T0350131

CAMBRIDGE STUDIES IN ADVANCED MATHEMATICS 185

FOUNDATIONS OF STABLE HOMOTOPY THEORY

The beginning graduate student in homotopy theory is confronted with a vast literature on spectra that is scattered across books, articles and decades. There is much folklore, but very few easy entry points. This comprehensive introduction to stable homotopy theory changes that. It presents the foundations of the subject together in one place for the first time, from the motivating phenomena to the modern theory, at a level suitable for those with only a first course in algebraic topology.

Starting from stable homotopy groups and (co)homology theories, the authors study the most important categories of spectra and the stable homotopy category, before moving on to computational aspects and more advanced topics such as monoidal structures, localisations and chromatic homotopy theory. The appendix containing essential facts on model categories, the numerous examples and the suggestions for further reading make this a friendly introduction to an often daunting subject.

David Barnes is a Senior Lecturer in Mathematics at Queen's University Belfast. His work centres on stable homotopy theory, usually with either a monoidal or equivariant focus, often using algebra to describe the structures in question.

Constanze Roitzheim is a Senior Lecturer in Mathematics at the University of Kent. Her work focuses on localisations of the stable homotopy category and related questions in algebra.

CAMBRIDGE STUDIES IN ADVANCED MATHEMATICS

All the titles listed below can be obtained from good booksellers or from Cambridge University Press. For a complete series listing, visit www.cambridge.org/mathematics.

Already Published

148 S. R. Garcia, J. Mashreghi & W. T. Ross *Introduction to Model Spaces and Their Operators*
149 C. Godsil & K. Meagher *Erdős–Ko–Rado Theorems: Algebraic Approaches*
150 P. Mattila *Fourier Analysis and Hausdorff Dimension*
151 M. Viana & K. Oliveira *Foundations of Ergodic Theory*
152 V. I. Paulsen & M. Raghupathi *An Introduction to the Theory of Reproducing Kernel Hilbert Spaces*
153 R. Beals & R. Wong *Special Functions and Orthogonal Polynomials*
154 V. Jurdjevic *Optimal Control and Geometry: Integrable Systems*
155 G. Pisier *Martingales in Banach Spaces*
156 C. T. C. Wall *Differential Topology*
157 J. C. Robinson, J. L. Rodrigo & W. Sadowski *The Three-Dimensional Navier–Stokes Equations*
158 D. Huybrechts *Lectures on K3 Surfaces*
159 H. Matsumoto & S. Taniguchi *Stochastic Analysis*
160 A. Borodin & G. Olshanski *Representations of the Infinite Symmetric Group*
161 P. Webb *Finite Group Representations for the Pure Mathematician*
162 C. J. Bishop & Y. Peres *Fractals in Probability and Analysis*
163 A. Bovier *Gaussian Processes on Trees*
164 P. Schneider *Galois Representations and* (φ, Γ)*-Modules*
165 P. Gille & T. Szamuely *Central Simple Algebras and Galois Cohomology (2nd Edition)*
166 D. Li & H. Queffelec *Introduction to Banach Spaces, I*
167 D. Li & H. Queffelec *Introduction to Banach Spaces, II*
168 J. Carlson, S. Müller-Stach & C. Peters *Period Mappings and Period Domains (2nd Edition)*
169 J. M. Landsberg *Geometry and Complexity Theory*
170 J. S. Milne *Algebraic Groups*
171 J. Gough & J. Kupsch *Quantum Fields and Processes*
172 T. Ceccherini-Silberstein, F. Scarabotti & F. Tolli *Discrete Harmonic Analysis*
173 P. Garrett *Modern Analysis of Automorphic Forms by Example, I*
174 P. Garrett *Modern Analysis of Automorphic Forms by Example, II*
175 G. Navarro *Character Theory and the McKay Conjecture*
176 P. Fleig, H. P. A. Gustafsson, A. Kleinschmidt & D. Persson *Eisenstein Series and Automorphic Representations*
177 E. Peterson *Formal Geometry and Bordism Operators*
178 A. Ogus *Lectures on Logarithmic Algebraic Geometry*
179 N. Nikolski *Hardy Spaces*
180 D.-C. Cisinski *Higher Categories and Homotopical Algebra*
181 A. Agrachev, D. Barilari & U. Boscain *A Comprehensive Introduction to Sub-Riemannian Geometry*
182 N. Nikolski *Toeplitz Matrices and Operators*
183 A. Yekutieli *Derived Categories*
184 C. Demeter *Fourier Restriction, Decoupling and Applications*
185 D. Barnes & C. Roitzheim *Foundations of Stable Homotopy Theory*
186 V. Vasyunin & A. Volberg *The Bellman Function Technique in Harmonic Analysis*
187 M. Geck & G. Malle *The Character Theory of Finite Groups of Lie Type*

Foundations of Stable Homotopy Theory

DAVID BARNES
Queen's University Belfast

CONSTANZE ROITZHEIM
University of Kent

CAMBRIDGE
UNIVERSITY PRESS

University Printing House, Cambridge CB2 8BS, United Kingdom

One Liberty Plaza, 20th Floor, New York, NY 10006, USA

477 Williamstown Road, Port Melbourne, VIC 3207, Australia

314–321, 3rd Floor, Plot 3, Splendor Forum, Jasola District Centre,
New Delhi – 110025, India

79 Anson Road, #06-04/06, Singapore 079906

Cambridge University Press is part of the University of Cambridge.

It furthers the University's mission by disseminating knowledge in the pursuit of education, learning, and research at the highest international levels of excellence.

www.cambridge.org
Information on this title: www.cambridge.org/9781108482783
DOI: 10.1017/9781108636575

First published 2020

A catalogue record for this publication is available from the British Library.

Library of Congress Cataloging-in-Publication Data
Names: Barnes, David, 1981- author.
Title: Foundations of stable homotopy theory / David Barnes, Queen's University Belfast, Constanze Roitzheim, University of Kent, Canterbury.
Description: Cambridge, United Kingdom ; New York, NY : Cambridge University Press, 2020. | Series: Cambridge studies in advanced mathematics | Includes bibliographical references and index.
Identifiers: LCCN 2019035713 | ISBN 9781108482783 (hardback) | ISBN 9781108636575 (ebook)
Subjects: LCSH: Homotopy theory.
Classification: LCC QA612.7 .B375 2020 | DDC 514/.24--dc23
LC record available at https://lccn.loc.gov/2019035713

ISBN 978-1-108-48278-3 Hardback

Contents

Introduction *page* 1

1 Basics of Stable Homotopy Theory 7
1.1 Stable Phenomena 7
1.2 The Spanier–Whitehead Category 28
1.3 A First Attempt at Spectra 32

2 Sequential Spectra and the Stable Homotopy Category 37
2.1 The Levelwise Model Structure 38
2.2 Homotopy Groups of Spectra 47
2.3 The Stable Model Structure 57
2.4 Explicit Fibrant Replacement 67
2.5 The Steenrod Algebra 71
2.6 The Adams Spectral Sequence 86

3 The Suspension and Loop Functors 93
3.1 Definition of the Functors 94
3.2 Stable Model Categories: A First Look 101
3.3 The Coaction of a Cofibre 105
3.4 Definition of Fibre and Cofibre Sequences 112
3.5 Shifting Fibre and Cofibre Sequences 115
3.6 The Long Exact Puppe Sequence 125

4 Triangulated Categories 128
4.1 Definition and Basic Properties 128
4.2 The Homotopy Category of a Stable Model Category 138
4.3 Comparison of Fibre and Cofibre Sequences 146
4.4 Consequences of Stability 149

4.5	Exact Functors and Quillen Functors	152
4.6	Toda Brackets	155
4.7	Muro's Exotic Triangulated Category	161
5	**Modern Categories of Spectra**	166
5.1	The Stable Homotopy Category – Revisited	167
5.2	Orthogonal Spectra	174
5.3	Symmetric Spectra	186
5.4	Properness of Spectra	210
5.5	Other Categories of Spectra	212
5.6	Compact Objects	222
5.7	Rigidity of Spectra	231
6	**Monoidal Structures**	234
6.1	Monoidal Model Categories	234
6.2	A Smash Product on the Stable Homotopy Category	251
6.3	Closed Monoidal Structures on Spectra	253
6.4	Monoidal Model Categories of Spectra	269
6.5	Spanier–Whitehead Duality	280
6.6	Ring Spectra and Modules	283
6.7	Commutative Ring Spectra	292
6.8	Applications of Monoidality	296
6.9	Homotopy Mapping Objects and Framings	298
7	**Left Bousfield Localisation**	317
7.1	General Localisation Techniques	317
7.2	Localisation of Stable Model Categories	325
7.3	Localisation of Spectra with Respect to Homology Theories	336
7.4	Examples of Left Localisation	343
Appendix	**Model Categories**	376
	References	411
	Index	418

Introduction

Aims and Intended Readership

The aim of this book is to be an accessible introduction to stable homotopy theory that novices, particularly graduate students, can use to learn the fundamentals of the subject. For the experts, we hope to have provided a useful compendium of results across the main areas of stable homotopy theory.

This book is not intended to replace any specific part of the existing literature, but instead to give a smoother, more coherent introduction to stable homotopy theory. We use modern techniques to give a streamlined development that avoids a number of outdated, and often over-complicated, constructions of a suitable stable homotopy category. We cover the most pressing topics for a novice and give a narrative to motivate the development. This narrative is missing from much of the current literature, which often assumes the reader already knows stable homotopy theory and hence understands why any given definition or result is important.

The majority of sections have been written to (hopefully) contain all details required for a graduate student. The remaining sections are intended to give an overview of more specialised or advanced topics, with references to the central texts for those areas. It is hoped that once the reader has read the chapters relevant to their research, they will be well prepared to dive into the rest of the literature and to know what to read next.

Prerequisites

Rather than rewrite many pages on model categories, category theory and unstable homotopy theory, we depend upon several excellent, and quite standard, references. As such, the reader should know a fair amount of point-set topology and algebraic topology. The standard texts are Gray [Gra75], Hatcher [Hat02],

May [May99a] and May and Ponto [MP12]. The reader should also know the basics of model categories. The best introductions are Dwyer and Spaliński [DS95] and the first chapters of Hovey [Hov99].

A certain amount of category theory is used throughout, the standard text is Mac Lane [Mac71]. For the chapters on the monoidal smash product, the reader will need some enriched category theory, easily obtained from Kelly [Kel05]. They may also like to have access to Borceaux [Bor94]. The chapter on localisations refers to Hirschhorn [Hir03] for some proofs and technical results, but the reader will not need to have read the book to follow the development.

A Historical Narrative

The book [Jam99] gives a treatment of the history of topology, while the chapter of May [May99b] (50 pages) covers stable homotopy theory from 1945 to 1966. Since then, the pace of development and publication has only quickened, a thorough history of stable homotopy theory would be a book by itself.

A basic problem in homotopy theory is the calculation of the homotopy groups of spheres. This problem is well known to be hopelessly difficult, but certain patterns in the homotopy groups were noticed. The Freudenthal Suspension Theorem gives a clear statement of a major pattern: the group $\pi_n(S^{k+n})$ is independent of n for $n > k + 1$. This and the suspension isomorphisms of homology and cohomology were a starting point of stable homotopy theory.

Calculations continued and the Spanier–Whitehead category was developed to study duality statements. It was also a useful category for the study of spaces under equivalences of stable homotopy groups. However, the Spanier–Whitehead category has some substantial drawbacks, in particular, it does not contain representatives for all reduced cohomology theories.

Several solutions to this were constructed, including Boardman's stable homotopy category, Lima's notion of spectra, Kan's semisimplicial spectra and Whitehead's developments of the notions of spectra. None of these categories were entirely satisfactory, so we jump ahead to Adams's construction of the stable homotopy category [Ada74], which was based on ideas of Boardman. This category contained the Spanier–Whitehead category, represented all cohomology theories and had a commutative smash product.

Having a good construction with sensible axioms allowed for further development of stable homotopy theory. A good notion of "categories of fractions", now known as Bousfield localisations at homology theories, greatly improved the ability to calculate stable homotopy groups via the Adams spectral sequence. Through work of Bousfield, these localisations were further developed

([Bou75], [Bou79]), and vast amounts of calculations were now possible. This led to the introduction of chromatic homotopy theory, which gives a framework for major structural results about the stable homotopy category (see [Rav84] and [Rav92a]), as well as techniques for even further calculations of stable homotopy groups of spheres [Rav86].

The lack of a good commutative monoidal point-set model for the stable homotopy category still held the subject area back. Brown representability posited the existence of function spectra and allowed for some homotopical calculations, but direct constructions were often impossible to give. The study of (commutative) ring spectra up to homotopy was difficult – keeping track of the homotopies and their coherence was particularly burdensome. Moreover, constructions up to homotopy prevented geometric constructions such as bundles of spectra or diagrams of spectra.

The development of coordinate–free spectra by May and others offered several improvements to the area. The use of operads to manage commutative multiplications up to homotopy allowed for serious study of derived algebra in spectra, the so-called "brave new algebra", see the work of May, Quinn and Ray [May77].

Coordinate–free spectra also led to the development of "spectrification" functors, which simplified the construction of maps between spectra. These functors played a central role in work of Lewis, May and Steinberger [LMSM86], which gave a construction of G-equivariant spectra and the G-equivariant stable homotopy category for G a compact Lie group.

While this technology did allow for a useful definition of an internal function object for spectra, the smash product was still only commutative and associative up to homotopy. The work of Lewis [Lew91] even suggested that there may be no commutative monoidal point-set model for the stable homotopy category, but this pessimism turned out to be unfounded.

Two independent solutions to the problem of commutative smash products came about in surprisingly quick succession: the S–modules of Elmendorff, Kriz, Mandell and May [EKMM97] and the symmetric spectra of Hovey, Shipley and Smith [HSS00]. These references gave closed symmetric monoidal model categories of spectra and model categories of (commutative) ring spectra. By using model categories, one had point-set level smash products and function objects which would have the correct homotopical properties after passing to the homotopy categories.

This reinvigorated the area and allowed for a great deal of further development in "brave new algebra", namely, the importing of statements from algebra into stable homotopy theory. For example, Hovey, Palmieri and Strickland [HPS97] were able to give an axiomatisation of stable homotopy theories.

Schwede [Sch01a] showed that the model categories of symmetric spectra and S–modules were Quillen equivalent. Yet more symmetric monoidal categories of spectra were constructed by Mandell, May, Schwede and Shipley [MMSS01]. These were all shown to be Quillen equivalent, and a particular highlight is the category of orthogonal spectra. Categories of spectra in categories other than simplicial sets or spaces were given in Schwede [Sch97] and Hovey [Hov01b].

All of these model categories are amenable to the theory of localisations as developed by Hirschhorn [Hir03], giving many point-set models for localisations of spectra and, in particular, those from chromatic homotopy theory.

We are now in the modern era of stable homotopy theory, with current topics such as topological modular forms and its variants, motivic stable homotopy theory, the study of commutative ring spectra and their localisations, Galois extensions of ring spectra and equivariant versions of most of those topics.

Explanation of Contents

We start with a study of stable phenomena, namely, the Freudenthal Suspension Theorem and the suspension isomorphisms of homology and cohomology. We discuss how this leads to the notion of a stable homotopy category and what axioms it should satisfy. We introduce the Spanier–Whitehead category and basic categories of spectra and show how these fail to satisfy those axioms. Using the benefit of hindsight, we then define the stable homotopy category to be the homotopy category of the stable model structure on sequential spectra. This approach avoids the difficulties of extending the Spanier–Whitehead category and the complicated constructions of maps and functions in Adams's category of spectra.

The category of sequential spectra evidently satisfies enough of the axioms to be worth studying further, but it will not be possible to give a category that satisfies all the axioms until after we have introduced the symmetric monoidal categories of symmetric and orthogonal spectra.

We digress from the further development of categories of spectra to ask about a formal framework in which those categories of spectra can be studied. The starting place is a suspension functor on general model categories and how it gives rise to cofibre and fibre sequences, leading to a notion of a stable model category. When the model category is stable, one can extend cofibre and fibre sequences in either direction and prove the fundamental statement: the homotopy category of a stable model category is triangulated. Working in this generality shows clear benefits of stability and simplifies the later work, where we can appeal to the triangulated arguments of these chapters.

The next task is to examine the generalisation of the smash product of the Spanier–Whitehead category to the stable homotopy category. Again, we want an approachable method, so we first show that the homotopy categories of symmetric spectra and orthogonal spectra are equivalent to the stable homotopy category. We then define the smash product and the internal function object of the stable homotopy category as coming from the smash products on orthogonal spectra and symmetric spectra. At this point, we have encountered three models for the stable homotopy category. Each has its own advantages. We have:

Sequential spectra: These are the simplest to define and can be motivated from a discussion of Brown representability. However, they do not have a commutative smash product. The weak equivalences are defined in terms of homotopy groups of spectra, a natural extension of the idea of stable homotopy groups.

Orthogonal spectra: These are slightly more complicated than sequential spectra and can be thought of as sequential spectra with extra structure. Their weak equivalences are defined by the forgetful functor to sequential spectra and, hence, are defined in terms of homotopy groups of spectra. The extra structure allows one to define a symmetric monoidal smash product and an internal function object.

Symmetric spectra: The final model is symmetric spectra (in either pointed topological spaces or pointed simplicial sets). This model is intermediate in its complexity, but the weak equivalences are harder to define. These spectra also have good monoidal properties. Their simplicity allows them to be described as "initial amongst stable model categories" in the sense of Sections 6.8 and 6.9.

The symmetric monoidal versions of spectra lead to the next important topic: spectra with (commutative) ring structures and Spanier–Whitehead duality for spectra, which is essentially a study of duality in the stable homotopy category.

We take the opportunity to consider framings and stable framings. This allows us to construct mapping spaces for an arbitrary model category and mapping spectra for an arbitrary stable model category.

We end the book with a chapter on Bousfield localisation, introducing and motivating the concept and proving a simple existence result for stable model categories. As an application, we discuss p-localisation, p-completion and localisation at complex topological K-theory.

Along the way, we furthermore include important results on stable homotopy theory and suggest further directions. The results include: rigidity and uniqueness of the monoidal structure, a description in terms of modules over spectrally enriched categories, the Adams spectral sequence and chromatic homotopy theory.

An appendix listing the results on model categories that are needed is included for easy reference. Some proofs are given, otherwise clear references are provided.

Omissions

An exhaustive treatment of stable homotopy theory would require several books and be impractical for the needs of many graduate students. Hence, certain topics have been omitted, a list is below. Reasons for the omission vary, from being somewhat outside the scope (stable infinity categories), being a topic that builds upon stable homotopy theory (equivariant or motivic stable homotopy theory), or having good textbooks already, albeit ones that assume a familiarity with stable homotopy theory.

- Infinite loop space machines and operads.
- Right Bousfield localisations.
- Comprehensive treatment of spectral sequences.
- Equivariant stable homotopy theory.
- Motivic stable homotopy theory.
- Stable infinity categories.
- In-depth treatment of K-theory, cobordism and formal group laws.
- The S–modules of Elmendorf, Kriz, Mandell and May.

Convention

Throughout the book we use the convention that the set of natural numbers $\mathbb{N}$ contains 0.

Acknowledgements

We are very grateful to the Isaac Newton Institute for Mathematical Sciences for their support and hospitality during the programme "Homotopy harnessing higher structures", during which a large part of this book was written. The programme was supported by EPSRC grant EP/R014604/1.

We also thank Andy Baker, Bob Bruner, Jocelyne Ishak, Magdalena Kędziorek, Tyler Lawson and Doug Ravenel for interesting discussions and inspiration.

1

Basics of Stable Homotopy Theory

In the crudest sense, stable homotopy theory is the study of those homotopy-invariant constructions of spaces which are preserved by suspension. In this chapter we show how there are naturally occurring situations which exhibit stable behaviour. We will discuss several historic attempts at constructing a "stable homotopy category" where this stable behaviour can be studied, and we relate these to the more developed notions of spectra and the Bousfield–Friedlander model structure.

Of course, if one wants to perform calculations of stable homotopy groups using spectral sequences, then one does not need much of the formalism of model categories of spectra. But as soon as one wishes to move away from those tasks and consider other stable homotopy theories (such as G-equivariant stable homotopy theory for some group G) or to make serious use of a symmetric monoidal smash product in the context of "Brave New Algebra", then the advantages of the more formal setup become overwhelming.

A more accurate title for the chapter might be along the lines of "First Encounters with Stability", as a detailed look at even the first question of stable homotopy theory – calculating the stable homotopy groups of spheres – is a book by itself.

1.1 Stable Phenomena

We introduce two standard results which exhibit stability, namely the Freudenthal Suspension Theorem and the interaction of homology with suspension. The proof of each result comes from an excision theorem and a long exact sequence. Our starting point is a quick introduction to the category of spaces that we use throughout the book.

1.1.1 Topological Spaces

We want to work in a closed monoidal category of topological spaces. We must therefore equip the set of continuous maps $\mathrm{Top}(A, B)$ with a topology such that there is a natural isomorphism

$$\mathrm{Top}(A \times B, C) \cong \mathrm{Top}(A, \mathrm{Top}(B, C))$$

for all spaces A, B and C in the category. As is well known, we cannot use the standard category of all topological spaces equipped with the Cartesian product and compact-open topology on the set of continuous maps. Instead we follow the standard pattern and work with compactly generated weak Hausdorff spaces. Recall that a space X is said to be *compactly generated* if a set $A \subseteq X$ is closed if and only if $A \cap K$ is closed in K for each compact subset K of X. A space X is said to be *weak Hausdorff* if for any compact space M and continuous map $f\colon M \longrightarrow X$, the image $f(M)$ is closed in X. More details can be found in Steenrod [Ste67], Hovey [Hov99, Sections 2.4 and 4.2], May [May99a, Chapter 5] and Schwede [Sch18, Appendix].

Definition 1.1.1 The *category of topological spaces* is the category of compactly generated weak Hausdorff topological spaces and continuous maps. We denote this by Top.

The category Top is a closed monoidal category with all small limits and colimits. We can also consider pointed spaces, which is also a closed monoidal category with all small limits and colimits.

Definition 1.1.2 The *category of pointed topological spaces* is the category of pointed, compactly generated weak Hausdorff spaces and continuous maps that preserve the basepoints. We denote this by Top_*.

We denote a pointed space as (X, x_0) or just X with an implicit basepoint. For topological spaces X and Y, we denote the set of homotopy classes of maps from X to Y by $[X, Y]$. If X and Y are pointed, we interpret $[X, Y]$ as the set of pointed homotopy classes of pointed maps from X to Y.

Monoidal products, function objects, limits and colimits can be a little mysterious in the categories Top and Top_*, but are mostly given by the expected constructions. For example, if X is Hausdorff, the monoidal product with X is the Cartesian product with the product topology, or the smash product in the pointed case. If X is compact and Hausdorff, then the topology on $\mathrm{Top}(X, Y)$ is the compact-open topology (similarly for $\mathrm{Top}_*(X, Y)$). A colimit of a sequential diagram of injections or a pushout of closed inclusions is given by the usual colimit of (pointed) spaces.

For general topological spaces, the monoidal product is the Cartesian product with the Kelly product topology (and the corresponding smash product for the pointed case). The topology on the set of maps from X to Y is given by the modified compact-open topology, which we now define. Let $a\colon K \to X$ be a continuous map from a compact Hausdorff space K to X. For U an open set of Y, let $C(a, U)$ be the set of continuous maps f from X to Y such that the image of $f \circ a\colon K \to Y$ lies in U. These sets $C(a, U)$ define a sub-basis for the modified compact-open topology on the set of continuous maps from X to Y.

Recall the Serre and Hurewicz model structures on Top from [DS95] or [MP12], see also Example A.1.5. The weak equivalences in the Serre model structure are the weak homotopy equivalences, and its fibrations are those maps with the right lifting property with respect to the inclusions

$$A \times \{0\} \longrightarrow A \times [0, 1]$$

for each CW-complex A. The cofibrations are called the q-cofibrations or Serre cofibrations.

The weak equivalences of the Hurewicz model structure are the homotopy equivalences, and its fibrations are those maps with the right lifting property with respect to the inclusions

$$A \times \{0\} \longrightarrow A \times [0, 1]$$

for each space A. The cofibrations are called the h-cofibrations or Hurewicz cofibrations. We see that every q-cofibration is a h-cofibration.

Both model structures extend to the category of pointed spaces, with weak equivalences, cofibrations and fibrations defined by forgetting the basepoint. Thus, a h-cofibration of pointed spaces means a basepoint-preserving map which is a h-cofibration in Top.

Every object of Top is cofibrant in the Hurewicz model structure, but this fails for Top_*. For pointed spaces, the homotopical behaviour of the inclusion of the basepoint into a space is a technical issue we will encounter several times.

Definition 1.1.3 We say that a pointed space (X, x_0) has a *non-degenerate basepoint* if the map $x_0 \longrightarrow X$ is a h-cofibration in Top. We say that a pointed space (X, x_0) is a *pointed CW-complex* if (X, x_0) is a CW-pair.

Many results about pointed spaces require the assumption of non-degenerate basepoints so that we can move to the simpler setting of unpointed spaces. For example, if X has a non-degenerate basepoint, then the map from the unreduced suspension of X to the reduced suspension of X is a homotopy equivalence. Hence statements about unreduced suspensions will extend to reduced suspensions in this case. Many spaces are easily seen to have non-degenerate basepoints, for example, the basepoint of any pointed CW-complex is non-degenerate. Any pointed space can be replaced by a space with a non-degenerate basepoint as follows. For a pointed space X, consider the space

$$wX = (X \amalg [0,1])/(x_0 \sim 0)$$

with the basepoint taken to be $1 \in [0,1]$. This space wX has a non-degenerate basepoint, and we may consider the construction $w(-)$ as a functor. Moreover, the contraction map $wX \to X$ is an (unpointed) homotopy equivalence and hence is a weak homotopy equivalence of pointed spaces.

1.1.2 The Freudenthal Suspension Theorem

The purpose of this subsection is to exhibit a pattern that occurs across many homotopy groups, namely that for a pointed CW-complex X, the homotopy groups of X and ΣX agree, up to a shift, over a range that depends on the connectivity of X. The precise statement is Theorem 1.1.10.

The starting point for this section is connectivity and its relation to suspensions. To prove our results we shall need a rather substantial ingredient: the homotopy excision theorem.

We follow the development of [May99a, Section 11]. Let us recall the notions of k-connected spaces, k-connected pairs of spaces and k-equivalences. Note that we will avoid defining or using $\pi_0(X, A)$ for spaces $A \subseteq X$. For a definition of $\pi_n(X, A)$, see [Hat02].

Definition 1.1.4 A pointed topological space (X, x_0) is *k-connected* if it is path-connected and $\pi_n(X, x_0) = 0$ for $1 \leqslant n \leqslant k$.

A pair $A \subseteq X$ is said to be *k-connected* if every path component of X intersects with A and $\pi_n(X, A) = 0$ for each $1 \leqslant n \leqslant k$.

A pointed map $f \colon X \longrightarrow Y$ is a *k-equivalence* of topological spaces if for all $x_0 \in X$, the map $\pi_n(f, x_0)$ is an isomorphism for $0 \leqslant n < k$ and surjective for $n = k$.

By convention, every pointed topological space is (-1)-connected.

It follows from the long exact Puppe sequence that a k-equivalence of pointed spaces has a $(k-1)$-connected homotopy fibre. We can also relate k-equivalences to the dimension of a CW-complex.

The following can be found in [Hat02].

Proposition 1.1.5 *Let $f\colon X \longrightarrow Y$ be a k-equivalence of pointed spaces and A be a pointed CW-complex. Then composition with f induces an isomorphism*

$$f_*\colon [A, X] \longrightarrow [A, Y]$$

if the dimension of A is less than k and a surjection if the dimension is equal to k.

Theorem 1.1.6 (Homotopy Excision) *Let X be a topological space with subsets A and B such that $A \cap B$ is non-empty and X is the union of the interiors of A and B.*

Assume that $(A, A \cap B)$ is m-connected for some $m \geqslant 1$, and that $(B, A \cap B)$ is n-connected for some $n \in \mathbb{N}$. Then the map induced by inclusion

$$\pi_a(A, A \cap B) \longrightarrow \pi_a(X, B)$$

is an isomorphism for $a < m + n$ and a surjection when $a = m + n$. Moreover, the square

$$\begin{array}{ccc} \pi_0(A \cap B) & \longrightarrow & \pi_0(A) \\ \downarrow & & \downarrow \\ \pi_0(B) & \longrightarrow & \pi_0(X) \end{array}$$

of maps induced by the inclusions is a pullback.

We can use homotopy excision to calculate some of the homotopy groups of the homotopy cofibre (mapping cone) of a map in terms of relative homotopy groups.

Definition 1.1.7 Let $f\colon X \longrightarrow Y$ be a map in Top. Then the *mapping cylinder* Mf of f is the topological space defined as the pushout

$$\begin{array}{ccc} X & \xrightarrow{i_0} & X \times [0,1] \\ {\scriptstyle f}\downarrow & & \downarrow \\ Y & \longrightarrow & Mf. \end{array}$$

The *mapping cone* Cf of f is the quotient $Mf/(X \times \{1\})$.

Proposition 1.1.8 *Let $f\colon X \longrightarrow Y$ be a k-equivalence in* Top *between $(k-1)$-connected spaces for $k \geqslant 1$. Then the projection map*

$$(Mf, X) \longrightarrow (Cf, *)$$

induces an isomorphism on π_n for $n < 2k$ and a surjection on π_n when $n = 2k$.

Proof Identifying $X \subset Mf$ as the collection of points $(x, 1) \in Mf$ for $x \in X$, we define two subsets of Cf

$$A = Y \cup (X \times [0, 2/3]) \quad B = (X \times [1/3, 1])/X.$$

Their intersection is a cylinder on X and the union of their interiors is Cf. We also see that

$$(A, A \cap B) \simeq (Mf, X), \ (Cf, B) \simeq (Cf, *) \text{ and } (B, A \cap B) \simeq (Cf, X).$$

Moreover, the quotient map

$$(Mf, X) \longrightarrow (Cf, *)$$

is homotopic to the inclusion

$$(A, A \cap B) \longrightarrow (Cf, B).$$

The long exact sequence of the homotopy groups of a pair implies that $(A, A \cap B)$ is k-connected and that $(B, A \cap B)$ is k-connected. The result then follows from homotopy excision. □

We may extend this result to the general case of a quotient map.

Corollary 1.1.9 *Let $i \colon A \longrightarrow X$ be a h-cofibration of $(k-1)$-connected spaces in* Top *that is a k-equivalence, $k \geqslant 1$. Then the quotient map*

$$(X, A) \longrightarrow (X/A, *)$$

induces an isomorphism on π_n for $n < 2k$ and a surjection on π_{2k}. The result also holds in Top_* *if, in addition, A and X have non-degenerate basepoints.*

Proof Since the map i is a h-cofibration, $Ci \simeq X/A$ by Lemma A.5.6. The result follows from the commutative diagram

$$\begin{array}{ccc} (Mi, A) & \longrightarrow & (Ci, *) \\ \downarrow \simeq & & \downarrow \simeq \\ (X, A) & \longrightarrow & (X/A, *). \end{array}$$

For the last statement we use the reduced mapping cylinder and reduced cofibre. Since the basepoints are non-degenerate, the quotient maps from the unreduced to reduced versions are homotopy equivalences, hence the pointed version follows from the unpointed version. □

We are now ready to prove the Freudenthal Suspension Theorem. Our method is to apply Corollary 1.1.9 to the map $(CX, X) \longrightarrow (\Sigma X, *)$ and carefully examine the relative homotopy groups. As the spaces involved are connected by assumption, we do not specify a basepoint.

Theorem 1.1.10 (Freudenthal Suspension Theorem) *Let $k \in \mathbb{N}$ and let X be a k-connected space with a non-degenerate basepoint. The map*

$$\pi_n(X) = [S^n, X] \xrightarrow{\Sigma} [\Sigma S^n, \Sigma X] \cong [S^{n+1}, \Sigma X] = \pi_{n+1}(\Sigma X)$$

$$[f] \longmapsto [\Sigma f]$$

is an isomorphism if $n < 2k+1$ and a surjection if $n = 2k+1$.

Proof In order to be consistent with the standard definition of relative homotopy groups, we let $C'X = X \wedge I$, with the basepoint of $I = [0,1]$ taken to be 0. Thus

$$C'X = (X \times I)/(X \times \{0\} \cup \{x_0\} \times I).$$

Consider a map $f\colon (I^n, \partial I^n) \longrightarrow (X, x_0)$ representing some element of $\pi_n(X, x_0)$. Then $f \times \mathrm{Id}$ induces a map of triples

$$(I^{n+1}, \partial I^{n+1}, J^q) \longrightarrow (C'X, X, x_0)$$

whose restriction to $I^n \times \{1\}$ is f. Taking the quotient by $X \times \{1\}$, we obtain Σf. It follows that the diagram below commutes, where ∂ is the connecting map and $\rho\colon C'X \longrightarrow \Sigma X$ is the quotient.

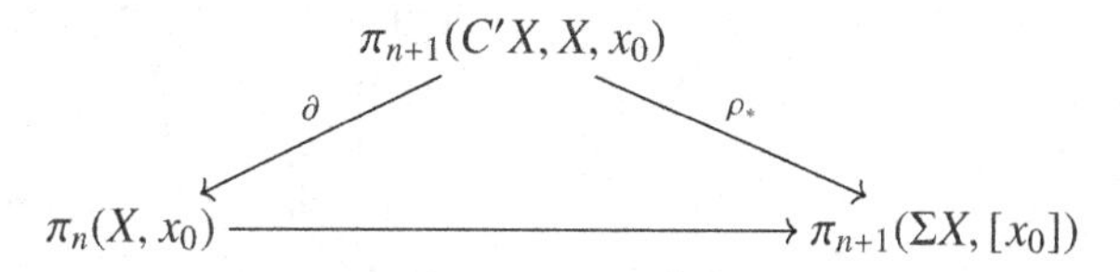

The inclusion $X \longrightarrow C'X$ is a h-cofibration and a $(k+1)$-equivalence of k-connected spaces, so Corollary 1.1.9 tells us that ρ_* is an isomorphism for $n+1 < 2k+2$ and a surjection for $n+1 = 2k+2$. As $C'X$ is contractible, ∂ is an isomorphism and the result follows. □

The map Σ could also be defined via the composite

$$[S^n, X] \longrightarrow [S^n, \Omega(\Sigma X)] \cong [S^{n+1}, \Sigma X],$$

where the first map is the unit of the (Σ, Ω)-adjunction on pointed spaces. Checking that this definition agrees with $[f] \mapsto [\Sigma f]$ is an exercise in relating the counit of an adjunction to the action of the functor on sets of maps. A similar exercise shows that the composite of suspension followed by the counit

$$[S^{n-1}, \Omega X] \longrightarrow [S^n, \Sigma(\Omega X)] \longrightarrow [S^n, X]$$

is given by the adjunction isomorphism.

We see that suspension increases connectivity of pointed spaces and that the unit of the (Σ, Ω)-adjunction induces isomorphisms on a range of homotopy groups.

Corollary 1.1.11 *Let $k \geqslant 1$, and let X be a k-connected pointed topological space with a non-degenerate basepoint. Then ΣX is $(k+1)$-connected and the unit map*

$$\eta\colon X \longrightarrow \Omega\Sigma X$$

is a $(2k+1)$-equivalence. Furthermore, the counit map

$$\varepsilon\colon \Sigma\Omega X \longrightarrow X$$

is a $2k$-equivalence.

Proof Given the above discussion, we only need to consider the counit ε. We want to know that the basepoint of ΩX (the constant path at x_0) is non-degenerate. This follows from a more general statement: if $A \longrightarrow B$ is the inclusion of a closed subset and a h-cofibration, then $\Omega A \to \Omega B$ is a h-cofibration.

By [May99a, Section 6.4], if $A \longrightarrow B$ is the inclusion of closed subset, then it is a h-cofibration if and only if it is an NDR pair (neighbourhood deformation retract pair). One can show that a presentation of $A \longrightarrow B$ as an NDR pair can be extended to give a presentation of $\Omega A \to \Omega B$ as an NDR pair.

As X is k-connected and has a non-degenerate basepoint, it follows that ΩX is $(k-1)$-connected and has a non-degenerate basepoint. Using the Freudenthal Suspension Theorem on ΩX and the commutative diagram

$$\begin{array}{ccc} \pi_{n-1}(\Omega X) & \xrightarrow{\Sigma} & \pi_n(\Sigma\Omega X) \\ & \searrow^{\cong} & \downarrow \varepsilon \\ & & \pi_n(X), \end{array}$$

we see that ε is a $2k$-equivalence. □

The Freudenthal Suspension Theorem can be used to calculate some homotopy groups of spheres. For example, we can prove that

$$\pi_n(S^n) \cong \mathbb{Z}$$

for each $n \geqslant 1$. We know the cases of $n = 1$ and $n = 2$ by the long exact sequence of the Hopf fibration. For $n > 2$ the result follows directly from the suspension theorem, as the suspension maps

$$\pi_n(S^n) \longrightarrow \pi_{n+1}(S^{n+1})$$

are isomorphisms. In fact, this isomorphism also holds for $n = 1$, as the only surjections $\mathbb{Z} \longrightarrow \mathbb{Z}$ are isomorphisms.

The ranges for the suspension theorem are usually sharp. Consider the suspension maps

$$\pi_2(S^1) \longrightarrow \pi_3(S^2) \longrightarrow \pi_4(S^3)$$

$$0 \longrightarrow \mathbb{Z} \longrightarrow \mathbb{Z}/2.$$

The first cannot be surjective and the second cannot be an isomorphism.

We restate the Freudenthal Suspension Theorem in terms of iterated suspensions.

Corollary 1.1.12 *Let X be a topological space with non-degenerate basepoint and let a and b be natural numbers with $b < a - 1$. Then the suspension map*

$$\pi_{a+b}(\Sigma^a X) \longrightarrow \pi_{a+b+1}(\Sigma^{a+1} X)$$

is an isomorphism. □

If we fix b in the above corollary and allow a to increase, we see that $\pi_{a+b}(\Sigma^a X)$ can take many different values until $a > b + 1$. Then, for larger a each homotopy group in the sequence is the same up to isomorphism. It is natural to study the eventual behaviour of these homotopy groups. We restrict our attention to pointed CW-complexes, as these are the spaces of primary interest.

Definition 1.1.13 For a pointed CW-complex X and $n \in \mathbb{N}$, the nth *stable homotopy group* of X is

$$\pi_n^{\text{stable}}(X) := \operatorname{colim}_a \pi_{n+a}(\Sigma^a X) = \pi_{2n+2}(\Sigma^{n+2} X).$$

We could define these groups for all $n \in \mathbb{Z}$ by simply ignoring those terms in the colimit where $n + a < 0$. However, we see that these groups are all zero. We also see that stable homotopy groups are stable under suspension, that is,

$$\pi_n^{\text{stable}}(X) \cong \pi_{n+1}^{\text{stable}}(\Sigma X).$$

There is no need to mention basepoints when using stable homotopy groups, as, aside from the first term, the spaces $\Sigma^a X$ will all be path-connected.

We can relate stable homotopy groups of spaces to ordinary homotopy groups by a sequential homotopy colimit construction, see Example A.7.9.

Definition 1.1.14 We define a functor Q from the category of pointed CW-complexes to pointed topological spaces by

$$QX = \mathrm{hocolim}_n\, \Omega^n \Sigma^n X.$$

The maps of the diagram are given by the unit of the (Ω, Σ)-adjunction.

To form the homotopy colimit, we may either replace each map by a (weakly equivalent) h-cofibration or take the union of the mapping cylinders.

We see that

$$\pi_*(QX) = \pi_*^{\mathrm{stable}}(X),$$

as the homotopy groups of a sequential homotopy colimit are the colimit of the homotopy groups. Equally, if $f\colon X \longrightarrow Y$ is a map that induces isomorphisms on all stable homotopy groups, then

$$Qf\colon QX \longrightarrow QY$$

is a weak homotopy equivalence.

As a first example of extra structure on stable homotopy groups, we will now discuss the ring structure on the stable homotopy groups of spheres. Let

$$\pi_*^{\mathrm{stable}}(S^0) = \bigoplus_{n \geqslant 0} \pi_n^{\mathrm{stable}}(S^0).$$

We may construct a product $*$ making $\pi_*^{\mathrm{stable}}(S^0)$ into a commutative graded ring, satisfying

$$\alpha * \beta = (-1)^{mn} \beta * \alpha, \quad \alpha \in \pi_m^{\mathrm{stable}}(S^0),\ \beta \in \pi_n^{\mathrm{stable}}(S^0).$$

Recall the notion of *homological degree* of a map $f\colon S^n \to S^n$. Choosing a generator $a \in H_n(S^n)$, we say that f has homological degree $k \in \mathbb{Z}$ if $f_*(a) = ka$. The homological degree of a map is independent of the choice of generator. We will see that the $(-1)^{mn}$ of the formula above originates from the twist isomorphism

$$\tau_{p,q}\colon S^p \wedge S^q \longrightarrow S^q \wedge S^p,$$

which has homological degree $(-1)^{pq}$.

To define $*$, we choose representatives $f\colon S^{a+m} \longrightarrow S^a$ and $g\colon S^{a+n} \longrightarrow S^a$. Then we set

$$\alpha * \beta = [g \circ \Sigma^n f].$$

This operation is well-defined and associative.

We must prove that it is distributive, and that the commutativity condition holds. For the rest of this section, we specify $\Sigma^p X = X \wedge S^p$. Hence, Σ^p applied to $f\colon X \longrightarrow Y$ is given by

$$f \wedge \mathrm{Id}_p\colon \Sigma^p X \longrightarrow \Sigma^p Y.$$

Let $f, g\colon S^{m+n+a} \longrightarrow S^{n+a}$ and $h\colon S^{n+a} \longrightarrow S^a$. Addition $[f] + [g]$ is represented by the map

$$f + g\colon S^{m+n+a} \longrightarrow S^{n+a},$$

which is f on the upper hemisphere of S^{n+m+a} and g on the lower hemisphere. Thus,

$$[h] * ([f] + [g]) = [h \circ (f + g)] = [h \circ f] + [h \circ g] = [h] * [f] + [h] * [g].$$

This is one of the equations for distributivity, the other one will follow from the first one plus commutativity, which we will show now.

To prove the commutativity statement, we show that the composition product $*$ can be defined in terms of the smash product. Let $f\colon S^{a+m} \longrightarrow S^a$ and $g\colon S^{a+n} \longrightarrow S^a$ with a even. Then

$$g \circ \Sigma^n f\colon S^{n+m+a} \longrightarrow S^a \quad \text{and} \quad \Sigma^a g \circ \Sigma^{a+n} f\colon S^{n+m+2a} \longrightarrow S^{2a}$$

represent the same class in the stable homotopy group $\pi^{\text{stable}}_{n+m}(S^0)$. The second representative is the top map in the following diagram.

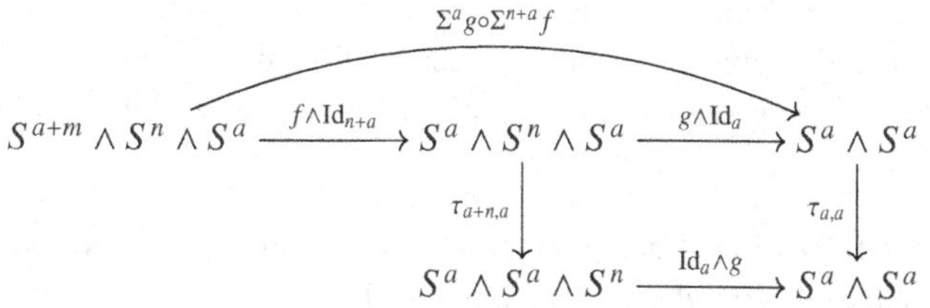

$$\begin{array}{ccccc}
 & & \xrightarrow{\Sigma^a g \circ \Sigma^{n+a} f} & & \\
S^{a+m} \wedge S^n \wedge S^a & \xrightarrow{f \wedge \mathrm{Id}_{n+a}} & S^a \wedge S^n \wedge S^a & \xrightarrow{g \wedge \mathrm{Id}_a} & S^a \wedge S^a \\
 & & \downarrow{\scriptstyle \tau_{a+n,a}} & & \downarrow{\scriptstyle \tau_{a,a}} \\
 & & S^a \wedge S^a \wedge S^n & \xrightarrow{\mathrm{Id}_a \wedge g} & S^a \wedge S^a
\end{array}$$

As a is even, each of the twist maps have homological degree 1 and so are homotopic to the identity. The composite along the lower path is therefore

$$(f \wedge \mathrm{Id}_{n+a}) \circ \tau_{a+n,a} \circ (\mathrm{Id}_a \wedge g) \simeq f \wedge g.$$

Hence

$$[g * f] = [\Sigma^a g * \Sigma^a f] = [f \wedge g].$$

To complete the proof of commutativity, we prove that

$$[f \wedge g] = (-1)^{mn}[g \wedge f].$$

This follows from the commutativity of the diagram

$$\begin{array}{ccc} S^{a+m} \wedge S^{a+n} & \xrightarrow{f \wedge g} & S^a \wedge S^a \\ \downarrow \tau_{a+m,a+n} & & \downarrow \tau_{a,a} \\ S^{a+n} \wedge S^{a+m} & \xrightarrow{g \wedge f} & S^a \wedge S^a \end{array}$$

and the fact that for even a, $\tau_{a+m,a+n}$ is of homological degree $(-1)^{mn}$. Hence

$$[g * f] = [f \wedge g] = (-1)^{mn}[g \wedge f] = (-1)^{mn}[f * g].$$

We can use our understanding of stable homotopy groups of spheres to prove a more general version of the Freudenthal Suspension Theorem.

Theorem 1.1.15 *Suppose X and Y are pointed CW-complexes with Y k-connected. Then the suspension map*

$$\Sigma\colon [X, Y] \longrightarrow [\Sigma X, \Sigma Y]$$

is surjective if X is of dimension $2k + 1$ and a bijection if X has dimension less than $2k + 1$.

Proof By Corollary 1.1.11, the map $Y \longrightarrow \Omega\Sigma Y$ is a $(2k + 1)$-equivalence, hence the result follows by Proposition 1.1.5. □

Let X and Y be finite CW-complexes and $a \geqslant 2$. As with stable homotopy groups, after enough suspensions ($a > \dim X$), the collection of abelian groups $[\Sigma^a X, \Sigma^a Y]$ are all isomorphic under the suspension map. Following the pattern of stable homotopy groups when X and Y are not finite, we take a colimit over suspensions.

Definition 1.1.16 Let X and Y be pointed CW-complexes. The set of *stable homotopy classes of pointed maps* from X to Y is

$$[X, Y]^s := \operatorname{colim}_a [\Sigma^a X, \Sigma^a Y].$$

When X is also compact,

$$[X, QY]_* = [X, \operatorname{hocolim}_n \Omega^n \Sigma^n Y]_* \cong [X, Y]^s_*$$

by the same argument as for homotopy groups.

This leads to the natural question: Can we make a category with objects given by the class of pointed topological spaces and morphisms given by the class of stable homotopy classes of pointed maps? We will see that this is possible in Section 1.2 but that the resulting category has some serious failings.

1.1.3 Homology and Cohomology

Homology and cohomology theories are closely linked to stable homotopy theory. We show that reduced (co)homology theories lead to a key idea for stable homotopy theory: spectra. We begin with a set of axioms for homology that are equivalent to the Eilenberg–Steenrod axioms.

Recall that a CW-pair (X, A) is a CW-complex X with a subcomplex A. In particular, there is an open neighbourhood of A in X which is a deformation retract of A.

Definition 1.1.17 A *reduced homology theory* is a functor $\widetilde{E}_*$ from pointed CW-complexes to graded abelian groups Ab_* satisfying the list of axioms below. For a map $f\colon X \longrightarrow Y$ of pointed CW-complexes, we write

$$f_* = \widetilde{E}_*(f)\colon \widetilde{E}_*(X) \longrightarrow \widetilde{E}_*(Y).$$

Let f_n be the nth level of this map of graded abelian groups.

1. If $f \simeq g$, then $f_* = g_*$.
2. For each CW-pair (X, A) there is a *boundary map* $\partial_*\colon \widetilde{E}_*(X/A) \longrightarrow \widetilde{E}_{*-1}(A)$, which is natural in pointed CW-pairs.
3. If (X, A) is a CW-pair, then the inclusion map $i\colon A \longrightarrow X$, the quotient map $q\colon X \longrightarrow X/A$ and the boundary map ∂_* give a natural long exact sequence
$$\cdots \xrightarrow{\partial_{n+1}} \widetilde{E}_n(A) \xrightarrow{i_n} \widetilde{E}_n(X) \xrightarrow{q_n} \widetilde{E}_n(X/A) \xrightarrow{\partial_n} \widetilde{E}_{n-1}(A) \xrightarrow{i_{n-1}} \cdots .$$
4. For a set of spaces $\{X_\alpha\}_\alpha$ the maps $i_\alpha\colon X_\alpha \longrightarrow \bigvee_\alpha X_\alpha$ induce isomorphisms
$$\bigoplus_\alpha (i_\alpha)_*\colon \bigoplus_\alpha \widetilde{E}_*(X_\alpha) \longrightarrow \widetilde{E}_*(\bigvee_\alpha X_\alpha).$$

The functoriality of a reduced homology theory $\widetilde{E}_*$ along with the long exact sequence forces $\widetilde{E}_*(*) = 0$. Since $\widetilde{E}_*$ sends homotopic maps to equal maps, it follows that $\widetilde{E}_*(Z) = 0$ for any $Z \simeq *$. These simple consequences of the axioms allow us to prove that reduced homology theories are "stable under suspension" in the sense of the following lemma.

Lemma 1.1.18 *Let $\widetilde{E}_*$ denote a reduced homology theory, then for any pointed CW-complex X there is a natural isomorphism*

$$\widetilde{E}_{*+1}(\Sigma X) \cong \widetilde{E}_*(X).$$

Proof Consider the CW-pair (CX, X) and the induced natural long exact sequence

$$\cdots \xrightarrow{\partial_{n+1}} \widetilde{E}_n(X) \xrightarrow{i_n} \widetilde{E}_n(CX) \xrightarrow{q_n} \widetilde{E}_n(CX/X) \xrightarrow{\partial_n} \widetilde{E}_{n-1}(X) \xrightarrow{i_{n-1}} \widetilde{E}_{n-1}(CX) \longrightarrow \cdots .$$

Since X has a non-degenerate basepoint, the reduced and unreduced suspensions are naturally homotopy equivalent. The unreduced suspension of X is naturally isomorphic to CX/X, and $\widetilde{E}_*(CX) = 0$ for all n. Thus we have the desired natural isomorphism

$$\widetilde{E}_{*+1}(\Sigma X) \cong \widetilde{E}_{*+1}(CX/X) \xrightarrow{\cong} \widetilde{E}_*(X). \qquad \square$$

The standard example of a reduced homology theory is reduced singular homology (or reduced cellular homology), which is defined as

$$\widetilde{H}_*(X) = \ker(H_*(X) \longrightarrow H_*(*)).$$

We have another example linking homology theories to stable homotopy theory.

Example 1.1.19 Stable homotopy groups define a reduced homology theory. The key point of attention is proving that the long exact sequence of stable homotopy groups of a CW-pair (X, A) is of the same form as in the long exact sequence axiom for reduced homology theories. Corollary 1.1.9 applied to the stable setting proves that $\pi_*^{\text{stable}}(X, A) \cong \pi_*^{\text{stable}}(X/A)$. This result combined with the isomorphism between $\Sigma(X/A)$ and $\Sigma X/\Sigma A$ gives the boundary maps and the long exact sequence.

Unstable homotopy groups do not form a homology theory. In general, $\pi_1(X)$ is not an abelian group. More substantially, the long exact sequence of a CW-pair cannot be used to define boundary maps. For example,

$$\pi_3(D^2, S^1) = 0 \;\text{ but }\; \pi_3(D^2/S^1) = \pi_3(S^2) = \mathbb{Z}.$$

A related idea is that of reduced cohomology theories. The axioms are dual to that of reduced homology theories.

Definition 1.1.20 A *reduced cohomology theory* is a contravariant functor $\widetilde{E}^*$ from pointed CW-complexes to graded abelian groups Ab_* satisfying the list of axioms that follow. For a map $f\colon X \longrightarrow Y$ of pointed CW-complexes, we write

$$f^* = \widetilde{E}^*(f)\colon \widetilde{E}^*(Y) \longrightarrow \widetilde{E}^*(X)$$

and let f^n be the nth level of this map of graded abelian groups.

1. If $f \simeq g$, then $f^* = g^*$.
2. For each CW-pair (X, A) there is a *coboundary map* $\partial^*\colon \widetilde{E}^*(A) \longrightarrow \widetilde{E}^{*+1}(X/A)$, which is natural in pointed CW-pairs.

3. If (X, A) is a CW-pair, then the inclusion map $i\colon A \longrightarrow X$, the quotient map $q\colon X \longrightarrow X/A$ and the coboundary map ∂^* give a natural long exact sequence
$$\cdots \xrightarrow{\partial^{n-1}} \widetilde{E}^n(X/A) \xrightarrow{q^n} \widetilde{E}^n(X) \xrightarrow{i^n} \widetilde{E}^n(A) \xrightarrow{\partial^n} \widetilde{E}^{n+1}(X/A) \xrightarrow{i^{n+1}} \cdots.$$
4. For a set of spaces $\{X_\alpha\}_\alpha$, the maps $i_\alpha : X_\alpha \longrightarrow \bigvee_\alpha X_\alpha$ induce isomorphisms
$$\prod_\alpha (i_\alpha)^*\colon \widetilde{E}^*(\bigvee_\alpha X_\alpha) \longrightarrow \prod_\alpha \widetilde{E}^*(X_\alpha).$$

We have an analogous result showing how cohomology and suspension interact.

Lemma 1.1.21 *Let $\widetilde{E}^*$ denote a reduced cohomology theory, then for any pointed CW-complex X there is a natural isomorphism*
$$\widetilde{E}^*(X) \longrightarrow \widetilde{E}^{*+1}(\Sigma X)$$
induced by the coboundary map. □

Remark 1.1.22 We say that two cohomology theories $\widetilde{E}^*$ and $\hat{E}^*$ are *isomorphic* if there is a bijective natural transformation $\widetilde{E}^* \longrightarrow \hat{E}^*$ which is compatible with the coboundary maps.

This extra requirement is to ensure that we have commutative diagrams
$$\begin{array}{ccc} \widetilde{E}^n(A) & \xrightarrow{\partial^n} & \widetilde{E}^{n+1}(X/A) \\ \downarrow & & \downarrow \\ \hat{E}^n(A) & \xrightarrow{\partial^n} & \hat{E}^{n+1}(X/A), \end{array}$$
which we would not obtain from just a natural transformation, as the boundary maps ∂^n are not induced by morphisms of spaces.

A major result about reduced cohomology theories of pointed spaces is that each level of a reduced cohomology theory $\widetilde{E}^*$ can be *represented* by a connected pointed CW-complex. The proof constructs a CW-complex with the correct property by adding cells according to the data of $\widetilde{E}^n(S^k)$ for varying k. The assumption of connectedness is necessary for uniqueness. For example, if X is a connected pointed space, Z is a pointed space, and $Z \amalg *$ is the disjoint union of Z with a point, then in the homotopy category of pointed spaces we have
$$[X, Z] \cong [X, Z \amalg *].$$

Theorem 1.1.23 (Brown Representability) *Let $\widetilde{E}^*$ be a reduced cohomology theory. Then for each $n \in \mathbb{Z}$ there is a connected pointed CW-complex K_n such that for each connected pointed CW-complex X there is a natural isomorphism*

$$\widetilde{E}^n(X) \cong [X, K_n],$$

where the right hand side denotes maps in the homotopy category of pointed topological spaces. Moreover, the spaces K_n are unique up to homotopy equivalence.

However, the sequence of spaces $\{K_n\}_{n\in\mathbb{Z}}$ does not determine $\widetilde{E}^n$. That is, a collection of pointed CW-complexes does not determine a reduced cohomology theory, as there is no good way to construct coboundary maps. Thus, we need to add some more structure to the sequence of spaces in order to recover cohomology theories.

Combining the above theorem with our lemma on suspensions, we have the following.

Corollary 1.1.24 *Let $\widetilde{E}^*$ be a reduced cohomology theory represented by the connected pointed CW-complexes $\{K_n\}_{n\in\mathbb{Z}}$. Then for any pointed connected CW-complex A we have isomorphisms*

$$[A, K_n] \cong \widetilde{E}^n(A) \cong \widetilde{E}^{n+1}(\Sigma A) \cong [\Sigma A, K_{n+1}] \cong [A, \Omega K_{n+1}],$$

which are natural in maps of pointed CW-complexes A. □

If we let $A = K_n$, then the image of the identity map of K_n gives us maps

$$\alpha_n : K_n \longrightarrow \Omega K_{n+1}$$

for each n, which we call the *structure maps*. Naturality then implies that the image of some $f : A \longrightarrow K_n$ under the isomorphism

$$[A, K_n] \cong \widetilde{E}^n(A) \cong \widetilde{E}^{n+1}(\Sigma A) \cong [\Sigma A, K_{n+1}] \cong [A, \Omega K_{n+1}]$$

is $\alpha_n \circ f$. The CW-complexes K_n are connected, so they are path-connected. Hence, if we let $A = S^k$ for varying $k \in \mathbb{N}$, we see that α_n is a weak homotopy equivalence. We now prove that the spaces K_n and the maps α_n are sufficient to determine a reduced cohomology theory.

Proposition 1.1.25 *A sequence of pointed CW-complexes $\{K_n\}_{n\in\mathbb{Z}}$ with weak homotopy equivalences $\alpha_n : K_n \longrightarrow \Omega K_{n+1}$ for each $n \in \mathbb{Z}$ determines a reduced cohomology theory $\widetilde{E}^*$ by*

$$\widetilde{E}^n(X) = [X, K_n],$$

with coboundary maps induced by

$$\alpha_n \colon K_n \longrightarrow \Omega K_{n+1}.$$

Let $\{K'_n, \alpha'_n\}_{n\in\mathbb{Z}}$ be another collection of pointed CW-complexes and weak homotopy equivalences determining a reduced cohomology theory $\hat{E}^*$. For each $n \in \mathbb{Z}$, let $f_n \colon K_n \longrightarrow K'_n$ be a pointed weak homotopy equivalence such that the square

$$\begin{array}{ccc} K_n & \xrightarrow{\alpha_n} & \Omega K_{n+1} \\ \downarrow{\scriptstyle f_n} & & \downarrow{\scriptstyle \Omega f_{n+1}} \\ K'_n & \xrightarrow{\alpha'_n} & \Omega K'_{n+1} \end{array}$$

commutes for all n. Then the maps f_n induce a natural isomorphism of cohomology theories $\widetilde{E}^* \longrightarrow \hat{E}^*$.

Proof We prove that the data of the spaces K_n together with the maps α_n define a reduced cohomology theory. The functoriality axiom, homotopy axiom and the wedge axiom all follow immediately from the definition. All that is left is to define the coboundary maps and show they form part of a long exact sequence.

For a CW-pair (X, A), consider the cofibration or Puppe sequence (using reduced mapping cones and reduced suspensions)

$$A \xrightarrow{i} X \xrightarrow{q} X/A \xrightarrow{d} \Sigma A \xrightarrow{\Sigma i} \Sigma X \xrightarrow{\Sigma q} \Sigma(X/A) \longrightarrow \cdots$$

given by iterating the mapping cone construction. Since $K_n \simeq \Omega^2 K_{n+2}$, applying the functor $[-, K_n]$ gives an exact sequence of abelian groups

$$[A, K_n] \longleftarrow [X, K_n] \longleftarrow [X/A, K_n] \longleftarrow [\Sigma A, K_n] \longleftarrow [\Sigma X, K_n] \longleftarrow \cdots.$$

This works for each $n \in \mathbb{Z}$. As

$$[\Sigma A, K_n] \cong [A, K_{n-1}],$$

the collection of exact sequences for varying n can be patched together using the maps α_n to give an exact sequence

$$\cdots \longleftarrow [A, K_n] \longleftarrow [X, K_n] \longleftarrow [X/A, K_n] \longleftarrow [A, K_{n+1}] \longleftarrow [X, K_{n+1}] \longleftarrow \cdots$$

and the first statement is complete.

The second statement follows straight from the construction. The commuting squares involving the structure maps α_n and α'_n ensure that one obtains an isomorphism of cohomology theories rather than simply an isomorphism of each level. □

We may call a sequence of spaces $\{K_n\}_{n\in\mathbb{Z}}$ with structure maps

$$\alpha_n\colon K_n \longrightarrow \Omega K_{n+1}$$

that are weak homotopy equivalences a "spectrum" (we discuss these objects further in Section 1.3). The following result is a formal statement of the slogan "spectra represent cohomology theories".

Corollary 1.1.26 *A reduced cohomology theory $\widetilde{E}^*$ determines, and is determined by, a sequence of connected pointed CW-complexes $\{K_n\}_{n\in\mathbb{Z}}$ with pointed weak homotopy equivalences $\alpha_n\colon K_n \longrightarrow \Omega K_{n+1}$. The spaces K_n and the maps α_n are unique up to homotopy equivalence.*

Proof One direction is given by Proposition 1.1.25. For the converse, let

$$\{K_n, \alpha_n\}_{n\in\mathbb{Z}} \quad \text{and} \quad \{K'_n, \alpha'_n\}_{n\in\mathbb{Z}}$$

be two collections of connected pointed CW-complexes and weak homotopy equivalences that represent $\widetilde{E}^*$ on connected pointed CW-complexes from Theorem 1.1.23.

We first show that for any pointed CW-complex A, $\widetilde{E}^n(A) \cong [A, K_n]$ (and similarly for K'_n). That is, we may now remove the connectedness assumption on the input space A. Using Corollary 1.1.24, we have isomorphisms

$$\widetilde{E}^n(A) \cong \widetilde{E}^{n+1}(\Sigma A) \cong [\Sigma A, K_{n+1}] \cong [A, \Omega K_{n+1}] \cong [A, K_n].$$

Secondly, we compare the spaces K_n and K'_n, and the maps α_n and α'_n. Theorem 1.1.23 gives us homotopy equivalences $p_n\colon K_n \to K'_n$. Since the α_n are constructed from the suspension isomorphisms of the cohomology theory $\widetilde{E}^*$, the square

$$\begin{array}{ccc} [A, K_n] & \xrightarrow{(\alpha_n)_*} & [A, \Omega K_{n+1}] \\ \downarrow{\scriptstyle (p_n)_*} & & \downarrow{\scriptstyle (\Omega p_{n+1})_*} \\ [A, K'_n] & \xrightarrow{(\alpha'_n)_*} & [A, \Omega K'_{n+1}] \end{array}$$

commutes for any pointed CW-complex A. If we set $A = K_n$, then the identity map in the top left corner is sent to

$$\alpha'_n \circ p_n \simeq \Omega p_{n+1} \circ \alpha_n$$

in the bottom right corner. Hence the square that follows commutes up to homotopy.

$$\begin{array}{ccc} K_n & \xrightarrow{\alpha_n} & \Omega K_n \\ {\scriptstyle p_n}\downarrow & & \downarrow{\scriptstyle \Omega p_{n+1}} \\ K'_n & \xrightarrow{\alpha'_n} & \Omega K'_n \end{array}$$

The map p_n is a homotopy equivalence, and Ωp_{n+1} is a weak homotopy equivalence between spaces of the homotopy type of a CW-complex, hence it is also a homotopy equivalence. Hence α_n and α'_n are unique up to homotopy equivalence. □

Example 1.1.27 The spaces that correspond to reduced singular cohomology with coefficients in the abelian group G are the Eilenberg–Mac Lane spaces $K(G, n)$ for $n \geqslant 0$ and $*$ for negative n.

The structure maps come from the universal properties of the spaces. That is, $\Omega K(G, n)$ is an Eilenberg–Mac Lane space for G of level $n-1$ and hence is weakly homotopy equivalent to $K(G, n-1)$.

Example 1.1.28 The spaces that correspond to reduced complex K-theory are $BU \times \mathbb{Z}$ in even degrees and U in odd degrees. This periodicity reflects Bott periodicity. We will introduce K-theory in more detail in Subsection 7.4.2.

A sequence of CW-complexes together with structure maps behaves very well with respect to stable homotopy groups. Consider a sequence of connected pointed CW-complexes $\{K_n\}_{n \in \mathbb{Z}}$ with pointed weak homotopy equivalences

$$\alpha_n \colon K_n \longrightarrow \Omega K_{n+1}.$$

It follows that we have maps $\sigma_n \colon \Sigma K_n \longrightarrow K_{n+1}$ and a commuting diagram

$$\begin{array}{ccccc} [S^{k+a}, K_a] & \xrightarrow{\Sigma} & [S^{k+a+1}, \Sigma K_a] & \xrightarrow{(\sigma_a)_*} & [S^{k+a+1}, K_{a+1}] \\ & \searrow^{(\alpha_a)_*} & & & \downarrow \cong \\ & & & & [S^{k+a}, \Omega K_{a+1}]. \end{array}$$

As the α_n are weak homotopy equivalences, we obtain that the nth stable homotopy group of K_0 is

$$\pi_n^{\text{stable}}(K_0) = \operatorname{colim}\,(\pi_n(K_0) \xrightarrow{(\alpha_0)_*} \pi_n(\Omega K_1) \xrightarrow{(\alpha_1)_*} \pi_n(\Omega^2 K_2) \xrightarrow{(\alpha_2)_*} \cdots) = \pi_n(K_0).$$

Hence the stable homotopy groups of K_0 are simply the homotopy groups of K_0. A similar calculation for stable homotopy classes of maps (Definition 1.1.16) shows that

$$[X, K_n]^s = [X, K_n] = \widetilde{E}^n(X).$$

The results of this subsection lead to a natural question: Can we make a category with objects given by "spectra", sequences of connected pointed CW-complexes $\{K_n\}_{n\in\mathbb{Z}}$ together with maps

$$\alpha_n \colon K_n \longrightarrow \Omega K_{n+1}$$

that are pointed weak homotopy equivalences? The morphisms would be sequences of maps compatible with the structure maps up to some notion of homotopy. We would then hope that this category would be strongly related to the category of reduced cohomology theories. If we are able to construct such a category, we would then want to know how it relates to the (theorised) category of CW-complexes described the end of Subsection 1.1.2, whose morphisms are the stable homotopy classes of maps.

We will see in Section 1.3 that we can make a category of "spectra". Unfortunately, the most straightforward notion of homotopy equivalence in this category is poorly behaved. Example 1.3.6 gives two spectra which will represent the same cohomology theory but are not homotopy equivalent.

1.1.4 Properties of a Stable Homotopy Category

Having seen some instances of stability, we would like to have a category in which to study those phenomena. More precisely, we would like to have a category satisfying the following list of properties. We will call this category the *stable homotopy category*, $\mathcal{SHC}$, and use $[-,-]$ to denote maps in this category.

We would like the stable homotopy category to satisfy the following list of conditions and properties.

1. There is an adjunction

$$\Sigma^\infty \colon \mathrm{Ho}(\mathrm{Top}_*) \rightleftarrows \mathcal{SHC} \colon \Omega^\infty.$$

2. Let A and B be pointed CW-complexes. If A has only finitely many cells, there is a natural isomorphism

$$[\Sigma^\infty A, \Sigma^\infty B] \cong [A, B]^s$$

where $[A, B]^s$ is the set of stable homotopy classes of maps of spaces from Definition 1.1.16.

3. The sets of maps in $\mathcal{SHC}$ can be equipped with the structure of graded abelian groups, and composition is bilinear. We will use $[-,-]_*$ for the graded set of maps.
4. The stable homotopy category has arbitrary (small) products and coproducts. Finite products and coproducts coincide.
5. For A a pointed CW-complex and $X, Y \in \mathcal{SHC}$, there are objects $X \wedge A$ and $F(A, Y)$ in $\mathcal{SHC}$ such that
$$[X \wedge A, Y] \cong [X, F(A, Y)],$$
and for any pointed CW-complex B, there is an isomorphism in $\mathcal{SHC}$
$$(\Sigma^\infty B) \wedge A \cong \Sigma^\infty(B \wedge A).$$
6. The functor
$$(-) \wedge S^1 \colon \mathcal{SHC} \longrightarrow \mathcal{SHC}$$
is an equivalence of categories.
7. Given a reduced cohomology theory $\widetilde{E}^*$, there is an object $E \in \mathcal{SHC}$ such that
$$\widetilde{E}^*(A) = [\Sigma^\infty A, E]_{-*}$$
for any pointed CW-complex A. Moreover, the object E is unique up to isomorphism. We say that E *represents* $\widetilde{E}^*$.
8. Every $E \in \mathcal{SHC}$ defines a reduced cohomology theory on pointed CW-complexes by
$$A \mapsto [\Sigma^\infty A, E]_* = \widetilde{E}^*(A).$$
9. A map $E_1 \longrightarrow E_2$ in $\mathcal{SHC}$ induces a map of cohomology theories
$$\widetilde{E}_1^*(X) \longrightarrow \widetilde{E}_2^*(X).$$
10. There is a monoidal product $\wedge$ on $\mathcal{SHC}$ with
$$\Sigma^\infty A \wedge \Sigma^\infty B \cong \Sigma^\infty(A \wedge B)$$
for all pointed CW-complexes A and B.
11. There is an internal function object $F(-,-)$ on $\mathcal{SHC}$ so that for $X, Y, Z \in \mathcal{SHC}$
$$[X \wedge Y, Z] \cong [X, F(Y, Z)].$$
12. Every $E \in \mathcal{SHC}$ defines a reduced homology theory via
$$E_n(X) = \pi_n(E \wedge X).$$

In Sections 1.2 and 1.3, we will construct some categories that satisfy a few of these properties to indicate the difficulty of achieving all properties at once. We will see our first construction of the stable homotopy category in Chapter 2, however, the proof that $\mathcal{SHC}$ satisfies the last three points will require modern categories of spectra as in Chapter 5. In Section 5.7 we will justify calling it *the* stable homotopy category by giving a uniqueness result.

Once we have a monoidal product on $\mathcal{SHC}$, we can define the homology of objects $X \in \mathcal{SHC}$ as

$$E_*(X) = \pi_*(E \wedge X).$$

We will show in Proposition 5.1.7 that this generalises the characterisation of homology of spaces given in the preceding list.

1.2 The Spanier–Whitehead Category

We introduce the Spanier–Whitehead category as a first attempt at making the stable homotopy category. The original intention for this construction was to have a good place to study Spanier–Whitehead duality [SW55].

As a candidate for $\mathcal{SHC}$, this category has a bilinear addition on its sets of maps and the suspension functor is an equivalence, but the Spanier–Whitehead category does not satisfy all items of Subsection 1.1.4. For example, it does not have countable coproducts, and there are reduced cohomology theories that it cannot represent. We can think of this as the category not having enough objects.

Definition 1.2.1 The *Spanier–Whitehead category* $\mathcal{SW}$ is defined as follows. Its objects are the pointed finite CW-complexes, and morphisms are given by

$$[X, Y]^s := \operatorname{colim}_a[\Sigma^a X, \Sigma^a Y].$$

We may further define the set of graded maps in the Spanier–Whitehead category.

$$[X, Y]^s_q := \begin{cases} \operatorname{colim}_a[\Sigma^{a+q} X, \Sigma^a Y] & \text{if } q \geqslant 0 \\ \operatorname{colim}_a[\Sigma^a X, \Sigma^{a-q} Y] & \text{if } q < 0 \end{cases}$$

The composite of maps $f \in [X, Y]^s$ and $g \in [Y, Z]^s$ is given by choosing representatives $f_a \colon \Sigma^a X \longrightarrow \Sigma^a Y$ and $g_b \colon \Sigma^b Y \longrightarrow \Sigma^b Z$ and defining

$$g \circ f = \Sigma^a g_b \circ \Sigma^b f_a.$$

One can check that this composition is well-defined, associative, unital and extends to graded maps.

A minor issue with the Spanier–Whitehead category is that defining Ω on $\mathcal{SW}$ is more complicated, as the loop functor does not preserve CW-complexes. Hence, for X a pointed CW-complex, we interpret $\Omega X \in \mathcal{SW}$ to be a CW-approximation to the loop space of X. On the positive side, this category has many useful properties related to stable homotopy theory. Firstly, we can naturally define an addition on the sets of maps. This result does not need our CW-complexes to be finite.

Lemma 1.2.2 *For pointed CW-complexes A and B, $[A, B]^s$ is naturally a (graded) abelian group and composition is bilinear.*

Proof For pointed CW-complexes A and B, the set of pointed homotopy classes of maps $[\Sigma^2 A, B]$ is an abelian group in a natural manner compatible with composition of maps. It follows that

$$[A, B]^s = \mathrm{colim}_a[\Sigma^a A, \Sigma^a B]$$

has the same properties, and this extends to graded maps. □

The second useful property is that suspension is an equivalence on $\mathcal{SW}$. This property requires that the CW-complexes are finite.

Proposition 1.2.3 *If $f\colon X \longrightarrow Y$ is a map of finite pointed CW-complexes that induces an isomorphism on stable homotopy groups, then $[f] \in [X, Y]^s$ is an isomorphism.*

The suspension functor $\Sigma\colon \mathcal{SW} \longrightarrow \mathcal{SW}$ induces isomorphisms on sets of maps, that is, it is a full and faithful functor.

Proof The last statement follows from the definition of sets, of maps in terms of colimits.

Consider a map $f\colon X \longrightarrow Y$ that induces isomorphisms on all stable homotopy groups, then $Qf\colon QX \longrightarrow QY$ (Definition 1.1.14) is a weak homotopy equivalence of pointed spaces. If A is a finite pointed CW-complex, it is a compact space, and so a map from A into a sequential homotopy colimit is given by a map from A into some term of that colimit. Thus we have isomorphisms

$$[A, QX]_* = [A, \mathrm{hocolim}_a\, \Omega^a\Sigma^a X]_* \cong \mathrm{colim}_a[A, \Omega^a\Sigma^a X]_* \cong [A, X]^s_*,$$

and it follows that we have an isomorphism

$$[f]_*\colon [A, X]^s_* \longrightarrow [A, Y]^s_*.$$

When $A = Y$, we use the surjectivity of $[f]_*$ to obtain a right homotopy inverse g to f. When $A = X$, the injectivity of $[f]_*$ shows that $[g]$ and $[f]$ are inverse isomorphisms in $\mathcal{SW}$. □

We give a similar construction to $\mathcal{SW}$ that is also referred to as the Spanier–Whitehead category. This category has no finiteness assumptions on the CW-complexes.

Definition 1.2.4 The category $\widehat{\mathcal{SW}}$ is defined as follows. Its objects are pairs (X, m) for X a CW-complex and $m \in \mathbb{Z}$, and morphisms are given by

$$\{(X, m), (Y, n)\} = \mathrm{colim}_a[\Sigma^{m+a} X, \Sigma^{n+a} Y].$$

Similarly, graded maps are given by the extension

$$\{(X, m), (Y, n)\}_b = \{(X, m + b), (Y, n)\},$$

where $b \in \mathbb{Z}$. We may define a shift suspension functor s by $s(X, n) = (X, n+1)$ on objects. If $\alpha\colon (X, m) \longrightarrow (Y, n)$ is represented by

$$f\colon \Sigma^{a+m} X \to \Sigma^{a+n} Y,$$

then $s\alpha$ is defined to be the equivalence class of Σf in the colimit.

One can check that the shift suspension is naturally equivalent to the topological suspension $(\Sigma X, n)$ and that each object (X, n) has an obvious desuspension, namely $(X, n - 1)$. Hence we have the following.

Lemma 1.2.5 *The suspension functor is an equivalence on* $\widehat{\mathcal{SW}}$. □

The category $\widehat{\mathcal{SW}}$ is richly structured by Margolis [Mar83, Chapter 1, Theorem 7], $\widehat{\mathcal{SW}}$ is an example of a *triangulated category*, the subject of Chapter 4.

There is a full and faithful functor from $\mathcal{SW}$ to $\widehat{\mathcal{SW}}$ given by sending a finite CW-complex X to $(X, 0)$ and acting as the identity on sets of maps. As noted in [Mar83, Remark, Page 9], this is not an equivalence, as $\mathcal{SW}$ is not closed under desuspension.

We may relate the categories $\mathcal{SW}$ and $\widehat{\mathcal{SW}}$ to stable homotopy groups and cohomology theories.

Theorem 1.2.6 *For X a finite pointed CW-complex, the functor*

$$\widetilde{X}^* = [-, X]^s_{-*} = \{(-, 0), (X, 0)\}_{-*}$$

from pointed CW-complexes to graded abelian groups defines a reduced cohomology theory, except for the wedge axiom.

There is a natural isomorphism of graded abelian groups

$$[S^0, X]^s_* \cong \pi^{\mathrm{stable}}_*(X).$$

Proof By Lemma 1.2.2, we know that the functor $\widetilde{X}^*$ takes values in graded abelian groups. If f and g are homotopic maps of CW-complexes, then $f^* = g^*$.

We construct the coboundary map and long exact sequence together. Consider a CW-pair (B, A). Applying $\widetilde{X}^n$ to the Puppe sequence (cofibration sequence)

$$A \longrightarrow B \longrightarrow B/A \longrightarrow \Sigma A \longrightarrow \cdots$$

gives an exact sequence

$$\cdots \longrightarrow \widetilde{X}^n(\Sigma A) \longrightarrow \widetilde{X}^n(B/A) \longrightarrow \widetilde{X}^n(B) \longrightarrow \widetilde{X}^n(A)$$

for each integer n. We want to patch these exact sequences together similarly to the proof of Proposition 1.1.25. To that end, for $n \geqslant 0$ we have isomorphisms

$$\begin{aligned} \widetilde{X}^n(\Sigma A) = [\Sigma A, X]^s_{-n} &= \operatorname{colim}_a [\Sigma^{1+a} A, \Sigma^{a+n} X]^s \\ &\cong \operatorname{colim}_b [\Sigma^b A, \Sigma^{b+n-1} X]^s \\ &= [A, X]^s_{-n+1} \\ &= \widetilde{X}^{n-1}(A). \end{aligned}$$

Similar statements apply to negative n. Define the coboundary to be the composite of this isomorphism with the map induced by $B/A \to \Sigma A$:

$$\widetilde{X}^{n-1}(A) \cong \widetilde{X}^n(\Sigma A) \longrightarrow \widetilde{X}^n(B/A).$$

The exactness of the sequences for $\widetilde{X}^n$ and these coboundary maps give the desired long exact sequence

$$\cdots \longrightarrow \widetilde{X}^{n-1}(A) \longrightarrow \widetilde{X}^n(B/A) \longrightarrow \widetilde{X}^n(B) \longrightarrow \widetilde{X}^n(A) \longrightarrow \cdots.$$

The statement about stable homotopy groups is immediate. □

For a finite wedge $\bigvee_{i=1}^n A_i$ of pointed CW-complexes, we have a natural isomorphism

$$\bigoplus_{i=1}^n [A_i, X]^s \longrightarrow [\bigvee_{i=1}^n A_i, X]^s.$$

It follows that $\widetilde{X}^*$ will send finite coproducts to finite direct sums. However, countable coproducts do not exist in $\mathcal{SW}$ or $\widehat{\mathcal{SW}}$ in general. For the first, this is simply that countable coproducts of finite CW-complexes will usually not be finite CW-complexes. For the second, the natural definition to attempt is to extend the countable coproduct of pointed spaces. This construction will not be the categorical coproduct, as the next example shows.

Example 1.2.7 The abelian group

$$\prod_{i \geqslant 0} \{(S^i, 0), (S^0, 0)\} = \prod_{i \geqslant 0} [S^i, S^0]^s = \prod_{i \geqslant 0} \pi_i^{\text{stable}}(S^0)$$

is not equal to

$$\left\{(\bigvee_{i\geqslant 0} S^i, 0), (S^0, 0)\right\} = \operatorname{colim}_a \left[\bigvee_{i\geqslant 0} S^{a+i}, S^a\right]$$
$$\cong \operatorname{colim}_a \prod\nolimits_{i\geqslant 0} [S^{a+i}, S^a]$$
$$= \operatorname{colim}_a \prod\nolimits_{i\geqslant 0} \pi_{a+i}(S^a).$$

In the last group, an element is an equivalence class of a sequence of homotopy class of maps $f_i \colon S^{a+i} \longrightarrow S^a$. Hence infinitely many terms f_i will be outside the stable range altogether.

A further major problem with $\mathcal{SW}$ and $\widehat{\mathcal{SW}}$ is that the cohomology theories of the form

$$\widetilde{X}^*(-) = [-, X]^s_{-*} = \{(-, 0), (X, 0)\}_{-*}$$

are by our convention bounded above on finite CW-complexes. Unpicking the definition, for $n \in \mathbb{N}$ and A a finite CW-complex, we have

$$\widetilde{X}^n(A) = [A, X]^s_{-n} = [A, \Sigma^n X]^s,$$

which is zero for n larger than the highest dimension cell of A. Hence we cannot represent, for example, K-theory in terms of $\mathcal{SW}$ or $\widehat{\mathcal{SW}}$, as it is periodic and therefore has cohomology groups in negative dimensions. This is expected, as we have already seen in Subsection 1.1.3 that more information is needed to represent a cohomology theory than simply a single topological space (and an integer).

Boardman gave a solution to this problem, see the lecture notes written by Vogt [Vog70]. The method is to formally add directed colimits to the Spanier–Whitehead category. This gives one of the first constructions of the stable homotopy category, but it is not a convenient category to work in. We find it preferable to construct the stable homotopy category using spectra.

We will relate $\mathcal{SW}$ and $\widehat{\mathcal{SW}}$ to stable homotopy category in Lemmas 5.1.2 and 5.1.3.

1.3 A First Attempt at Spectra

We give a second attempt at constructing the stable homotopy category. Given that the Spanier–Whitehead category does not have enough objects to represent all reduced cohomology theories, we try to make a homotopy category based on spectra.

Lima [Lim59] was the first to define spectra, and the definition has been revised many times. For this section we take our definition from G. Whitehead, [Whi62]. We will see that the most straightforward definition of homotopy on this category does not give a good candidate for $\mathcal{SHC}$. In particular, there will be spectra that represent the same reduced cohomology theory but are not homotopy equivalent. We can think of this as the category not having enough maps.

Definition 1.3.1 A *spectrum* X is a collection of pointed topological spaces X_n for $n \in \mathbb{N}$, with maps

$$\sigma_n^X : \Sigma X_n \longrightarrow X_{n+1}$$

called the *structure maps* of X. An Ω*-spectrum* Z is a spectrum such that the adjoint structure maps

$$\widetilde{\sigma}_n^Z : Z_n \longrightarrow \Omega Z_{n+1}$$

are weak homotopy equivalences.

A *map of spectra* $f : X \longrightarrow Y$ is a collection of continuous maps of pointed topological spaces $f_n : X_n \longrightarrow Y_n$, such that the square that follows commutes, up to homotopy, for each $n \in \mathbb{N}$.

$$\begin{array}{ccc} \Sigma X_n & \xrightarrow{\Sigma f_n} & \Sigma Y_n \\ \downarrow{\scriptstyle \sigma_n^X} & & \downarrow{\scriptstyle \sigma_n^Y} \\ X_{n+1} & \xrightarrow{f_{n+1}} & Y_{n+1} \end{array}$$

This definition is rather natural, given the earlier results on representing cohomology theories by sequences of spaces. By Corollary 1.1.26, a cohomology theory is determined by an Ω-spectrum whose spaces are connected CW-complexes. We could try to restrict to such spectra, but this excludes a number of simple and important examples, such as the sphere spectrum that we define below.

Example 1.3.2 The *sphere spectrum* $\mathbb{S}$ is defined by $\mathbb{S}_n = S^n$, with structure maps given by the canonical maps $\Sigma S^n \longrightarrow S^{n+1}$.

Example 1.3.3 The *suspension spectrum* $\Sigma^\infty K$ of a pointed topological space K is given by $S^n \wedge K$ in level n, with the structure maps being the canonical maps. Hence $\mathbb{S} = \Sigma^\infty S^0$.

Example 1.3.4 The *Eilenberg–Mac Lane spectrum* HG for the abelian group G is the spectrum with level n given by the space $K(G, n)$. The adjoint structure maps are given by (a choice of) the weak homotopy equivalence

$$K(G, n) \longrightarrow \Omega K(G, n + 1).$$

Given a space A and a spectrum X, we can make a new spectrum $X \wedge A$ with level n given by $X_n \wedge A$ and structure maps given by smashing the structure maps of X with A. Hence $\mathbb{S} \wedge A = \Sigma^\infty A$.

Given a space A and a spectrum X, we can make a new spectrum $F(A, X)$ with level n given by $\mathrm{Top}_*(A, X_n)$ and adjoint structure maps given by

$$\mathrm{Top}_*(A, X_n) \xrightarrow{\mathrm{Top}_*(A, \widetilde{\sigma}_n^X)} \mathrm{Top}_*(A, \Omega X_{n+1}) \cong \Omega \mathrm{Top}_*(A, X_{n+1}).$$

The last isomorphism is our first encounter with a recurring issue in spectra, namely the twist isomorphism

$$\tau_{A,B} \colon A \wedge B \longrightarrow B \wedge A.$$

For example, if we wanted to define the smash product of a spectrum X with a space A as $A \wedge X$, then we would need to define the structure maps of $A \wedge X$ by

$$\Sigma(A \wedge X_n) = S^1 \wedge A \wedge X_n \xrightarrow{\tau_{S^1 \wedge A, X_n}} A \wedge S^1 \wedge X_n \xrightarrow{A \wedge \sigma_n^X} A \wedge X_{n+1}.$$

Since the twist maps are not always homotopic to the identity (τ_{S^1,S^1}, for example) we will need to keep track of where they are needed.

The functors $(- \wedge A, F(A, -))$ demonstrate why it is difficult to only consider Ω-spectra: while $F(A, -)$ preserves Ω-spectra, the functor $- \wedge A$ will not, in general. Hence we need a way to turn a spectrum into an Ω-spectrum. We will discuss this in Section 2.4. For now, we simply assume that for every spectrum X there is an Ω-spectrum RX and a levelwise weak homotopy equivalence $RX \xrightarrow{\sim} X$.

Using the smash product of spectra with pointed spaces, we can define homotopy between maps of spectra in a entirely analogous way to homotopy of pointed spaces.

Definition 1.3.5 Let X and Y be spectra. A *homotopy* between two maps $f, g \colon X \longrightarrow Y$ of spectra is a map

$$H \colon X \wedge [0, 1]_+ \longrightarrow Y$$

such that $H \circ i_0 = f$ and $H \circ i_1 = g$, where i_0 and i_1 are the two endpoint inclusions of S^0 into $[0, 1]_+$.

Hence, a homotopy between maps of spectra f and g is a collection of homotopies H_n between f_n and g_n which are compatible with the structure maps of X and Y.

We may define a category whose object class is given by the class of spectra, and morphism class given by the class of homotopy classes of maps of spectra.

We have already seen that such objects determine reduced cohomology theories in Proposition 1.1.25. However, this category does not have enough maps to be a good candidate for the stable homotopy category. The following example illustrates this by giving two spectra which are not homotopy equivalent but represent the same cohomology theory.

Example 1.3.6 Consider the spectrum $\mathbb{S}^{(1)}$, which is the sphere spectrum at all levels except 0, where it is a point. The structure maps are the canonical maps and $* \longrightarrow S^1$ in degree 0. There is a levelwise inclusion

$$\mathbb{S}^{(1)} \longrightarrow \mathbb{S},$$

which becomes a levelwise weak homotopy equivalence

$$R\mathbb{S}^{(1)} \longrightarrow R\mathbb{S}.$$

By Proposition 1.1.25, these two Ω-spectra represent the same cohomology theory. (Here, we assume that the homotopy colimit functor defining R takes values in CW-complexes.)

However, each map

$$f\colon \mathbb{S} \longrightarrow \mathbb{S}^{(1)}$$

is levelwise contractible. At level 0,

$$f_0\colon S^0 \longrightarrow *$$

is the constant map to the one-point space. At level 1, the homotopy commuting square forces f_1 to be homotopic to the composite

$$S^1 \longrightarrow * \longrightarrow S^1.$$

By induction, each f_n is homotopic to a constant map.

The example also makes it clear that homotopy equivalence is too strong a condition for a suitable notion of weak equivalence of spectra. One obvious candidate would be inducing an isomorphism on stable homotopy groups: for X a spectrum and $n \in \mathbb{Z}$, the structure maps of X allow us to define

$$\begin{aligned}\pi_n(X) &= \operatorname{colim}_a \pi_{n+a}(X_a)\\ &= \operatorname{colim}(\cdots \to [S^{n+a}, X_a] \xrightarrow{\Sigma} [S^{n+a+1}, \Sigma X_a] \xrightarrow{\sigma_*} [S^{n+a+1}, X_{a+1}] \to \cdots),\end{aligned}$$

see also Definition 2.2.1. It is evident that $\mathbb{S}^{(1)} \longrightarrow \mathbb{S}$ induces an isomorphism on all stable homotopy groups, but we do not have a suitable map of spectra inducing an inverse.

There are a number of solutions to this lack of maps from one spectrum to another. Adams [Ada74] defines the complicated notion of a cofinal subspectrum and considers (equivalences classes of) maps from cofinal subspectra of X to Y. Composition is then quite complicated, as one must find suitably compatible cofinal subspectra.

Another approach is to recognise that a similar problem occurs in topological spaces and simplicial sets. The functor $\mathrm{Top}_*(A, -)$ is homotopically poorly-behaved unless A is a CW-complex, and the functor $\mathrm{sSet}_*(-, Z)$ requires Z to be a Kan complex to be a useful homotopical functor. Hence to get the correct set of maps from X to Y in the homotopy category of spectra, we should require X to be a "CW-spectrum" and Y to be an Ω-spectrum. We then face the unpleasant choice between restricting to Ω-spectra that are also "CW-spectra", or defining a suitable CW-approximation functor and using both it and the functor R to define composition. A notable approach along these lines is that of Lewis, May and Steinberger [LMSM86], which defines a highly structured version of R that is a right adjoint and outputs an Ω-spectrum whose adjoint structure maps are homeomorphisms.

The preferred solution of the authors is to use model categories. This allows us to have a point-set model of spectra with a rich homotopy theory, whose homotopy category satisfies the list of conditions and properties of the stable homotopy category of Subsection 1.1.4. In particular, the objects of $\mathcal{SHC}$ will represent reduced cohomology theories, and $\mathcal{SHC}$ will contain the Spanier–Whitehead category $\mathcal{SW}$.

The machinery of model categories essentially gives us the functor R and a CW-approximation functor. Furthermore, it takes care of when these functors need to be applied. Another advantage of the model category approach is that (with a suitable base category) it will allow for a good definition of the smash product of spectra, which is associative and commutative before passage to the level of homotopy categories. This avoids numerous problems with the "handicrafted smash product" that is only useful at the homotopy category level. See Chapter 6 for more details.

2

Sequential Spectra and the Stable Homotopy Category

Bousfield and Friedlander defined the stable homotopy category in terms of the homotopy category of a model category of spectra [BF78]. We will construct this model category following an approach similar to [MMSS01]. We do this in two stages. First, we construct a levelwise model structure, and then build the desired stable model structure from the levelwise model structure. We then formally define the stable homotopy category to be the homotopy category of sequential spectra.

A sequential spectrum is a sequence of pointed topological spaces (and structure maps), thus, a natural candidate for an analogue of weak homotopy equivalences are those maps of spectra inducing a weak homotopy equivalence at every level. However, we will see that these levelwise weak homotopy equivalences are not sufficient to define a class of weak equivalences leading to a meaningful stable homotopy theory. A key ingredient is the definition of homotopy groups of spectra $\pi_*(X)$ and π_*-isomorphisms. This generalises the notion of stable homotopy groups of topological spaces that we encountered earlier. Making the π_*-isomorphisms the weak equivalences of sequential spectra will give us a construction of our desired stable homotopy category.

The task of much of the later chapters is to prove that the homotopy category of sequential spectra satisfies the desired properties of the stable homotopy category (see Subsection 1.1.4). To achieve this, we will introduce the language of stable model categories and triangulated categories more formally in Chapters 3 and 4. Unfortunately, sequential spectra are not a monoidal model category, making it very difficult to prove directly that the homotopy category of sequential spectra has a suitable smash product. We resolve this in Chapter 6, using the more highly structured versions of spectra from Chapter 5.

Nevertheless, we will know enough about spectra and the stable homotopy category by the end of this chapter to introduce the reader to the most powerful tool for calculating the stable homotopy groups of spheres, namely, the Adams spectral sequence.

2.1 The Levelwise Model Structure

We discussed the motivation behind spectra in Chapter 1. In this chapter we define the category of sequential spectra, whose objects are similar to the spectra we have seen previously. The morphisms of sequential spectra are sequences of maps commuting strictly with the structure maps, rather than commuting up to homotopy.

We give a levelwise model structure on the category of sequential spectra, whose primary purpose is to help us construct the stable model structure of Section 2.3.

For this section, we take the suspension functor Σ to be $S^1 \wedge -$.

Definition 2.1.1 A *sequential spectrum* X is a sequence of pointed topological spaces X_n, $n \in \mathbb{N}$ with *structure maps*

$$\sigma_n^X : \Sigma X_n \longrightarrow X_{n+1}.$$

We denote the adjoints of the structure maps by

$$\widetilde{\sigma}_n^X : X_n \longrightarrow \Omega X_{n+1},$$

and call them the *adjoint structure maps*. A spectrum X is called an Ω-*spectrum* if the adjoint structure maps are weak homotopy equivalences. The *category of sequential spectra* $\mathcal{S}^{\mathbb{N}}$ is given by the following. The objects are the class of sequential spectra. A morphism

$$f : X \longrightarrow Y$$

in $\mathcal{S}^{\mathbb{N}}$ is a sequence of pointed maps of topological spaces $f_n : X_n \longrightarrow Y_n$ such that for each $n \in \mathbb{N}$, the square below commutes.

$$\begin{array}{ccc} \Sigma X_n & \xrightarrow{\mathrm{Id} \wedge f_n} & \Sigma Y_n \\ \downarrow \sigma_n^X & & \downarrow \sigma_n^Y \\ X_{n+1} & \xrightarrow{f_{n+1}} & Y_{n+1} \end{array}$$

We will often shorten "sequential spectrum" to "spectrum" and omit the superscript X from the (adjoint) structure maps.

Example 2.1.2 The *sphere spectrum* $\mathbb{S}$ is defined by $\mathbb{S}_n = S^n$ with structure maps given by the canonical maps $\Sigma S^n \longrightarrow S^{n+1}$.

For $n \in \mathbb{N}$, we write $\mathbb{S}^n = \Sigma^n \mathbb{S}$ and $\mathbb{S}^{-n} = \mathrm{F}_n^{\mathbb{N}} S^0$, with the latter defined in the next example.

Example 2.1.3 The *shifted suspension spectrum* $\mathrm{F}_d^{\mathbb{N}} K$ of a pointed topological space K with $d \in \mathbb{N}$ is defined by

$$(\mathrm{F}_d^{\mathbb{N}} K)_n = \begin{cases} S^{n-d} \wedge K & n \geqslant d \\ * & n < d. \end{cases}$$

The structure maps are given by the canonical maps and by the inclusion of a point for $n = d - 1$. To match other notations, we often write Σ^∞ for $\mathrm{F}_0^{\mathbb{N}}$.

The shifted suspension spectrum of a pointed space is part of an adjoint functor pair with the right adjoint sending a spectrum X to the space X_d

$$\mathrm{F}_d^{\mathbb{N}} : \mathrm{Top}_* \rightleftarrows \mathcal{S}^{\mathbb{N}} : \mathrm{Ev}_d^{\mathbb{N}}.$$

Example 2.1.4 Given a space A and a spectrum X, we can make a new spectrum $X \wedge A$ with level n given by $X_n \wedge A$ and structure maps given by smashing the structure maps of X with A. We may also define a spectrum $A \wedge X$ with level n given by $A \wedge X_n$ and structure maps given by smashing with A and using the twist map τ to exchange S^1 and A. The twist map induces an isomorphism of sequential spectra between $X \wedge A$ and $A \wedge X$.

Given a space A and a sequential spectrum X, we can make a new spectrum $\mathrm{Top}_*(A, X)$ with level n given by $\mathrm{Top}_*(A, X_n)$ and adjoint structure maps given by

$$\mathrm{Top}_*(A, X_n) \xrightarrow{\mathrm{Top}_*(A, \tilde{\sigma}_n^X)} \mathrm{Top}_*(A, \Omega X_{n+1}) \cong \Omega \mathrm{Top}_*(A, X_{n+1}).$$

We write ΣX for $S^1 \wedge X$ and ΩX for $\mathrm{Top}_*(S^1, X)$.

The point-set definition of the category of sequential spectra allows us to define products, coproducts, limits and colimits in a straightforward manner. The definition demonstrates how certain constructions are easier to define using the structure maps and others using the adjoint structure maps.

Definition 2.1.5 Given a diagram of spectra $\{X^{(i)}, \alpha_{i,j} \colon X^{(i)} \longrightarrow X^{(j)}\}$, we may form the colimit $\mathrm{colim}_i X^{(i)}$ and the limit $\lim_i X^{(i)}$ levelwise. The structure maps for the colimit and limit are

$$\Sigma \,\mathrm{colim}_i X^{(i)} \cong \mathrm{colim}_i \Sigma X^{(i)} \xrightarrow{\mathrm{colim}_i \sigma_i^X} \mathrm{colim}_i \Sigma X_{i+1}$$

$$\lim_i X^{(i)} \xrightarrow{\lim_i \tilde{\sigma}_i^X} \lim_i \Omega X^{(i)} \cong \Omega \lim_i X^{(i)}.$$

The category of sequential spectra has all small limits and colimits, as the category of pointed spaces does. The initial and terminal objects are $* = \Sigma^\infty(\text{point})$.

We now construct the levelwise model structure on spectra. It is a useful stepping stone to the stable model structure.

Proposition 2.1.6 *There is a* levelwise model structure *on sequential spectra, where the weak equivalences are the levelwise weak homotopy equivalences of pointed spaces. The fibrations are the class of levelwise Serre fibrations of pointed spaces. The cofibrations are called the* q-cofibrations.

This model structure is cofibrantly generated with generating sets given by

$$\begin{aligned} I^{\mathbb{N}}_{\text{level}} &= \{\mathrm{F}^{\mathbb{N}}_d S^{n-1}_+ \longrightarrow \mathrm{F}^{\mathbb{N}}_d D^n_+ \mid n, d \in \mathbb{N}\} \\ J^{\mathbb{N}}_{\text{level}} &= \{\mathrm{F}^{\mathbb{N}}_d D^n_+ \longrightarrow \mathrm{F}^{\mathbb{N}}_d (D^n \times [0,1])_+ \mid n, d \in \mathbb{N}\}. \end{aligned}$$

The q-cofibrations are, in particular, levelwise q-cofibrations of pointed topological spaces.

Proof We follow the Recognition Theorem for cofibrantly generated model structures, Theorem A.6.9.

1. The levelwise weak homotopy equivalences satisfy the two-out-of-three condition.
2. The domains of the generating sets are small with respect to the class of levelwise cofibrations of spaces.
3. A map in $J^{\mathbb{N}}_{\text{level}}$-cell is a levelwise weak equivalence and an $I^{\mathbb{N}}_{\text{level}}$-cofibration.
4. The class of maps with the right lifting property with respect to $I^{\mathbb{N}}_{\text{level}}$ is exactly the class of levelwise weak equivalences that also have the right lifting property with respect to $J^{\mathbb{N}}_{\text{level}}$.

The two-out-of-three property for the weak equivalences is evident.

As colimits are defined levelwise, a sequential colimit of pushouts of maps in $I^{\mathbb{N}}_{\text{level}}$ or $J^{\mathbb{N}}_{\text{level}}$ is in each level a sequential colimit of pushouts of q-cofibrations of pointed spaces, see Example A.1.5. Hence, each such map is a levelwise q-cofibration of pointed spaces.

For the smallness conditions, we prove that the domains of $I^{\mathbb{N}}_{\text{level}}$ are small with respect to the class of levelwise q-cofibrations. Consider a sequential diagram

$$Y^0 \longrightarrow Y^1 \longrightarrow Y^2 \longrightarrow \cdots$$

of spectra with each map a levelwise q-cofibration. The adjunction $(\mathrm{F}_d^{\mathbb{N}}, \mathrm{Ev}_d^{\mathbb{N}})$ for $d \in \mathbb{N}$ induces the first and last isomorphism of

$$\begin{aligned}
\operatorname{colim}_i \mathcal{S}^{\mathbb{N}}(\mathrm{F}_d^{\mathbb{N}} S_+^{n-1}, Y^i) &\cong \operatorname{colim}_i \mathrm{Top}_*(S_+^{n-1}, Y_d^i) \\
&\cong \mathrm{Top}_*(S_+^{n-1}, \operatorname{colim}_i Y_d^i) \\
&= \mathrm{Top}_*(S_+^{n-1}, \mathrm{Ev}_d^{\mathbb{N}} \operatorname{colim}_i Y^i) \\
&\cong \mathcal{S}^{\mathbb{N}}(\mathrm{F}_d^{\mathbb{N}} S_+^{n-1}, \operatorname{colim}_i Y^i),
\end{aligned}$$

and the second isomorphism comes from the smallness of S_+^{n-1} with respect to q-cofibrations of pointed spaces (see also Corollary A.7.10). A similar argument shows that the domains of $J_{\text{level}}^{\mathbb{N}}$ are small with respect to the class of levelwise acyclic q-cofibrations.

The lifting properties relating the generating sets and the (acyclic) fibrations also follow from this adjunction and the corresponding properties of the Serre model structure on Top_*. For example, consider the two lifting squares below, where X and Y are spectra, A and B are pointed spaces and $d \in \mathbb{N}$. The first square below has a lift if and only if the other does.

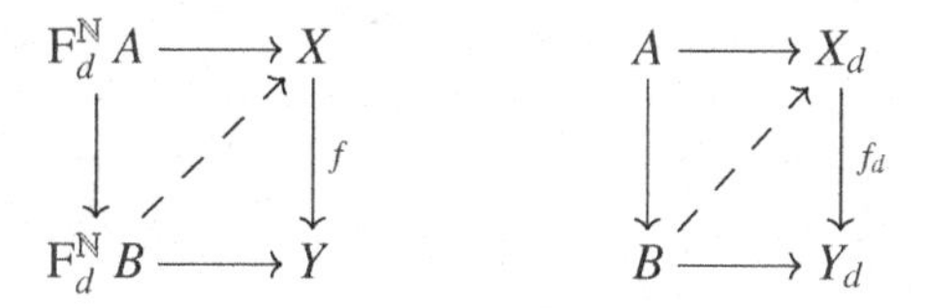

Assume that $f\colon X \to Y$ has the right lifting property with respect to $I_{\text{level}}^{\mathbb{N}}$. By the equivalence of the lifting diagrams above, it follows that each

$$f_d\colon X_d \longrightarrow Y_d$$

has the right lifting property with respect to $S_+^{n-1} \longrightarrow D_+^n$. Hence, each f_d is a levelwise acyclic fibration. The converse also holds, as does the equivalent statement for $J_{\text{level}}^{\mathbb{N}}$ and its converse. It follows that the class of levelwise weak homotopy equivalences with the right lifting property with respect to $J_{\text{level}}^{\mathbb{N}}$ is exactly the class of maps with the right lifting property with respect to $I_{\text{level}}^{\mathbb{N}}$.

One must also consider the set of maps $J_{\text{level}}^{\mathbb{N}}$-cell. A sequential colimit of pushouts of maps in $J_{\text{level}}^{\mathbb{N}}$ is in each level a sequential colimit of pushouts of acyclic cofibrations of the Serre model structure on pointed spaces. Hence, each such map is a levelwise weak homotopy equivalence by the cofibrant generation of the Serre model structure.

Since the maps in $J_{\text{level}}^{\mathbb{N}}$ have the left lifting property with respect to levelwise acyclic fibrations, it follows from Lemma A.6.11 that sequential colimits of pushouts of maps in $J_{\text{level}}^{\mathbb{N}}$ also have this left lifting property and hence are q-cofibrations. □

We can give an explicit description of the q-cofibrations. It may be useful to compare the following with Theorem A.7.1 and Example A.7.9.

Theorem 2.1.7 *Let X and Y be sequential spectra. A map $i\colon X \longrightarrow Y$ in the levelwise model structure is a q-cofibration if and only if*

$$i_0\colon X_0 \longrightarrow Y_0$$

is a q-cofibration and the induced map $\hat{\sigma}_n$ in the pushout diagram below

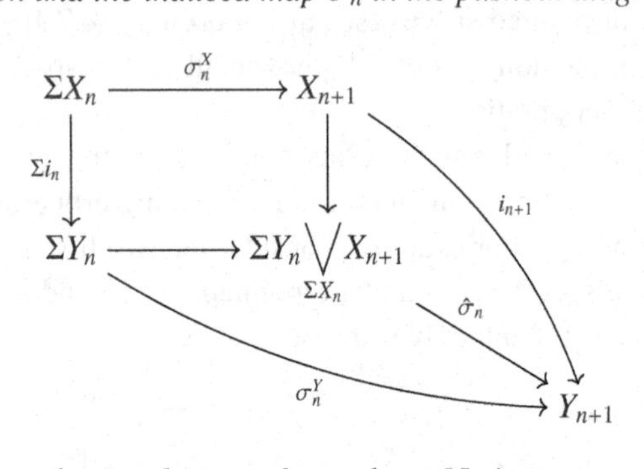

is a q-cofibration of pointed spaces for each $n \in \mathbb{N}$. A spectrum Z is q-cofibrant in the levelwise model structure if and only if Z_0 is a q-cofibrant pointed space and the maps $\Sigma Z_n \longrightarrow Z_{n+1}$ are q-cofibrations.

Proof Let $i\colon X \longrightarrow Y$ be a map of spectra such that i_0 is a q-cofibration and $\hat{\sigma}_n$ is a q-cofibration in Top_* for each $n \in \mathbb{N}$. We show that i is a levelwise q-cofibration in Top_*. Assume we have shown that

$$i_n\colon X_n \longrightarrow Y_n$$

is a q-cofibration. Then the left-hand map in the pushout square

$$\begin{array}{ccc} \Sigma X_n & \xrightarrow{\sigma_n^X} & X_{n+1} \\ {\scriptstyle \Sigma i_n}\downarrow & & \downarrow{\scriptstyle a_n} \\ \Sigma Y_n & \longrightarrow & \Sigma Y_n \bigvee_{\Sigma X_n} X_{n+1} \end{array}$$

is a q-cofibration, hence so is the right-hand map a_n. The map i_{n+1} is the composite $\hat{\sigma}_n \circ a_n$ and hence is the composite of two q-cofibrations.

To prove that i is a q-cofibration in the levelwise model structure, consider a lifting square in spectra

$$\begin{array}{ccc} X & \xrightarrow{f} & P \\ {\scriptstyle i}\downarrow & & \downarrow{\scriptstyle p} \\ Y & \xrightarrow[g]{} & Q \end{array}$$

with $p\colon P \longrightarrow Q$ a levelwise acyclic fibration. We want to construct a map of spectra $h\colon Y \longrightarrow P$ making the two triangles commute. At level 0, we can choose some lift h_0 in the Serre model structure on pointed spaces, as $X_0 \longrightarrow Y_0$ is a q-cofibration and $P_0 \longrightarrow Q_0$ is an acyclic fibration. Therefore, we obtain

$$\begin{array}{ccc} X_0 & \xrightarrow{f_0} & P_0 \\ {\scriptstyle i_0}\downarrow & \overset{h_0}{\nearrow} & \downarrow{\scriptstyle p_0} \\ Y_0 & \xrightarrow[g_0]{} & Q_0. \end{array}$$

At level 1, we must construct a lift h_1 that is compatible with our choice of h_0. The maps $\sigma_0^P \circ \Sigma h_0$ and f_1 induce a map

$$k_0\colon \Sigma Y_0 \bigvee_{\Sigma X_0} X_1 \longrightarrow P_1.$$

This map fits into the commuting square below, where we can choose a lift h_1, as $\hat{\sigma}_0$ is a q-cofibration and p_1 is an acyclic fibration.

$$\begin{array}{ccc} \Sigma Y_0 \bigvee_{\Sigma X_0} X_1 & \xrightarrow{k_0} & P_1 \\ {\scriptstyle \hat{\sigma}_0}\downarrow & \overset{h_1}{\nearrow} & \downarrow{\scriptstyle p_1} \\ Y_1 & \xrightarrow[g_1]{} & Q_1 \end{array}$$

Let b_0 be the inclusion $\Sigma Y_0 \longrightarrow \Sigma Y_0 \bigvee_{\Sigma X_0} X_1$. Then, the commutativity of the upper-left triangle implies

$$h_1 \circ \sigma_0^Y = h_1 \circ \hat{\sigma}_0 \circ b_0 = k_0 \circ b_0 = \sigma_0^P \circ \Sigma h_0,$$

so that h_1 is compatible with our choice of h_0. Continuing inductively, we obtain a lift h. Hence, i has the left lifting property with respect to acyclic fibrations and hence is a q-cofibration.

We now must prove the converse, namely, that the map $\hat{\sigma}_n$ induced by a q-cofibration of spectra

$$i\colon X \longrightarrow Y$$

is a q-cofibration in Top_* for each $n \in \mathbb{N}$ and that i_0 is a q-cofibration. The second statement is immediate: each level of i is a q-cofibration of Top_*. For the first, we argue via cofibrant generation of the levelwise model structure, see Section A.6. Every q-cofibration is a retract of a sequential colimit of pushouts of maps in

$$I^{\mathbb{N}}_{\text{level}} = \{\mathrm{F}^{\mathbb{N}}_d S^{n-1}_+ \longrightarrow \mathrm{F}^{\mathbb{N}}_d D^n_+ \mid n, d \in \mathbb{N}\}.$$

Consider a pushout square of spectra

$$\begin{array}{ccc} \mathrm{F}^{\mathbb{N}}_d A & \xrightarrow{f} & X \\ {\scriptstyle \mathrm{F}^{\mathbb{N}}_d j}\big\downarrow & & \big\downarrow{\scriptstyle i} \\ \mathrm{F}^{\mathbb{N}}_d B & \xrightarrow[g]{} & Y, \end{array}$$

where $j\colon A \longrightarrow B$ is a cofibration in Top_*. For $n \neq d-1$, the structure maps $\Sigma(\mathrm{F}^{\mathbb{N}}_d B)_n \longrightarrow (\mathrm{F}^{\mathbb{N}}_d B)_{n+1}$ are isomorphisms. Thus, we have isomorphisms

$$\begin{aligned} \Sigma Y_n \bigvee_{\Sigma X_n} X_{n+1} &\cong \left(\Sigma(\mathrm{F}^{\mathbb{N}}_d B)_n \bigvee_{\Sigma(\mathrm{F}^{\mathbb{N}}_d A)_n} \Sigma X_n \right) \bigvee_{\Sigma X_n} X_{n+1} \\ &\cong \Sigma(\mathrm{F}^{\mathbb{N}}_d B)_n \bigvee_{\Sigma(\mathrm{F}^{\mathbb{N}}_d A)_n} X_{n+1} \\ &\cong (\mathrm{F}^{\mathbb{N}}_d B)_{n+1} \bigvee_{(\mathrm{F}^{\mathbb{N}}_d A)_{n+1}} X_{n+1} \\ &\cong Y_{n+1}. \end{aligned}$$

The composite of these isomorphisms is exactly

$$\hat{\sigma}_n \colon \Sigma Y_n \bigvee_{\Sigma X_n} X_{n+1} \longrightarrow Y_{n+1}.$$

Thus, $\hat{\sigma}_n$ is a homeomorphism, and hence a q-cofibration, when $n \neq d-1$. When $n = d-1$, $\hat{\sigma}_n$ is given by

$$\Sigma Y_{d-1} \bigvee_{\Sigma X_{d-1}} X_d \cong * \bigvee_{*} \Sigma X_{d-1} \bigvee_{\Sigma X_{d-1}} X_d \cong X_d \xrightarrow{i_d} Y_d,$$

and we know that i_d is a q-cofibration for all d. Hence, for each n, $\hat{\sigma}_n$ is a q-cofibration.

Now consider a sequential diagram

$$(X = X^0 \xrightarrow{i^0} X^1 \xrightarrow{i^1} \cdots) \longrightarrow \operatorname{colim}_i X^i = Y,$$

where each map $i^k \colon X^k \longrightarrow X^{k+1}$ is given by a wedge of pushouts of maps in $I^{\mathbb{N}}_{\mathrm{level}}$. We know that each $\hat{\sigma}^k_n$ is a q-cofibration, and we want to prove that

$$\Sigma Y_n \bigvee_{\Sigma X_n} X_{n+1} \longrightarrow Y_{n+1}$$

is a q-cofibration for each $n \in \mathbb{N}$. Writing out the colimits defining Y, the above map is

$$\operatorname{colim}_k \Sigma X^k_n \bigvee_{\Sigma X_n} X_{n+1} \longrightarrow \operatorname{colim}_k X^k_{n+1}.$$

It suffices to prove that in the projective model structure on sequential diagrams in Top_*, the map

$$\Sigma X^\bullet_n \bigvee_{\Sigma X_n} X_{n+1} \longrightarrow X^\bullet_{n+1}$$

induced by the diagram below is a cofibration (as the colimit is a left Quillen functor with this model structure and hence preserves cofibrations).

$$\begin{array}{ccc}
\Sigma X^0_n \vee_{\Sigma X_n} X_{n+1} & \longrightarrow & X^0_{n+1} \\
\downarrow & & \downarrow \\
\Sigma X^1_n \vee_{\Sigma X_n} X_{n+1} & \longrightarrow & X^1_{n+1} \\
\downarrow & & \downarrow \\
\Sigma X^2_n \vee_{\Sigma X_n} X_{n+1} & \longrightarrow & X^2_{n+1} \\
\downarrow & & \downarrow \\
\vdots & & \vdots
\end{array}$$

This amounts to proving that the first map is a q-cofibration in Top_* (in our case it is the identity) and that for each square of the diagram, the map

$$\left(\Sigma X^{k+1}_n \bigvee_{\Sigma X_n} X_{n+1}\right) \bigvee_{\left(\Sigma X^k_n \bigvee_{\Sigma X_n} X_{n+1}\right)} X^k_{n+1} \longrightarrow X^{k+1}_{n+1}$$

is a q-cofibration in Top_*. By studying the domain, we see this map is isomorphic to

$$\Sigma X_n^{k+1} \bigvee_{\Sigma X_n^k} X_{n+1}^k \longrightarrow X_{n+1}^{k+1},$$

which we already know to be a q-cofibration. Hence, we see that the colimit is a q-cofibration. Therefore, we proved our statement for sequential colimits of pushouts of maps in $I^{\mathbb{N}}_{\text{level}}$.

The general case of a q-cofibration $i\colon X \longrightarrow Y$ is a retract of a sequential colimit of pushouts of maps in $I^{\mathbb{N}}_{\text{level}}$. Since retracts preserve cofibrations, we see that our i has the desired property, namely, that the induced $\hat{\sigma}_n$ are q-cofibrations of topological spaces.

The statement about cofibrant objects follows from setting $X = *$. □

This condition on cofibrant objects of the levelwise model structure has appeared in various forms before (see, for example, [LMSM86], or [EKMM97]), and these spectra have been called Σ-cofibrant spectra. A related definition in the early literature [Ada74] is that of a CW-spectrum and its stable cells. Using these ideas, one can extend many of the cell–based arguments on spaces to spectra.

Definition 2.1.8 A sequential spectrum X is a *CW-spectrum* if each X_n is a pointed CW-complex and each structure map $\Sigma X_n \to X_{n+1}$ is an isomorphism onto a subcomplex of X_{n+1}.

For a d-cell c of the CW-complex X_n, the suspension of c is a $(d+1)$-cell Σc of X_{n+1} via the structure map. Let C_n be the set of cells of X_n, then we have maps

$$s_n\colon C_n \to C_{n+1}.$$

The set of *stable cells* is the colimit of the C_n under the maps s_n. A *stable d-cell* is a stable cell represented by an $(n+d)$-cell of X_n for some n.

Note that the definition implies that the structure maps are, in particular, cellular maps. Moreover, a CW-spectrum is cofibrant. The following result gives the converse to this statement and is analogous to the situation for spaces.

Proposition 2.1.9 *For a pointed CW-complex B and a subcomplex A, the spectra $\mathrm{F}^{\mathbb{N}}_d A$ and $\mathrm{F}^{\mathbb{N}}_d B$ are CW-spectra, and the map*

$$\mathrm{F}^{\mathbb{N}}_d A \longrightarrow \mathrm{F}^{\mathbb{N}}_d B$$

is an inclusion of a sub-CW-spectrum.

If Y is a cofibrant spectrum, then there is a CW-spectrum X and a map $X \longrightarrow Y$ that is a levelwise homotopy equivalence.

Proof The first and second statements are evident by inspection of the levels and the structure maps.

CW-complexes are preserved by pushouts along cellular maps and by sequential colimits of cellular maps. Furthermore, a retract of a CW-complex has the homotopy type of a CW-complex. Thus, we see that the levels of a cofibrant spectrum Y have the homotopy type of CW-complexes.

We now choose a CW-replacement

$$f_0 \colon X_0 \longrightarrow Y_0.$$

This is a homotopy equivalence, as Y_0 has the homotopy type of a CW-complex. The structure maps of Y are q-cofibrations of pointed spaces by Theorem 2.1.7. Hence, they are monomorphisms, and we may choose a CW-replacement

$$f_1 : X_1 \longrightarrow Y_1$$

containing ΣX_0 as a subcomplex so that the square involving structure maps commutes. Continuing inductively gives a map $X \longrightarrow Y$ which at each level is a homotopy equivalence. □

2.2 Homotopy Groups of Spectra

The levelwise weak equivalences are not the correct weak equivalences for the stable model structure. Recalling Example 1.3.6, we would like the map

$$\mathrm{F}_1^{\mathbb{N}} S^1 = \mathbb{S}^{(1)} \longrightarrow \mathbb{S}$$

to be a weak equivalence in the stable model structure. Our earlier investigations indicate that these weak equivalences should be related to stable homotopy groups. For a spectrum X, $n \in \mathbb{Z}$ and $a \in \mathbb{N}$ such that $n + a > 1$, we have maps of abelian groups

$$\pi_{n+a}(X_a) \xrightarrow{\Sigma} \pi_{n+a+1}(\Sigma X_a) \xrightarrow{\sigma_a^X} \pi_{n+a+1}(X_{a+1}) \xrightarrow{\Sigma} \pi_{n+a+2}(\Sigma X_{a+1}) \longrightarrow \cdots$$

which in the case of $X = \Sigma^\infty A$ is the system of groups used to define $\pi_n^{\mathrm{stable}}(A)$. The following definition is then a natural continuation.

Definition 2.2.1 The *homotopy groups of a spectrum* X are defined to be

$$\pi_n(X) = \mathrm{colim}_a\, \pi_{n+a}(X_a)$$

for $n \in \mathbb{Z}$. A map $f \colon X \longrightarrow Y$ is a *π_*-isomorphism* if the induced map

$$\pi_n(f) \colon \pi_n(X) \longrightarrow \pi_n(Y)$$

is an isomorphism for each $n \in \mathbb{Z}$. We regard π_* as a functor from spectra to graded abelian groups. We say that spectra X and Y are *π_*-isomorphic* if there is a sequence of π_*-isomorphisms relating X and Y.

Note that, again, this definition of homotopy groups does not depend on the choice of a basepoint.

Example 2.2.2 The homotopy groups of the shifted suspension spectrum $F_d^{\mathbb{N}} A$ are given by

$$\pi_n(F_d^{\mathbb{N}} A) = \pi_{n+d}^{\text{stable}}(A),$$

hence, the homotopy groups of $\Sigma^\infty A$ are the stable homotopy groups of the pointed topological space A. In particular, the homotopy groups of the sphere spectrum $\mathbb{S} = \Sigma^\infty S^0$ are the stable homotopy groups of spheres

$$\pi_n(\mathbb{S}) = \pi_n^{\text{stable}}(S^0).$$

The adjoint of the identity map $S^{n+1} \longrightarrow S^{n+1} = (F_0^{\mathbb{N}} S^n)_1$ is a map

$$\lambda_1 \colon F_1^{\mathbb{N}} S^{n+1} \longrightarrow F_0^{\mathbb{N}} S^n = \Sigma^\infty S^n.$$

This map is a π_*-isomorphism, as after the first level, the two spectra are the same. In general, we may replace the first n levels of a spectrum X with $*$ obtaining a spectrum $X^{(n)}$, and the resulting inclusion

$$X^{(n)} \longrightarrow X$$

will be a π_*-isomorphism. Similarly, the map

$$\lambda_n \colon F_{n+1}^{\mathbb{N}} S^1 \longrightarrow F_n^{\mathbb{N}} S^0$$

is a π_*-isomorphism, as it is the identity above degree n.

Thus, we see that spectra can have non-trivial negative homotopy groups as, for example,

$$\pi_{-3}(F_4^{\mathbb{N}} S^0) = \pi_1^{\text{stable}}(S^0).$$

Example 2.2.3 The map $\eta \colon X \longrightarrow \Omega\Sigma X$ is a π_*-isomorphism. The adjoint structure maps of the spectrum $\Omega\Sigma X$ are given by

$$\Omega\Sigma X_a \xrightarrow{\Omega\Sigma\tilde{\sigma}_a^X} \Omega\Sigma\Omega X_a \longrightarrow \Omega\Omega\Sigma X_a \xrightarrow{\tau} \Omega(\Omega\Sigma X_a),$$

where the unlabelled map is induced by the natural transformation $\Sigma\Omega \longrightarrow \Omega\Sigma$ and τ swaps the two copies of Ω. From this description, we can check that η induces a map of sequential spectra. As $\pi_*(\eta)$ is part of the colimit sequence defining homotopy groups of spectra

$$\pi_{n+a}(X_a) \longrightarrow \pi_{n+a}(\Omega\Sigma X_a) \cong \pi_{n+a+1}(\Sigma X_a) \longrightarrow \pi_{n+a+1}(X_{a+1}),$$

it follows that η is a π_*-isomorphism.

Example 2.2.4 For G an abelian group, the homotopy groups of the Eilenberg–Mac Lane spectrum HG are

$$\pi_k(HG) = \mathrm{colim}_n\, \pi_{k+n}(K(G,n)) = \begin{cases} G & \text{if } k = 0 \\ 0 & \text{otherwise.} \end{cases}$$

From the uniqueness of Eilenberg–Mac Lane spaces and the above calculation, it follows that any two Eilenberg–Mac Lane spectra for G are π_*-isomorphic.

Remark 2.2.5 Similarly to Eilenberg–Mac Lane spectra, given an abelian group G, we can construct a spectrum whose homology is concentrated in degree zero, where it takes value G. These spectra are known as *Moore spectra* by analogy to Moore spaces. We construct a Moore spectrum for $\mathbb{Z}/n$ in Example 2.5.6. The general case needs substantially more technology and is given in Example 7.4.7.

Example 2.2.6 Let us have a look at the homotopy groups of the spectrum representing complex topological K-theory, see Subsection 7.4.2 for a more detailed introduction. The unitary group $U(n)$ may be considered as a subgroup of $U(n+1)$ by acting by the identity on the last coordinate. Taking the union over all n gives the infinite unitary group

$$U = \cup_{n \geqslant 0} U(n).$$

The loop space of U is $BU \times \mathbb{Z}$ and $\Omega^2 U \simeq U$ via Bott periodicity. It follows that we can define an Ω-spectrum K (often denoted KU in literature):

$$K_n = \begin{cases} BU \times \mathbb{Z} & \text{if } n \text{ is even} \\ U & \text{if } n \text{ is odd.} \end{cases}$$

Let X be a finite CW-complex (in particular compact and Hausdorff). We then have

$$[\Sigma^\infty X, K]_* \cong [X, K_0]_* \cong \widetilde{K}_*(X).$$

It follows that

$$\pi_n(K) = [\Sigma^\infty S^0, K]_n \cong \widetilde{K}_n(S^0) = \begin{cases} \mathbb{Z} & \text{if } n \text{ is even} \\ 0 & \text{if } n \text{ is odd.} \end{cases}$$

Hence, we have a spectrum with infinitely many non-zero negative homotopy groups.

It follows from the definition of π_*-isomorphisms that (small) coproducts or products of π_*-isomorphisms are π_*-isomorphisms. We also show that π_*-isomorphisms are preserved by suspension and loops.

Lemma 2.2.7 *A map of spectra $f\colon X \longrightarrow Y$ is a π_*-isomorphism if and only if*

$$\Sigma^k f\colon \Sigma^k X \longrightarrow \Sigma^k Y$$

is a π_*-isomorphism for all $k \in \mathbb{N}$.

A map of spectra $f\colon X \longrightarrow Y$ is a π_*-isomorphism if and only if

$$\Omega^k f\colon \Omega^k X \longrightarrow \Omega^k Y$$

is a π_*-isomorphism for all $k \in \mathbb{N}$.

Proof The result follows from the formulae

$$\pi_n(\Sigma^k X) \cong \pi_{n-k}(X), \qquad \pi_n(\Omega^k X) \cong \pi_{n+k}(X),$$

which hold for all $n \in \mathbb{Z}$ and $k \in \mathbb{N}$. We prove the first for $k = 1$, the second is similar. We have a diagram as below where τ is the twist map which swaps the two copies of S^1 in the double suspension. The left-hand square commutes and the right-hand square commutes up to a factor of -1, which corresponds to the effect of the twist map τ on homotopy groups. The colimit of the top row is $\pi_n(\Sigma X)$, and the colimit of the second row is $\pi_{n-1}(X)$.

$$\begin{array}{ccccccccc}
\cdots \longrightarrow & \pi_{n+a}(\Sigma X_a) & \xrightarrow{\Sigma} & \pi_{n+a+1}(\Sigma\Sigma X_a) & \xrightarrow{\Sigma\sigma_a^X \circ \tau} & \pi_{n+a+1}(\Sigma X_{a+1}) & \longrightarrow \cdots \\
& \downarrow{\scriptstyle \sigma_a^X} & & \downarrow{\scriptstyle \Sigma(\sigma_a^X)} & \ominus & \downarrow{\scriptstyle \sigma_{a+1}^X} & \\
\cdots \longrightarrow & \pi_{n+a}(X_{a+1}) & \xrightarrow[\Sigma]{} & \pi_{n+a+1}(\Sigma X_{a+1}) & \xrightarrow[\sigma_{a+1}^X]{} & \pi_{n+a+1}(X_{a+2}) & \longrightarrow \cdots
\end{array}$$

Since the structure maps appear in both rows and the horizontal maps, a diagram chase shows that the induced map on colimits is an isomorphism.

The series of isomorphisms below completes the proof.

$$\begin{aligned}
\pi_n(\Sigma X) = \operatorname{colim}_a \pi_{n+a}(\Sigma X_a) &\cong \operatorname{colim}_a \pi_{n+a}(X_{a+1}) \\
&= \operatorname{colim}_a \pi_{n-1+1+a}(X_{a+1}) \\
&\cong \operatorname{colim}_b \pi_{n-1+b}(X_b) = \pi_{n-1}(X)
\end{aligned}$$ □

Remark 2.2.8 Note how the above result does not assume anything about the basepoints of the levels of the spectrum. Given how much Subsection 1.1.2 relies on non-degenerate basepoints, this result should be considered a little surprising. Moreover, it allows us to prove that when A is a pointed CW-complex, $A \wedge -$ preserves π_*-isomorphisms, see Proposition 2.2.14. Thus, we see that the weak equivalences of sequential spectra are better behaved than those of pointed spaces.

Since homotopy groups of spaces interact well with coproducts, we have the following result for homotopy groups of spectra.

Lemma 2.2.9 *For a set of spectra* $\{X_i \mid i \in I\}$, *there is a natural isomorphism*

$$\bigoplus_{i\in I} \pi_n(X_i) \longrightarrow \pi_n(\bigvee_{i\in I} X_i). \qquad \square$$

Let us recall the notion of homotopy cofibres and homotopy fibres of maps of pointed spaces. Let $f\colon A \longrightarrow B$ be a map in Top_*. The *homotopy cofibre* of f is the pushout of the diagram

$$CA = A \wedge [0,1] \xleftarrow{i_0} A \xrightarrow{f} B,$$

and the *homotopy fibre* of f is the pullback of the diagram

$$A \xrightarrow{f} B \xleftarrow{p_0} \mathrm{Top}_*([0,1], B) = PB,$$

where $[0,1]$ has basepoint 1. If f is a h-cofibration of pointed spaces, then the natural quotient map

$$Cf \longrightarrow B/A$$

to the cokernel of f is a homotopy equivalence by Lemma A.5.6. Similarly, if f is a Serre fibration of pointed spaces, Ff is homotopy equivalent to $f^{-1}(*)$ by the dual statement. We define the homotopy cofibre and homotopy fibre of a map g of spectra analogously, so that level n is the homotopy (co)fibre of the nth level g_n of g. A *homotopy fibre sequence* is then a sequence of maps in spectra

$$A \longrightarrow B \longrightarrow C$$

with A weakly homotopy equivalent to the homotopy fibre of

$$B \longrightarrow C.$$

We define the *homotopy cofibre sequence* dually.

Proposition 2.2.10 *A map of spectra* $f\colon X \longrightarrow Y$ *induces long exact sequences of homotopy groups as below*

$$\cdots \longrightarrow \pi_{n+1}(Y) \to \pi_n(Ff) \to \pi_n(X) \to \pi_n(Y) \to \pi_{n-1}(Ff) \to \cdots$$
$$\cdots \to \pi_{n+1}(Cf) \to \pi_n(X) \to \pi_n(Y) \to \pi_n(Cf) \to \pi_{n-1}(X) \to \cdots,$$

where Cf *denotes the homotopy cofibre of* f, *and* Ff *is the homotopy fibre of* f.

Proof The first sequence is obtained from looking at the levelwise long exact sequence of homotopy groups of a Serre fibration. For all a, the map $f_a \colon X_a \longrightarrow Y_a$ is a map in Top_*, so we can construct the long exact sequence of homotopy groups

$$\cdots \to \pi_{n+a}(\Omega X_a) \longrightarrow \pi_{n+a}(\Omega Y_a) \longrightarrow \pi_{n+a}(Ff_a) \longrightarrow \pi_{n+a}(X_a) \longrightarrow \cdots$$

from the Puppe sequence of f_a (see Theorem 3.6.1). Here, n can be any integer such that $n + a > 1$. Taking colimits over a and using Lemma 2.2.7 gives the result.

For the second sequence, we can construct the Puppe sequence of pointed spaces

$$X_n \xrightarrow{f_n} Y_n \xrightarrow{i_n} Cf_n \xrightarrow{\delta_n} \Sigma X_n \xrightarrow{-\Sigma f_n} \Sigma Y_n \longrightarrow \cdots .$$

Recall that with the given sign convention, any three consecutive terms in the sequence is a homotopy cofibre sequence. Applying suspension and using the structure maps gives the diagram

$$\begin{array}{ccccccccccc} \Sigma X_n & \xrightarrow{f_n} & \Sigma Y_n & \xrightarrow{i_n} & \Sigma Cf_n & \xrightarrow{\delta_n} & \Sigma^2 X_n & \xrightarrow{-\Sigma^2 f_n} & \Sigma^2 Y_n & \longrightarrow & \cdots \\ \downarrow & & \downarrow & & \downarrow & & \downarrow & & \downarrow & & \\ X_{n+1} & \xrightarrow{f_{n+1}} & Y_{n+1} & \xrightarrow{i_{n+1}} & Cf_{n+1} & \xrightarrow{\delta_{n+1}} & \Sigma X_{n+1} & \xrightarrow{-\Sigma f_{n+1}} & \Sigma Y_{n+1} & \longrightarrow & \cdots , \end{array}$$

hence, we have a levelwise homotopy cofibre sequence of spectra

$$X \xrightarrow{f} Y \xrightarrow{i} Cf \xrightarrow{\delta} \Sigma X \xrightarrow{-\Sigma f} \Sigma Y \longrightarrow \cdots .$$

The homotopy group functor π_n gives a sequence of abelian groups

$$\cdots \longrightarrow \pi_n(X) \xrightarrow{f_*} \pi_n(Y) \xrightarrow{i_*} \pi_n(Cf) \xrightarrow{\partial} \pi_n(\Sigma X) \longrightarrow \cdots .$$

If we know that this sequence is exact, we can apply Lemma 2.2.7 and patch together these sequences for varying n to obtain the desired long exact sequence. Since

$$Y \longrightarrow Cf \longrightarrow \Sigma X \text{ and } Cf \longrightarrow \Sigma X \longrightarrow \Sigma Y$$

are homotopy cofibre sequences, it suffices to show exactness at $\pi_n(Y)$. The composite $i_* \circ f_*$ is zero. Consider an element α of $\pi_n(Y)$ which maps to zero in $\pi_n(Cf)$. We can represent α as the homotopy class of a map

$$g \colon S^{n+a} \longrightarrow Y_a,$$

where $i_a \circ g$ is homotopic to zero. This homotopy is represented by a map out of a cone

$$h\colon CS^{n+a} \longrightarrow Cf_a.$$

Representing this information as a diagram of homotopy cofibre sequences, we obtain the following diagram. Hence, we may choose a map k that makes the squares commute.

$$\begin{array}{ccccccccc} S^{n+a} & \longrightarrow & CS^{n+a} & \longrightarrow & \Sigma S^{n+a} & \longrightarrow & \Sigma S^{n+a} & \longrightarrow & \cdots \\ \downarrow g & & \downarrow h & & \downarrow k & & \downarrow g & & \\ Y_a & \longrightarrow & Cf_a & \longrightarrow & \Sigma X_a & \xrightarrow{\Sigma f_a} & \Sigma Y_a & \longrightarrow & \cdots \end{array}$$

The homotopy class of k is an element of $\pi_{n+a+1}(\Sigma X_a)$ and hence an element $\beta \in \pi_{n+1}(\Sigma X)$. The commutativity of the diagram and the formula

$$\pi_{n+1}(\Sigma X) \cong \pi_n(X)$$

shows that $f_*\beta = \alpha$. □

The proof shows nicely how stability is used to construct this long exact sequence, just as it was needed to construct the homological long exact sequence of stable homotopy groups of Example 1.1.19.

Corollary 2.2.11 *Let $f\colon X \longrightarrow Y$ be a map of spectra, let Cf be the homotopy cofibre of f, and let Ff be the homotopy fibre of f. Then the following are equivalent:*

- *The map f is a π_*-isomorphism.*
- *The homotopy cofibre Cf is π_*-isomorphic to $*$.*
- *The homotopy fibre Ff is π_*-isomorphic to $*$.* □

Given the similarity of the two long exact sequences of Proposition 2.2.10, one may ask if Ff and Cf are related. In fact there is a natural map

$$Ff \longrightarrow \Omega Cf$$

that is a π_*-isomorphism. A direct construction of $Ff \longrightarrow \Omega Cf$ is as follows. An element of Ff is a point $x \in X$ and a path σ from $f(x)$ to $* \in Y$. In Cf, we have a path τ_x from $[x,0] = [f(x)]$ to $* = [x,1]$ given by $t \mapsto [x,t]$. Taking the reverse of σ concatenated with τ_x gives a loop in Cf starting and ending at $*$. To see that this map is continuous, we may include Ff into $X \times PY$ and map this to $PCX \times PCf$, where $X \longrightarrow PCX$ sends $x \in X$ to τ_x. Reversing σ and concatenating with τ gives the map to PCf, and since the path is closed, we have the desired map $Ff \longrightarrow \Omega Cf$.

The easiest way to see that this is a π_*-isomorphism is to wait until we have our stable model structure on sequential spectra and use the technology of triangulated categories from Chapter 4. Specifically, Lemma 4.4.3 gives the following.

Corollary 2.2.12 *For any map of spectra f, the natural map $Ff \longrightarrow \Omega Cf$ is a π_*-isomorphism.* □

As one should expect, these long exact sequences are a fundamental tool for calculating the homotopy groups of spectra. They also allow us to construct new π_*-isomorphisms from certain colimits of π_*-isomorphisms.

Lemma 2.2.13 *Several standard operations preserve π_*-isomorphisms.*

1. *If $g\colon X \longrightarrow Y$ is a levelwise h-cofibration and a π_*-isomorphism, then the pushout of g along another map of spectra is also a π_*-isomorphism.*
2. *Given a diagram as below, where i and i′ are levelwise h-cofibrations and the vertical maps are all π_*-isomorphisms*

$$\begin{array}{ccccc} B & \xleftarrow{i} & A & \longrightarrow & C \\ \downarrow \simeq & & \downarrow \simeq & & \downarrow \simeq \\ B' & \xleftarrow{i'} & A' & \longrightarrow & C', \end{array}$$

   then the induced map from the pushout P of the top row to the pushout of the second row P′ is a π_*-isomorphism.
3. *If $f^i\colon X^i \longrightarrow X^{i+1}$ for $i \in \mathbb{N}$ is a collection of levelwise h-cofibrations and π_*-isomorphisms, then the map from the initial object into the colimit*

$$X^0 \longrightarrow \mathrm{colim}_i X^i$$

   is a π_*-isomorphism and a levelwise h-cofibration.

Proof We must show that the map h in the pushout square below is a π_*-isomorphism.

$$\begin{array}{ccc} X & \longrightarrow & P \\ \downarrow g & & \downarrow h \\ Y & \longrightarrow & Q. \end{array}$$

Since g and h are levelwise h-cofibrations, the homotopy cofibres of g and h are levelwise weakly equivalent to the cokernels Y/X and Q/P by Lemma A.5.6. The induced map

$$Y/X \longrightarrow P/Q$$

is an isomorphism since we have a pushout square. As g is a π_*-isomorphism, Y/X is π_*-isomorphic to a point, hence, so is P/Q. By Corollary 2.2.11, h is a π_*-isomorphism as claimed.

For the second statement, the same argument as above shows that B/A and B'/A' are the cofibres of i and i'. They are π_*-isomorphic by Proposition 2.2.10 and the Five Lemma. Since

$$B/A \cong P/C \text{ and } B'/A' \cong P'/C',$$

we see that P/C and P'/C' are π_*-isomorphic. The Five Lemma then implies that the natural map $P \longrightarrow P'$ is a π_*-isomorphism.

For the third statement, as the maps are levelwise h-cofibrations, Corollary A.7.10 implies that the colimit is levelwise homotopy equivalent to the sequential homotopy colimit of the maps f_i, see Example A.7.9. Here, we work in the Hurewicz model structure on topological spaces. Corollary A.7.10 also implies that the natural map

$$\operatorname{colim}_i \pi_{n+k}(X^i_k) \longrightarrow \pi_{n+k}(\operatorname{colim}_i X^i_k)$$

is an isomorphism for all k and n. The result then follows by the definition of π_*-isomorphisms. □

We may use these rules to prove that smashing with a pointed CW-complex preserves all π_*-isomorphisms. Again we note that no assumptions on the basepoints of the spectra are required.

Proposition 2.2.14 *Let $f: X \longrightarrow Y$ be a π_*-isomorphism of spectra and A a pointed CW-complex. Then the map*

$$f \wedge \mathrm{Id}: X \wedge A \longrightarrow Y \wedge A$$

is a π_*-isomorphism.

Proof We want to argue via the CW-structure of A, so we begin by proving that $- \wedge D^n_+$ and $- \wedge S^n_+$ preserve all π_*-isomorphisms.

Let X be a spectrum, then $X \wedge D^n_+$ at level d is given by the quotient space

$$X_d \times D^n/(\{x_d\} \times D_n),$$

where x_d is the basepoint of X_d. The contraction of D_n to a point gives a homotopy equivalence of spectra

$$X \wedge D^n_+ \simeq X$$

by [Rot88, Chapter 8, Lemma 8.9] which deals with the point set issues of identification spaces and products, see also [Sch18, Propositions A.2 and A.3].

Now consider the maps of homotopy cofibre sequences

$$\begin{array}{ccccc} X \wedge S^{n-1}_+ & \longrightarrow & X \wedge D^n_+ & \longrightarrow & X \wedge S^n \\ \downarrow & & \downarrow & & \downarrow \\ Y \wedge S^{n-1}_+ & \longrightarrow & Y \wedge D^n_+ & \longrightarrow & Y \wedge S^n. \end{array}$$

We have seen that the second and third vertical arrows are π_*-isomorphisms. Proposition 2.2.10 and the Five Lemma imply that so is the first vertical arrow.

To deal with a general CW-complex A, we use the cellular filtration of A

$$A = \operatorname{colim}(A_0 \longrightarrow A_1 \longrightarrow A_2 \longrightarrow \cdots),$$

where A_{n+1} is formed from A_n by a pushout

$$\begin{array}{ccc} \bigvee S^{n-1}_+ & \longrightarrow & A_n \\ \downarrow & & \downarrow \\ \bigvee D^n_+ & \longrightarrow & A_{n+1}. \end{array}$$

Consider the diagram below. Since π_*-isomorphisms are preserved by coproducts and smashing with spheres and discs, we may inductively assume that each vertical arrow is a π_*-isomorphism.

$$\begin{array}{ccccc} X \wedge \bigvee D^n_+ & \longleftarrow & X \wedge \bigvee S^{n-1}_+ & \longrightarrow & X \wedge A_n \\ \downarrow & & \downarrow & & \downarrow \\ Y \wedge \bigvee D^n_+ & \longleftarrow & Y \wedge \bigvee S^{n-1}_+ & \longrightarrow & Y \wedge A_n \end{array}$$

Lemma 2.2.13 states that the induced map on pushouts is a π_*-isomorphism. We now have a diagram of the form

$$\begin{array}{ccccccccc} X \wedge A_0 & \longrightarrow & X \wedge A_1 & \longrightarrow & X \wedge A_2 & \longrightarrow & X \wedge A_3 & \longrightarrow & \cdots \\ \downarrow & & \downarrow & & \downarrow & & \downarrow & & \\ Y \wedge A_0 & \longrightarrow & Y \wedge A_1 & \longrightarrow & Y \wedge A_2 & \longrightarrow & Y \wedge A_3 & \longrightarrow & \cdots, \end{array}$$

where each horizontal map is a levelwise h-cofibration in Top_*. Taking the homotopy cofibre of each vertical map gives a sequence of levelwise h-cofibrations

$$Z \wedge A_0 \longrightarrow Z \wedge A_1 \longrightarrow Z \wedge A_2 \longrightarrow Z \wedge A_3 \longrightarrow \cdots$$

between spectra that are π_*-isomorphic to a point. Lemma 2.2.13 implies that there is a π_*-isomorphism

$$Z \wedge A_0 \longrightarrow \mathrm{colim}_n(Z \wedge A_n) \cong Z \wedge A,$$

and hence $Z \wedge A$ is π_*-isomorphic to a point. Thus,

$$f \wedge \mathrm{Id}\colon X \wedge A \longrightarrow Y \wedge A$$

is a π_*-isomorphism by Corollary 2.2.11. □

2.3 The Stable Model Structure

Our goal is to construct a model structure on sequential spectra which "models" the behaviour of cohomology theories as outlined earlier as well as mimicking the behaviour of topological spaces. This will be called the "stable model structure". Its weak equivalences should be the π_*-isomorphisms. As fibrant objects, we would like the Ω-spectra. Hence, we want fewer fibrations than in the levelwise model structure, which can be achieved by adding more maps to the set of generating acyclic cofibrations. We know that these new maps should be π_*-isomorphisms, as they will be weak equivalences in the new model structure. While we will construct everything directly, it should be noted that this is exactly the context of a left Bousfield localisation, see Chapter 7.

We will need to have a much better understanding of how the levelwise model structure on spectra interacts with the Serre model structure on pointed spaces to construct the stable model structure and to understand the stable fibrations. For this, we need a topological version of Quillen's axiom SM7 [Qui67].

Earlier on, we defined a functor

$$- \wedge -\colon \mathcal{S}^{\mathbb{N}} \times \mathrm{Top}_* \longrightarrow \mathcal{S}^{\mathbb{N}},$$

which sends (X, A) to the spectrum $X \wedge A$ with level n being $X_n \wedge A$ and with structure maps

$$\sigma_n^X \wedge \mathrm{Id}_A\colon \Sigma X_n \wedge A \longrightarrow X_{n+1} \wedge A.$$

We refer to this functor as the *tensor* of spaces with spectra. We would also like an enrichment and a cotensor functor satisfying the expected adjointness properties.

Definition 2.3.1 The adjoints to $- \wedge -$ are

$$\mathcal{S}^{\mathbb{N}}(-,-)\colon (\mathcal{S}^{\mathbb{N}})^{op} \times \mathcal{S}^{\mathbb{N}} \longrightarrow \mathrm{Top}_* \qquad \mathrm{Top}_*(-,-)\colon \mathrm{Top}_*^{op} \times \mathcal{S}^{\mathbb{N}} \longrightarrow \mathcal{S}^{\mathbb{N}}.$$

For spectra X and Y, the pointed space $\mathcal{S}^{\mathbb{N}}(X,Y)$ is the subspace of

$$\prod_{n\geqslant 0} \mathrm{Top}_*(X_n, Y_n)$$

consisting of maps of spectra. Its basepoint is the constant map at $*$. We refer to this functor as the *enrichment* of spectra in spaces.

For a pointed space A, the *cotensor* is the spectrum $\mathrm{Top}_*(A,X)$ whose level n is given by the space of maps $\mathrm{Top}_*(A,X_n)$ and adjoint structure maps given by

$$\mathrm{Top}_*(A,X_n) \longrightarrow \mathrm{Top}_*(A,\Omega X_{n+1}) \cong \Omega \mathrm{Top}_*(A,X_{n+1}).$$

Just as with spaces, there is no danger of confusing the set of maps of spectra with the space of maps of spectra. We also note that the adjunction

$$\mathrm{F}_d^{\mathbb{N}}\colon \mathrm{Top}_* \rightleftarrows \mathcal{S}^{\mathbb{N}} : \mathrm{Ev}_d^{\mathbb{N}}$$

induces an isomorphism

$$\mathcal{S}^{\mathbb{N}}(\mathrm{F}_d^{\mathbb{N}} A, X) \cong \mathrm{Top}_*(A, X_d).$$

In particular,

$$\mathcal{S}^{\mathbb{N}}(\mathrm{F}_d^{\mathbb{N}} S^0, X) \cong X_d \quad \text{and} \quad \mathcal{S}^{\mathbb{N}}(\mathrm{F}_d^{\mathbb{N}} S^1, X) \cong \Omega X_d$$

as pointed spaces.

Looking at the adjunction levelwise, we obtain the expected relation between these three functors.

Lemma 2.3.2 *For a pointed space A and spectra X and Y, there are natural isomorphisms of pointed spaces*

$$\mathrm{Top}_*(A, \mathcal{S}^{\mathbb{N}}(X,Y)) \cong \mathcal{S}^{\mathbb{N}}(X \wedge A, Y) \cong \mathcal{S}^{\mathbb{N}}(X, \mathrm{Top}_*(A,Y)). \qquad \square$$

We now prove that sequential spectra with the levelwise model structure is a "topological model category", see Definition 6.1.28

Proposition 2.3.3 *Let $f\colon X \longrightarrow Y$ be a q-cofibration of spectra, $i\colon A \longrightarrow B$ a q-cofibration of pointed spaces, and $p\colon P \longrightarrow Q$ a levelwise fibration of spectra. Then the natural map induced by f and p*

$$\mathrm{hom}_{\square}(f,p)\colon \mathcal{S}^{\mathbb{N}}(Y,P) \longrightarrow \mathcal{S}^{\mathbb{N}}(X,P) \underset{\mathcal{S}^{\mathbb{N}}(X,Q)}{\times} \mathcal{S}^{\mathbb{N}}(Y,Q)$$

is a fibration of pointed spaces. Moreover, if one of f or p is a levelwise weak equivalence, the map $\mathrm{hom}_\square(f, p)$ *is a weak homotopy equivalence.*

The natural map

$$f \square i\colon Y \wedge A \bigvee_{X \wedge A} X \wedge B \longrightarrow Y \wedge B$$

is a q-cofibration of spectra. Moreover, if i is a weak homotopy equivalence or f a levelwise weak equivalence, then $f \square i$ is a levelwise weak equivalence.

The natural map

$$\mathrm{hom}_\square(i, p)\colon \mathrm{Top}_*(B, P) \longrightarrow \mathrm{Top}_*(A, P) \times_{\mathrm{Top}_*(A,Q)} \mathrm{Top}_*(B, Q)$$

is a levelwise fibration of spectra. Moreover, if i is a weak homotopy equivalence or p is a levelwise weak equivalence, then $\mathrm{hom}_\square(i, p)$ *is a levelwise weak equivalence.*

Proof The three statements are equivalent. The proof of this equivalence is an exercise similar to Lemma 6.1.8.

The last statement is the easiest to prove: it follows directly from levelwise statements about pointed topological spaces (specifically, they are a monoidal model category and hence satisfy the pushout product axiom). □

We list some consequences of the previous proposition which we will need to construct the stable model structure on spectra and to describe its fibrations. Each is proven by taking some suitable version of Proposition 2.3.3 and letting one of the objects be the trivial spectrum $*$.

Corollary 2.3.4 *Let X and Y be spectra. The enrichment functor $\mathcal{S}^{\mathbb{N}}(-, Y)$ sends q-cofibrations to fibrations and acyclic q-cofibrations to acyclic fibrations of pointed spaces. In particular, it sends levelwise weak equivalences between q-cofibrant spectra to weak homotopy equivalences.*

If X is a q-cofibrant spectrum, the functor $\mathcal{S}^{\mathbb{N}}(X, -)$ is a right Quillen functor from the levelwise model structure on spectra to pointed spaces. Its left adjoint is $X \wedge -$. □

We will need a homotopy–invariant version of pullbacks to define the fibrations of the stable model structure.

Definition 2.3.5 The *levelwise homotopy pullback* of a diagram of spectra $X \longrightarrow Z \longleftarrow Y$ is the pullback of the diagram

$$X \longrightarrow Z \longleftarrow Y',$$

where $Y \longrightarrow Y' \longrightarrow Z$ is a factorisation of $Y \longrightarrow Z$ into an acyclic q-cofibration followed by a levelwise fibration.

A diagram of spectra

$$\begin{array}{ccc} W & \xrightarrow{f} & X \\ {\scriptstyle p}\downarrow & & \downarrow{\scriptstyle q} \\ Y & \xrightarrow[g]{} & Z \end{array}$$

is said to be a *levelwise homotopy pullback square* if the map induced by f and p from W to the levelwise homotopy pullback of $X \longrightarrow Z \longleftarrow Y$ is a levelwise weak homotopy equivalence.

This definition of levelwise homotopy pullback is the same as a homotopy pullback in the level model structure, see Section A.7. Homotopy pullback squares are also known as homotopy cartesian squares. Note that the definition of a homotopy pullback usually requires both maps in the square to be factored as in the construction, but in this case, one will suffice, see Lemma A.7.20.

We have already seen that the square

$$\begin{array}{ccc} \mathcal{S}^{\mathbb{N}}(Y,P) & \longrightarrow & \mathcal{S}^{\mathbb{N}}(X,P) \\ \downarrow & & \downarrow \\ \mathcal{S}^{\mathbb{N}}(Y,Q) & \longrightarrow & \mathcal{S}^{\mathbb{N}}(X,Q) \end{array}$$

is a levelwise homotopy pullback square when the map $X \longrightarrow Y$ is a q-cofibration, the map $P \longrightarrow Q$ is a fibration, and one of those maps is a levelwise weak equivalence.

We can extend the standard results about homotopy pullback squares of pointed spaces to the levelwise model structure on spectra.

Proposition 2.3.6 *Consider a commutative square $\mathfrak{S}$ of spectra*

$$\begin{array}{ccc} W & \xrightarrow{f} & X \\ {\scriptstyle p}\downarrow & & \downarrow{\scriptstyle q} \\ Y & \xrightarrow{g} & Z. \end{array}$$

If one of q or g is a levelwise fibration, then the square is a levelwise homotopy pullback square if and only if the map induced by f and p

$$W \longrightarrow Y \times_Z X$$

is a levelwise weak homotopy equivalence.

Consider a map of squares $\alpha\colon \mathfrak{S} \longrightarrow \mathfrak{S}'$ (a cube) such that the component maps of α are levelwise weak equivalences.

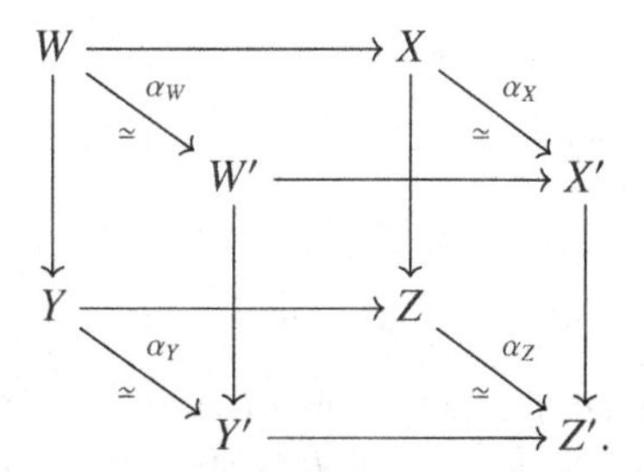

Then α induces a levelwise weak homotopy equivalence between the homotopy pullbacks of the punctured squares

$$Y \longrightarrow Z \longleftarrow X \quad \text{and} \quad Y' \longrightarrow Z' \longleftarrow X'.$$

Moreover, the back square $\mathfrak{S}$ is a levelwise homotopy pullback square if and only if the front square $\mathfrak{S}'$ is a levelwise homotopy pullback square. □

With these preliminaries understood, we can start defining the additional acyclic cofibrations needed to make the stable model structure.

Definition 2.3.7 Let $\lambda_n\colon \mathrm{F}^{\mathbb{N}}_{n+1} S^1 \longrightarrow \mathrm{F}^{\mathbb{N}}_n S^0$ be the adjoint of the identity map

$$S^1 \longrightarrow \mathrm{Ev}^{\mathbb{N}}_{n+1} \mathrm{F}^{\mathbb{N}}_n S^0 = S^1.$$

This map is the identity in every level except n, where it is the inclusion

$$* \longrightarrow S^0.$$

We have seen in Example 2.2.2 that λ_n is a π_*-isomorphism. While λ_n is a levelwise h-cofibration, it is not a q-cofibration. Hence, we need to alter the λ_n before we can add them to our set of generating acyclic cofibrations.

Definition 2.3.8 Define $M\lambda_n$ to be the mapping cylinder of λ_n, that is, the pushout of

$$\begin{array}{ccc} \mathrm{F}^{\mathbb{N}}_{n+1} S^1 & \xrightarrow{\lambda_n} & \mathrm{F}^{\mathbb{N}}_n S^0 \\ \downarrow{\scriptstyle i_1} & & \downarrow{\scriptstyle s_n} \\ (\mathrm{F}^{\mathbb{N}}_{n+1} S^1) \wedge [0,1]_+ & \xrightarrow{t_n} & M\lambda_n, \end{array}$$

where i_1 is the inclusion into the one end of the interval. We then have a map

$$k_n\colon \mathrm{F}^{\mathbb{N}}_{n+1} S^1 \longrightarrow M\lambda_n$$

coming from the inclusion into the zero end of the interval (so $k_n = t_n \circ i_0$). We also have a deformation retraction

$$r_n \colon M\lambda_n \longrightarrow \mathrm{F}_n^{\mathbb{N}} S^0$$

induced by the collapse map

$$(\mathrm{F}_{n+1}^{\mathbb{N}} S^1) \wedge [0,1]_+ \longrightarrow \mathrm{F}_{n+1}^{\mathbb{N}} S^1.$$

The composition $r_n \circ k_n$ is the original map λ_n.

The maps i_0 and i_1 are q-cofibrations of spectra, as any q-cofibration of pointed spaces smashed with a q-cofibrant spectrum is a q-cofibration. The map k_n is a π_*-isomorphism, as both r_n and λ_n are as well. It is also a q-cofibration of spectra by Lemma A.7.7 (the Patching Lemma).

Recall the generating sets for the levelwise model structure on sequential spectra.

$$\begin{aligned} I^{\mathbb{N}}_{\text{level}} &= \{\mathrm{F}_d^{\mathbb{N}}(S^{a-1}_+ \to D^a_+) \mid a, d \in \mathbb{N}\} \\ J^{\mathbb{N}}_{\text{level}} &= \{\mathrm{F}_d^{\mathbb{N}}(D^a_+ \to (D^a \times [0,1])_+) \mid a, d \in \mathbb{N}\} \end{aligned}$$

Definition 2.3.9 The following sets will be the generating sets of the *stable model structure* on sequential spectra.

$$\begin{aligned} I^{\mathbb{N}}_{\text{stable}} &= I^{\mathbb{N}}_{\text{level}} \\ J^{\mathbb{N}}_{\text{stable}} &= J^{\mathbb{N}}_{\text{level}} \cup \{k_n \,\square\, (S^{a-1}_+ \to D^a_+) \mid a, n \in \mathbb{N}\} \end{aligned}$$

We can characterise those maps with the right lifting property with respect to $J^{\mathbb{N}}_{\text{stable}}$. These maps will be the fibrations of our stable model structure. In particular, the following result makes it clear that the fibrant objects will be the Ω-spectra.

Proposition 2.3.10 *A map of spectra $f \colon X \longrightarrow Y$ has the right lifting property with respect to $J^{\mathbb{N}}_{\text{stable}}$ if and only if f is a levelwise fibration of spaces and for each $n \in \mathbb{N}$, the map*

$$X_n \longrightarrow Y_n \underset{\Omega Y_{n+1}}{\times} \Omega X_{n+1}$$

induced by $\widetilde{\sigma}_n^X$ and f is a weak homotopy equivalence. If $Y = *$, then X has this right lifting property if and only if it is an Ω-spectrum.

Proof A map $f \colon X \longrightarrow Y$ has the right lifting property with respect to $J^{\mathbb{N}}_{\text{stable}}$ if and only if f is a levelwise fibration of spaces and it has the right lifting property with respect to

$$\{k_n \,\square\, (S^{a-1}_+ \to D^a_+) \mid a, n \in \mathbb{N}\}.$$

The adjunctions between $-\Box-$ and $\hom_\Box(-,-)$ of Proposition 2.3.3 show that a levelwise fibration f of spectra has a lift in the first square if and only if the map of pointed spaces $\hom_\Box(k_n^*, f_*)$ has a lift in the second square.

$$\begin{array}{ccc} M\lambda_n \wedge S^{a-1}_+ \bigvee_{\mathrm{F}^{\mathbb{N}}_{n+1} S^1 \wedge S^{a-1}_+} \mathrm{F}^{\mathbb{N}}_{n+1} S^1 \wedge D^a_+ & \longrightarrow & X \\ {\scriptstyle k_n \Box i}\big\downarrow & & \big\downarrow{\scriptstyle f} \\ M\lambda_n \wedge D^a_+ & \longrightarrow & Y \end{array}$$

$$\begin{array}{ccc} S^{a-1}_+ & \longrightarrow & \mathcal{S}^{\mathbb{N}}(M\lambda_n, X) \\ {\scriptstyle i}\big\downarrow & & \big\downarrow{\scriptstyle \hom_\Box(k_n^*, f_*)} \\ D^a_+ & \longrightarrow & \mathcal{S}^{\mathbb{N}}(\mathrm{F}^{\mathbb{N}}_{n+1} S^1, X) \times_{\mathcal{S}^{\mathbb{N}}(\mathrm{F}^{\mathbb{N}}_{n+1} S^1, Y)} \mathcal{S}^{\mathbb{N}}(M\lambda_n, Y) \end{array}$$

Hence, f has the right lifting property with respect to $J^{\mathbb{N}}_{\text{stable}}$ if and only if f is a levelwise fibration of spaces and $\hom_\Box(k_n^*, f_*)$ is an acyclic fibration of pointed spaces for all n. We already know from Proposition 2.3.3 that $\hom_\Box(k_n^*, f_*)$ will be a fibration when f is a levelwise fibration, so the condition is now that f is a levelwise fibration and $\hom_\Box(k_n^*, f_*)$ is a weak homotopy equivalence.

Now assuming that f is a levelwise fibration, consider the map of squares (a cube) induced by the homotopy equivalence r_n^*

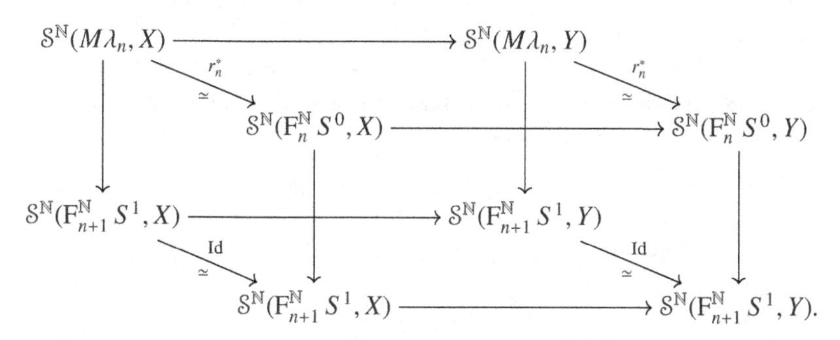

By the adjunction between $\mathrm{F}^{\mathbb{N}}_d$ and $\mathrm{Ev}^{\mathbb{N}}_d$, the front square is isomorphic to the square

$$\begin{array}{ccc} X_n & \xrightarrow{\ f\ } & Y_n \\ {\scriptstyle \widetilde{\sigma}^X_n}\big\downarrow & & \big\downarrow{\scriptstyle \widetilde{\sigma}^Y_n} \\ \Omega X_{n+1} & \xrightarrow[\Omega f_{n+1}]{} & \Omega Y_{n+1}. \end{array}$$

By Proposition 2.3.6, the above square is a homotopy pullback square if and only if the back square of the cube is a homotopy pullback. Moreover, both right-hand vertical maps are fibrations, hence $\mathrm{hom}_\square(k_n^*, f_*)$ is a weak homotopy equivalence if and only if

$$X_n \longrightarrow Y_n \underset{\Omega Y_{n+1}}{\times} \Omega X_{n+1}$$

is a weak homotopy equivalence for all $n \in \mathbb{N}$.

The second statement follows from the first. □

Corollary 2.3.11 *Let $f\colon X \longrightarrow Y$ be a levelwise fibration between Ω-spectra. Then f has the right lifting property with respect to $J^{\mathbb{N}}_{\text{stable}}$.* □

Theorem 2.3.12 *The* stable model structure *on sequential spectra is defined by the three classes below.*

- *The weak equivalences are the π_*-isomorphisms.*
- *The cofibrations are the q-cofibrations.*
- *The fibrations are given by Proposition 2.3.10 and are called the* stable fibrations.

The fibrant spectra are the Ω-spectra. The model structure is cofibrantly generated with generating sets given by Definition 2.3.9.

Proof The weak equivalences have the two-out-of-three property. The generating sets $I^{\mathbb{N}}_{\text{level}}$ and $J^{\mathbb{N}}_{\text{stable}}$ have small codomains, as pushouts of diagrams of small objects are small by Lemma A.6.5.

A map $f\colon X \longrightarrow Y$ with the right lifting property with respect to $I^{\mathbb{N}}_{\text{level}}$ is a levelwise acyclic fibration. We must show it is also a stable fibration. Letting $n \in \mathbb{N}$, we must show that the square

$$\begin{array}{ccc} X_n & \xrightarrow[\simeq]{f} & Y_n \\ \big\downarrow{\scriptstyle \tilde{\sigma}_n^X} & & \big\downarrow{\scriptstyle \tilde{\sigma}_n^Y} \\ \Omega X_{n+1} & \xrightarrow[\Omega f_{n+1}]{\simeq} & \Omega Y_{n+1} \end{array}$$

is a levelwise homotopy pullback square. We know that Ωf_{n+1} is an acyclic fibration, so it suffices to prove that

$$X_n \longrightarrow Y_n \underset{\Omega Y_{n+1}}{\times} \Omega X_{n+1}$$

is a weak homotopy equivalence of pointed spaces. The projection map

$$Y_n \underset{\Omega Y_{n+1}}{\times} \Omega X_{n+1} \longrightarrow Y_n$$

is an acyclic fibration, as these are preserved by pullbacks, and $X_n \longrightarrow Y_n$ is a weak equivalence, so the result follows by the two-out-of-three property.

Now consider a stable fibration $f\colon X \longrightarrow Y$ that is a π_*-isomorphism. We must prove it is a levelwise acyclic fibration. We know it is a levelwise fibration, so we must prove that each f_n is a weak homotopy equivalence. Since each f_n is a fibration, we know that the homotopy fibre of f is given by the levelwise fibre $(Ff)_n = f_n^{-1}(*)$. Moreover, the map $Ff \longrightarrow *$ is an acyclic stable fibration, as it is the pullback of one. Thus, Ff is an Ω-spectrum by our characterisation of fibrations, Proposition 2.3.10, and has trivial homotopy groups. Since the maps

$$Ff_n \longrightarrow \Omega Ff_{n+1}$$

are weak homotopy equivalences, the homotopy groups of the spectrum Ff are given by levelwise homotopy groups

$$0 = \pi_n(Ff) \cong \pi_{n+a}(Ff_a), \qquad a \in \mathbb{N}.$$

Hence, each level of Ff is weakly contractible, and

$$\pi_q(X_n) \longrightarrow \pi_q(Y_n)$$

is an isomorphism for each $q > 0$ and $n \in \mathbb{N}$ by the long exact sequences of homotopy groups. The map

$$\Omega f_n\colon \Omega X_n \longrightarrow \Omega Y_n$$

is therefore a weak homotopy equivalence for all $n \in \mathbb{N}$. (This is to be expected, as ΩX_n only depends on the basepoint component of X_n.) We can write f_n as the composite

$$X_n \longrightarrow Y_n \underset{\Omega Y_{n+1}}{\times} \Omega X_{n+1} \longrightarrow Y_n.$$

As the first map is a weak homotopy equivalence (f is a stable fibration) and the second map is the pullback of the acyclic fibration of spaces Ωf_{n+1}, we see that f_n is a weak equivalence.

Finally, we must show that a sequential colimit of pushouts of maps in $J^{\mathbb{N}}_{\text{stable}}$ is a π_*-isomorphism and a q-cofibration. Every map in $J^{\mathbb{N}}_{\text{stable}}$ is a q-cofibration, and these are preserved by pushouts and sequential colimits. Moreover, every q-cofibration is a levelwise q-cofibration, so Lemma 2.2.13 completes this part of the argument. □

By general model category theory, the homotopy category of sequential spectra with the stable model structure may be described as spectra with inverses to the π_*-isomorphisms added (taking care of the lack of maps mentioned previously) or as homotopy classes of maps from a cofibrant spectrum to a fibrant spectrum. This last description implies that we are looking at homotopy classes of maps from CW-spectra to Ω-spectra, as on [Ada74, Page 141]. It follows that maps in this homotopy category agree with the more classical approaches.

Definition 2.3.13 The *stable homotopy category* is the homotopy category of sequential spectra equipped with the stable model structure

$$\mathcal{SHC} = \mathrm{Ho}(\mathcal{S}^{\mathbb{N}}).$$

We shall see in Chapter 4 that the stable homotopy category is *triangulated*, an important structural property that gives many useful results. We need one more piece of information for this to hold, namely, we need to know that the stable homotopy category is "stable with respect to Σ", that is, Σ induces an equivalence on the stable homotopy category.

Theorem 2.3.14 *The loop–suspension adjunction*

$$\Sigma \colon \mathcal{S}^{\mathbb{N}} \rightleftarrows \mathcal{S}^{\mathbb{N}} : \Omega$$

on sequential spectra equipped with the stable model structure is a Quillen equivalence. Hence, for any spectra X and Y, the suspension functor induces an isomorphism of maps in the stable homotopy category

$$[X, Y]^{\mathcal{S}^{\mathbb{N}}} \cong [\Sigma X, \Sigma Y]^{\mathcal{S}^{\mathbb{N}}}.$$

Proof The functor Σ preserves q-cofibrations, and by Lemma 2.2.7, it preserves π_*-isomorphisms. Hence, it is a left Quillen functor. To prove that it is a Quillen equivalence, we note that Ω preserves and detects all π_*-isomorphisms by that same lemma, and that the map $X \longrightarrow \Omega\Sigma X$ is a π_*-isomorphism for all spectra X by Example 2.2.3. □

We can relate the stable model structure on sequential spectra to the levelwise model structure. Let us temporarily use $\mathcal{S}^{\mathbb{N}}_l$ to denote sequential spectra with the levelwise model structure and $[-,-]^l$ for maps in the corresponding homotopy category.

Lemma 2.3.15 *The identity functor from the levelwise structure on sequential spectra to the stable model structure is a left Quillen functor*

$$\mathrm{Id} \colon \mathcal{S}^{\mathbb{N}}_l \rightleftarrows \mathcal{S}^{\mathbb{N}} : \mathrm{Id}.$$

If E is an Ω-spectrum, then

$$[X, E] \cong [X, E]^l.$$

Proof Since the generating sets for the levelwise model structure are contained in the generating sets for the stable model structure, we have the Quillen adjunction as claimed. The second statement follows by looking at the derived adjunction. □

Lemma 2.3.16 *For $d \in \mathbb{N}$, the shifted suspension spectrum functor $F_d^{\mathbb{N}}$ and the evaluation functor $Ev_d^{\mathbb{N}}$ form a Quillen adjunction*

$$F_d^{\mathbb{N}} : \mathrm{Top}_* \rightleftarrows \mathcal{S}^{\mathbb{N}} : Ev_d^{\mathbb{N}} .$$

In particular, there is a Quillen adjunction

$$\Sigma^\infty : \mathrm{Top}_* \rightleftarrows \mathcal{S}^{\mathbb{N}} : Ev_0^{\mathbb{N}} .$$

Proof This follows from inspecting the generating sets for the Serre model structure and for the stable model structure (Definition 2.3.9), which are given in terms of the functors $F_d^{\mathbb{N}}$ for $d \in \mathbb{N}$. □

We will discuss the derived functor of $Ev_0^{\mathbb{N}}$ in Section 2.4.

A common question is whether it is useful to index spectra over the integers rather than the natural numbers. The method of this section can be used to put a stable model structure on integer-indexed spectra. If X is a spectrum indexed over the natural numbers, we can always extend it to an integer-indexed spectrum EX by putting a point in negative degrees. A spectrum Z indexed over the integers can be truncated to give a spectrum UZ indexed over the natural numbers. The map $X \longrightarrow UEX$ is an isomorphism, and the inclusion EUZ into Z is a π_*-isomorphism.

We see that the resulting functors (E, U) form a Quillen equivalence, and hence both categories are models for the stable homotopy category. In general, the "smaller" category is used rather than integer-indexed spectra. One possible reason is that it is harder to perform inductive constructions over integer-indexed spectra.

2.4 Explicit Fibrant Replacement

Now that we have a stable model structure on spectra, we can investigate how the stable homotopy category relates to pointed spaces. For this, we will need an explicit construction of fibrant replacement in the stable model structure of sequential spectra.

Let X be a sequential spectrum. We want to define an Ω-spectrum $R_\infty X$ with a π_*-isomorphism $X \longrightarrow R_\infty X$. Let $R_0 X = X$. For $k \geqslant 1$, define $R_k X$ by

$$(R_k X)_n = \Omega^k X_{n+k}$$

with structure map given by

$$(R_k X)_n = \Omega^k X_{n+k} \xrightarrow{\Omega^k \widetilde{\sigma}^X_{n+k}} \Omega^k \Omega X_{n+k+1} \xrightarrow{\cong} \Omega \Omega^k X_{n+k+1},$$

where the last isomorphism is an associativity isomorphism $S^1 \wedge S^k \cong S^k \wedge S^1$ without reordering the coordinates of the sphere.

We have a map $R_k X \longrightarrow R_{k+1} X$ induced by the adjoint structure maps

$$\begin{array}{ccc}
(R_k X)_n = \Omega^k X_{n+k} & \xrightarrow{\Omega^k \widetilde{\sigma}_{n+k}} & \Omega^{k+1} X_{n+k+1} = (R_{k+1} X)_n \\
\Big\downarrow {\scriptstyle \Omega^k \widetilde{\sigma}_{n+k}} & & \Big\downarrow {\scriptstyle \Omega^{k+1} \widetilde{\sigma}_{n+k+1}} \\
\Omega (R_k X)_{n+1} = \Omega^{k+1} X_{n+k+1} & \xrightarrow{\Omega^{k+1} \widetilde{\sigma}_{n+k+1}} & \Omega^{k+2} X_{n+k+2} = \Omega (R_{k+1} X)_{n+1}.
\end{array}$$

Each map $R_k X \longrightarrow R_{k+1}$ is a π_*-isomorphism, as the adjoint structure maps are "inverted" by the colimit defining the homotopy groups of a spectrum.

We may then define $R_\infty X = \operatorname{hocolim}_k R_k X$. If we wanted a functorial definition, then we could choose a functorial construction of homotopy colimits. This is done in [Sch97, Lemma 2.1.3] using the functorial factorisations of the stable model structure on sequential spectra. For now, we just want *some* construction.

We claim that $R_\infty X$ is an Ω-spectrum. Level n of the spectrum is given by

$$(R_\infty X)_n = (\operatorname{hocolim}_k R_k X)_n = \operatorname{hocolim}_k \Omega^k X_{n+k}$$

with structure map

$$\begin{array}{rcl}
\operatorname{hocolim}_k \Omega^k X_{n+k} & \xrightarrow{\operatorname{hocolim}_k \Omega^k \widetilde{\sigma}^X_{n+k}} & \operatorname{hocolim}_k \Omega^k \Omega X_{n+k+1} \\
& \xrightarrow{\cong} & \operatorname{hocolim}_k \Omega \Omega^k X_{n+k+1} \\
& \xrightarrow{\simeq} & \Omega \operatorname{hocolim}_k \Omega^k X_{n+k+1}.
\end{array}$$

The last map is a weak homotopy equivalence, as the loop functor commutes with sequential homotopy colimits of pointed spaces, see Corollary A.7.10. The second map is an instance of the associativity isomorphism, and the first map is a weak homotopy equivalence, as the maps defining the homotopy colimit are the adjoint structure maps.

It follows that $X \longrightarrow R_\infty X$ is a π_*-isomorphism to an Ω-spectrum. Hence, we may take this to be fibrant replacement in the stable model structure on sequential spectra.

Definition 2.4.1 For a sequential spectrum X, we define a pointed topological space

$$\Omega^\infty X = \mathbb{R}\,\mathrm{Ev}_0^{\mathbb{N}} X = \mathrm{Ev}_0^{\mathbb{N}} X^{fib} \simeq \mathrm{hocolim}_n \Omega^n X_n.$$

We use the convention that Ω^∞ is the derived functor of evaluation at 0 so as to match the notation of many older texts which only consider Ω^∞. In particular, we see that for A a CW-complex,

$$\Omega^\infty \Sigma^\infty A = \mathrm{Ev}_0^{\mathbb{N}}(\Sigma^\infty A)^{fib} \simeq \mathrm{hocolim}_n \Omega^n \Sigma^n A = QA,$$

where the functor Q was introduced in Definition 1.1.14.

The spaces in the image of Ω^∞ have many interesting properties. In particular, they satisfy the following definition.

Definition 2.4.2 An *infinite loop space* is a pointed topological space Z such that for each n, there is a space Z_n and a weak homotopy equivalence

$$Z \longrightarrow \Omega^n Z_n.$$

We caution the reader that there are many competing definitions of an infinite loop space – we have chosen the simplest and least structured.

Proposition 2.4.3 *For a sequential spectrum X, the space $\Omega^\infty X$ is an infinite loop space.*

Proof The fibrant replacement $Y = X^{fib}$ of a spectrum X is an Ω-spectrum. We then have maps $Y_n \simeq \Omega Y_{n+1}$ for $n \in \mathbb{N}$. Thus, $\Omega^\infty X = Y_0$ is an infinite loop space. □

One can also prove this from the definition of the homotopy colimit definition of the fibrant replacement. This requires care when commuting the various copies of Ω.

In the reverse direction, we construct a spectrum from an infinite loop space.

Definition 2.4.4 A spectrum whose negative homotopy groups are all trivial is called a *connected spectrum*.

A spectrum with only finitely many homotopy groups in negative degrees is called a *connective spectrum*.

The suspension spectrum of a space is a connected spectrum by Example 2.2.2.

Proposition 2.4.5 *Every infinite loop space Z defines a connected Ω-spectrum $\bar{Z}$ with $Z \simeq \Omega^\infty \bar{Z}$.*

Proof Let $Z = Z_0$ be an infinite loop space. Choose a delooping Z_1 so that $Z \simeq \Omega Z_1$. We may choose Z_1 to be path-connected, as Ω only depends on the basepoint component. Continuing in this way gives connected spaces Z_n and weak homotopy equivalences

$$Z_n \longrightarrow \Omega Z_{n+1},$$

thus, we have an Ω-spectrum $\bar{Z}$. Having $\bar{Z}_0 = Z$ implies the last statement.

The assumption that the spaces Z_n, $n > 0$, are connected ensures that the homotopy groups of each Z_n are entirely determined by Z. Moreover, for $k < n$,

$$\pi_k(Z_n) \cong \pi_0(\Omega^k Z_n) \cong \pi_0(Z_{n-k}) = 0.$$

The spectrum $\bar{Z}$ is connected, as for negative a one has

$$\pi_a(\bar{Z}) = \operatorname{colim}_k \pi_{k+a}(Z_k) = 0.$$ □

In positive degrees, the homotopy groups of the spectrum $\bar{Z}$ are given by the homotopy groups of the space Z.

To illustrate, we may build a connective spectrum ku representing connective complex K-theory from the infinite loop space $BU \times \mathbb{Z}$. We start with

$$ku_0 = BU \times \mathbb{Z}.$$

The space ku_1 should be a connected space satisfying

$$\Omega ku_1 \simeq ku_0.$$

Hence, we may take $ku_1 = U = KU_1$ as it is connected. For ku_2, we take $ku_2 = BU$ rather than $KU_2 = BU \times \mathbb{Z}$, as it also needs to be connected. The third space must satisfy $\Omega ku_3 \simeq BU$ and be 2-connected, so we may take $ku_3 = SU$. The fourth space is then the classifying space of the group SU, $ku_4 = BSU$. From here on, the spaces no longer have particularly meaningful names, other than being certain Postnikov–type constructions on U and BU.

Remark 2.4.6 We would like to give a formal statement and proof of the slogan "up to homotopy, infinite loop spaces are connected spectra". But, as we have defined it, the construction of $\bar{Z}$ from an infinite loop space Z does not need to be unique or functorial.

Resolving these problems is a substantial task which we leave to the numerous references in the literature, for example, the work of Adams [Ada78] and the extensive work of May on the subject and its relations to operads [May77], [May09a] and [May09b].

With this section complete, we have an excellent point-set model $\mathcal{S}^{\mathbb{N}}$ for the stable homotopy category $\mathcal{SHC}$. It is easy to construct objects and maps in this category, it represents cohomology theories by the Brown Representability Theorem 1.1.23, and it is nicely related to topological spaces by Σ^∞ and Ω^∞. Moreover, we have an explicit description of cofibrations, fibrations and fibrant replacements, as well as a theory of CW-objects similar to that of spaces.

There is one major problem with this model category: it is unable to model the smash product of the stable homotopy category. We discuss this issue (and give definitions and solutions) in Chapters 5 and 6.

2.5 The Steenrod Algebra

In this section, we assume that the homotopy category of sequential spectra satisfies the list of properties from Subsection 1.1.4 and examine some of the consequences of this structure.

For a space X, we know that its cohomology $\widetilde{H}^*(X;\mathbb{F}_p)$ is a module over $\mathbb{F}_p$ for a prime p, but sometimes this is not enough structure for our purposes. We will see in this section that $\widetilde{H}^*(X;\mathbb{F}_p)$ is in addition a module over a more structured algebra, the *Steenrod algebra* $\mathcal{A} = \mathcal{A}^*$. The Steenrod algebra is necessary for discussing the Adams spectral sequence in Section 2.6, and it is also one of the first examples of a Hopf algebra naturally occurring in topology. Therefore, let us go through some of the properties of $\mathcal{A}$, which can be found in, for example, [Ste62].

Throughout this section, let p denote a prime. For a pointed topological space X, we know that we have an isomorphism between the reduced $\mathbb{F}_p$–cohomology of X and maps in the stable homotopy category from the suspension spectrum of X to the Eilenberg–Mac Lane spectrum for $\mathbb{F}_p$:

$$\widetilde{H}^*(X;\mathbb{F}_p) \cong [\Sigma^\infty X, H\mathbb{F}_p]_{-*}.$$

As $[H\mathbb{F}_p, H\mathbb{F}_p]_*$ is a (graded) ring under addition and composition, it follows that $\widetilde{H}^*(X;\mathbb{F}_p)$ is a (graded) module over $[H\mathbb{F}_p, H\mathbb{F}_p]_*$ under composition.

For this section, we assume that the stable homotopy category has a commutative smash product and that $H\mathbb{F}_p$ is a commutative ring object in the stable homotopy category. The multiplication map (in the stable homotopy category) $H\mathbb{F}_p \wedge H\mathbb{F}_p \longrightarrow H\mathbb{F}_p$ gives the second map in

$$[\mathbb{S}, H\mathbb{F}_p]_* \otimes [H\mathbb{F}_p, H\mathbb{F}_p]_* \longrightarrow [H\mathbb{F}_p, H\mathbb{F}_p \wedge H\mathbb{F}_p]_* \longrightarrow [H\mathbb{F}_p, H\mathbb{F}_p]_*.$$

We know that $[\mathbb{S}, H\mathbb{F}_p]_*$ is $\mathbb{F}_p$ in degree 0 and that the multiplicative structure on $\mathbb{F}_p$ corresponds to the smash product induced multiplication on $[\mathbb{S}, H\mathbb{F}_p]_*$. It follows that $[H\mathbb{F}_p, H\mathbb{F}_p]_*$ and $[X, H\mathbb{F}_p]_*$ are $\mathbb{F}_p$–modules.

The definition of $\mathbb{F}_p$–cohomology can be extended from spaces to spectra quite easily using the stable homotopy category. For a spectrum Z, define

$$\widetilde{H}^*(Z;\mathbb{F}_p) \cong [Z, H\mathbb{F}_p]_{-*}.$$

This is again a $[H\mathbb{F}_p, H\mathbb{F}_p]_*$–module by the argument above.

This module structure amounts to a natural transformation of functors

$$\theta\colon \widetilde{H}^*(-,\mathbb{F}_p) \longrightarrow \widetilde{H}^{*+a}(-;\mathbb{F}_p)$$

for each $\theta \in [H\mathbb{F}_p, H\mathbb{F}_p]_a$. These natural transformations must also commute with the coboundary maps of $\widetilde{H}^*$, as in Remark 1.1.22. As a consequence, θ will be compatible with the long exact sequence of a pair and with the suspension isomorphism of $\widetilde{H}^*$. We call these natural transformations θ *cohomology operations* and can therefore think of $[H\mathbb{F}_p, H\mathbb{F}_p]_a$ as the algebra of cohomology operations.

Definition 2.5.1 A *cohomology operation* is a natural transformation of functors

$$\theta\colon \widetilde{H}^*(-,\mathbb{F}_p) \longrightarrow \widetilde{H}^*(-;\mathbb{F}_p),$$

where $\widetilde{H}^*$ denotes reduced singular cohomology of topological spaces.

Using the composition of natural transformations as well as the additive structure on morphisms in $\mathcal{SHC}$, the cohomology operations form a $\mathbb{F}_p$–algebra.

Definition 2.5.2 The mod-p *Steenrod algebra* $\mathcal{A} = \mathcal{A}^*$ is the $\mathbb{F}_p$–algebra of cohomology operations

$$\widetilde{H}^*(-;\mathbb{F}_p) \longrightarrow \widetilde{H}^*(-;\mathbb{F}_p).$$

Equivalently, we can define

$$\mathcal{A} = \mathcal{A}^* = [H\mathbb{F}_p, H\mathbb{F}_p]_*.$$

In the literature, the mod-p Steenrod algebra is often denoted by $\mathcal{A}_p$ to recognise that this definition depends on the prime p. As we will only use $p = 2$ in this section from this point onwards, we omit the subscript and mod-p from our notation. We refer the interested reader to [Ste62] for the odd primary analogues of the results.

The definition of the Steenrod algebra is often given via this equivalent, axiomatic characterisation.

Theorem 2.5.3 *The Steenrod algebra $\mathcal{A}$ is generated by additive homomorphisms*

$$Sq^n\colon \widetilde{H}^m(X,\mathbb{F}_2) \to \widetilde{H}^{m+n}(X,\mathbb{F}_2)$$

natural in X satisfying the following.

1. *Sq^0 is the identity.*
2. *For $x \in \widetilde{H}^n(X;\mathbb{F}_2)$, $Sq^n(x) = x^2$.*
3. *For $x \in \widetilde{H}^m(X;\mathbb{F}_2)$ with $m < n$, $Sq^n(x) = 0$.*
4. *$Sq^n(x \cup y) = \sum_{i+j=n} Sq^i(x) \cup Sq^j(y)$, which is known as the* Cartan formula.
5. *Sq^1 is the Bockstein homomorphism β associated to the short exact sequence*

$$0 \longrightarrow \mathbb{Z}/2 \longrightarrow \mathbb{Z}/4 \longrightarrow \mathbb{Z}/2 \longrightarrow 0.$$

6. *If $0 < a < 2b$, the* Adem relations *hold, that is,*

$$Sq^a Sq^b = \sum_{i=0}^{\lfloor a/2 \rfloor} \binom{b-1-i}{a-2i} Sq^{a+b-i} Sq^i.$$

Because of the second point, the elements Sq^n are often referred to as "Steenrod squares".

We will not give the proof that the two definitions of the Steenrod algebra are indeed equivalent. The construction of the Sq^n and the result that an algebra satisfying (1.) – (6.) is unique are performed in the setting of power operations in cohomology, see, for example, [May70]. The axiomatic set-up of power operations is indeed a very strong tool, leading to Steenrod squares, Dyer–Lashof operations and other, more algebraic applications. However, it is also lengthy and technical, which is why we refer the reader to the literature, for example, [Ste62] and [Hat02].

Remark 2.5.4 The product $\cup$ in the Cartan formula is the cup product

$$\cup\colon \widetilde{H}^i(X;\mathbb{F}_2) \otimes \widetilde{H}^j(X;\mathbb{F}_2) \longrightarrow \widetilde{H}^{i+j}(X;\mathbb{F}_2).$$

There is an equivalent version of the Cartan formula, namely,

$$Sq^n(x \otimes y) = \sum_{i+j=n} Sq^i(x) \otimes Sq^j(y),$$

where

$$x \in \widetilde{H}^*(X;\mathbb{F}_2), \quad y \in \widetilde{H}^*(Y;\mathbb{F}_2),$$

and $x \otimes y$ is viewed as an element in

$$x \otimes y \in \widetilde{H}^*(X;\mathbb{F}_2) \otimes \widetilde{H}^*(Y;\mathbb{F}_2) \cong \widetilde{H}^*(X \times Y;\mathbb{F}_2).$$

Assume that the cup product version of the Cartan formula holds. Let x, y as above, $(x \otimes 1), (1 \otimes y) \in \widetilde{H}^*(X \times Y;\mathbb{F}_2)$, and let

$$p_X\colon X \times Y \longrightarrow X \text{ and } p_Y\colon X \times Y \longrightarrow Y$$

be the respective projections. Then,

$$\begin{aligned}
Sq^n(x \otimes y) &= Sq^n((x \otimes 1) \cup (1 \otimes y)) \\
&= \sum_i Sq^i(x \otimes 1) \cup Sq^{n-i}(1 \otimes y) \\
&= \sum_i Sq^i(p_X^*(x)) \cup Sq^{n-i}(p_Y^*(y)) \\
&= \sum_i p_X^* Sq^i(x) \cup p_Y^* Sq^{n-i}(y) \\
&= \sum_i (Sq^i(x) \otimes 1) \cup (1 \otimes Sq^{n-i}(y)) \\
&= \sum_i Sq^i(x) \otimes Sq^{n-i}(y).
\end{aligned}$$

Conversely, as the cup product is the composition of the Künneth isomorphism with the morphism induced by the diagonal map, the cup product version of the Cartan formula follows from the new version.

The $\mathcal{A}$–module structure of $\widetilde{H}^*(X; \mathbb{F}_2)$ gives the cohomology considerable additional structure. For example, we know that $\widetilde{H}^*(\mathbb{R}P^\infty_+; \mathbb{F}_2)$ is a polynomial algebra over $\mathbb{F}_2$ with one generator in degree 1, using the characterisation of Theorem 2.5.3 we can also describe its $\mathcal{A}$–module structure.

Corollary 2.5.5 *On the mod-2 cohomology of the infinite-dimensional real projective space, we have*

$$\widetilde{H}^i(\mathbb{R}P^\infty_+; \mathbb{F}_2) = \mathbb{F}_2[u], \quad |u| = 1,$$

with the Steenrod algebra action given by

$$Sq^j(u^{2^k}) = \begin{cases} u^{2^k} & \text{if } j = 0 \\ u^{2^{k+1}} & \text{if } j = 2^k \\ 0 & \text{otherwise.} \end{cases}$$ □

The above corollary is particularly useful, as we can show many of the algebraic properties of $\mathcal{A}$ by evaluating Steenrod squares on powers of u.

The $\mathcal{A}$–module structure also helps us distinguish between spectra, as we demonstrate in the following example and lemma.

Example 2.5.6 Since we are assuming that the homotopy category of sequential spectra satisfies the list of properties from Subsection 1.1.4, we see that $[\mathbb{S}, \mathbb{S}] = \mathbb{Z}$. The self map of the sphere spectrum corresponding to n is called *multiplication by n*. Choosing a representative map

$$f\colon \mathbb{S}^{cof} \longrightarrow \mathbb{S}^{fib}$$

of multiplication by n in sequential spectra, we call the homotopy cofibre of f the *mod-n Moore spectrum* $M(\mathbb{Z}/n)$. We may calculate the homology of $M(\mathbb{Z}/n)$ using the long exact homology sequence and see that it is concentrated in degree zero, where it takes value $\mathbb{Z}/n$. For a construction of Moore spectra for arbitrary abelian groups G, see Example 7.4.7.

Lemma 2.5.7 *Let M denote the mod-2 Moore spectrum, that is, the homotopy cofibre of multiplication by 2 on the sphere spectrum $\mathbb{S}$. Then we have*

$$\widetilde{H}^*(M \wedge M; \mathbb{F}_2) \cong \widetilde{H}^*(M \vee \Sigma M; \mathbb{F}_2) \text{ as graded } \mathbb{F}_2\text{–modules,}$$

but

$$\widetilde{H}^*(M \wedge M; \mathbb{F}_2) \ncong \widetilde{H}^*(M \vee \Sigma M; \mathbb{F}_2) \text{ as } \mathcal{A}\text{–modules.}$$

Proof The long exact cohomology sequence associated to the homotopy cofibre sequence

$$\mathbb{S} \xrightarrow{2} \mathbb{S} \longrightarrow M$$

shows that the mod-2 cohomology of M is

$$\widetilde{H}^i(M; \mathbb{F}_2) = \begin{cases} \mathbb{F}_2 & \text{if } i = 0, 1 \\ 0 & \text{otherwise.} \end{cases}$$

We would like to determine the structure of $\widetilde{H}^*(M; \mathbb{F}_2)$ as an $\mathcal{A}$–module. For degree reasons, the only non-trivial option for this is to check whether $Sq^1(x_0) = x_1$ or $Sq^1(x_0) = 0$, where x_i denotes the $\mathbb{F}_2$–generator in $\widetilde{H}^i(M; \mathbb{F}_2)$. We show that $Sq^1(x_0) = x_1$.

By Theorem 2.5.3, Sq^1 is the Bockstein homomorphism associated to the short exact sequence of coefficients

$$0 \longrightarrow \mathbb{Z}/2 \xrightarrow{i} \mathbb{Z}/4 \xrightarrow{p} \mathbb{Z}/2 \longrightarrow 0.$$

Applying cohomology to these coefficients gives us

$$\cdots \to \widetilde{H}^0(M; \mathbb{Z}/4) \xrightarrow{p_*} \widetilde{H}^0(M; \mathbb{F}_2) \xrightarrow{\beta} \widetilde{H}^1(M; \mathbb{F}_2) \xrightarrow{i_*} \widetilde{H}^1(M; \mathbb{Z}/4) \to \cdots.$$

To work out the terms in the sequence, we use the Universal Coefficient Theorem, which says that for a group G, the sequence

$$0 \to \mathrm{Ext}^1_{\mathbb{Z}}(H_{n-1}(X,\mathbb{Z}),G) \to \widetilde{H}^n(X,G) \to \mathrm{Hom}_{\mathbb{Z}}(H_n(X,\mathbb{Z}),G) \longrightarrow 0$$

is exact. Using $\widetilde{H}_0(M,\mathbb{Z}) = \mathbb{F}_2$ and $\widetilde{H}_1(M,\mathbb{Z}) = 0$, one obtains

$$\widetilde{H}^0(M;\mathbb{Z}/4) \cong \mathrm{Hom}_{\mathbb{Z}}(\mathbb{F}_2;\mathbb{Z}/4) \cong \mathbb{F}_2 \quad \text{and} \quad \widetilde{H}^1(M;\mathbb{Z}/4) = 0.$$

Therefore, the above long exact cohomology sequence is

$$\cdots \longrightarrow \mathbb{F}_2 \xrightarrow{p_*} \mathbb{F}_2 \xrightarrow{\beta} \mathbb{F}_2 \longrightarrow 0 \longrightarrow \cdots.$$

The cyclic generator of $\widetilde{H}^0(M;\mathbb{Z}/4) \cong \mathrm{Hom}_{\mathbb{Z}}(\mathbb{F}_2;\mathbb{Z}/4)$ is precisely

$$i\colon \mathbb{F}_2 \longrightarrow \mathbb{Z}/4$$

from the original short exact coefficient sequence. Consequently, the map p_* is zero as $p \circ i = 0$. Thus,

$$Sq^1 = \beta\colon \widetilde{H}^0(M;\mathbb{F}_2) \longrightarrow \widetilde{H}^1(M;\mathbb{F}_2)$$

is an isomorphism. So altogether, as a module over $\mathbb{F}_2$, $\widetilde{H}^*(M;\mathbb{F}_2)$ is generated by two elements x_0 and x_1 in degrees 0 and 1, respectively, and the $\mathcal{A}$–module structure is $Sq^1(x_0) = x_1$, see Theorem 2.5.3.

Consequently, $\widetilde{H}^*(M \vee \Sigma M;\mathbb{F}_2)$ is generated over $\mathbb{F}_2$ by elements x_0, x_1, y_1 and y_2 in respective degrees 0, 1, 1 and 2. The Steenrod algebra action is

$$Sq^1(x_0) = x_1 \quad \text{and} \quad Sq^1(y_1) = y_2$$

and is trivial in all other cases. We can illustrate $\widetilde{H}^*(M \vee \Sigma M;\mathbb{F}_2)$ as an $\mathcal{A}$–module in the following picture.

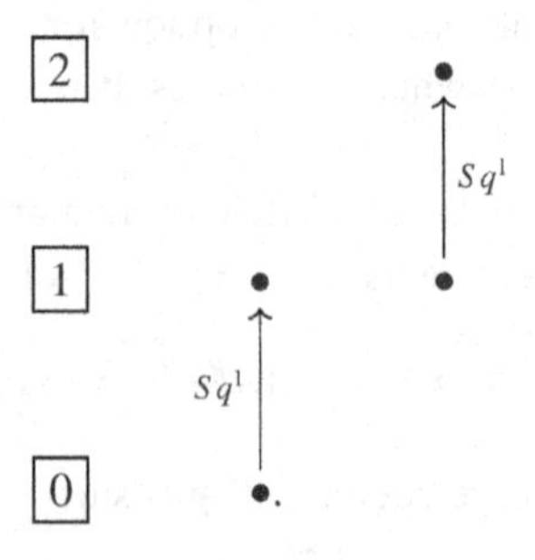

Meanwhile, by the Künneth isomorphism we have

$$\widetilde{H}^*(M \wedge M;\mathbb{F}_2) \cong \widetilde{H}^*(M;\mathbb{F}_2) \otimes \widetilde{H}^*(M;\mathbb{F}_2),$$

which as a graded $\mathbb{F}_2$–module is generated by the elements

$$\{x_0 \otimes x_0,\ x_0 \otimes x_1,\ x_0 \otimes x_1,\ x_1 \otimes x_1\}$$

and therefore isomorphic to $\widetilde{H}^*(M \vee \Sigma M; \mathbb{F}_2)$, as a graded $\mathbb{F}_2$–module. Another possible basis is

$$\{x_0 \otimes x_0,\ x_0 \otimes x_1 + x_1 \otimes x_0,\ x_1 \otimes x_0,\ x_1 \otimes x_1\}.$$

The Cartan formula tells us that

$Sq^1(x_0 \otimes x_0) = x_0 \otimes x_1 + x_1 \otimes x_0$, $Sq^1(x_0 \otimes x_1) = x_1 \otimes x_1$ and $Sq^2(x_0 \otimes x_0) = x_1 \otimes x_1$,

as illustrated below.

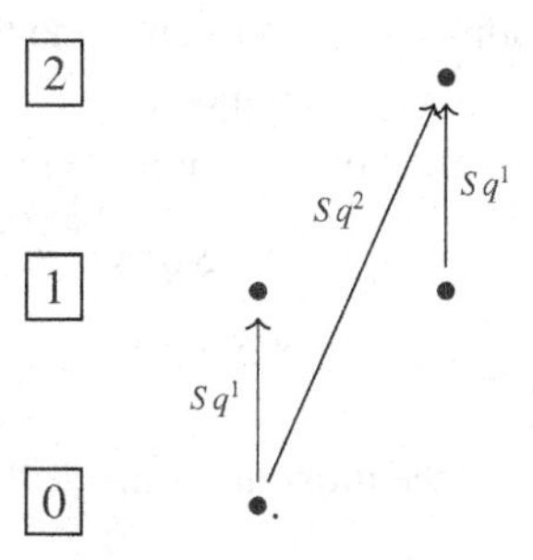

As $\widetilde{H}^*(M \vee \Sigma M; \mathbb{F}_2)$ does not possess any non-trivial action by Sq^2, it therefore cannot be isomorphic to $\widetilde{H}^*(M \wedge M; \mathbb{F}_2)$ as $\mathcal{A}$–module. □

Taking a closer look at the Adem relations, we see that not all possible products of the Sq^n are necessary to generate $\mathcal{A}$ as a graded module over $\mathbb{F}_2$. In fact, we can give a minimal generating set.

Definition 2.5.8 A monomial $Sq^{i_1} Sq^{i_2} \cdots Sp^{i_k}$ in $\mathcal{A}$ is *admissible* if $i_k \geqslant 1$ and $i_{r-1} \geqslant 2i_r$ for $k \geqslant r \geqslant 2$.

Proposition 2.5.9 *As a module over $\mathbb{F}_2$, the admissible monomials form a basis of the Steenrod algebra $\mathcal{A}$.*

Proof We have to show that every monomial $Sq^{i_1} Sq^{i_2} \cdots Sp^{i_k}$ can be uniquely written as a sum of admissible monomials. We use the notation

$$Sq^I = Sq^{i_1} \cdots Sq^{i_k} \text{ for } I = (i_1, \cdots, i_k).$$

We speak of k as the length of I, and the degree of I is $i_1 + \cdots + i_k$. To show the generating statement, one looks at the Adem relations

$$Sq^a Sq^b = \sum_{i=0}^{\lfloor a/2 \rfloor} \binom{b-1-i}{a-2i} Sq^{a+b-i} Sq^i$$

to see that every monomial can be rearranged into a sum of admissibles. The following terminology will allow for a smooth inductive argument. We define the *moment* of the monomial Sq^I to be

$$m(I) := \sum_{s=1}^{k} s i_s.$$

The idea is that the monomials with a small moment are those with relatively large i_j for j small, and those with a big moment have large i_j for j large. So to get a monomial closer to being admissible, one has to decrease its moment, which we can do with the Adem relations.

Assume that every monomial with moment smaller than the moment of Sq^I can be written as a sum of admissibles. Also, assume that Sq^I is not an admissible monomial itself, that is, there is an index r with $i_r < 2i_{r+1}$. Applying the Adem relations to that place in the monomial gives us

$$\begin{aligned} Sq^I &= Sq^{(i_1,\cdots,i_{r-1})} Sq^{i_r} Sq^{i_{r+1}} Sq^{(i_{r+2},\cdots,i_k)} \\ &= \sum_{j=0}^{\lfloor i_r/2 \rfloor} \chi_j \, Sq^{(i_1,\cdots,i_{r-1})} Sq^{i_r+i_{r+1}-j} Sq^j Sq^{(i_{r+2},\cdots,i_k)}, \end{aligned}$$

where $\chi_j \in \mathbb{F}_2$. The moment of the monomials inside the sum on the right-hand side is

$$\sum_{s=1}^{r-1} s i_s + r(i_r + i_{r+1} - j) + (r+1)j + \sum_{s=r+2}^{k} s i_s.$$

This is smaller than the moment of Sq^I, because

$$r i_r + (r+1) i_{r+1} > r(i_r + i_{r+1} - j) + (r+1)j$$

as

$$0 \leqslant j \leqslant i_r/2 < i_{r+1}.$$

Thus, by induction,

$$Sq^{(i_1,\cdots,i_{r-1})} Sq^{i_r+i_{r+1}-j} Sq^j Sq^{(i_{r+2},\cdots,i_k)}$$

can be written as a sum of admissibles, and therefore, also Sq^I.

Now we have to show that the admissible monomials are linearly independent. We show this by evaluating them on

$$u^{\otimes n} \in \widetilde{H}^n((\mathbb{R}P^\infty_+)^{\wedge n}; \mathbb{F}_2),$$

where u is the polynomial generator of the cohomology ring of $\mathbb{R}P^\infty_+$. We will do this by induction on n.

Let $0 \leqslant d \leqslant n$. Consider a linear combination

$$\sum \chi_I \cdot Sq^I(u^{\otimes n}) = 0, \tag{2.1}$$

where $\chi_I \in \mathbb{F}_2$ and the sum is taken over all I for which Sq^I is admissible and of degree $d \leqslant n$. We would like to show that $\chi_I = 0$ for all I, which we again show by (decreasing) induction on the length of I. We note that we can start this induction because no monomial of degree d (not containing any Sq^0) can have length greater than d, particularly not an admissible one. The main idea of the overall claim is splitting our linear combination

$$\sum \chi_I \cdot Sq^I(u^{\otimes n})$$

into sums involving $u^{\otimes j}$ for $j < n$ using the Cartan formula and our induction assumption: let us assume that if

$$\sum \chi_I \cdot Sq^I(u^{\otimes n-1}) = 0, \tag{2.2}$$

where the sum runs over all admissible monomials of degree d where $d \leqslant n-1$, then all the χ_I are zero. Furthermore, assume that all χ_I in (2.1) are zero if the length of I is greater than m.

The Künneth isomorphism gives us

$$\begin{aligned} Sq^I(u^{\otimes n}) &\in \widetilde{H}^{d+n}((\mathbb{R}P^\infty_+)^{\wedge n}; \mathbb{F}_2) \\ &\cong \bigoplus_r \widetilde{H}^r(\mathbb{R}P^\infty_+; \mathbb{F}_2) \otimes \widetilde{H}^{d+n-r}((\mathbb{R}P^\infty_+)^{\wedge(n-1)}; \mathbb{F}_2). \end{aligned}$$

Applying the Cartan formula to this results in

$$Sq^I(u^{\otimes n}) = \sum_{J \leqslant I} Sq^J(u) \otimes Sq^{I-J}(u^{\otimes n-1}).$$

In this notation, the difference $I - J$ is taken componentwise, and $J \leqslant I$ means that $j_r \leqslant i_r$ for all r. Let us take a closer look at the summand belonging to $r = 2^m$ in the Künneth formula by considering the projection pr onto this summand.

We saw in Corollary 2.5.5 that

$$Sq^J(u) = \begin{cases} u^{2^k} & \text{if } J = J_k := (2^{k-1}, 2^{k-2}, \cdots, 2, 1) \\ 0 & \text{otherwise.} \end{cases}$$

As Sq^{J_m} is the only monomial in the summand $r = 2^m$ acting non-trivially on u, we obtain

$$pr(Sq^I(u^{\otimes n})) = \begin{cases} u^{2^m} \otimes Sq^{I-J_m}(u^{\otimes n-1}) & \text{if the length of } I \text{ is } m \\ 0 & \text{otherwise.} \end{cases}$$

Now applying the projection pr to the original linear combination (2.1) gives us

$$pr\Big(\sum \chi_I Sq^I(u^{\otimes n})\Big) = \sum_{\text{length}(I)=m} \chi_I \cdot pr(Sq^I(u^{\otimes n})) + \sum_{\text{length}(I)<m} \chi_I \cdot pr(Sq^I(u^{\otimes n}))$$
$$= u^{2^m} \otimes \sum_{\text{length}(I)=m} \chi_I \cdot Sq^{I-J_m}(u^{\otimes n-1}) = 0.$$

The sets of all monomials of the form $I - J_m$, where I is admissible of length m and degree d, is in one-to-one correspondence with the set of admissible monomials of length m or less and degree $d - 2^m + 1$. Therefore, the final sum in the previous formula is the same as (2.2), which we assumed to imply that $\chi_I = 0$ for all I. Thus, we have proved that $\chi_I = 0$ for all admissible I of length m and degree $d \leqslant n$, which was precisely our claim.

Hence, the map

$$e_n \colon \mathcal{A} \longrightarrow \widetilde{H}^n((\mathbb{R}P^\infty_+)^{\wedge n}; \mathbb{F}_2)$$

given by evaluation on $u^{\otimes n}$ sends the set of admissible monomials to a linearly independent set. In particular, it is an isomorphism in degrees n and smaller. As a consequence, the set of admissible monomials must be linearly independent itself. □

Definition 2.5.10 Let $\bar{\mathcal{A}}$ denote the ideal of all positive degree elements in $\mathcal{A}$. An element of $\mathcal{A}$ is *decomposable* if it is in the image of $\bar{\mathcal{A}} \otimes \bar{\mathcal{A}}$ under the multiplication map

$$m \colon \mathcal{A} \otimes \mathcal{A} \longrightarrow \mathcal{A}.$$

If an element of $\mathcal{A}$ is not in this image, it is called *indecomposable.*

In other words, an element is indecomposable if it cannot be written as a (sum of a) product of two other elements. In fact, we can say what the indecomposable elements of $\mathcal{A}$ are.

Lemma 2.5.11 *For $p = 2$, the element Sq^i is indecomposable if and only if i is a power of 2.*

Proof Let $i = 2^k$. Assume that Sq^i is decomposable, that is, there are $m_j \in \mathcal{A}$ with

$$Sq^{2^k} = \sum_{j=1}^{2^k-1} m_j \cdot Sq^j.$$

Evaluated on u^{2^k}, with u the polynomial generator of $\widetilde{H}^*(\mathbb{R}P^\infty_+; \mathbb{F}_2)$, this would be

$$u^{2^{k+1}} = Sq^{2^k}(u^{2^k}) = \sum_{j=1}^{2^k-1} m_j \cdot Sq^j(u^{2^k}) = \sum_{j=1}^{2^k-1} m_j \cdot 0 = 0$$

because of Corollary 2.5.5. This contradiction shows that Sq^{2^k} must be indecomposable.

Now let $i \neq 2^k$. We can write $i = a + b$, where $0 < a < 2^k$ and $b = 2^k$ for a maximal k. Then the Adem relations give us

$$\binom{b-1}{a} Sq^{a+b} = Sq^a Sq^b + \sum_{j \geqslant 0} \binom{b-1-j}{a-2j} Sq^{a+b-j} Sq^j.$$

As $b - 1 = 1 + \cdots + 2^{k-2} + 2^{k-1}$, induction shows that

$$\binom{b-1}{a} \equiv 1 \bmod 2,$$

which means that $Sq^{a+b} = Sq^i$ is decomposable as claimed. □

Corollary 2.5.12 *The elements Sq^{2^k} generate the mod-2 Steenrod algebra $\mathcal{A}$ as an algebra.* □

A useful algebraic property of $\mathcal{A}$ is that it has the structure of a *Hopf algebra*, meaning that it is a bialgebra with an antipode and compatibility conditions, see, for example, [Wei94].

Definition 2.5.13 A *Hopf algebra* is a (graded) algebra A over a ring R together with

- an augmentation $\epsilon\colon A \longrightarrow R$,
- an R-algebra homomorphism

$$\Delta\colon A \longrightarrow A \otimes A,$$

- an antipode $c\colon A \longrightarrow A$ which is an R–module isomorphism

satisfying the following properties.

- With Δ and ϵ, A is a cocommutative coalgebra.
- The composites

$$A \xrightarrow{\Delta} A \otimes A \xrightarrow{c \otimes 1} A \otimes A \xrightarrow{m} A$$

and

$$A \xrightarrow{\Delta} A \otimes A \xrightarrow{1 \otimes c} A \otimes A \xrightarrow{m} A,$$

where m denotes the multiplication in A, equal the composite

$$A \xrightarrow{\epsilon} R \xrightarrow{\iota} A.$$

Recall that a graded $\mathbb{F}_2$–algebra A is connected if it is 0 in negative gradings and $A_0 = \mathbb{F}_2$.

Remark 2.5.14 Work of Milnor and Moore [MM65], proves the following result. If we have a graded connected bialgebra A with comultiplication of the form

$$\Delta(x) = x \otimes 1 + 1 \otimes x + \sum_i y_i \otimes z_i,$$

where y_i and z_i are elements of positive degree, then there is a unique R–module homomorphism $c\colon A \longrightarrow A$ satisfying $c(1) = 1$, c is an anti-automorphism, meaning that it is a linear isomorphism satisfying $c(x \cdot y) = (-1)^{|x||y|}c(y)c(x)$, if x has positive degree and $\Delta(x) = \sum_i a_i \otimes b_i$, then $\sum_i a_i \cdot c(b_i) = 0$, and c^2 is the identity.

Furthermore, the last point is equivalent to

$$\sum_i c(a_i) \cdot b_i = 0,$$

which follows from applying c to the third point and using $c^2 = \mathrm{Id}$. In other words, in this situation, A is a Hopf algebra with antipode c.

Theorem 2.5.15 *The Steenrod algebra $\mathcal{A}$ is a* Hopf algebra *over $\mathbb{F}_2$ with comultiplication given by*

$$\Delta(Sq^k) = \sum_{i=0}^{k} Sq^i \otimes Sq^{k-i}$$

and antipode defined by

$$c(Sq^0) = Sq^0, \quad \sum_{i=0}^{n} Sq^i \cdot c(Sq^{n-i}) = 0 \ \textit{for} \ n \geqslant 1.$$

Proof We first have to show that

$$\Delta\colon \mathcal{A} \longrightarrow \mathcal{A} \otimes \mathcal{A}$$

is an algebra homomorphism. Let F denote the free graded algebra generated by elements Sq^n of degree $n \in \mathbb{N}$ ($Sq^0 = \mathrm{Id}$), and let

$$pr\colon F \longrightarrow \mathcal{A}$$

denote the canonical projection. Then, the kernel of pr is exactly the Adem relations. As F is free, $\Delta \circ pr$ induces an algebra homomorphism

$$\bar{\Delta}\colon F \longrightarrow \mathcal{A} \otimes \mathcal{A}.$$

To show that Δ itself is also an algebra homomorphism, we need to show that $\bar{\Delta}$ sends the Adem relations to zero.

We have seen at the end of Proposition 2.5.9 that the map

$$e_n\colon \mathcal{A} \longrightarrow \widetilde{H}^*((\mathbb{R}P^\infty_+)^{\wedge n}; \mathbb{F}_2)$$

given by evaluating an element of the Steenrod algebra on the n-fold tensor product $u^{\otimes n}$ of the polynomial generator $u \in \widetilde{H}^1((\mathbb{R}P^\infty_+)^{\wedge n}; \mathbb{F}_2)$ is an isomorphism in degrees less than or equal to n. Therefore,

$$e_n \otimes e_n\colon \mathcal{A} \otimes \mathcal{A} \longrightarrow \widetilde{H}^*((\mathbb{R}P^\infty_+)^{\wedge n}; \mathbb{F}_2) \otimes \widetilde{H}^*((\mathbb{R}P^\infty_+)^{\wedge n}; \mathbb{F}_2)$$

is also an isomorphism in degrees less than or equal to n. Similarly,

$$e_{2n}\colon \mathcal{A} \longrightarrow \widetilde{H}^*(((\mathbb{R}P^\infty_+)^{\wedge n})^2; \mathbb{F}_2)$$

induces an isomorphism in degrees less than or equal to $2n$. We can fit all these maps into a diagram, which the Cartan formula tells us will be commutative.

$$\begin{array}{ccccc}
F & & \xrightarrow{\quad pr \quad} & & \mathcal{A} \\
\downarrow{\scriptstyle \bar{\Delta}} & & & & \downarrow{\scriptstyle e_{2n}} \\
\mathcal{A} \otimes \mathcal{A} & \xrightarrow[e_n \otimes e_n]{} & \widetilde{H}^*((\mathbb{R}P^\infty_+)^{\wedge n}; \mathbb{F}_2)^{\otimes 2} & \xrightarrow[k]{\cong} & \widetilde{H}^*(((\mathbb{R}P^\infty_+)^{\wedge n})^2; \mathbb{F}_2)
\end{array}$$

In other words, the two F–module structures $\widetilde{H}^*((\mathbb{R}P^\infty_+)^{\wedge n})^2; \mathbb{F}_2)$ inherits from the diagram are identical. So now let S be an element of the kernel of pr of degree n or less. As the diagram commutes, this means that

$$(k \circ (e_n \otimes e_n) \circ \bar{\Delta})(S) = 0.$$

But k and $e_n \otimes e_n$ are isomorphisms, meaning that $\bar{\Delta}(S)$ must already be zero. As this holds for all n, we have proved our claim that $\bar{\Delta}$ sends the Adem relations to zero, meaning that the algebra homomorphism $\bar{\Delta}$ induces the algebra homomorphism

$$\Delta\colon \mathcal{A} \longrightarrow \mathcal{A} \otimes \mathcal{A}$$

as claimed.

The next claim on our list is that Δ is a cocommutative, coassociative coalgebra with counit ϵ. This is a straightforward calculation using the explicit formulas of Δ and ϵ, so we are not going to spell this out here.

Finally, we have to deal with the antipode $c\colon \mathcal{A} \longrightarrow \mathcal{A}$. We are in precisely the situation of Remark 2.5.14, thereby finishing the proof that $\mathcal{A}$ is a Hopf algebra. □

It is also worth considering the dual of $\mathcal{A}$ due to its comparatively simple structure.

Definition 2.5.16 The *dual* of the Steenrod algebra $\mathcal{A}$ is defined by

$$\mathcal{A}_* = \mathrm{Hom}_{\mathbb{F}_p}(\mathcal{A}, \mathbb{F}_p).$$

Its grading is given by

$$\mathcal{A}_n = \mathrm{Hom}_{\mathbb{F}_p}(\mathcal{A}^n, \mathbb{F}_p).$$

The dual of $\mathcal{A}$ is in fact traditionally denoted $\mathcal{A}_*$ rather than $\mathcal{A}^*$ – the latter is often used for the Steenrod algebra itself.

Proposition 2.5.17 *For $p = 2$, the dual $\mathcal{A}_*$ of the Steenrod algebra $\mathcal{A}$ is the polynomial algebra*

$$\mathbb{F}_2[\xi_1, \xi_2, \cdots],$$

where ξ_k is the dual of the admissible monomial $Sq^{2^{k-1}} Sq^{2^{k-2}} \cdots Sq^2 Sq^1$.

Proof Let $I = (i_1, \cdots, i_n)$ be a sequence of non-negative integers, and let

$$Sq^I = Sq^{i_1} \cdots Sq^{i_n}.$$

Then we define $\xi_k \in \mathcal{A}_*$ to be the dual of Sq^I with

$$I = (2^{k-1}, 2^{k-2}, \cdots, 2, 1)$$

in the following sense. We recall that by Proposition 2.5.9, a graded $\mathbb{F}_2$–vector space basis of $\mathcal{A}$ is given by those admissible Sq^I, so for an admissible $a \in \mathcal{A}$, we define

$$\langle \xi_k, a \rangle = \begin{cases} 1 & \text{if } a = Sq^{(2^{k-1}, 2^{k-2}, \cdots, 2, 1)} \\ 0 & \text{otherwise.} \end{cases}$$

Our goal is to show that the standard map

$$\Phi\colon \mathbb{F}_2[\xi_1, \xi_2, \cdots] \longrightarrow \mathcal{A}_*, \quad \xi_i \mapsto \xi_i$$

is an isomorphism of algebras. We will first show that it is an epimorphism, and then that the dimensions of the source and target over $\mathbb{F}_2$ agree in every degree.

Firstly, let us introduce some notation. Let us inductively define

$$\xi_I = \xi_i \text{ if } I = (i) \quad \text{and} \quad \xi_I = \xi_{i_1} \cdot \xi_{(i_2, \cdots, i_n)} \in \mathcal{A}_*$$

if the length of I is greater than 1, where the multiplication $\cdot$ in $\mathcal{A}_*$ is dual to the comultiplication in $\mathcal{A}$

$$\xi_I \cdot \xi_J = \Delta^*(\xi_I \otimes \xi_J).$$

By definition, these elements ξ_I lie in the image of Φ.

We prove the following claim: given an element $a \in \mathcal{A}$, if $\langle \xi_I, a\rangle = 0$ for all possible I, then $a = 0$. This is shown by using the $\mathcal{A}$–module structure on $\widetilde{H}^*((\mathbb{R}P^\infty_+)^{\wedge n}; \mathbb{F}_2)$. Let u be the polynomial generator of $\widetilde{H}^*(\mathbb{R}P^\infty_+; \mathbb{F}_2)$. One can prove by induction that

$$a(u^{\otimes n}) = \sum_{\text{length}(I)=n} \langle \xi_I, a\rangle u(I),$$

where $u(I)$ is some word in tensor powers of u. Therefore, if $\langle \xi_I, a\rangle$ was always trivial, then the action of $\mathcal{A}$ on $u^{\otimes n}$ would also be trivial. But we have seen in the proof of Proposition 2.5.9 that it is not. So, $\langle \xi_I, a\rangle = 0$ for all I implies that $a = 0$. However, this means that the ξ_I (and therefore the whole image of Φ) must already make up the whole of $\mathcal{A}_*$, as otherwise we could construct an I and an a such that $\langle \xi(I), a\rangle = 1$. Therefore, Φ is an epimorphism.

Now we will show that

$$\Phi \colon \mathbb{F}_2[\xi_1, \xi_2, \cdots] \longrightarrow \mathcal{A}_*$$

is not just an epimorphism, but an isomorphism by proving that the dimension of $\mathbb{F}_2[\xi_1, \xi_2, \cdots]$ as a $\mathbb{F}_2$–vector space is the same as the dimension of $\mathcal{A}_*$ in each degree. This is purely combinatorial.

Monomials in $\mathbb{F}_2[\xi_1, \xi_2, \cdots]$ correspond to sequences of non-negative integers

$$I = (i_1, i_2, \cdots, i_n, 0, 0, \cdots)$$

with finitely many non-zero entries. On the right-hand side, additive generators of $\mathcal{A}_*$ correspond, via duality, to admissible monomials in $\mathcal{A}$ (Proposition 2.5.9), which in turn correspond to sequences of non-negative integers

$$J = (j_1, j_2, \cdots, j_n, 0, \cdots), \quad j_k \geqslant 2j_{k+1}, \quad j_n \geqslant 1.$$

We can set up a bijection between the two respective sets of integer sequences as follows. Let

$$I_k := (0, \cdots, 0, 1, 0, \cdots) \text{ (non-zero entry in position } k)$$

and

$$J_k := (2^{k-1}, 2^{k-2}, \cdots, 2, 1, 0, \cdots).$$

We define a map on sequences by sending I_k to J_k and asking for this map to be additive with respect to componentwise addition. It is now straightforward to check that an admissible sequence

$$J = (j_1, j_2, \cdots, j_n, 0, \cdots)$$

has the unique preimage

$$I = (i_1, i_2, \cdots, i_n, 0, \cdots) \text{ where } i_k = j_k - 2j_{k+1}.$$

This completes the proof that Φ is an algebra isomorphism between the polynomial algebra $\mathbb{F}_2[\xi_1, \xi_2, \cdots]$ and the dual of the Steenrod algebra. □

As a consequence of Theorem 2.5.15, we get the following.

Corollary 2.5.18 *The dual $\mathcal{A}_*$ of the Steenrod algebra $\mathcal{A}$ is a Hopf algebra.*

Proof This is standard because $\mathcal{A}$ is finite in every degree. The multiplication in $\mathcal{A}_*$ is then the dual of the comultiplication in $\mathcal{A}$ and vice versa. □

Remark 2.5.19 The dual map of the multiplication on $\mathcal{A}$

$$\Delta\colon \mathcal{A}_* \longrightarrow \mathcal{A}_* \otimes \mathcal{A}_*$$

is given by

$$\psi(\xi_k) = \sum_{i=0}^{k} \xi_{k-i}^{2^i} \otimes \xi_i.$$

This is again shown by using the cohomology of $(\mathbb{R}P^\infty_+)^{\wedge n}$, see [Ste62] or [Mil58]. One can furthermore derive a closed formula for the antipode on $\mathcal{A}_*$ using the methods of Remark 2.5.14.

2.6 The Adams Spectral Sequence

The Adams spectral sequence is not only one of the most powerful tools to calculate the stable homotopy groups of spheres (and of other spectra), but it is also related to other phenomena in stable homotopy theory such as power operations. We will give a brief overview of the construction of the spectral sequence and its properties. This is by no means a replacement for a rigorous discussion, such as Bruner [Bru09] and Rognes [Rog12], but we hope that it will serve as a gentle introduction which may inspire further reading. We assume that the reader is familiar with the basic terminology of spectral sequences, see McCleary [McC01].

As with the previous section, we assume that the homotopy category of sequential spectra satisfies the list of properties from Subsection 1.1.4.

2.6.1 Construction of the Spectral Sequence

In this section, p will denote a fixed prime, $\mathbb{Z}_p^\wedge$ will be the p-adic integers and $\mathbb{Z}_{(p)}$ will be the p-local integers

$$\begin{aligned} \mathbb{Z}_p^\wedge &= \lim_{n\geqslant 0} \mathbb{Z}/p^n \\ \mathbb{Z}_{(p)} &= \{\tfrac{a}{b} \mid p \nmid b\} \subseteq \mathbb{Q}. \end{aligned}$$

Given two spectra X and Y, applying mod-p singular cohomology H^* gives us a map

$$[X, Y] \xrightarrow{H^*} \mathrm{Hom}_{\mathcal{A}}(H^*(Y), H^*(X)),$$

where $\mathcal{A}$ denotes the mod-p Steenrod algebra and $\mathrm{Hom}_{\mathcal{A}}(-, -)$ denotes morphisms in the category of modules over $\mathcal{A}$. The Adams spectral sequence is an attempt to reverse this, to find out about $[X, Y]$ by knowing the cohomology of X and Y as modules over $\mathcal{A}$.

By a spectrum of "finite type", we mean a spectrum whose cohomology is finitely generated in each degree. For reasons of convergence, we also assume our spectra to have bounded below homotopy groups. The example we have in mind is $X = Y = \mathbb{S}$.

Theorem 2.6.1 (The Adams spectral sequence) *Let X and Y be spectra of finite type, and let p be a prime. Furthermore, assume that the homotopy groups of X and Y are bounded below. Then there is a spectral sequence*

$$E_2^{s,t} = \mathrm{Ext}_{\mathcal{A}}^{s,t}(H^*(Y), H^*(X)) \Longrightarrow [X, Y_p^\wedge]_{t-s},$$

where $Y_p^\wedge$ denotes the p-completion of Y.

For details on the p-completion of a spectrum, see Section 7.4. Alternatively, the reader may just bear in mind that in the case of $X = Y = \mathbb{S}$, one has

$$[\mathbb{S}, \mathbb{S}_p^\wedge] = [\mathbb{S}, \mathbb{S}] \otimes \mathbb{Z}_p^\wedge.$$

Remark 2.6.2 For each r, the E_r-term of the spectral sequence carries a bilinear pairing

$$E_r^{s,t}(X, Y) \otimes E_r^{s',t'}(Y, Z) \longrightarrow E_r^{s+s',t+t'}(X, Z),$$

which is natural in X, Y and Z, and which converges to the pairing given by composing morphisms of spectra. For $r = 2$, this pairing coincides with the Yoneda product on Ext groups.

Remark 2.6.3 As the dual $\mathcal{A}_*$ of the Steenrod algebra is so much simpler algebraically than the Steenrod algebra itself, it is preferable in some situations to use the homological variant of the Adams spectral sequence, which is

$$E^2_{s,t} = \mathrm{Ext}^{s,t}_{\mathcal{A}_*}(H_*(X), H_*(Y)) \Longrightarrow [X, Y]_{t-s} \otimes \mathbb{Z}_{(p)}.$$

Let us now look into the construction of the spectral sequence.

Definition 2.6.4 An *Adams tower* for a spectrum Y is a diagram in $\mathcal{SHC}$

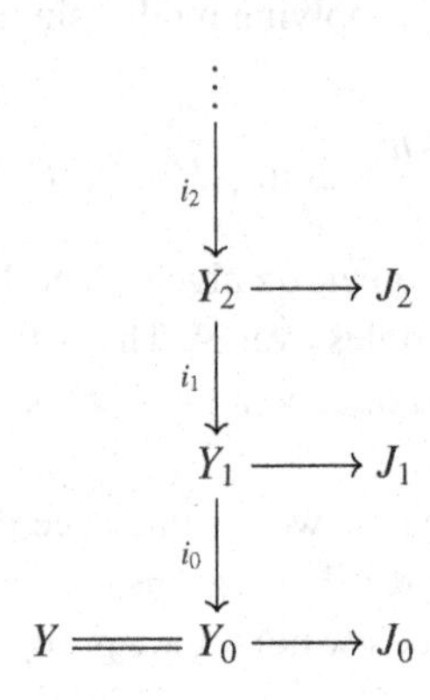

such that

- J_n is the homotopy cofibre of i_n,
- $H^*(i_n) = 0$ for all n,
- $H^*(J_n)$ is a projective $\mathcal{A}$–module for all n,
- the map

  $$[X, J_n]_* \longrightarrow \mathrm{Hom}^*_{\mathcal{A}}(H^*(J_n), H^*(X))$$

  given by $[f] \mapsto H^*(f)$ is an isomorphism for all n and all X (i.e. J_n is an "$\mathcal{A}$-Eilenberg–Mac Lane object").

For any spectrum Y, there is an Adams tower with the above properties. For the sphere spectrum, we can construct an Adams tower as follows. Let $H\mathbb{F}_p$ denote the Eilenberg–Mac Lane spectrum of $\mathbb{F}_p$, and let $\overline{H\mathbb{F}_p}$ be the homotopy cofibre of the Hurewicz map

$$\mathbb{S} \xrightarrow{hur} H\mathbb{F}_p \longrightarrow \overline{H\mathbb{F}_p}$$

which is the inclusion of $\mathbb{S}$ into the bottom cell of $H\mathbb{F}_p$. Then we define

$$J_n := H\mathbb{F}_p \wedge \overline{H\mathbb{F}_p}^{\wedge(n-1)}$$

and

$$Y_{n+1} := \overline{H\mathbb{F}_p}^{\wedge n},$$

where $(-)^{\wedge(n-1)}$ denotes the $(n-1)$–fold smash product. This gives us an Adams tower, the "standard Adams tower".

Applying $[X,-]_*$ to an Adams tower gives us an exact couple

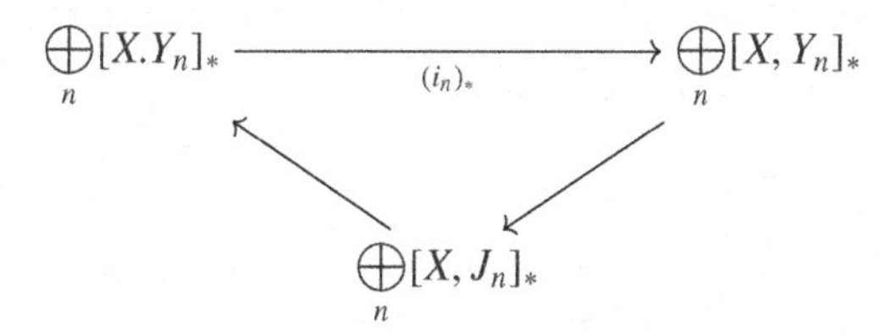

by using the Puppe exact sequence for spectra. Generally, having an exact couple

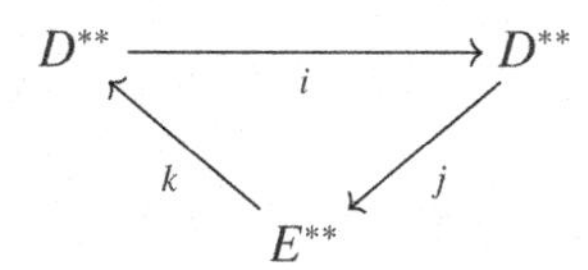

induces the derived exact couple

$$E' := \ker(j \circ k)/\operatorname{Im}(j \circ k), \quad D' := \ker(j), \quad d' := j \circ k$$

with j' and k' being the maps induced by j and k. By inductively defining

$$E_r := E'_{r-1}, \quad d_r := j'_{r-1} \circ k'_{r-1},$$

we obtain a spectral sequence $\{E_r^{**}, d_r\}$.

Remark 2.6.5 One can show that any Adams tower for Y will give us the same spectral sequence (up to isomorphism).

In any case, the exact couple yields

$$E_1^{s,t} = [X, \Sigma^s J_s]_t = [\Sigma^t X, \Sigma^s J_s],$$

where the right-hand side denotes ungraded morphisms in $\mathcal{SHC}$.

We obtain a diagram where the two rows are exact sequences.

$$\begin{array}{ccccccc} [X,\Sigma^n Y_n]_t \xrightarrow{j_n} [X,\Sigma^n J_n]_t \xrightarrow{k_n} & [X,\Sigma^{n+1} Y_{n+1}]_t & \xrightarrow{i_n} [X,\Sigma^{n+1} Y_n]_t & \\ & \| & & \\ [X,\Sigma^{n+1} Y_{n+2}]_t \xrightarrow{i_{n+1}} & [X,\Sigma^{n+1} Y_{n+1}]_t & \xrightarrow{j_{n+1}} [X,\Sigma^{n+1} J_{n+1}]_t \xrightarrow{k_{n+1}} [X,\Sigma^{n+2} Y_{n+2}]_t. & \end{array}$$

(By abuse of notation, we write $i_n = (i_n)_*$, where the first i_n is part of the exact couple, and the second i_n is part of the Adams tower.)

The differential on the E_1-term is given by

$$d_1 = j_{n+1} \circ k_n \colon [X, \Sigma^n J_n] \longrightarrow [X, \Sigma^{n+1} J_{n+1}].$$

It satisfies $d_1 \circ d_1 = 0$, as this composite factors over $k_{n+1} \circ j_{n+1}$, which is zero. In fact, the differential d_1 is induced by a morphism $\partial\colon J_n \longrightarrow \Sigma J_{n+1}$ in $\mathcal{SHC}$ which satisfies $\partial \circ \partial = 0$ by the properties of the Adams tower. Also, by the assumptions on the Adams tower, each $H^*(J_n)$ is a projective $\mathcal{A}$–module, so applying H^* to the sequence

$$Y \xrightarrow{\partial} J_0 \xrightarrow{\partial} \Sigma^1 J_1 \xrightarrow{\partial} \Sigma^2 J_2 \xrightarrow{\partial} \cdots$$

results in a projective resolution of $\mathcal{A}$–modules

$$H^*(Y) \xleftarrow{\partial^*} H^*(J_0) \xleftarrow{\partial^*} H^*(\Sigma^1 J_1) \xleftarrow{\partial^*} H^*(\Sigma^2 J_2) \xleftarrow{\partial^*} \cdots .$$

Recall that the E_1-term of our spectral sequence was

$$[X, J_n]_* \xrightarrow{\cong} \mathrm{Hom}^*_{\mathcal{A}}(H^*(J_n), H^*(X)).$$

This means that the E_2-term of our spectral sequence is

$$E_2^{s,t} = H_s(\mathrm{Hom}(H^*(\Sigma^n J_n), H^{*+t}(X)), \partial_*) = \mathrm{Ext}^{s,t}_{\mathcal{A}}(H^*(Y), H^*(X)).$$

We will not discuss the convergence of this spectral sequence and only mention that it converges strongly with respect to the following filtration. We say that an element f of $[X, Y]$ has filtration k if it lifts over the kth level of the Adams tower

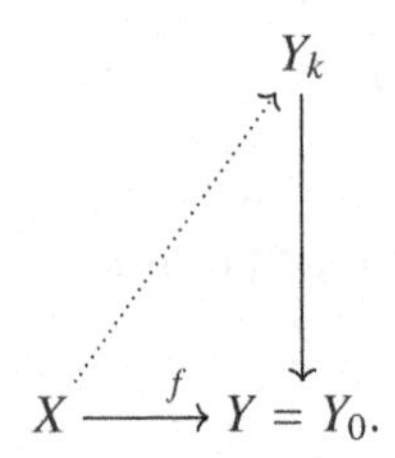

In other words,

$$F^k[X, Y] = \mathrm{Im}([X, Y_k] \longrightarrow [X, Y]).$$

Remark 2.6.6 With the method described in this section, it is possible to construct a spectral sequence for a fixed homology theory E,

$$E_2^{s,t} = \mathrm{Ext}^{s,t}_{E^*E}(E^*(Y), E^*(X)) \Longrightarrow [X, L_E Y]_{t-s},$$

see Chapter 7 for the definition of $L_E Y$. However, convergence becomes a separate question depending on E.

For the Johnson–Wilson theories $E = E(n)$ (see Definition 7.4.43), this has been discussed by Hovey and Strickland [HS99, Proposition 6.5]. For complex cobordism MU or its p-local version, the Brown–Peterson spectrum BP, this is known as the *Adams–Novikov spectral sequence*, whose technical advantages have been discussed in, for example, [Rav78]. The Thom reduction map

$$\Phi \colon BP \longrightarrow H\mathbb{Z}/p$$

induces a morphism of spectral sequences between the Adams–Novikov spectral sequence and the Adams spectral sequence. Applying it to the respective Adams towers, we see that if a morphism has Adams filtration s, it has Adams–Novikov filtration at most s.

2.6.2 A Look at the E_2-Term

The most frequently discussed calculation of the Adams spectral sequence is $X = Y = \mathbb{S}$:

$$E_2^{s,t} = \mathrm{Ext}_{\mathcal{A}}^{s,t}(\mathbb{F}_2, \mathbb{F}_2) \Longrightarrow \pi_{t-s}(\mathbb{S}) \otimes \mathbb{Z}_2^{\wedge}.$$

First, let us look at the 1-line of the E_2-term. We will just talk about the case $p = 2$, as the odd primary case is very similar.

Lemma 2.6.7 *For $p = 2$,*

$$\mathrm{Ext}_{\mathcal{A}}^{1,t}(\mathbb{F}_2, \mathbb{F}_2) = \begin{cases} \mathbb{F}_2\{h_i\} & \text{if } t = 2^i \\ 0 & \text{if } t \neq 2^i. \end{cases}$$

Proof An element in $\mathrm{Ext}_{\mathcal{A}}^{1,t}(\mathbb{F}_2, \mathbb{F}_2)$ is represented by a short exact sequence

$$0 \longrightarrow \mathbb{F}_2[t] \longrightarrow M \longrightarrow \mathbb{F}_2[0] \longrightarrow 0,$$

where M is a module over $\mathcal{A}$, and $\mathbb{F}_p[s]$ denotes one copy of $\mathbb{F}_2$ in degree s generated over $\mathbb{F}_2$ by the element x_s.

As an underlying graded $\mathbb{F}_2$–module, $M = \mathbb{F}_2[t] \oplus \mathbb{F}_2[0]$ generated by x_0 and x_t, but we are interested in M as an $\mathcal{A}$–module. Therefore, the different possibilities for M in the short exact sequence are encoded by those elements $Q \in \mathcal{A}$ of degree t, where $Qx_0 = x_t$.

This means that Q is an *indecomposable* element of $\mathcal{A}$, that is, Q cannot be written as a product $Q = Q_1 \cdot Q_2$, see Definition 2.5.10. If $Q = Q_1 \cdot Q_2$, then the underlying $\mathbb{F}_2$–module would have to have at least rank 3 to accommodate $Q_2 x_0$.

By Lemma 2.5.11, the indecomposable elements of $\mathcal{A}$ are given by

$$\{Sq^{2^i} \mid i \in \mathbb{N}\},$$

so there is a bijection between $\mathrm{Ext}_{\mathcal{A}}^{1,t}(\mathbb{F}_2, \mathbb{F}_2)$ and the set of indecomposables as above. The Ext class represented by the short exact sequence corresponding to Sq^{2^i} is traditionally denoted by h_i. This completes our proof. □

The next question is of course if any of the h_i survive the spectral sequence, that is, if they are permanent cycles. It turns out that the only permanent cycles are h_0, h_1, h_2 and h_3, giving rise to the Hopf elements

$$\begin{aligned} 2 &\in \pi_0(\mathbb{S}) \otimes \mathbb{Z}_2^\wedge = \mathbb{Z}_2^\wedge, \\ \eta &\in \pi_1(\mathbb{S}) \otimes \mathbb{Z}_2^\wedge = \mathbb{Z}/2, \\ \nu &\in \pi_3(\mathbb{S}) \otimes \mathbb{Z}_2^\wedge = \mathbb{Z}/8, \\ \sigma &\in \pi_7(\mathbb{S}) \otimes \mathbb{Z}_2^\wedge = \mathbb{Z}/16. \end{aligned}$$

The fact that these elements survive to the E_∞-term is related to the classical geometric question of whether S^n is parallelisable, see, for example, [Ada60] or [Ste62].

Of course, the 1-line is just the start of many calculations, relations and structural results that one can compute in the E_2-term. The E_2-term

$$E_2^{s,t} = \mathrm{Ext}_{\mathcal{A}}^{s,t}(\mathbb{F}_p, \mathbb{F}_p)$$

can be calculated manually in small degrees by writing down an explicit projective resolution of $\mathbb{F}_p$ as $\mathcal{A}$–module, which in practice makes heavy use of the Adem relations.

For example, one can use a minimal resolution [Ben98] or bar resolution [Wei94]. The minimal resolution has, unsurprisingly, the advantage that it is small in terms of generators used, whereas the bar resolution allows us to detect structure of $\pi_*(\mathbb{S})$ already on the E_2-term, such as the product structure or Toda brackets (see Section 4.6), as well as a wealth of other patterns.

As mentioned earlier, the relative simplicity of the dual $\mathcal{A}_*$ of $\mathcal{A}$ means that in some situations it is preferable to use the homological Adams spectral sequence and a cobar resolution instead [Rav86].

In any case, the Adams spectral sequence provides a very useful method for calculating morphisms in $\mathcal{SHC}$ by using cohomology and cohomology operations (or, respectively, endomorphisms of $H\mathbb{F}_p$), which is why it is a fitting application to conclude this chapter.

3

The Suspension and Loop Functors

In this chapter, we introduce the notion of pointed model categories and show that the homotopy category of a pointed model category has a suspension functor with an adjoint called the loop functor. This suspension functor is a generalisation of the standard notion of (reduced) suspension of pointed topological spaces. We shall also see that in the case of chain complexes over a ring, this suspension functor is modelled by the shift functor. With these constructions in place, we can define the notion of a stable model category.

The suspension and loop functors allow us to define cofibre and fibre sequences in an arbitrary pointed model category. These sequences are a generalisation of cofibre and fibre sequences for pointed spaces and are a useful aid to calculations. When the model category is also stable, these cofibre and fibre sequences form the basis of important additional structure on the homotopy category, as we examine in Chapter 4.

It would be tempting to think that a cofibre sequence in $\mathrm{Ho}(\mathcal{C})$ simply consists of a sequence of maps

$$A \xrightarrow{f} B \longrightarrow C,$$

where C is the cofibre of f or that a fibre sequence is

$$F \longrightarrow E \xrightarrow{p} B,$$

where F is the fibre of p. However, some additional information is needed to obtain the properties that are enjoyed by fibre and cofibre sequences in topological spaces.

Specifically, a cofibre sequence comes with a coaction of the suspension of A on the cofibre of $f\colon B \longrightarrow C$, and part of the information of a fibre sequence with fibre F is an action of the loops of the base space ΩB on the fibre. These actions are standard in topological spaces, see [Bre97], but we will obtain a well-defined action and coaction using nothing but the definition of a model category and some other basic category theory similar to [Qui67].

We will also examine some key properties of fibre and cofibre sequences, including that they can be "shifted" to the left or to the right to obtain a new fibre or cofibre sequence, that $[X, -]$ takes fibre sequences to long exact sequences of groups and that $[-, X]$ takes cofibre sequences to long exact sequences of groups.

3.1 Definition of the Functors

In Chapter 1, we worked primarily in the category of pointed topological spaces with basepoint–preserving maps as morphisms. The basepoint plays a role similar to that of the zero object in $\mathrm{Ch}(\mathcal{A})$, chain complexes in an abelian category $\mathcal{A}$ (or in other "algebraic" model categories). This notion can be generalised to the model category context as follows.

Definition 3.1.1 A model category $\mathcal{C}$ is *pointed* if the unique map from the initial object to the terminal object is an isomorphism. This object is denoted $*$ and called the *basepoint*.

Given any two objects A and B in $\mathcal{C}$, the composite map $A \longrightarrow * \longrightarrow B$ is called the *zero map* from A to B.

The composite of any map with a zero map is again a zero map.

Recall that a cylinder object for $X \in \mathcal{C}$ is a factorisation of the fold map $X \amalg X \longrightarrow X$

$$X \amalg X \xrightarrow{(i_0, i_1)} \mathrm{Cyl}(X) \xrightarrow{\sim} X.$$

Similarly, a path object for X is a factorisation of the diagonal $X \longrightarrow X \times X$

$$X \xrightarrow{\sim} PX \xrightarrow{(p_0, p_1)} X \times X,$$

see Definition A.2.1. We assume without loss of generality that our cylinder and path objects are "very good" (see Section A.2), so the first map is a cofibration and the second map a fibration.

Definition 3.1.2 Let $\mathcal{C}$ be a pointed model category, and let $X \in \mathcal{C}$ be cofibrant. The *suspension* ΣX *of* X is defined as the pushout of the diagram

$$* \longleftarrow X \amalg X \xrightarrow{(i_0, i_1)} \mathrm{Cyl}(X).$$

Dually, let $Y \in \mathcal{C}$ be fibrant. The *loop object ΩY of Y* is defined as the pullback of the map

$$PY \xrightarrow{(p_0,p_1)} Y \times Y \longleftarrow *.$$

The above definition does not depend on the choice of cylinder or path object in the following sense.

Proposition 3.1.3 *Let $\mathcal{C}$ be a pointed model category. The suspension and loop constructions define functors*

$$\Sigma\colon \operatorname{Ho}(\mathcal{C}) \longrightarrow \operatorname{Ho}(\mathcal{C}) \text{ and } \Omega\colon \operatorname{Ho}(\mathcal{C}) \longrightarrow \operatorname{Ho}(\mathcal{C}).$$

Proof Let $\operatorname{Cyl}(X)$ and $\operatorname{Cyl}(X)'$ denote two (very good) path objects for a cofibrant X. Then, by Section A.2 there are comparison maps

$$c_1\colon \operatorname{Cyl}(X) \longrightarrow \operatorname{Cyl}(X)' \text{ and } c_2\colon \operatorname{Cyl}(X)' \longrightarrow \operatorname{Cyl}(X)$$

which are weak equivalences and mutually homotopy inverse. Denote the respective pushouts by

$$p_1\colon \operatorname{Cyl}(X) \longrightarrow \Sigma X \text{ and } p_2\colon \operatorname{Cyl}(X)' \longrightarrow (\Sigma X)'.$$

Now, a map $f\colon \operatorname{Cyl}(X) \longrightarrow A$ which is trivial on both "ends" of the cylinder induces a unique map $\Sigma X \longrightarrow A$, but also a unique map $(\Sigma X)' \longrightarrow A$ by precomposing f with c_2. Conversely, any $g\colon \operatorname{Cyl}(X)' \longrightarrow A$ inducing a map $(\Sigma X)' \longrightarrow A$ also induces a map $\Sigma X \longrightarrow A$.

Thus, we have commutative diagrams

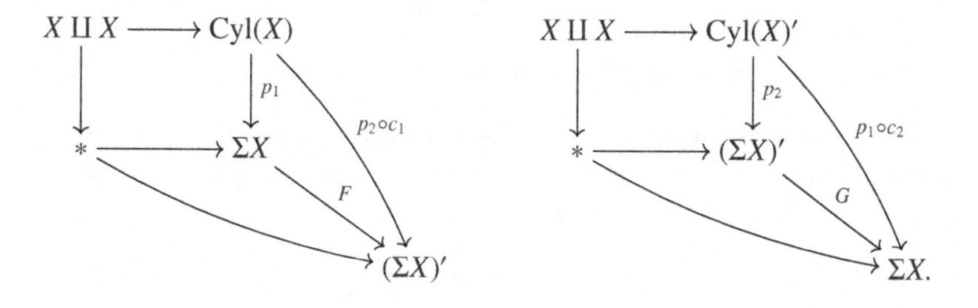

Putting them together gives a big diagram

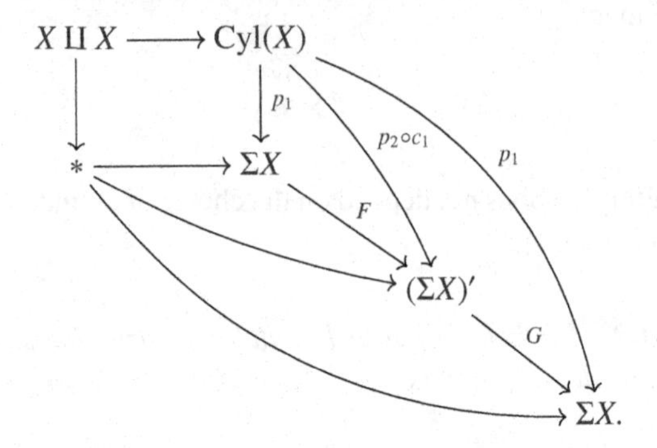

This diagram does not commute strictly, but as $p_1 \circ c_2 \circ c_1 \simeq p_1$, it commutes in $\mathrm{Ho}(\mathcal{C})$. Thus, in $\mathrm{Ho}(\mathcal{C})$, we have $G \circ F = \mathrm{Id}$ and, repeating the process with $\mathrm{Cyl}(X)$ and $\mathrm{Cyl}(X)'$ interchanged, $F \circ G = \mathrm{Id}$, so ΣX and $(\Sigma X)'$ are canonically isomorphic in $\mathrm{Ho}(\mathcal{C})$.

Now let A and B be both fibrant and cofibrant in $\mathcal{C}$. By definition, ΣA is the pushout of the diagram

$$* \longleftarrow A \amalg A \longrightarrow \mathrm{Cyl}(A),$$

and ΣB is given by the pushout

$$* \longleftarrow B \amalg B \longrightarrow \mathrm{Cyl}(B).$$

A morphism $f\colon A \longrightarrow B$ in $\mathcal{C}$ therefore gives a commutative diagram

$$\begin{array}{ccccc} * & \longleftarrow & A \amalg A & \longrightarrow & \mathrm{Cyl}(A) \\ \big\downarrow & & \big\downarrow{\scriptstyle f \amalg f} & & \big\downarrow{\scriptstyle \mathrm{Cyl}(f)} \\ * & \longleftarrow & B \amalg B & \longrightarrow & \mathrm{Cyl}(B). \end{array}$$

Note that by Remark A.3.4, the map $\mathrm{Cyl}(f)$ is unique in $\mathrm{Ho}(\mathcal{C})$. The universal property of pushouts now gives us a morphism

$$\Sigma f\colon \Sigma A \longrightarrow \Sigma B.$$

An analogous argument holds for the loop construction. □

The derived left adjoint of a Quillen adjunction preserves coproducts, pushouts and cylinder objects, hence, it commutes with suspensions. Similarly, the derived right adjoint will commute with loops.

Corollary 3.1.4 *Let $\mathcal{C}$ and $\mathcal{D}$ be model categories, and let*

$$F\colon \mathcal{C} \rightleftarrows \mathcal{D} : G$$

be a Quillen adjunction. Then for $X \in \mathcal{C}$ and $Y \in \mathcal{D}$, there are natural isomorphisms

$$\mathbb{L}F(\Sigma X) \cong \Sigma \mathbb{L}F(X) \qquad \mathbb{R}G(\Omega Y) \cong \Omega \mathbb{R}G(Y),$$

where

$$\mathbb{L}F\colon \mathrm{Ho}(\mathcal{C}) \longrightarrow \mathrm{Ho}(\mathcal{D}) \quad \text{and} \quad \mathbb{R}G\colon \mathrm{Ho}(\mathcal{D}) \longrightarrow \mathrm{Ho}(\mathcal{C})$$

denote the respective derived functors of F and G. □

Example 3.1.5 The definition gives the usual suspension and loop constructions on pointed topological spaces and pointed simplicial sets. For example, if X is a cofibrant pointed topological space, we can take the cylinder object on X to be $X \wedge [0,1]_+$ with $X \vee X \to X \wedge [0,1]_+$ being the inclusion of the ends of the cylinder. We then form the pushout diagram

$$\begin{array}{ccc} X \vee X & \longrightarrow & X \wedge [0,1]_+ \\ \downarrow & & \downarrow \\ * & \longrightarrow & \Sigma X. \end{array}$$

For a cofibrant sequential spectrum Z, we can perform the above constructions levelwise to obtain $\mathrm{Cyl}(Z)$ and ΣZ. The spectrum ΣZ is defined by $(\Sigma Z)_n = \Sigma Z_n$ and the structure maps are given by

$$\Sigma\Sigma Z_n \xrightarrow{\tau \wedge \mathrm{Id}} \Sigma\Sigma Z_n \xrightarrow{\Sigma\sigma_n} \Sigma Z_{n+1}.$$

Note the use of the twist map. Its necessity can be seen more clearly by looking at the structure map of $A \wedge Z$ for a pointed space A, see Example 2.1.4.

Example 3.1.6 For chain complexes $\mathrm{Ch}(\mathcal{A})$, one obtains the following. Let $A_* \in \mathrm{Ch}(\mathcal{A})$. A cylinder object for A_* is given by

$$\mathrm{Cyl}(A_*)_n = A_n \oplus A_{n-1} \oplus A_n, \quad \partial(a,b,c) = (d(a)+b, -d(b), d(c)-b),$$

where d denotes the differential in A_*. One can verify that the inclusion

$$i\colon A_* \oplus A_* \longrightarrow \mathrm{Cyl}(A_*),$$

which sends (a,b) to $(a,0,b)$, and the projection

$$p\colon \mathrm{Cyl}(A_*) \longrightarrow A_*$$

given by $p(a,b,c) = a + c$ are chain maps. Their composite is the fold map $A_* \oplus A_* \longrightarrow A_*$, and p_* is a homology isomorphism. Furthermore, we see that the cokernel of i and thus the suspension of A_* is simply given by the degree shift

$$(\Sigma A)_n = A_{n-1}, \quad d_{\Sigma A}(x) = -d_A(x).$$

Similarly, a path object for $A_* \in \mathrm{Ch}(\mathcal{A})$ is given by

$$PA_n = A_n \times A_{n+1} \times A_n, \quad \partial(a,b,c) = (d(a), -d(b) - a + c, d(c)).$$

The maps $i\colon A_* \longrightarrow PA_*$ given by $i(x) = (x, 0, x)$ and

$$p\colon PA_* \longrightarrow A_* \times A_*$$

given by $p(a,b,c) = (a,c)$ are chain maps, their composition is the diagonal map $A_* \longrightarrow A_* \times A_*$, and i is a homology isomorphism. We can then see that the kernel of i is ΩA_* with

$$\Omega A_n = A_{n+1}, \quad d_{\Omega A}(x) = -d_A(x),$$

that is, the loop object is given by the degree shift in the opposite direction.

Note that these are not automatically very good cylinder objects in the projective or injective model structure, but they are both very good in either model structure if A_* is fibrant and cofibrant.

Proposition 3.1.7 *Let $\mathcal{C}$ be a pointed model category. Then the suspension and loop construction induce an adjoint functor pair*

$$\Sigma\colon \mathrm{Ho}(\mathcal{C}) \rightleftarrows \mathrm{Ho}(\mathcal{C}) : \Omega,$$

hence, there are natural isomorphisms

$$\rho_{A,B}\colon [\Sigma A, B] \xrightarrow{\cong} [A, \Omega B]$$

for $A, B \in \mathrm{Ho}(\mathcal{C})$.

Proof We saw that Σ and Ω are functors in Proposition 3.1.3, so we now have to deal with the adjointness.

Definition of the Isomorphism ρ

By definition of the suspension, a map $[f] \in [\Sigma A, B]$ is represented by a map

$$F\colon \mathrm{Cyl}(A) \longrightarrow B$$

in $\mathcal{C}$ such that the composite

$$A \amalg A \longrightarrow \mathrm{Cyl}(A) \xrightarrow{F} B$$

is the zero map $A \amalg A \longrightarrow * \longrightarrow B$. In other words, F is a left homotopy between two copies of the zero map $A \to B$. In a similar way, elements $[g] \in [A, \Omega B]$ are represented by right homotopies between two copies of the zero map from A to B. As A and B can be chosen to be both fibrant and cofibrant, being left homotopic is equivalent to being right homotopic (see Lemma A.2.6), so $[\Sigma A, B]$ and $[A, \Omega B]$ agree. We will spell out the details below.

As A and B are both fibrant and cofibrant, a left homotopy

$$F\colon \mathrm{Cyl}(A) \longrightarrow B$$

gives rise to a right homotopy G between two copies of the zero map, namely, $G = H \circ i_1 : A \longrightarrow PB$ with H a lift in the commutative diagram below

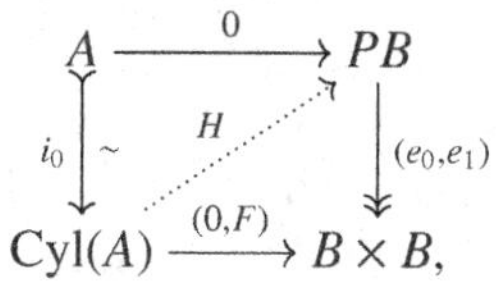

$$\begin{array}{ccc} A & \xrightarrow{\quad 0 \quad} & PB \\ {\scriptstyle i_0}\downarrow{\scriptstyle \sim} & \overset{H}{\nearrow} & \downarrow{\scriptstyle (e_0,e_1)} \\ \mathrm{Cyl}(A) & \xrightarrow{(0,F)} & B \times B, \end{array}$$

see Lemma A.2.6. Note that because A is cofibrant, i_0 is an acyclic cofibration see Section A.2. As

$$(e_0, e_1) \circ G = (e_0 \circ H \circ i_1, e_1 \circ H \circ i_1) = (0 \circ i_1, F \circ i_1) = (0, 0),$$

the map G factors over the pullback of

$$* \longrightarrow B \times B \longleftarrow PB,$$

thus gives us a unique map $g\colon A \longrightarrow \Omega B$. Furthermore, the choice of H is unique in $\mathrm{Ho}(\mathcal{C})$, meaning that a different choice of lift H would result in the same $[g] \in [A, \Omega B]$.

ρ is Well-Defined

This part of the proof boils down to the following statement: if

$$F\colon \mathrm{Cyl}(A) \longrightarrow B \text{ and } F'\colon \mathrm{Cyl}(A) \longrightarrow B$$

are right homotopic, then their corresponding right homotopies

$$G\colon A \longrightarrow PB \text{ and } G'\colon A \longrightarrow PB$$

are left homotopic. So to show that the assignment

$$\rho\colon [\Sigma A, B] \longrightarrow [A, \Omega B], \quad [f] \mapsto [g]$$

is well-defined we must show that it does not depend on the choice of representative F of $[f]$. Let $[f] = [f'] \in [\Sigma A, B]$. Then f and f' are related by a right homotopy

$$K\colon \Sigma A \longrightarrow PB.$$

We can consider K as a map

$$K\colon \mathrm{Cyl}(A) \longrightarrow PB$$

with $K \circ i_0 = K \circ i_1 = 0$. If we pick a different representative

$$F'\colon \mathrm{Cyl}(A) \longrightarrow B$$

for $[f] \in [\Sigma X, Y]$, then F and F' are related by the homotopy K, with $e_0 \circ K = F$ and $e_1 \circ K = F'$.

Performing the same construction with F' as we did with F earlier, we obtain a lift $H'\colon \mathrm{Cyl}(A) \longrightarrow PB$ and $G' := H' \circ i_1$ which factors over $g'\colon A \longrightarrow \Omega B$. We now have to show that G and G' are homotopic. We can concatenate the homotopies H and K to obtain

$$K * H\colon \mathrm{Cyl}(A) \longrightarrow PB$$

using the construction given in Section A.2. Note that we can concatenate K and H both as left and as right homotopies, and the results are the same in the homotopy category. We are allowed to perform this construction because

$$e_1 \circ H = F = e_0 \circ K,$$

that is, the homotopies "fit together".

We can choose $H' = H * K$ because

$$e_0 \circ (H * K) = e_0 \circ H = 0$$

and

$$e_1 \circ (H * K) = e_1 \circ K = F'$$

as well as

$$(H * K) \circ i_0 = (H \circ i_0) * (K \circ i_0) = 0 * 0 = 0.$$

By definition,

$$(H * K) \circ i_1 = (H \circ i_1) * (K \circ i_1) = G * 0,$$

which is homotopic to G. Therefore, we have just shown that the left homotopy F' corresponds to both the right homotopy G' as well as to $G * 0$. As correspondence between left and right homotopy is a bijection up to homotopy, G' and $G * 0$ (and consequently G' and G) are homotopic, and therefore, $[g] = [g']$.

This proves that our construction of $\rho\colon [\Sigma X, Y] \longrightarrow [X, \Omega Y]$ is indeed well-defined.

ρ is Bijective

The previous part showed that if two left homotopies $F\colon \mathrm{Cyl}(A) \longrightarrow B$ and $F'\colon \mathrm{Cyl}(A) \longrightarrow B$ are right homotopic, then their corresponding right homotopies $G, G'\colon A \longrightarrow PB$ are left homotopic. For ρ to be bijective, we have to show the converse. The proof is exactly dual to the proof of ρ being well-defined, so we shall not repeat it here.

This finishes the proof that ρ is an isomorphism. □

3.2 Stable Model Categories: A First Look

With the technicalities of suspension and loop functors in place, we can now come to the central definition of this whole book.

Definition 3.2.1 A pointed model category $\mathcal{C}$ is *stable* if the suspension and loop functors are equivalences of categories from $\mathrm{Ho}(\mathcal{C})$ to itself.

Example 3.2.2 The model category of pointed topological spaces is not stable. Recall that $\pi_n(X) = [S^n, X]$ for a pointed space X, and that $\pi_2(S^1) = 0$, but $\pi_3(S^2) = \mathbb{Z}$ (generated by the Hopf map) and $\pi_4(S^3) = \mathbb{Z}/2$. It follows that the suspension functor is not an equivalence on the homotopy category.

Sequential spectra are stable by Theorem 2.3.14. Section 2.4 gives a nice illustration of stability on sets of maps between suspension spectra $\Sigma^\infty A$ and $\Sigma^\infty B$ of CW complexes, provided A is finite:

$$\begin{aligned}[\Sigma^\infty A, \Sigma^\infty B] &= \mathrm{colim}_n[\Sigma^n A, \Sigma^n B] \\ &= \mathrm{colim}_n[\Sigma^{n+1} A, \Sigma^{n+1} B] = [\Sigma(\Sigma^\infty A), \Sigma(\Sigma^\infty B)].\end{aligned}$$

Example 3.2.3 For $\mathcal{A}$ an abelian category, the injective and projective model structures on chain complexes in $\mathcal{A}$ are stable, whenever they exist. We have seen that suspension and loop functors are simply given by degree shifts. In particular, they are equivalences of categories and form a Quillen adjunction. Hence, they induce an equivalence of categories on the homotopy category.

Definition 3.2.4 For $\mathcal{C}$ a stable model category, the *graded set of maps* from X to Y in $\mathrm{Ho}(\mathcal{C})$, denoted $[X, Y]_*^{\mathcal{C}}$, is defined as

$$[X, Y]_n = \begin{cases} [\Sigma^n X, Y] & \text{if } n \geqslant 0 \\ [X, \Sigma^{-n} Y] & \text{if } n < 0. \end{cases}$$

A map in $[X, Y]_n$ is called a *map of degree n*.

It follows from stability that the composite of a map of degree n and a map of degree m is a map of degree $n + m$. We may also relate the grading to Ω via

$$[X, \Omega Y]_n = [\Sigma X, Y]_n = [X, Y]_{n+1} \quad \text{and} \quad [\Omega X, Y]_n \cong [\Sigma\Omega X, Y]_{n-1} \cong [X, Y]_{n-1}.$$

This grading convention agrees with that of homotopy groups of spectra. For a sequential spectrum X, we will see in Proposition 5.6.6 that

$$\pi_n(X) = [\mathbb{S}, X]_n^{\mathcal{S}^{\mathbb{N}}}.$$

Definition 3.2.5 A category $\mathcal{C}$ is *additive* if it has all finite products and coproducts and is enriched in the category of abelian groups. In other words, the morphism sets are abelian groups, and the composition of morphisms

$$\circ \colon \mathcal{C}(Y, Z) \times \mathcal{C}(X, Y) \longrightarrow \mathcal{C}(X, Z)$$

is bilinear. An *additive functor* is a functor

$$F \colon \mathcal{C} \longrightarrow \mathcal{D}$$

of additive categories such that

$$F \colon \mathcal{C}(X, Y) \longrightarrow \mathcal{D}(F(X), F(Y))$$

is a morphism of abelian groups.

A consequence of the definition is that finite products and coproducts coincide. Given $f \colon A \longrightarrow X$ and $g \colon A \longrightarrow Y$, we have a map

$$(f \amalg g) \circ (i_1 + i_2) \colon A \longrightarrow X \amalg Y$$

given by adding together the two inclusions $i_1,\ i_2 \colon A \longrightarrow A \amalg A$. Let

$$p_X \colon X \amalg Y \longrightarrow X \quad \text{and} \quad p_1 \colon A \amalg A \longrightarrow A$$

be the projections onto the first summands, and let

$$p_Y \colon X \amalg Y \longrightarrow Y$$

be the projection onto the second summand. Then

$$p_X \circ (f \amalg g) \circ (i_1 + i_2) = f \circ p_1 \circ (i_1 + i_2) = f$$
$$p_Y \circ (f \amalg g) \circ (i_1 + i_2) = g \circ p_2 \circ (i_1 + i_2) = g.$$

This proves that $X \amalg Y$ is also the product of X and Y. In an additive category, the product (or coproduct) is sometimes called the *biproduct* or the *direct sum*.

Our goal is to show that the homotopy category of a stable model category is additive and that a Quillen functor between stable model categories induces an additive functor. But let us first consider the following.

Lemma 3.2.6 *If $\mathcal{C}$ is a pointed model category, then for any $X, Y \in \mathcal{C}$, the sets $[\Sigma X, Y]$ and $[X, \Omega Y]$ are groups, and the adjunction isomorphism*

$$\rho \colon [\Sigma X, Y] \longrightarrow [X, \Omega Y]$$

is a group isomorphism.

Proof The group composition on $[\Sigma X, Y]$ is given by concatenating left homotopies as in Definition A.2.8. We explain at the end of Section A.2 that for fibrant and cofibrant A and B, homotopy classes of maps from A to B are the objects of a groupoid. Therefore, restricting to homotopies between the zero map and itself gives us a group. The isomorphism ρ is given by sending a left homotopy to its corresponding right homotopy, and concatenation and correspondence commute in the homotopy category, see again Section A.2. Thus, ρ is a group homomorphism. □

The Yoneda Lemma yields the following.

Corollary 3.2.7 *Let $\mathcal{C}$ be a pointed model category, and let $X \in \mathcal{C}$. Then ΣX is a cogroup object in* $\mathrm{Ho}(\mathcal{C})$, *and ΩX is a group object in* $\mathrm{Ho}(\mathcal{C})$. □

Lemma 3.2.8 *Let $\mathcal{C}$ be a pointed model category, and X, $Y \in \mathcal{C}$. Then the groups $[\Sigma^2 X, Y]$ and $[X, \Omega^2 Y]$ are abelian.*

Proof We only prove the first claim as the second claim is dual. Let

$$X \amalg X \longrightarrow \mathrm{Cyl}(X) \longrightarrow X$$

be a cylinder object for X. Then we can obtain a cylinder object for $\mathrm{Cyl}(X)$ either the usual way or by applying Cyl to the previous diagram (it is, after all, a functor on $\mathrm{Ho}(\mathcal{C})$, although not one on $\mathcal{C}$ itself). We denote the resulting double cylinder objects by $\mathrm{Cyl}'(\mathrm{Cyl}(X))$ and $\mathrm{Cyl}(\mathrm{Cyl}'(X))$.

Let α and β be representatives of elements in $[\Sigma^2 X, Y]$. We know that in $\mathrm{Ho}(\mathcal{C})$

$$\alpha * \beta = (\alpha * 0) * (0 * \beta) \colon \ \mathrm{Cyl}'(\mathrm{Cyl}(X)) \longrightarrow Y.$$

The concatenation $*$ within the brackets is the concatenation in the Cyl′-"direction", whereas the concatenation between the brackets is the concatenation on Cyl. This is illustrated by the first equivalence in the following diagram.

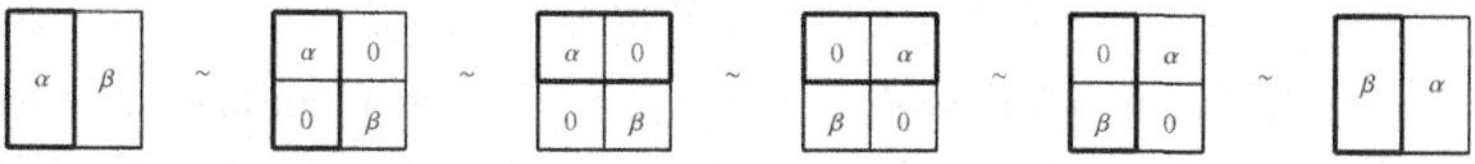

$$\boxed{\alpha \mid \beta} \sim \begin{array}{|c|c|}\hline \alpha & 0 \\ \hline 0 & \beta \\ \hline\end{array} \sim \begin{array}{|c|c|}\hline \alpha & 0 \\ \hline 0 & \beta \\ \hline\end{array} \sim \begin{array}{|c|c|}\hline 0 & \alpha \\ \hline \beta & 0 \\ \hline\end{array} \sim \begin{array}{|c|c|}\hline 0 & \alpha \\ \hline \beta & 0 \\ \hline\end{array} \sim \boxed{\beta \mid \alpha}$$

The second equivalence in the picture is given by the isomorphism

$$\mathrm{Cyl}'(\mathrm{Cyl}(X)) \cong \mathrm{Cyl}(\mathrm{Cyl}'(X)).$$

On the third object,

$$\alpha * 0 = 0 * \alpha \text{ and } 0 * \beta = \beta * 0,$$

which gives the third equivalence. The fourth equivalence is given by the inverse isomorphism $\mathrm{Cyl}(\mathrm{Cyl}'(X))$ to $\mathrm{Cyl}'(\mathrm{Cyl}(X))$, and the final equivalence is given by

$$(0 * \beta) * (\alpha * 0) = \beta * \alpha,$$

which is the result we wanted. □

Proposition 3.2.9 *Let $\mathcal{C}$ be a stable model category. Then its homotopy category $\mathrm{Ho}(\mathcal{C})$ is an additive category. If $\mathcal{C}$ and $\mathcal{D}$ are stable model categories and $F\colon \mathcal{C} \longrightarrow \mathcal{D}$ is a (left or right) Quillen functor, then F is an additive functor.*

Proof By stability, $[X, Y] \cong [\Sigma^2 X, \Sigma^2 Y]$, which is an abelian group by Lemmas 3.2.6 and 3.2.8.

Composing morphisms is bilinear, because Definition A.2.8 implies that in $\mathrm{Ho}(\mathcal{C})$,

$$f \circ (\alpha * \beta) = (f \circ \alpha) * (f \circ \beta) \text{ and } (\alpha * \beta) \circ g = (\alpha \circ g) * (\beta \circ g).$$

For the first equality, we use left homotopies and pushouts, and for the second equality, we use right homotopies and pullbacks.

A left or right Quillen functor F satisfies $F(\alpha * \beta) = F(\alpha) * F(\beta)$ because $F(\mathrm{Cyl}(X))$ is isomorphic to $\mathrm{Cyl}(F(X))$ in $\mathrm{Ho}(\mathcal{C})$ and $P(F(Y))$ is isomorphic to $F(P(Y))$. □

Corollary 3.2.10 *Let $\mathcal{C}$ and $\mathcal{D}$ be stable model categories with a Quillen adjunction*

$$F\colon \mathcal{C} \rightleftarrows \mathcal{D} :G.$$

Then the derived functors $\mathbb{L}F$ and $\mathbb{R}G$ commute with both Σ and Ω and preserve the grading on maps in the homotopy categories.

Proof By Corollary 3.1.4, $\mathbb{L}F$ commutes with Σ and $\mathbb{R}G$ commutes with Ω. Since $\mathcal{C}$ and $\mathcal{D}$ are stable model categories, $X \cong \Sigma\Omega X$ for any $X \in \mathrm{Ho}(\mathcal{C})$ and $Y \cong \Omega\Sigma Y$ for any Y in $\mathrm{Ho}(\mathcal{D})$. Thus,

$$\Omega\mathbb{L}F(X) \cong \Omega\mathbb{L}F(\Sigma\Omega X) \cong \Omega\Sigma\mathbb{L}F(\Omega X) \cong \mathbb{L}F(\Omega X).$$

The case of $\mathbb{R}G$ commuting with Σ is dual, and the statement about gradings follows from the definition. □

We will uncover a lot more of the structure of $\mathrm{Ho}(\mathcal{C})$ for stable $\mathcal{C}$ in the next sections and Chapter 4.

3.3 The Coaction of a Cofibre

We start by abstracting the coaction associated to a cofibre and the action associated to a fibre to the level of model categories. We need this coaction and action to give the definition of a cofibre sequence and a fibre sequence in a homotopy category.

Definition 3.3.1 Let $\mathcal{C}$ be a pointed model category and $f\colon A \longrightarrow B$ a map in $\mathcal{C}$.

The *cofibre* of f is the pushout of the diagram

$$* \longleftarrow A \xrightarrow{f} B.$$

The *fibre* of f is the pullback of the diagram

$$A \xrightarrow{f} B \longleftarrow *.$$

The cofibre of $f\colon A \to B$ is often written as a quotient B/A, and the fibre of f is sometimes written as $f^{-1}(*)$. Let

$$A \xrightarrow{f} B \xrightarrow{g} C$$

be a sequence of maps in a pointed model category $\mathcal{C}$, where C is the cofibre of f. Assume that f is a cofibration of cofibrant objects. We start by defining an action of the group $[\Sigma A, X]$ on $[C, X]$, where X is any (fibrant) object of $\mathcal{C}$. We know that an element of $[\Sigma A, X]$ can be thought of as a homotopy

$$h\colon A \longrightarrow PX$$

between the zero map $A \longrightarrow X$ and itself. We can pick the path object PX such that the canonical map

$$PX \xrightarrow{e_0} X$$

is an acyclic fibration. Furthermore, let

$$q\colon C \longrightarrow X$$

represent an element of $[C, X]$. Then the diagram

$$\begin{array}{ccc} A & \xrightarrow{h} & PX \\ {\scriptstyle f}\big\downarrow & & \big\downarrow{\scriptstyle \sim\, e_0} \\ B & \xrightarrow{q\circ g} & X \end{array}$$

commutes as both composites are zero. Because f is a cofibration and e_0 an acyclic fibration, there is a lift

$$\phi\colon B \longrightarrow PX$$

in this diagram. Because h is a right homotopy from zero to zero, this satisfies

$$e_1 \circ h = (e_1 \circ \phi) \circ f = 0.$$

This implies that $e_1 \circ \phi$ factors over the cofibre C of f, that is,

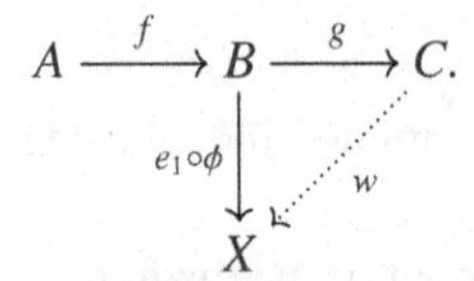

We now define $[q] \odot [h] := [w]$. We will see soon that this pairing is actually well-defined and gives a right action of $[\Sigma A, X]$ on $[C, X]$.

We can perform a dual construction involving the fibre of a map. Let

$$F \xrightarrow{i} E \xrightarrow{p} B$$

be a sequence of maps in $\mathcal{C}$, where p is a fibration between fibrant objects and F is the fibre of p. Furthermore, let A be any (cofibrant) object of $\mathcal{C}$. We saw that an element $[A, \Omega B]$ can be represented by a left homotopy

$$h\colon \mathrm{Cyl}(A) \longrightarrow B$$

between the zero map $A \longrightarrow B$ and itself. Lastly, let

$$f\colon A \longrightarrow F$$

represent an element of $[A, F]$. Then, the diagram

$$\begin{array}{ccc} A & \xrightarrow{i\circ f} & E \\ {\scriptstyle i_0}\big\downarrow{\scriptstyle \sim} & & \big\downarrow{\scriptstyle p} \\ \mathrm{Cyl}(A) & \xrightarrow{h} & B \end{array}$$

commutes because both composites are zero. Again, we can choose the cylinder object $\mathrm{Cyl}(A)$ so that the map i_0 is an acyclic cofibration. Thus, the diagram has a lift

$$\alpha\colon \mathrm{Cyl}(A) \longrightarrow E$$

satisfying

$$h \circ i_1 = p \circ (\alpha \circ i_1) = 0$$

because h is a homotopy between 0 and 0. This means that $\alpha \circ i_1$ factors over the fibre F of p

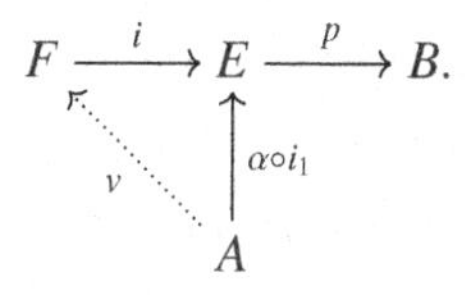

We define $[f] \odot [h] := [v]$.

Example 3.3.2 In the case of pointed topological spaces and $A = S^0$, this action corresponds to the classical action of the fundamental group of the base space on the fibre. A loop in B is represented by a continuous map

$$\omega\colon [0,1] \longrightarrow B$$

with basepoint x_0. For $f \in F$, the homotopy lifting property gives us a path

$$\tilde{\omega}\colon [0,1] \longrightarrow E$$

in the total space that lifts ω and starts at f. The action

$$\odot\colon \pi_0(F) \times \pi_0(\Omega B) \longrightarrow \pi_0(F)$$

assigns to a point $f \in F$ and a loop ω the end point of the lifted path $\tilde{\omega}$.

We can now show that this action and coaction are well-defined and natural. The proof of the following theorem will occupy the rest of this section, so some readers may want to simply read the statement and skip to Section 3.4.

Theorem 3.3.3 *Let $f\colon A \to B$ be a cofibration of cofibrant objects with cofibre C. The map*

$$\odot\colon [C,X] \times [\Sigma A, X] \longrightarrow [C,X]$$

is well-defined natural in $X \in \mathcal{C}$, associative and unital and thus defines a group action of $[\Sigma A, X]$ on $[C,X]$.

Now let $p\colon E \to B$ be a fibration of fibrant objects with fibre F. The map

$$\odot\colon [A,F] \times [A,\Omega B] \longrightarrow [A,F]$$

is well-defined natural in $A \in \mathcal{C}$, associative and unital and thus defines a group action of $[A,\Omega B]$ on $[A,F]$.

Proof We are going to prove the statement for the second action, using the notation from the definition.

The Operation $\odot$ is Well-Defined

First, we have to show that $\odot$ is well-defined by picking a representative

$$h'\colon \operatorname{Cyl}(A) \longrightarrow B \ \text{ for } \ [h] \in [A, \Omega B]$$

(possibly different from h) and

$$g\colon A \longrightarrow F \ \text{ representing } \ [f] \in [A, F].$$

Then, we pick a lift β in the diagram

$$\begin{array}{ccc} A & \xrightarrow{i \circ g} & E \\ {\scriptstyle i_0}\downarrow\wr & \overset{\beta}{\nearrow} & \downarrow{\scriptstyle p} \\ \operatorname{Cyl}(A) & \xrightarrow{h'} & B \end{array}$$

and a map v' factoring over F as

$$\begin{array}{ccccc} F & \xrightarrow{i} & E & \xrightarrow{p} & B. \\ & {\scriptstyle v'}\nwarrow & \uparrow{\scriptstyle \beta \circ i_1} & & \\ & & A & & \end{array}$$

We have to show that v' is homotopic to the original v obtained using h and f instead of h' and g.

As $[h] = [h']$, there is a right homotopy $\bar{H}\colon \Sigma A \longrightarrow PB$. This homotopy $\bar{H}$ is defined by a map

$$H\colon \operatorname{Cyl}(A) \longrightarrow PB \ \text{ with } \ e_0 \circ H = h, e_1 \circ H = h', H \circ i_0 = H \circ i_1 = 0.$$

As $[f] = [g]$, there is a right homotopy

$$K\colon A \longrightarrow PF$$

relating f and g in the usual way, $e_0 \circ K = f, e_1 \circ K = g$.

Let Q denote the pullback of the diagram

$$E \times E \xrightarrow{(p,p)} B \times B \xleftarrow{(e_0,e_1)} PB.$$

Then there are commutative diagrams

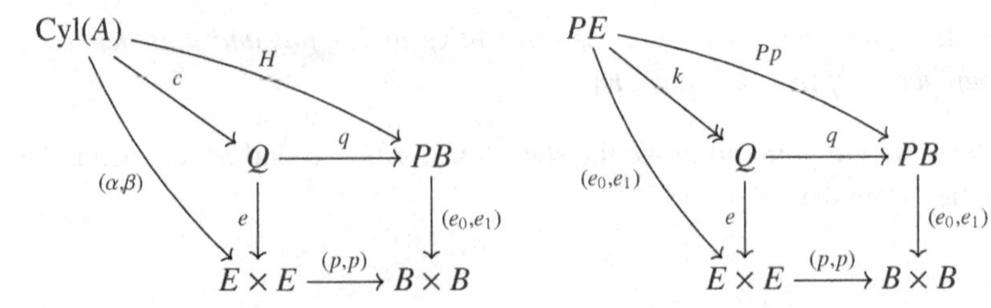

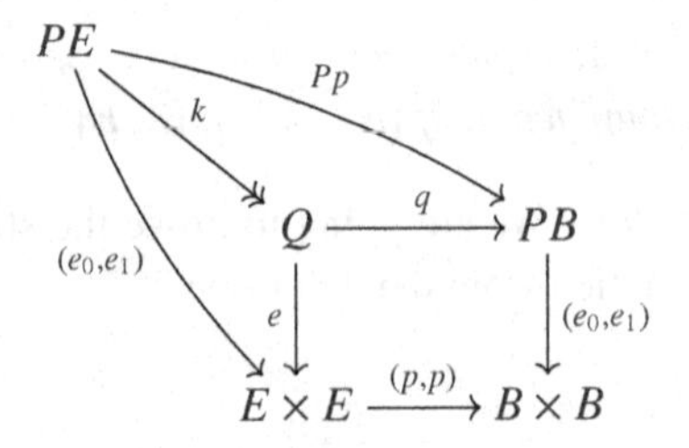

because $p \circ \beta = h'$, $p \circ \alpha = h$ and h is homotopic to h' via H. Without loss of generality, we can pick k to be a fibration. If not, then we can factor k as an acyclic cofibration and a fibration as follows

$$k\colon\ PE \overset{\sim}{\rightarrowtail} PE' \twoheadrightarrow Q$$

and use PE' as a path object instead of PE. Thus, there is a commutative diagram

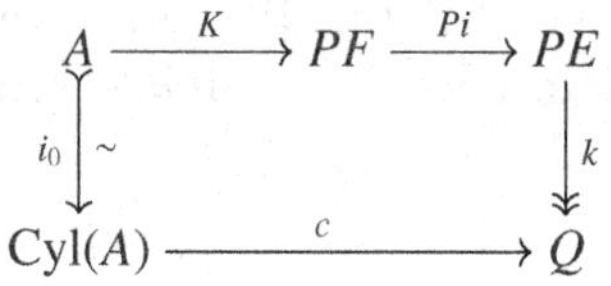

$$\begin{array}{ccccc} A & \xrightarrow{K} & PF & \xrightarrow{Pi} & PE \\ {\scriptstyle i_0}\downarrow{\scriptstyle \sim} & & & & \downarrow{\scriptstyle k} \\ \mathrm{Cyl}(A) & & \xrightarrow{c} & & Q \end{array}$$

because the composites

$$q \circ k \circ Pi \circ K\colon A \longrightarrow PB \text{ and } q \circ c \circ i_0\colon A \longrightarrow PB$$

are both zero and

$$e \circ k \circ Pi \circ K\colon A \longrightarrow E \times E \text{ and } e \circ c \circ i_0\colon A \longrightarrow E \times E$$

are both equal to $(i \circ f, i \circ g)$. Now there is a lift

$$\Gamma\colon\ \mathrm{Cyl}(A) \longrightarrow PE$$

in the above diagram. We have that

$$(e_0 \circ \Gamma, e_1 \circ \Gamma) = e \circ k \circ \Gamma = e \circ c = (\alpha, \beta),$$

that is, Γ is a right homotopy between α and β.

We have $Pp \circ \Gamma = q \circ k \circ \Gamma = q \circ c = H$ and thus,

$$Pp \circ \Gamma \circ i = q \circ k \circ \Gamma \circ i_1 = H \circ i_1 = 0.$$

Thus, $\Gamma \circ i_1$ factors over the fibre of $q \circ k = Pp$:

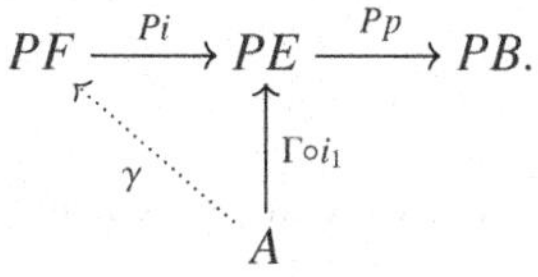

$$\begin{array}{ccccc} PF & \xrightarrow{Pi} & PE & \xrightarrow{Pp} & PB. \\ & {\scriptstyle \gamma}\nwarrow & \uparrow{\scriptstyle \Gamma \circ i_1} & & \\ & & A & & \end{array}$$

Recall that Γ is a homotopy between α and β, thus, $\Gamma \circ i_1$ is a homotopy between $\alpha \circ i_1$ and $\beta \circ i_1$, so $\gamma\colon A \longrightarrow PF$ is a homotopy between v and v', which is what we needed for $\odot$ to be a well-defined map.

The Operation $\odot$ is Natural in A

Let $\varphi\colon A' \longrightarrow A$ be a map of cofibrant objects. As before, let $h\colon \mathrm{Cyl}(A) \longrightarrow B$ represent an element of $[A, \Omega B]$ and f represent an element of $[A, F]$. Then

$$h \circ \mathrm{Cyl}(\varphi)\colon \mathrm{Cyl}(A') \longrightarrow B$$

represents an element of $[A', \Omega B]$ and

$$f \circ \varphi\colon A' \longrightarrow F$$

represents an element of $[A', F]$. We are now going to calculate $[f \circ \varphi] \odot [h \circ \mathrm{Cyl}(\varphi)]$. For that, we need a lift in the commutative square

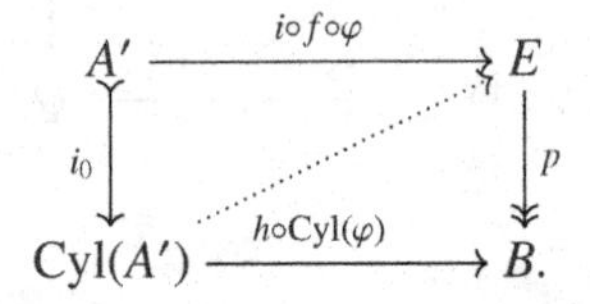

We can take this lift to be $\alpha \circ \mathrm{Cyl}(\varphi)$, where α is a lift in the second square

$$\begin{array}{ccccc} A' & \xrightarrow{\varphi} & A & \xrightarrow{i\circ f} & E \\ \Big\downarrow{\scriptstyle i_0} & & \Big\downarrow{\scriptstyle i_0} & \overset{\alpha}{\nearrow} & \Big\downarrow{\scriptstyle p} \\ \mathrm{Cyl}(A') & \xrightarrow{\mathrm{Cyl}(\varphi)} & \mathrm{Cyl}(A) & \xrightarrow{h} & B \end{array}$$

as before. So,

$$[w] = [f \circ \varphi] \odot [h \circ \mathrm{Cyl}(\varphi)],$$

where w is the unique map in the diagram below.

$$\begin{array}{ccccc} F & \xrightarrow{i} & E & \xrightarrow{p} & B \\ & {\scriptstyle w}\nwarrow & \Big\uparrow{\scriptstyle \alpha\circ\mathrm{Cyl}(\varphi)\circ i_1} & & \\ & & A' & & \end{array}$$

With our usual notation, $[v] = [f] \odot [h]$, where v is given by the lift

$$\begin{array}{ccccc} F & \xrightarrow{i} & E & \xrightarrow{p} & B. \\ & {\scriptstyle v}\nwarrow & \Big\uparrow{\scriptstyle \alpha\circ i_1} & & \\ & & A & & \end{array}$$

Comparing these two diagrams, we can see that w is given by $v \circ \varphi$, because $i_1 \circ \varphi = \mathrm{Cyl}(\varphi) \circ i_1$. Thus,

$$[v \circ \varphi] = [f \circ \varphi] \odot [h \circ \mathrm{Cyl}(\varphi)]$$

as claimed.

The Operation $\odot$ is Unital

The unit of $[A, \Omega B]$ is represented by the trivial homotopy $0\colon \operatorname{Cyl}(A) \longrightarrow B$. Denote by r the map given in the factorisation

$$A \amalg A \xrightarrow{(i_0,i_1)} \operatorname{Cyl}(A) \xrightarrow{r} A.$$

By definition, we have $r \circ i_0 = r \circ i_1 = \mathrm{Id}_A$. This means that in the diagram

$$\begin{array}{ccc} A & \xrightarrow{i \circ f} & E \\ {\scriptstyle i_0}\big\downarrow \sim & \nearrow & \big\downarrow {\scriptstyle p} \\ \operatorname{Cyl}(A) & \xrightarrow{0} & B, \end{array}$$

we can take our lift to be $i \circ f \circ r$ because $i \circ f \circ r \circ i_0 = i \circ f$ and $p \circ i \circ f \circ r = 0$. For our definition of $[v] = [f] \odot [0]$, we are now looking for a map v such that $i \circ v = i \circ f \circ r \circ i_1$. But we can simply take $v = f$. Thus, $\odot$ is unital.

The Operation $\odot$ is Associative

We start by calculating

$$[k] := ([f] \odot [h]) \odot [h']$$

and show that this equals $[f] \odot [h * h']$.

First of all, let $[v] = [f] \odot [h]$ be represented by $v\colon A \longrightarrow F$ defined via a lift α in the usual diagram. This lift satisfies $i \circ v = \alpha \circ i_1$. To define $[k]$, we need a lift in the diagram

$$\begin{array}{ccc} A & \xrightarrow{i \circ v} & E \\ {\scriptstyle i_0}\big\downarrow & \overset{\beta}{\nearrow} & \big\downarrow {\scriptstyle p} \\ \operatorname{Cyl}(A) & \xrightarrow{h'} & B. \end{array}$$

This means that

$$\beta \circ i_0 = \alpha \circ i_1,$$

and we define k to be the map such that $i \circ k = \beta \circ i_1$.

To calculate $[f] \odot [h * h']$, recall the notion of concatenation from Definition A.2.8. Because $\alpha \circ i_1 = \beta \circ i_0$, we may concatenate as below.

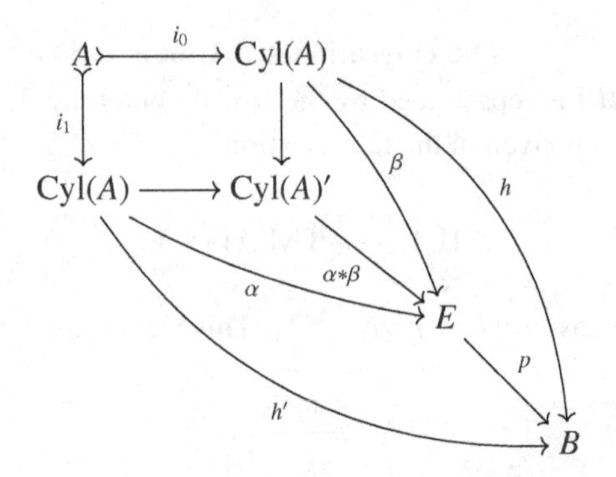

We see that $p \circ (\alpha * \beta)$ is a concatenation of h and h'. Moreover,

$$(\alpha * \beta) \circ i_1 = \beta \circ i_1 \quad \text{and} \quad (\alpha * \beta) \circ i_0 = \alpha \circ i_0 = i \circ f.$$

It follows that $\alpha * \beta$ is a lift for the commutative diagram

$$\begin{array}{ccc} A & \xrightarrow{i \circ f} & E \\ {\scriptstyle i_0}\big\downarrow & \nearrow & \big\downarrow{\scriptstyle p} \\ \mathrm{Cyl}(A)' & \xrightarrow{h * h'} & B. \end{array}$$

Because $(\alpha * \beta) \circ i_1 = \beta \circ i_1$, we may choose k to be the lift in the diagram

$$\begin{array}{ccccc} F & \xrightarrow{i} & E & \xrightarrow{p} & B, \\ & \nwarrow & \big\uparrow{\scriptstyle (\alpha*\beta)\circ i_1} & & \\ & & A & & \end{array}$$

and thus,

$$[k] = ([f] \odot [h]) \odot [h'] = [f] \odot [h * h'].$$

□

3.4 Definition of Fibre and Cofibre Sequences

We have seen in Section 3.3 that ΩC is a (unital) group object in $\mathrm{Ho}(\mathcal{C})$ for $C \in \mathrm{Ho}(\mathcal{C})$, and that ΣA is a (counital) cogroup object in $\mathrm{Ho}(\mathcal{C})$ for $A \in \mathrm{Ho}(\mathcal{C})$. Let $p\colon E \twoheadrightarrow B$ be a fibration of fibrant objects in $\mathcal{C}$ with fibre F.

The group action from Theorem 3.3.3 gives a map

$$[F \times \Omega B, F \times \Omega B] \cong [F \times \Omega B, F] \times [F \times \Omega B, \Omega B] \xrightarrow{\odot} [F \times \Omega B, F].$$

The identity map of $F \times \Omega B$ thus specifies a map in $\mathrm{Ho}(\mathcal{C})$

$$\odot\colon F \times \Omega B \longrightarrow F,$$

the same name is used for this map and the group action, as the context usually prevents confusion.

The Yoneda Lemma tells us that $\odot$ defines an action of the group object ΩB on F. That is, the following diagrams in $\mathrm{Ho}(\mathcal{C})$ commute, where the unique map $* \to \Omega B$ acts as the unit, and the composition of loops is denoted by m.

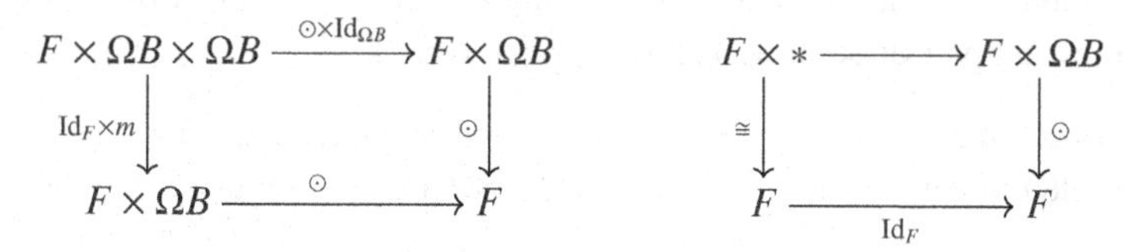

Dually, let $f\colon A \to B$ be a cofibration of cofibrant objects with cofibre C. The Yoneda Lemma tells us that the group action from Theorem 3.3.3 gives a map

$$\odot\colon C \longrightarrow C \amalg \Sigma A$$

that defines a coaction of the cogroup object ΣA on C in $\mathrm{Ho}(\mathcal{C})$. With this, we can now make our main definitions.

Definition 3.4.1 Let $\mathcal{C}$ be a pointed model category. A *cofibre sequence* in $\mathrm{Ho}(\mathcal{C})$ is a diagram

$$X \longrightarrow Y \longrightarrow Z$$

that is isomorphic in $\mathrm{Ho}(\mathcal{C})$ to a diagram

$$A \xrightarrow{f} B \xrightarrow{g} C,$$

where f is a cofibration between cofibrant objects and C is the cofibre of f. In addition, the diagram is equipped with a right coaction in $\mathrm{Ho}(\mathcal{C})$

$$Z \longrightarrow Z \amalg \Sigma X$$

that is isomorphic to the coaction from Theorem 3.3.3.

Definition 3.4.2 Let $\mathcal{C}$ be a pointed model category. A *fibre sequence* is a diagram

$$X \longrightarrow Y \longrightarrow Z$$

that is isomorphic in $\mathrm{Ho}(\mathcal{C})$ to a diagram

$$F \xrightarrow{i} E \xrightarrow{p} B,$$

where p is a fibration between fibrant objects and F is the fibre of p. Furthermore, the diagram is equipped with a right action in $\mathrm{Ho}(\mathcal{C})$

$$X \times \Omega Z \longrightarrow X$$

that is isomorphic to the action from Theorem 3.3.3.

It is important to remember that the (co)actions are all defined at the level of homotopy categories. One could pick a representative for the map $\odot$, but then one would have to take care over the (co)fibrancy of the objects. Consequently, the following pair of definitions give maps in $\mathrm{Ho}(\mathcal{C})$.

Definition 3.4.3 Let $X \longrightarrow Y \longrightarrow Z$ be a cofibre sequence in $\mathrm{Ho}(\mathcal{C})$, where $\mathcal{C}$ is a pointed model category. The *boundary map* is the composite

$$\partial \colon Z \xrightarrow{\odot} Z \amalg \Sigma X \xrightarrow{(0,\mathrm{Id})} \Sigma X,$$

where the first map is the coaction associated to the cofibre sequence.

Definition 3.4.4 Let $X \longrightarrow Y \longrightarrow Z$ be a fibre sequence in $\mathrm{Ho}(\mathcal{C})$, where $\mathcal{C}$ is a pointed model category. The *boundary map* is the composite

$$\partial \colon \Omega Z \xrightarrow{(0,\mathrm{Id})} X \times \Omega Z \xrightarrow{\odot} X,$$

where the second map is the action associated to the fibre sequence.

The remainder of this chapter focuses on the properties of these actions, coactions and the boundary maps. In the next section, we will use the boundary maps to "shift" cofibre and fibre sequences, and in Section 3.6, we will use the boundary maps to extend cofibre and fibre sequences indefinitely to obtain the Puppe sequences.

The interesting thing about the boundary map is that it contains the same information as the (co)action associated to a (co)fibre sequence: one uses the (co)action to define the boundary map, but can also recover the (co)action if given just the boundary map. More precisely, in the cofibre case, if

$$X \xrightarrow{f} Y \xrightarrow{g} Z$$

is a cofibre sequence with boundary map $\partial \colon Z \longrightarrow \Sigma X$ and there is a commutative diagram in $\mathrm{Ho}(\mathcal{C})$

$$\begin{array}{ccccccc} X & \xrightarrow{f} & Y & \xrightarrow{g} & Z & \xrightarrow{\partial} & \Sigma X \\ \downarrow & & \downarrow & & \downarrow & & \downarrow \\ A & \xrightarrow{u} & B & \xrightarrow{v} & C & \xrightarrow{w} & \Sigma A, \end{array}$$

where the vertical arrows are isomorphisms, then

$$A \xrightarrow{u} B \xrightarrow{v} C$$

is a cofibre sequence with boundary map $w\colon C \longrightarrow \Sigma A$. This is proven in full detail in Lemma 4.2.2 and Remark 4.2.3 and there is a dual statement.

3.5 Shifting Fibre and Cofibre Sequences

In this section, we show that a fibre sequence can be "shifted" to the left using its boundary map to obtain another fibre sequence. This result is a powerful tool that will make the subsequent results a lot easier. The proof will occupy the rest of this section, so some readers may wish to read the statement and skip ahead to Section 3.6.

Lemma 3.5.1 *Let $X \xrightarrow{f} Y \xrightarrow{g} Z$ be a fibre sequence in* $\mathrm{Ho}(\mathcal{C})$*, where $\mathcal{C}$ is a pointed model category. Then*

$$\Omega Z \xrightarrow{\partial} X \xrightarrow{f} Y$$

is a fibre sequence with the action

$$\odot\colon \Omega Z \times \Omega Y \longrightarrow \Omega Z$$

given by

$$\Omega Z \times \Omega Y \xrightarrow{(\mathrm{Id}, \Omega g)} \Omega Z \times \Omega Z \xrightarrow{(\mathrm{Id}, -1)} \Omega Z \times \Omega Z \xrightarrow{*} \Omega Z,$$

where $*$ is the group structure map of the loop object and -1 the group inverse. In other words,

$$[-((\Omega g) \circ h)] * [u] = [u] \odot [h],$$

where $u \in [A, \Omega Z]$ and $[h] \in [A, \Omega Y]$ for $A \in$ $\mathrm{Ho}(\mathcal{C})$.

Note that we write the inverse additively rather than multiplicatively, as in our stable setting all such groups will be abelian. On $[A, \Omega Z]$, the inverse is given by reversing homotopies and the group structure is given by composing homotopies. Note that when we write $f * g$, we read this group composition from left to right, that is, f is the homotopy that is being executed first. Because we are dealing with a group inverse, $[u]$ is on the right side in the term $[-((\Omega g) \circ h)] * [u]$ despite $[h]$ acting from the right.

Proof Our strategy is to construct a new fibre sequence, compare it to the original, and then show that this comparison is an isomorphism compatible with the actions of each fibre sequence.

Construction of the Fibre Sequence

Without loss of generality, let us assume that our starting fibre sequence is of the form $F \xrightarrow{i} E \xrightarrow{p} B$, where p is a fibration between fibrant objects in $\mathcal{C}$. We let ∂ be the boundary map. Now let F' be the pullback of

$$PB \xrightarrow{(e_0,e_1)} B \times B \xleftarrow{(p,0)} E.$$

We can extend this to a commutative diagram

$$\begin{array}{ccc} \Omega B & \longrightarrow & * \\ {\scriptstyle h'}\downarrow & & \downarrow \\ F' & \xrightarrow{i'} & E \\ {\scriptstyle \varphi}\downarrow & & \downarrow{\scriptstyle (p,0)} \\ PB & \xrightarrow{(e_0,e_1)} & B \times B. \end{array}$$

The outer square and the bottom square are pullback squares, thus, so is the top square [Str11, Theorem 2.42], and we have a fibre sequence

$$\Omega B \xrightarrow{h'} F' \xrightarrow{i'} E.$$

Our claim is that this fibre sequence is isomorphic to the one of the statement, $\Omega B \xrightarrow{\partial} F \xrightarrow{i} E$. We must prove the following points.

- There is a weak equivalence $\phi \colon F \longrightarrow F'$.
- This map satisfies $\phi \circ \partial = h'$ and $i' \circ \phi = i$ in the homotopy category.
- The action $\Omega B \times \Omega E \longrightarrow \Omega B$ associated to this fibre sequence is isomorphic to the action in the statement of the lemma.

First, we must obtain a map $\phi \colon F \longrightarrow F'$. Rewriting $(p, 0)$ as $(\mathrm{Id}, 0) \circ p$, we have the diagram

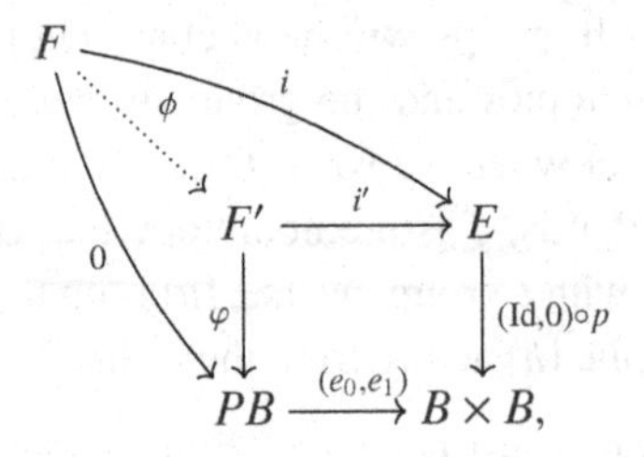

where the outer square commutes because $p \circ i = 0$. Thus, the universal property of the pullback gives us ϕ with $i' \circ \phi = i$ at the level of model categories.

Comparison of Fibres

To show that ϕ is a weak equivalence, we rewrite it as a map into a double pullback, followed by two maps of double pullbacks, each of which is a weak equivalence. The double pullback diagrams and maps between them are given below, with their limits on the right.

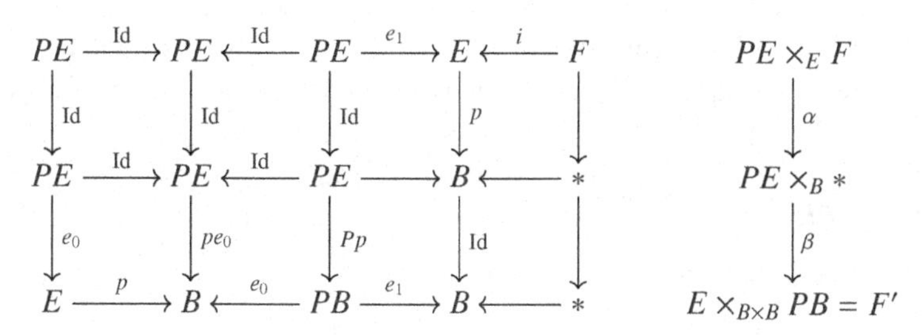

The map from F into the first row is induced by the identity of F and the composite $s \circ i$, where $s\colon E \longrightarrow PE$ is a weak equivalence that is part of the structure of the path object. Furthermore, note that

$$F' = E \times_{B\times B} PB \cong E \times_B PB \times_B *.$$

By considering the composite maps from F into the outer terms of the last pullback, we see that ϕ is the composite

$$F \xrightarrow{(s\circ i,\mathrm{Id})} PE \times_E F \xrightarrow{\alpha} PE \times_B * \xrightarrow{\beta} E \times_{B\times B} PB = F'.$$

As $e_0\colon PE \longrightarrow E$ is an acyclic fibration, so is the pullback of it along i, which is a map $PE \times_E F \longrightarrow F$. The two-out-of-three axiom implies that $(s \circ i, \mathrm{Id})$ is a weak equivalence as desired.

For the second map, we consider the commutative diagram

$$\begin{array}{ccccc} PE \times_E F & \longrightarrow & F & \longrightarrow & * \\ \big\downarrow{\scriptstyle pr} & & \big\downarrow{\scriptstyle i} & & \big\downarrow \\ PE & \xrightarrow{e_0} & E & \overset{p}{\twoheadrightarrow} & B. \end{array}$$

The two small squares are both pullback squares, thus, the outer rectangle is a pullback square too. In other words, the identity of PE, p and the zero map $F \longrightarrow *$ induce an isomorphism

$$\alpha\colon PE \times_E F \longrightarrow PE \times_B *.$$

We will prove that the third map β fits into a pullback square, with the right-hand vertical being an acyclic fibration.

$$\begin{array}{ccc} PE \times_B * & \xrightarrow{pr} & PE \\ \beta \downarrow & & \downarrow \sim (e_0, Pp) \\ E \times_{B\times B} PB & \xrightarrow{(\mathrm{Id},\mathrm{Id})} & E \times_B PB \end{array} \tag{3.1}$$

Since the pullback of an acyclic fibration is itself an acyclic fibration, so is β, and the result will be complete. To see that β is given by that pullback, we rewrite the above square as a commutative diagram in the diagram category $\mathcal{C}^{1\to 2\leftarrow 3\to 4\leftarrow 5}$.

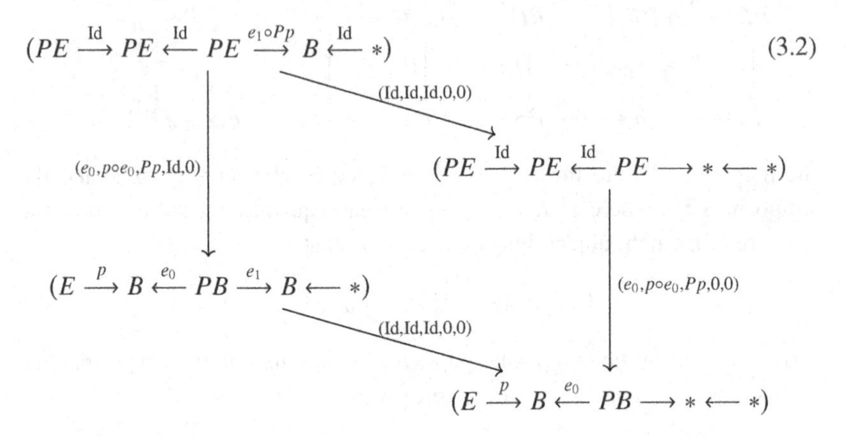

Evaluating the diagram (3.2) at each vertex of the index category

$$1 \to 2 \leftarrow 3 \to 4 \leftarrow 5$$

gives a pullback square. This means that the diagram (3.2) is a pullback square in $\mathcal{C}^{1\to 2\leftarrow 3\to 4\leftarrow 5}$, as limits are defined termwise. Consequently, applying the limit functor to each of the four corners of (3.2) gives another pullback square in $\mathcal{C}$. But that square is exactly the square (3.1).

It still remains to show that the map

$$\beta\colon PE \times_B * \longrightarrow E \times_{B\times B} PB = F'$$

is an acyclic fibration. Recall that β is the pullback of $PE \longrightarrow E \times_B PB$ along

$$(\mathrm{Id}, \mathrm{Id})\colon E \times_{B\times B} PB \longrightarrow E \times_B PB.$$

We construct a specific model for PE so that the map $PE \longrightarrow E \times_B PB$ induced by e_0 and Pp is an acyclic fibration. We first construct a double pullback

$$E \times_B PB \times_B E = \lim\left(E \xrightarrow{p} B \xleftarrow{e_0} PB \xrightarrow{e_1} B \xleftarrow{p} E\right).$$

We can view this as an iterated pullback, hence, we get a diagram as below.

$$\begin{array}{ccccc}
E \times_B PB \times_B E & \twoheadrightarrow & PB \times_B E & \xrightarrow[\sim]{} & E \\
\downarrow \lambda & & \downarrow & & \downarrow p \\
E \times_B PB & \twoheadrightarrow & PB & \xrightarrow[\sim]{e_1} & B \\
\downarrow \sim & & \downarrow e_0 \sim & & \\
E & \xrightarrow{p} & B & &
\end{array}$$

We may then recognise $E \times E$, $B \times B$ and PB in terms of limits of double pullback diagrams.

$$\begin{aligned}
E \times E &= \lim (E \longrightarrow * \longleftarrow * \longrightarrow * \longleftarrow E) \\
B \times B &= \lim (B \longrightarrow * \longleftarrow * \longrightarrow * \longleftarrow B) \\
PB &= \lim \left(B \xrightarrow{\mathrm{Id}} B \xleftarrow{e_0} PB \xrightarrow{e_1} B \xleftarrow{\mathrm{Id}} B \right).
\end{aligned}$$

The maps $(p, p)\colon E \times E \longrightarrow B \times B$ and $(e_0, e_1)\colon PB \to B \times B$ are recognisable as limits of maps of these diagrams. Specifically, (p, p) is induced by a map from the first diagram to the second, which is p on the two outermost terms and zero elsewhere. Furthermore, (e_0, e_1) is induced by a map from the third to the second, which is the identity map on the outermost terms and zero elsewhere.

Using these maps of diagrams, $E \times_B PB \times_B E$ can be recognised as the pullback of (p, p) along (e_0, e_1). This is similar to the earlier statements relating diagrams (3.1) and (3.2). Moreover, as $PB \to B \times B$ is a fibration, the map created from the pullback

$$(pr_0, pr_1)\colon E \times_B PB \times_B E \longrightarrow E \times E$$

is also a fibration.

The maps $\mathrm{Id}\colon E \longrightarrow E$, $s \circ p\colon E \longrightarrow PB$ and $\mathrm{Id}\colon E \longrightarrow E$ induce a map

$$\nu\colon E \longrightarrow E \times_B PB \times_B E$$

such that $(pr_0, pr_1) \circ \nu$ is the diagonal $E \longrightarrow E \times E$. It follows that factorising ν into a acyclic cofibration s followed by a fibration μ

$$E \xrightarrow{s} PE \xrightarrow{\mu} E \times_B PB \times_B E$$

gives a path object for E

$$E \xrightarrow{s} PE \xrightarrow{(pr_0, pr_1) \circ \mu} E \times E.$$

We call the last map (e_0, e_1) to match the standard notation. As before, $e_0 \circ s$ and $e_1 \circ s$ are the identity maps of E. Moreover, by definition of a map into a pullback, μ induces a map $PE \longrightarrow PB$, which we may call Pp as

$$e_0 \circ Pp = p \circ e_0 \quad \text{and} \quad e_1 \circ Pp = p \circ e_1.$$

The composite of the fibrations μ and λ from the pullback diagram explaining $E \times_B PB \times_B E$ gives a fibration

$$PE \xrightarrow{\mu} E \times_B PB \times_B E \xrightarrow{\lambda} E \times_B PB,$$

which is precisely that induced by e_0 and Pp. Since $\kappa \circ \lambda \circ \nu$ is the identity of E, κ is a weak equivalence and μ is a factorisation of ν, it follows that $\lambda \circ \mu$ is an acyclic fibration as required.

Comparison of the New Fibre Sequence with the Original Sequence
We now show that $\phi \circ \partial = h'$ to obtain a commutative diagram

$$\begin{array}{ccccc} \Omega B & \xrightarrow{\partial} & F & \xrightarrow{i} & E \\ \| & & \downarrow{\scriptstyle \phi} & & \| \\ \Omega B & \xrightarrow{h'} & F' & \xrightarrow{i'} & E, \end{array}$$

where all three vertical maps are weak equivalences. We are going to show that

$$[h' \circ f] = [\phi \circ \partial \circ f] \text{ for all } [f] \in [A, \Omega B].$$

Let $[f]$ be represented by both the right homotopy $j\colon A \longrightarrow PB$ and the left homotopy $k\colon \mathrm{Cyl}(A) \longrightarrow B$ which are related via the correspondence

$$\begin{array}{ccc} A \amalg A & \xrightarrow{(j,0)} & PB \\ {\scriptstyle (i_0,i_1)} \downarrow & \overset{K}{\nearrow} & \downarrow {\scriptstyle e_1} \\ \mathrm{Cyl}(A) & \xrightarrow{0} & B \end{array}$$

with

$$e_0 \circ K = k, \quad K \circ i_0 = j, \quad e_1 \circ K = 0 = K \circ i_1.$$

Firstly, $[h' \circ f] \in [A, F']$ is given by

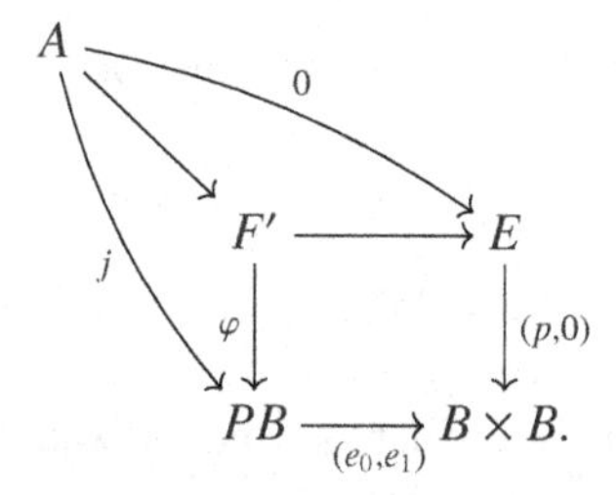

Now let us look at a representative for $[\phi \circ \partial \circ f]$. By definition of ∂, $[\partial \circ f] = [0] \odot [f]$, so

$$[\phi \circ \partial \circ f] = [\phi] \circ ([0] \odot [f]).$$

To compute $[0] \odot [f]$, we get a lift H in the commutative square

$$\begin{array}{ccc} A & \xrightarrow{0} & E \\ {\scriptstyle i_0}\downarrow & \overset{H}{\nearrow} & \downarrow{\scriptstyle p} \\ \mathrm{Cyl}(A) & \xrightarrow{k} & B. \end{array}$$

Then $[0] \odot [f]$ is represented by a map $v \colon A \longrightarrow F$ such that $i \circ v = H \circ i_1$. As $i' \circ \phi = i$, this means that $H \circ i_1$ represents $i' \circ \phi \circ \partial \circ f$, so $\phi \circ \partial \circ f$ is given by the commutative diagram

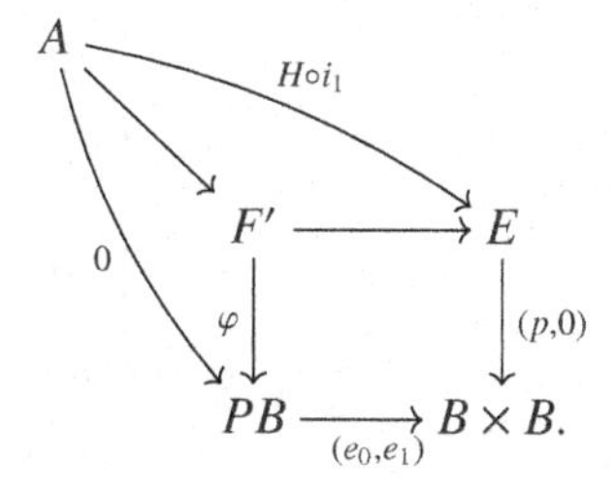

We now consider the map π induced by H and the homotopy correspondence K

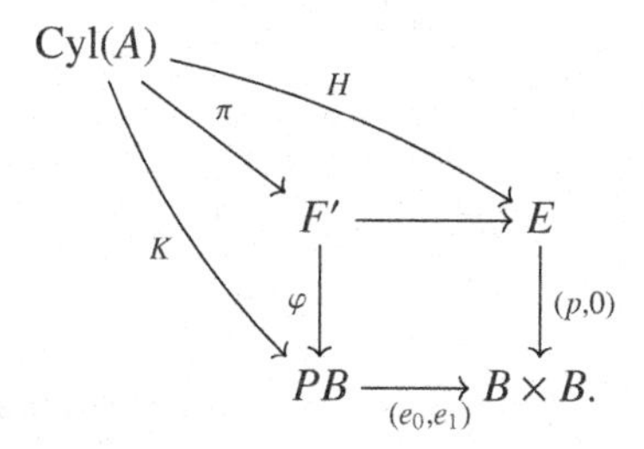

This map π is in fact a (left) homotopy between $h' \circ f$ and $\phi \circ \partial \circ f$ because

$$K \circ i_0 = j, \quad H \circ i_0 = 0 \quad \text{and thus,} \quad \pi \circ i_0 = h' \circ f,$$

as well as

$$K \circ i_1 = 0, \quad \text{and thus,} \quad \pi \circ i_1 = \phi \circ \partial \circ f,$$

meaning $[h' \circ f] = [\phi \circ \partial \circ f]$, which is what we wanted to prove.

Comparison of the Two Actions

Lastly, we compare the natural action of ΩE on ΩB as part of the fibre sequence

$$\Omega B \xrightarrow{h'} F' \xrightarrow{i'} E$$

with the action given in the statement of the lemma. For this, let $f\colon A \longrightarrow PB$ represent an element of $[A, \Omega B]$, and let $h\colon \mathrm{Cyl}(A) \longrightarrow E$ represent an element $[h]$ of $[A, \Omega E]$. By definition of the loop object, there is a commutative diagram

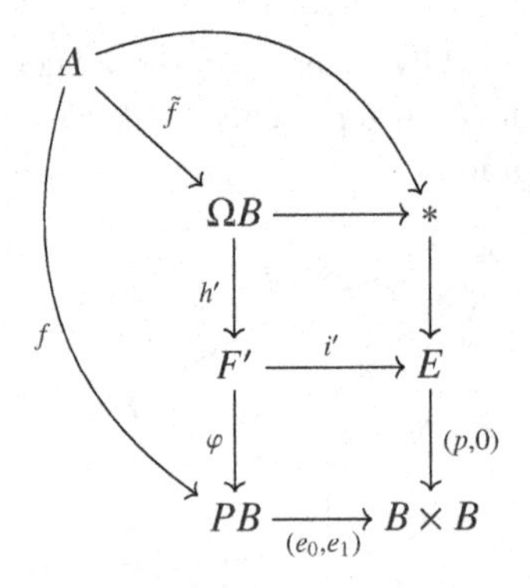

for some $\tilde{f}\colon A \longrightarrow \Omega B$. This gives us the usual commutative square from the definition of $\odot$

$$\begin{array}{ccc} A & \xrightarrow{h' \circ \tilde{f}} & F' \\ {\scriptstyle i_0}\downarrow & \nearrow & \downarrow{\scriptstyle i'} \\ \mathrm{Cyl}(A) & \xrightarrow{h} & E \end{array}$$

with a lift $\alpha\colon \mathrm{Cyl}(A) \longrightarrow F'$. As this α is a map into a pullback, this is equivalent to a commutative diagram

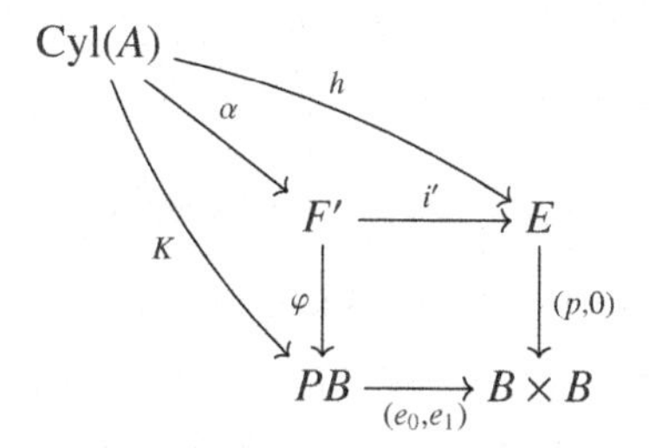

for some map $K\colon \mathrm{Cyl}(A) \longrightarrow PB$. This map K satisfies

$$K \circ i_0 = \varphi \circ \alpha \circ i_0 = \varphi \circ h' \circ \tilde{f} = f,$$
$$e_0 \circ K = p \circ h \quad \text{and} \quad e_1 \circ K = 0.$$

The element $[f] \odot [h] \in [A, \Omega B]$ is defined as the homotopy class of the map v in the commutative diagram below.

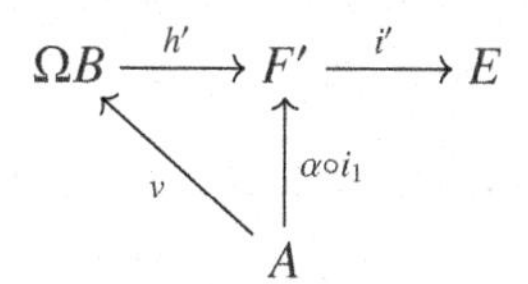

First, we notice that because

$$e_0 \circ K \circ i_1 = p \circ h \circ i_1 = 0 = e_1 \circ K \circ i_1,$$

the map $K \circ i_1 \colon A \longrightarrow PB$ factors over ΩB. We denote this map by $k\colon A \longrightarrow \Omega B$. Thus, we have

$$\varphi \circ h' \circ k = K \circ i_1 = \varphi \circ \alpha \circ i_1,$$

and we can conclude that $[v] = [f] \odot [h] = [k]$. This means that our claim amounts to

$$[k] = [(-\Omega p \circ h) * f],$$

where $*$ is the group multiplication (i.e. composition of homotopies) on $[A, \Omega B]$.

We look at $K \circ i_1 \colon A \longrightarrow PB$ as a right homotopy between 0 and 0. This corresponds to a left homotopy $\bar{h}\colon \mathrm{Cyl}(A) \longrightarrow B$ via a correspondence $\bar{H}$, that is, we have a commutative diagram

$$\begin{array}{ccc} A \amalg A & \xrightarrow{(K \circ i_1, 0)} & PB \\ {\scriptstyle (i_0, i_1)}\downarrow & \overset{\bar{H}}{\nearrow} & \downarrow{\scriptstyle e_1} \\ \mathrm{Cyl}(A) & \xrightarrow{0} & B \end{array}$$

with

$$\bar{H} \circ i_0 = K \circ i_1,\ \bar{H} \circ i_1 = 0,\ e_0 \circ \bar{H} = \bar{h} \text{ and } e_1 \circ \bar{H} = 0.$$

The first equation means that we can form $K * \bar{H}$. We therefore get

$$\begin{aligned}(K * \bar{H}) \circ i_1 &= \bar{H} \circ i_1 = 0,\\ (K * \bar{H}) \circ i_0 &= K \circ i_0 = f,\\ e_0 \circ (K * \bar{H}) &= (e_0 \circ K) * (e_0 \circ \bar{H}) = (p \circ h) * \bar{h},\\ e_1 \circ (K * \bar{H}) &= (e_1 \circ K) * (e_1 \circ \bar{H}) = 0 * 0 = 0.\end{aligned}$$

In other words, we have a commutative diagram

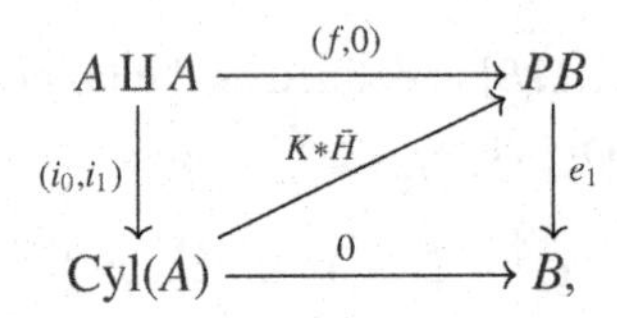

which means that $K * \bar{H}$ is a correspondence between f and

$$e_0 \circ (K * \bar{H}) = (p \circ h) * \bar{h},$$

so $[f] = [p \circ h] * [\bar{h}]$. Now $\bar{h}$ is in the same homotopy class as $K \circ i_1$ which represents $[k] = [f] \odot [h]$, so

$$[f] = [p \circ h] * [\bar{h}] = [p \circ h] * ([f] \odot [h]).$$

By slight abuse of notation, $h\colon \operatorname{Cyl}(A) \longrightarrow B$ is $h\colon A \longrightarrow \Omega B$, so

$$[f] = [\Omega p \circ h] * ([f] \odot [h]).$$

Rearranging the above equation then gives

$$[-(\Omega p) \circ h] * [f] = [f] \odot [h]$$

as required. □

Corollary 3.5.2 *With the previous notation, if* $F \xrightarrow{i} E \xrightarrow{p} B$ *is a fibre sequence, then so is*

$$\Omega E \xrightarrow{-\Omega p} \Omega B \xrightarrow{\partial} F.$$

Proof Using the action $\odot$ of ΩE on ΩB given in Lemma 3.5.1 as well as Definition 3.4.4, we see that the boundary map associated to $\Omega B \longrightarrow F \longrightarrow E$ is given by

$$[X, \Omega E] \longrightarrow [X, \Omega B] \times [X, \Omega E] \xrightarrow{\quad \odot \quad} [X, \Omega B],$$

$$[f] \longmapsto ([0], [f]) \longmapsto [0] * [-\Omega p \circ f] = [-\Omega p \circ f]. \quad \square$$

The proof of this corollary very neatly justifies the minus sign appearing in the shifted fibre sequence, which plays an important role in the following long exact sequence and will do so again in Definition 4.1.2. For topological spaces, this is known as the Puppe sequence or simply the long exact fibre sequence, and it also holds in a general model category environment.

3.6 The Long Exact Puppe Sequence

Because we have already done so much preparation work, the proof of the long exact sequence is now relatively brief. In the following result and Theorem 3.6.4, note that the signs $(-1)^n$ can be removed while keeping the sequences exact.

Theorem 3.6.1 *Let $X \xrightarrow{f} Y \xrightarrow{g} Z$ be a fibre sequence in* $\mathrm{Ho}(\mathcal{C})$, *where $\mathcal{C}$ is a pointed model category, and let $\partial \colon \Omega Z \longrightarrow X$ be the boundary map from Definition 3.4.4. Then the sequence*

$$\cdots \to [A, \Omega^{n+1} Z] \xrightarrow{(-1)^n (\Omega^n \partial)_*} [A, \Omega^n X] \xrightarrow{(-1)^n (\Omega^n f)_*} [A, \Omega^n Y] \xrightarrow{(-1)^n (\Omega^n g)_*} \cdots$$
$$\cdots \to [A, \Omega Z] \xrightarrow{\partial_*} [A, X] \xrightarrow{f_*} [A, Y] \xrightarrow{g_*} [A, Z]$$

is exact for any $A \in \mathrm{Ho}(\mathcal{C})$.

Of course, the homotopy category of a pointed model category $\mathcal{C}$ is not necessarily additive, so unless, for example, $\mathcal{C}$ is stable or A is a cogroup object, the last three terms of the sequence need not carry any group structure. Thus, exactness at these terms should be considered as exactness in pointed sets.

Proof Because of Lemma 3.5.1 and Corollary 3.5.2, it is sufficient to show exactness of

$$[A, X] \xrightarrow{f_*} [A, Y] \xrightarrow{g_*} [A, Z]$$

only. Note that Corollary 3.5.2 is important here because Lemma 3.5.1 alone does not tell us that the boundary map in the twice shifted fibre sequence is actually what one expects. Without loss of generality, let us assume that our fibre sequence is of the form

$$F \xrightarrow{i} E \xrightarrow{p} B,$$

where p is a fibration of fibrant objects, so we will show that

$$[A, F] \xrightarrow{i_*} [A, E] \xrightarrow{p_*} [A, B]$$

is exact. Because $p \circ i = 0$ by definition, we have that the image of (i_*) is in the kernel of p_*. Now let $u\colon A \longrightarrow E$ satisfy $[p \circ u] = 0$. We would like to show that u is in the image of i_*. As $[p \circ u] = 0$, there is a homotopy $h\colon \mathrm{Cyl}(A) \longrightarrow B$ with $h \circ i_0 = p \circ u$ and $h \circ i_1 = 0$. As p is a fibration, there is a lift in the following commutative square.

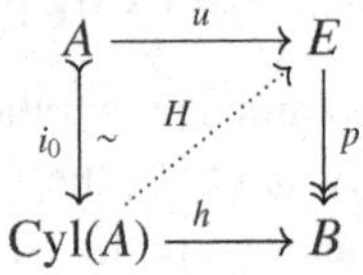

$$\begin{array}{ccc} A & \xrightarrow{u} & E \\ {\scriptstyle i_0}\big\downarrow{\scriptstyle \sim} & \overset{H}{\nearrow} & \big\downarrow{\scriptstyle p} \\ \mathrm{Cyl}(A) & \xrightarrow{h} & B \end{array}$$

Because $p \circ (H \circ i_1) = h \circ i_1 = 0$, the map $H \circ i_1$ lifts over the fibre of p, meaning that there is

$$v\colon A \longrightarrow F \quad \text{such that} \quad i \circ v = H \circ i_1.$$

Since H is a homotopy between $H \circ i_1$ and $H \circ i_0$, they represent the same homotopy class, hence,

$$i_*[v] = [i \circ v] = [H \circ i_1] = [H \circ i_0] = [u],$$

thus, $[u]$ is in the image of i_* as required. □

There are dual statements of the previous results for cofibre sequences.

Lemma 3.6.2 *Let $X \xrightarrow{f} Y \xrightarrow{g} Z$ be a cofibre sequence in* $\mathrm{Ho}(\mathcal{C})$, *where $\mathcal{C}$ is a pointed model category. Then*

$$Y \xrightarrow{g} Z \xrightarrow{\partial} \Sigma X$$

is a cofibre sequence with the coaction

$$\odot\colon \Sigma X \longrightarrow \Sigma X \amalg \Sigma Y$$

given by

$$[u] \odot [h] = [u] * [-h \circ \Sigma f].$$ □

Corollary 3.6.3 *If $X \xrightarrow{f} Y \xrightarrow{g} Z$ is a cofibre sequence, then so is*

$$Z \xrightarrow{\partial} \Sigma X \xrightarrow{-\Sigma f} \Sigma Y.$$

□

Theorem 3.6.4 *Let $X \xrightarrow{f} Y \xrightarrow{g} Z$ be a cofibre sequence in* $\mathrm{Ho}(\mathcal{C})$*, where $\mathcal{C}$ is a pointed model category, and let $\partial\colon Z \longrightarrow \Sigma X$ be its boundary map. Then the sequence*

$$\cdots \to [\Sigma^{n+1}X, A] \xrightarrow{(-1)^n(\Sigma^n\partial)^*} [\Sigma^n Z, A] \xrightarrow{(-1)^n(\Sigma^n g)^*} [\Sigma^n Y, A] \to \cdots$$

$$\cdots \xrightarrow{\partial^*} [Z, A] \xrightarrow{g^*} [Y, A] \xrightarrow{f^*} [X, A]$$

is exact for any $A \in \mathrm{Ho}(\mathcal{C})$. □

Indeed, all standard properties of fibre sequences can be reformulated for cofibre sequences and vice versa. The proofs are identical, one just has to exchange limits for colimits, left homotopies for right homotopies etc. In the next chapter, the duality between fibre and cofibre sequences for *stable* model categories is highlighted even more strongly.

4

Triangulated Categories

The aim of this chapter is to state the definition of a triangulated category and show that the homotopy category of a stable model category is a triangulated category, see Theorem 4.2.1. Triangulated categories were developed to axiomatise the structure of the derived category of an abelian category. This structure comes in the form of exact triangles, a replacement for the short exact sequences (and kernels and cokernels) of an abelian category. The exact triangles of a stable model category are defined in terms of cofibre sequences.

Once we have proven the main theorem, we will investigate some of the consequences for the homotopy category of a stable model category. These include the agreement between fibre and cofibre sequences and finite products equalling finite coproducts. Next, we will consider the notion of an exact functor, which is a functor that is compatible with the structures of triangulated categories. We will show that a Quillen functor of stable model categories induces an exact functor of the respective homotopy categories.

We end the chapter with two overview sections. The first introduces the concept of Toda brackets, an important method of calculation in triangulated categories, and applies the theory to the stable homotopy category. The second gives an example of a triangulated category that does not arise from a stable model category.

4.1 Definition and Basic Properties

Let us begin with the axioms, which arise naturally from some of the properties of fibre and cofibre sequences, as well as from the algebraic structures found on the derived category $D(\mathcal{A})$ of an abelian category $\mathcal{A}$. The version we give has some redundancies, but has been chosen for its ease of use. For example, axiom (T3) follows from (T1) and (T4), but is the first step in proving that the homotopy category of a stable model category is triangulated.

Recall that an additive category is a category whose sets of morphisms come with abelian group structures such that composition is bilinear and the category has all finite coproducts and products (which coincide and are called direct sums), see Definition 3.2.5. Note that the zero object is the direct sum of an empty collection.

Definition 4.1.1 Let $\mathcal{T}$ be an additive category equipped with an additive self-equivalence $\Sigma\colon \mathcal{T} \longrightarrow \mathcal{T}$. A *triangle* in $\mathcal{T}$ is a sequence of morphisms

$$X \xrightarrow{f_1} Y \xrightarrow{f_2} Z \xrightarrow{f_3} \Sigma X.$$

A *morphism of triangles* from

$$X \to Y \to Z \to \Sigma X \text{ to } X' \to Y' \to Z' \to \Sigma X'$$

is a commutative diagram

$$\begin{array}{ccccccc} X & \xrightarrow{f_1} & Y & \xrightarrow{f_2} & Z & \xrightarrow{f_3} & \Sigma X \\ \downarrow{\scriptstyle \phi_1} & & \downarrow{\scriptstyle \phi_2} & & \downarrow{\scriptstyle \phi_3} & & \downarrow{\scriptstyle \Sigma\phi_1} \\ X' & \xrightarrow{f_1'} & Y' & \xrightarrow{f_2'} & Z' & \xrightarrow{f_3'} & \Sigma X'. \end{array}$$

Given what we know about fibres and cofibres, we might expect the definition of a triangle to require that the composites $f_2 \circ f_1$ and $f_3 \circ f_2$ are the zero map. However, we shall shortly see that this follows from the axioms of a triangulated category.

Definition 4.1.2 Let $\mathcal{T}$ be an additive category equipped with an additive self-equivalence $\Sigma\colon \mathcal{T} \longrightarrow \mathcal{T}$. (This functor is often called the "shift functor" as a concession to the most common examples.)

We say that $\mathcal{T}$ is a *triangulated category* if there is a class of distinguished triangles called *exact triangles* satisfying the following axioms.

(T1) The triangle

$$* \longrightarrow X = X \longrightarrow *$$

is exact for every $X \in \mathcal{T}$. A triangle isomorphic to an exact triangle is exact. Every morphism $f\colon X \longrightarrow Y$ fits into some exact triangle

$$X \xrightarrow{f} Y \xrightarrow{g} Z \xrightarrow{u} \Sigma X.$$

(T2) The triangle

$$X \xrightarrow{f} Y \xrightarrow{g} Z \xrightarrow{u} \Sigma X$$

is exact if and only if the triangle

$$Y \xrightarrow{g} Z \xrightarrow{u} \Sigma X \xrightarrow{-\Sigma f} \Sigma Y$$

is exact.

(T3) Let

$$\begin{array}{ccccccc} X & \xrightarrow{f_1} & Y & \xrightarrow{f_2} & Z & \xrightarrow{f_3} & \Sigma X \\ \downarrow{\scriptstyle\phi_1} & & \downarrow{\scriptstyle\phi_2} & & & & \downarrow{\scriptstyle\Sigma\phi_1} \\ X' & \xrightarrow{f_1'} & Y' & \xrightarrow{f_2'} & Z' & \xrightarrow{f_3'} & \Sigma X' \end{array}$$

be a diagram such that the two rows are exact triangles and the left square commutes. Then one can add a morphism $\phi_3 \colon Z \longrightarrow Z'$ to this diagram such that the resulting second and third square commute.

(T4) Let

$$\begin{array}{ccccccc} X & \xrightarrow{f_1} & Y & \xrightarrow{f_2} & Z & \xrightarrow{f_3} & \Sigma X \\ \| & & \downarrow{\scriptstyle u_1} & & & & \| \\ X & \xrightarrow{g_1} & U & \xrightarrow{g_2} & V & \xrightarrow{g_3} & \Sigma X \\ & & \downarrow{\scriptstyle u_2} & & & & \\ & & W & & & & \\ & & \downarrow{\scriptstyle u_3} & & & & \\ & & \Sigma Y & & & & \end{array}$$

be a commutative diagram such that the column and two rows are exact triangles. Then there is an exact triangle

$$Z \xrightarrow{v_1} V \xrightarrow{v_2} W \xrightarrow{v_3} \Sigma Z$$

that can be added to the first diagram to obtain the commutative diagram

$$\begin{array}{ccccccc} X & \xrightarrow{f_1} & Y & \xrightarrow{f_2} & Z & \xrightarrow{f_3} & \Sigma X \\ \| & & \downarrow{\scriptstyle u_1} & & \downarrow{\scriptstyle v_1} & & \| \\ X & \xrightarrow{g_1} & U & \xrightarrow{g_2} & V & \xrightarrow{g_3} & \Sigma X \\ & & \downarrow{\scriptstyle u_2} & & \downarrow{\scriptstyle v_2} & & \\ & & W & = & W & & \\ & & \downarrow{\scriptstyle u_3} & & \downarrow{\scriptstyle v_3} & & \\ & & \Sigma Y & \xrightarrow{\Sigma f_2} & \Sigma Z. & & \end{array}$$

First, we note that because of (T2), the fill-in axiom (T3) could equivalently be reformulated with ϕ_2 instead of ϕ_3 and asking for the right square to commute.

The axiom (T4) is known as the *octahedral axiom*. This is due to its original form concerning the four exact triangles arranged in the shape of an octahedron. The version we stated above is equivalent to the original octahedral axiom [Kra07, Appendix] and has established itself as a more convenient version in practice.

Rewriting the object Z in the first row of the diagrams in (T4) as "$Z = Y/X$", the octahedral axiom then has the appearance of the Third Isomorphism Theorem

$$(Y/X)\big/(U/X) = Y/U.$$

It is somewhat unclear why "triangulated" is the name for this kind of category. A possible explanation lies in the following. One could rewrite an exact triangle

$$X \xrightarrow{f} Y \xrightarrow{g} Z \xrightarrow{u} \Sigma X$$

as below, where u is considered as a morphism of degree 1 (indicated by a special arrow $\multimap\!\!\rightarrow$)

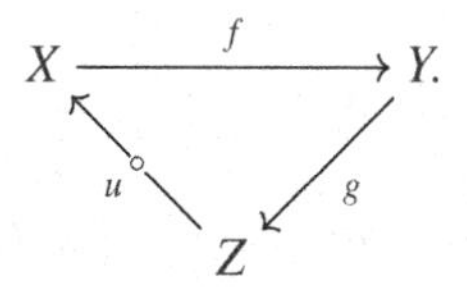

The axiom (T2) would then say that one can "rotate" such a triangle either way and still obtain a triangle. The octahedral axiom then says that the diagram

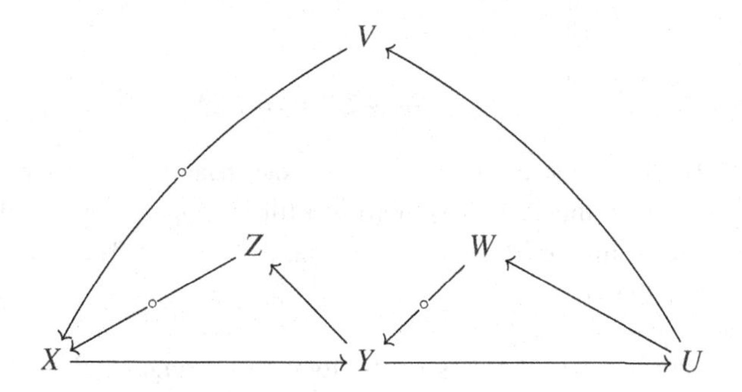

(where the bottom two triangles are exact and the big outer triangle is exact) can be filled in so that the top central triangle is also exact.

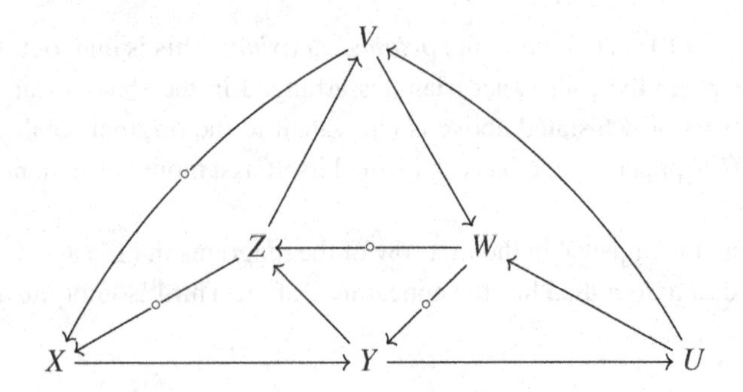

(The other triangle shapes appearing in the second diagram are not exact triangles, which is already evident from the direction of the arrows and degrees of the maps.)

Definition 4.1.3 A subcategory $\mathcal{T}'$ of a triangulated category $\mathcal{T}$ is a *triangulated subcategory* if it is a triangulated category with shift and exact triangles inherited from $\mathcal{T}$.

Let us look at some of the immediate consequences of the definition.

Lemma 4.1.4 *Let*

$$X \xrightarrow{f} Y \xrightarrow{g} Z \xrightarrow{u} \Sigma X$$

be an exact triangle in a triangulated category $\mathcal{T}$. Then $g \circ f = 0$ and $u \circ g = 0$.

Proof We have the commutative diagram

$$\begin{array}{ccccccc} X & \xrightarrow{f} & Y & \xrightarrow{g} & Z & \xrightarrow{u} & \Sigma X \\ \| & & & & \downarrow{\scriptstyle u} & & \| \\ X & \longrightarrow & * & \longrightarrow & \Sigma X & = & \Sigma X. \end{array}$$

(T1) and (T2) tell us that the two rows are exact triangles. The right square evidently commutes, thus, by (T3) there is a fill-in map $Y \longrightarrow *$ making the resulting square commute. But this means that $u \circ g = 0$. The case $g \circ f$ is similar, but uses the fact that $h = 0$ if and only if $\Sigma h = 0$. $\square$

The following is one of the most useful tools in a triangulated category.

Proposition 4.1.5 *Let $\mathcal{T}$ be a triangulated category, $X \xrightarrow{f_1} Y \xrightarrow{f_2} Z \xrightarrow{f_3} \Sigma X$ an exact triangle in $\mathcal{T}$ and A an object in $\mathcal{T}$. Then the two sequences of abelian groups*

$$\mathcal{T}(A,X) \xrightarrow{f_{1*}} \mathcal{T}(A,Y) \xrightarrow{f_{2*}} \mathcal{T}(A,Z) \xrightarrow{f_{3*}} \mathcal{T}(A,\Sigma X)$$

$$\mathcal{T}(\Sigma X,A) \xrightarrow{f_3^*} \mathcal{T}(Z,A) \xrightarrow{f_2^*} \mathcal{T}(Y,A) \xrightarrow{f_1^*} \mathcal{T}(X,A)$$

are exact.

Proof We will only show exactness for the first sequence. By (T2), it is sufficient to show exactness of

$$\mathcal{T}(A,X) \xrightarrow{f_{1*}} \mathcal{T}(A,Y) \xrightarrow{f_{2*}} \mathcal{T}(A,Z).$$

Take $g\colon A \longrightarrow Y$ and the diagram

$$\begin{array}{ccccccc} A & = & A & \longrightarrow & * & \longrightarrow & \Sigma A \\ & & \downarrow{\scriptstyle g} & & & & \\ X & \xrightarrow{f_1} & Y & \xrightarrow{f_2} & Z & \xrightarrow{f_3} & \Sigma X, \end{array}$$

where the two rows are exact triangles. If g is in the image of f_{1*}, this means that there is a map $u\colon A \longrightarrow X$ giving the first and last vertical arrow in the commutative diagram

$$\begin{array}{ccccccc} A & = & A & \longrightarrow & * & \longrightarrow & \Sigma A \\ \downarrow{\scriptstyle u} & & \downarrow{\scriptstyle g} & & \vdots & & \downarrow{\scriptstyle \Sigma u} \\ X & \xrightarrow{f_1} & Y & \xrightarrow{f_2} & Z & \xrightarrow{f_3} & \Sigma X. \end{array}$$

By (T3), we can fill in the third vertical arrow, which implies that $g \in \ker(f_{2*})$ because $f_2 \circ g$ factors over $*$.

Conversely, if we assume that $g \in \ker(f_{2*})$, we have the second and third vertical arrow in the diagram, so (T3) gives us a map $u\colon A \longrightarrow X$ to make the first square commute, thus, $g \in \operatorname{Im}(f_{1*})$. □

Remark 4.1.6 Proposition 4.1.5 will play an important role when we study the homotopy category of a stable model category. Theorems 3.6.1 and 3.6.4 give similar long exact sequences in the homotopy category of a pointed model category, one for fibre sequences and one for cofibre sequences. Once we know that the homotopy category of a triangulated model category is triangulated, we can actually compare them and show that in a stable model category, cofibre sequences and fibre sequences agree. We discuss this in detail in Section 4.3.

Definition 4.1.7 A functor from a triangulated category to an abelian category that sends exact triangles to exact sequences as in Proposition 4.1.5 is called a *cohomological functor*.

Applying the Five Lemma to Proposition 4.1.5 gives the following.

Corollary 4.1.8 *Let*

$$\begin{array}{ccccccc} X & \xrightarrow{f_1} & Y & \xrightarrow{f_2} & Z & \xrightarrow{f_3} & \Sigma X \\ \downarrow{\scriptstyle \phi_1} & & \downarrow{\scriptstyle \phi_2} & & \downarrow{\scriptstyle \phi_3} & & \downarrow{\scriptstyle \Sigma\phi_1} \\ X' & \xrightarrow{f_1'} & Y' & \xrightarrow{f_2'} & Z' & \xrightarrow{f_3'} & \Sigma X' \end{array}$$

be a morphism of exact triangles. If two of the ϕ_i are isomorphisms, then so is the third. □

Let us continue with some more properties of exact triangles.

Lemma 4.1.9 *Exact triangles are closed under products and coproducts.*

Proof We only show the proof for coproducts, as the other case is very similar. Let

$$X_i \xrightarrow{f_i} Y_i \xrightarrow{g_i} Z_i \xrightarrow{u_i} \Sigma X_i$$

be an exact triangle for $i \in \mathfrak{I}$. We want to show that

$$\coprod_{i\in\mathfrak{I}} X_i \xrightarrow{\coprod_{i\in\mathfrak{I}} f_i} \coprod_{i\in\mathfrak{I}} Y_i \xrightarrow{\coprod_{i\in\mathfrak{I}} g_i} \coprod_{i\in\mathfrak{I}} Z_i \xrightarrow{\coprod_{i\in\mathfrak{I}} u_i} \Sigma \coprod_{i\in\mathfrak{I}} X_i$$

is exact by showing that it is isomorphic to an exact triangle. (Note that the shift Σ commutes with coproducts as it is an equivalence of categories.) By (T1), the morphism

$$\coprod_{i\in\mathfrak{I}} f_i \colon \coprod_{i\in\mathfrak{I}} X_i \longrightarrow \coprod_{i\in\mathfrak{I}} Y_i$$

can be completed to an exact triangle

$$\coprod_{i\in\mathfrak{I}} X_i \xrightarrow{\coprod_{i\in\mathfrak{I}} f_i} \coprod_{i\in\mathfrak{I}} Y_i \longrightarrow W \longrightarrow \Sigma \coprod_{i\in\mathfrak{I}} X_i.$$

Furthermore, by (T3), we have a morphism of exact triangles

$$\begin{array}{ccccccc} X_i & \xrightarrow{f_i} & Y_i & \xrightarrow{g_i} & Z_i & \xrightarrow{u_i} & \Sigma X_i \\ \downarrow & & \downarrow & & \downarrow & & \downarrow \\ \coprod_{i\in\mathcal{I}} X_i & \xrightarrow{\coprod_{i\in\mathcal{I}} f_i} & \coprod_{i\in\mathcal{I}} Y_i & \longrightarrow & W & \longrightarrow & \Sigma \coprod_{i\in\mathcal{I}} X_i. \end{array}$$

The universal property of the coproduct now implies that there is a morphism

$$\coprod_{i\in\mathcal{I}} Z_i \longrightarrow W,$$

which is part of a commutative diagram

$$\begin{array}{ccccccc} \coprod_{i\in\mathcal{I}} X_i & \longrightarrow & \coprod_{i\in\mathcal{I}} Y_i & \longrightarrow & \coprod_{i\in\mathcal{I}} Z_i & \longrightarrow & \Sigma \coprod_{i\in\mathcal{I}} X_i \\ \downarrow & & \downarrow & & \downarrow & & \downarrow \\ \coprod_{i\in\mathcal{I}} X_i & \longrightarrow & \coprod_{i\in\mathcal{I}} Y_i & \longrightarrow & W & \longrightarrow & \Sigma \coprod_{i\in\mathcal{I}} X_i. \end{array}$$

Let $A \in \mathcal{T}$. Applying the cohomological functor $\mathcal{T}(-, A)$ to the exact triangle

$$X_i \xrightarrow{f_i} Y_i \xrightarrow{g_i} Z_i \xrightarrow{u_i} \Sigma X_i$$

gives us a long exact sequence of abelian groups by Proposition 4.1.5. Taking the coproduct of these results in the long exact sequence

$$\cdots \longrightarrow \bigoplus_{i\in\mathcal{I}} \mathcal{T}(Z_i, A) \longrightarrow \bigoplus_{i\in\mathcal{I}} \mathcal{T}(Y_i, A) \longrightarrow \bigoplus_{i\in\mathcal{I}} \mathcal{T}(X_i, A) \longrightarrow \cdots,$$

which by definition of the coproduct is isomorphic to the long exact sequence

$$\cdots \longrightarrow \mathcal{T}(\coprod_{i\in\mathcal{I}} Z_i, A) \longrightarrow \mathcal{T}(\coprod_{i\in\mathcal{I}} Y_i, A) \longrightarrow \mathcal{T}(\coprod_{i\in\mathcal{I}} X_i, A) \longrightarrow \cdots.$$

Applying $\mathcal{T}(-, A)$ to the exact triangle

$$\coprod_{i\in\mathcal{I}} X_i \xrightarrow{\coprod_{i\in\mathcal{I}} f_i} \coprod_{i\in\mathcal{I}} Y_i \longrightarrow W \longrightarrow \Sigma \coprod_{i\in\mathcal{I}} X_i$$

also gives a long exact sequence. Comparing this sequence to the previous exact sequence using the Five Lemma gives us a canonical isomorphism

$$\mathcal{T}(W, A) \longrightarrow \mathcal{T}(\coprod_{i\in\mathcal{I}} Z_i, A),$$

which, by the Yoneda Lemma, means that the morphism

$$\coprod_{i\in\mathcal{I}} Z_i \longrightarrow W$$

is in fact an isomorphism. We now have a commutative diagram

$$\begin{array}{ccccccc}
\coprod_{i\in\mathcal{I}} X_i & \longrightarrow & \coprod_{i\in\mathcal{I}} Y_i & \longrightarrow & \coprod_{i\in\mathcal{I}} Z_i & \longrightarrow & \Sigma \coprod_{i\in\mathcal{I}} X_i \\
\downarrow & & \downarrow & & \downarrow & & \downarrow \\
\coprod_{i\in\mathcal{I}} X_i & \longrightarrow & \coprod_{i\in\mathcal{I}} Y_i & \longrightarrow & W & \longrightarrow & \Sigma \coprod_{i\in\mathcal{I}} X_i,
\end{array}$$

where the bottom row is an exact triangle and all the vertical arrows are isomorphisms, so (T1) tells us that the top row is an exact triangle too, which is what we wanted to show. □

Remark 4.1.10 The axiom (T3) does not specify that the fill-in is unique. In general, there will be several suitable maps. For example, consider the diagram

$$\begin{array}{ccccccc}
\Sigma^{-1}X & \xrightarrow{0} & Y & \longrightarrow & X \coprod Y & \longrightarrow & X \\
\| & & \| & & & & \| \\
\Sigma^{-1}X & \xrightarrow{0} & Y & \longrightarrow & X \coprod Y & \longrightarrow & X,
\end{array}$$

where the rows are the coproduct of (shifts of) the triangles

$$* \longrightarrow X = X \longrightarrow *$$

and

$$* \longrightarrow Y = Y \longrightarrow *.$$

Let $f\colon X \longrightarrow Y$ be a map in $\mathcal{T}$. Then the map

$$\bar{f}\colon X \coprod Y \longrightarrow X \coprod Y$$

induced by

$$(\mathrm{Id}, f)\colon X \longrightarrow X \coprod Y \quad \text{and} \quad (0, \mathrm{Id})\colon Y \longrightarrow X \coprod Y$$

is a possible fill-in for the diagram. As this is true for any $f\colon X \longrightarrow Y$, the fill-in thus need not be unique.

Definition 4.1.11 An exact triangle

$$X \xrightarrow{f_1} Y \xrightarrow{f_2} Z \xrightarrow{f_3} \Sigma X$$

in a triangulated category $\mathcal{T}$ is *split* if one of the maps f_1, f_2, or f_3 is zero.

Lemma 4.1.12 *If*

$$X \xrightarrow{f_1} Y \xrightarrow{f_2} Z \xrightarrow{f_3} \Sigma X$$

is a split exact triangle in a triangulated category $\mathcal{T}$, then $Y \cong X \coprod Z$ in $\mathcal{T}$.

Proof We prove the result in the case $f_3 = 0$. By the axiom (T1),

$$X = X \longrightarrow * \longrightarrow \Sigma X$$

and

$$* \longrightarrow Z = Z \longrightarrow *$$

are exact triangles. Therefore, by Lemma 4.1.9, their coproduct

$$X \longrightarrow X \coprod Z \longrightarrow Z \xrightarrow{0} \Sigma X$$

is an exact triangle. We also have a commutative diagram

$$\begin{array}{ccccccc} X & \longrightarrow & Y & \longrightarrow & Z & \xrightarrow{0} & \Sigma X \\ \| & & & & \| & & \| \\ X & \longrightarrow & X \coprod Z & \longrightarrow & Z & \xrightarrow{0} & \Sigma X, \end{array}$$

which by (T3) can be completed to a morphism of exact triangles

$$\begin{array}{ccccccc} X & \longrightarrow & Y & \longrightarrow & Z & \xrightarrow{0} & \Sigma X \\ \| & & \downarrow & & \| & & \| \\ X & \longrightarrow & X \coprod Z & \longrightarrow & Z & \xrightarrow{0} & \Sigma X. \end{array}$$

By Corollary 4.1.8, the resulting map $Y \longrightarrow X \coprod Z$ is an isomorphism. □

The following result may be thought of as saying that the cofibres of a map of exact triangles form an exact triangle. The proof is quite lengthy, so we leave it to the references [BBD82, Proposition 1.1.11] or [May01, Lemma 2.6].

Lemma 4.1.13 (3 × 3 Lemma) *A commutative square*

$$\begin{array}{ccc} X_1 & \xrightarrow{f_1} & Y_1 \\ {\scriptstyle a_1}\downarrow & & \downarrow{\scriptstyle b_1} \\ X_2 & \xrightarrow{f_2} & Y_2 \end{array}$$

can be extended to the diagram below, with all rows and columns triangles.

$$\begin{array}{ccccccc}
X_1 & \xrightarrow{f_1} & Y_1 & \xrightarrow{g_1} & Z_1 & \xrightarrow{h_1} & \Sigma X_1 \\
\downarrow{\scriptstyle a_1} & & \downarrow{\scriptstyle b_1} & & \downarrow{\scriptstyle c_1} & & \downarrow{\scriptstyle \Sigma a_1} \\
X_2 & \xrightarrow{f_2} & Y_2 & \xrightarrow{g_2} & Z_2 & \xrightarrow{h_2} & \Sigma X_2 \\
\downarrow{\scriptstyle a_2} & & \downarrow{\scriptstyle b_2} & & \downarrow{\scriptstyle c_2} & & \downarrow{\scriptstyle \Sigma a_2} \\
X_3 & \xrightarrow{f_3} & Y_3 & \xrightarrow{g_3} & Z_3 & \xrightarrow{h_3} & \Sigma X_3 \\
\downarrow{\scriptstyle a_3} & & \downarrow{\scriptstyle b_3} & & \downarrow{\scriptstyle c_3} & \ominus & \downarrow{\scriptstyle \Sigma a_3} \\
\Sigma X_1 & \xrightarrow{\Sigma f_1} & \Sigma Y_1 & \xrightarrow{\Sigma g_1} & \Sigma Z_1 & \xrightarrow{\Sigma h_1} & \Sigma^2 X_1 .
\end{array}$$

The squares of this diagram commute, except the bottom-right square, which commutes up to sign

$$\Sigma a_3 \circ h_3 = -\Sigma h_1 \circ c_3.$$

4.2 The Homotopy Category of a Stable Model Category

The main goal of this chapter is the following.

Theorem 4.2.1 *Let $\mathcal{C}$ be a stable model category. Then its homotopy category* $\mathrm{Ho}(\mathcal{C})$ *is a triangulated category where the exact triangles are given by cofibre sequences with their boundary maps*

$$X \longrightarrow Y \longrightarrow Z \xrightarrow{\partial} \Sigma X.$$

We are going to present the proof of the theorem in a separate lemma for each axiom. (T1) follows immediately from the definition of a cofibre sequence.

We are going to start with (T3) because, although the axioms of a triangulated category are traditionally listed in the given order, the proof of (T2) actually needs (T3).

Lemma 4.2.2 (Proof of (T3)) *Let $\mathcal{C}$ be a stable model category. Assume that we have a commutative diagram in* $\mathrm{Ho}(\mathcal{C})$

$$\begin{array}{ccccccc}
X' & \xrightarrow{f'} & Y' & \xrightarrow{g'} & Z' & \xrightarrow{u'} & \Sigma X' \\
\downarrow{\scriptstyle \phi_1} & & \downarrow{\scriptstyle \phi_2} & & & & \downarrow{\scriptstyle \Sigma \phi_1} \\
X & \xrightarrow{f} & Y & \xrightarrow{g} & Z & \xrightarrow{u} & \Sigma X,
\end{array}$$

where the two rows are cofibre sequences with their respective boundary maps. Then there is a map $\phi_3 \colon Z' \longrightarrow Z$ making the resulting second and third square commute.

Proof This is a direct calculation using the definition of the coaction. Without loss of generality, let f and f' be cofibrations between cofibrant objects, Z' the cofibre of f' and Z the cofibre of f. That is, Z and Z' are given by the pushouts of the diagrams

$$* \longleftarrow X' \xrightarrow{f'} Y' \quad \text{and} \quad * \longleftarrow X \xrightarrow{f} Y.$$

There is a map ϕ_3 making the following diagram commute

$$\begin{array}{ccccc} X' & \xrightarrow{f'} & Y' & \xrightarrow{g'} & Z' \\ \downarrow{\scriptstyle \phi_1} & & \downarrow{\scriptstyle \phi_2} & & \vdots{\scriptstyle \phi_3} \\ X & \xrightarrow{f} & Y & \xrightarrow{g} & Z \end{array}$$

because

$$g \circ \phi_2 \circ f' = g \circ f \circ \phi_1 = 0,$$

so $g \circ \phi_2$ factors over the cofibre of f'. To show that $\Sigma\phi_1 \circ u' = \phi_3 \circ u$, we are going to show that ϕ_3 is compatible with the coaction of the two cofibre sequences because the boundary maps u and u' are defined using those. This means that we have to show that

$$\begin{array}{ccc} Z' & \xrightarrow{\odot} & Z' \coprod \Sigma X' \\ \downarrow{\scriptstyle \phi_3} & & \downarrow{\scriptstyle \phi_3 \coprod \Sigma\phi_1} \\ Z & \xrightarrow{\odot} & Z \coprod \Sigma X \end{array}$$

commutes. That is, for $w\colon Z \longrightarrow A$ and for $h\colon X \longrightarrow PA$ representing an element of $[\Sigma X, A]$, we show that

$$[w \circ \phi_3] \odot [h \circ \phi_1] = ([w] \odot [h]) \circ \phi_3.$$

Recall that $[w] \odot [h]$ is constructed as follows: we have a commutative diagram

$$\begin{array}{ccc} X & \xrightarrow{h} & PA \\ \downarrow{\scriptstyle f} & & \downarrow{\scriptstyle \sim\, e_0} \\ Y & \xrightarrow{w \circ g} & A \end{array}$$

with a lift $\alpha\colon Y \longrightarrow PA$. The element $[w] \odot [h]$ is represented by the map v such that $v \circ g = e_1 \circ \alpha$ as below.

$$\begin{array}{ccccc} X & \xrightarrow{f} & Y & \xrightarrow{g} & Z \\ & & \downarrow{\scriptstyle e_1 \circ \alpha} & \swarrow{\scriptstyle v} & \\ & & A & & \end{array}$$

Similarly, for $[v'] = [w \circ \phi_3] \odot [h \circ \phi_1]$, we start with a commutative diagram

$$\begin{array}{ccc} X' & \xrightarrow{h\circ\phi_1} & PA \\ {\scriptstyle f'}\big\downarrow & & \big\downarrow{\scriptstyle e_0}\ \sim \\ Y' & \xrightarrow{w\circ\phi_3\circ g'} & A \end{array}$$

in which there is a lift $\beta\colon Y' \longrightarrow PA$, and we get

$$\begin{array}{ccccc} X' & \xrightarrow{f'} & Y' & \xrightarrow{g'} & Z. \\ & & {\scriptstyle e_1\circ\beta}\big\downarrow & \swarrow{\scriptstyle v'} & \\ & & A & & \end{array}$$

But we can take $\beta = \alpha \circ \phi_2$ because

$$\alpha \circ \phi_2 \circ f' = \alpha \circ f \circ \phi_1 = h \circ \phi_1$$

and

$$e_0 \circ \alpha \circ \phi_2 = w \circ g \circ \phi_2 = w \circ \phi_3 \circ g'.$$

This means that $v' = v \circ \phi_3$ because

$$v' \circ g' = v \circ \phi_3 \circ g' = v \circ g \circ \phi_2 = e_1 \circ \alpha \circ \phi_2 = e_1 \circ \beta$$

as required.

So we have shown that the coactions of the two cofibre sequences are compatible. By definition of the boundary maps, $\phi_3 \circ u' = u \circ \phi_3$ is equivalent to having a commutative diagram

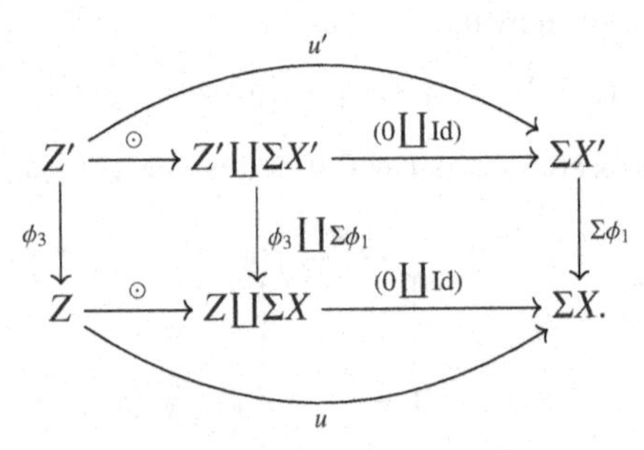

It is evident that the second square commutes, and we have just shown that the first square commutes, which completes the proof. □

Remark 4.2.3 The proof of this lemma also shows the following: let $\mathcal{C}$ be a pointed model category and let

$$X \xrightarrow{f} Y \xrightarrow{g} Z$$

be a cofibre sequence with boundary map $\partial\colon Z \longrightarrow \Sigma X$. If there is a commutative diagram in $\mathrm{Ho}(\mathcal{C})$

$$\begin{array}{ccccccc} X & \xrightarrow{f} & Y & \xrightarrow{g} & Z & \xrightarrow{\partial} & \Sigma X \\ \downarrow & & \downarrow & & \downarrow & & \downarrow \\ A & \xrightarrow{u} & B & \xrightarrow{v} & C & \xrightarrow{w} & \Sigma A, \end{array}$$

where the vertical arrows are isomorphisms, then

$$A \xrightarrow{u} B \xrightarrow{v} C$$

is a cofibre sequence with boundary map $w\colon C \longrightarrow \Sigma A$. We will make use of this in the proof of (T2).

Remark 4.2.4 We would also like to point out that in the case of $\mathrm{Ho}(\mathcal{C})$ for a stable model category $\mathcal{C}$, there is a canonical choice of fill-in map in (T3) coming from the induced map of cofibres. However, as in the general case of Remark 4.1.10, there can still be other possible fill-in maps.

The statement and proof of (T3) only used properties of cofibre sequences in the homotopy category of a pointed model category and did not require the model category to be stable. However, stability comes into play now. Let

$$\epsilon_X\colon \Sigma\Omega X \longrightarrow X \quad \text{and} \quad \eta_X\colon X \longrightarrow \Omega\Sigma X$$

denote the counit and unit of the loop-suspension adjunction on $\mathrm{Ho}(\mathcal{C})$.

By Lemma 3.6.2, we can shift exact triangles of cofibre sequence to the right, giving one half of (T2). The proof of (T2) shows that in a stable model category, one can shift cofibre sequences to the left and not just to the right.

Lemma 4.2.5 (Proof of (T2)) *Let $\mathcal{C}$ be a stable model category. Then*

$$X \xrightarrow{f} Y \xrightarrow{g} Z$$

is a cofibre sequence in $\mathrm{Ho}(\mathcal{C})$ *with boundary map $u\colon Z \longrightarrow \Sigma X$ if and only if*

$$\Omega Z \xrightarrow{-\eta_X^{-1}\circ\Omega u} X \xrightarrow{f} Y$$

is a cofibre sequence with boundary map $\epsilon_Z^{-1}\circ g\colon Y \longrightarrow \Sigma(\Omega Z)$.

Proof Assume that

$$\Omega X \xrightarrow{-\eta_X^{-1}\circ\Omega u} X \xrightarrow{f} Y$$

is a cofibre sequence with boundary map $\epsilon_Z^{-1} \circ g$. Then by Lemma 3.6.2,

$$X \xrightarrow{f} Y \xrightarrow{\epsilon_Z^{-1} \circ g} Z$$

is a cofibre sequence with boundary map

$$-\Sigma(-\eta_X^{-1} \circ \Omega u) = \Sigma \eta_X^{-1} \circ \Sigma\Omega u.$$

The unit of the loop-suspension adjunction ϵ_X is natural in X, so we have

$$u \circ \epsilon_Z = \epsilon_{\Sigma X} \circ \Sigma\Omega u.$$

But $\epsilon_{\Sigma X} = \Sigma(\eta_X^{-1}) = (\Sigma\eta_X)^{-1}$, which means that we have the following commutative diagram

$$\begin{array}{ccccccc}
X & \xrightarrow{f} & Y & \xrightarrow{\epsilon_Z^{-1} \circ g} & \Sigma\Omega Z & \xrightarrow{\Sigma\eta_X^{-1} \circ \Sigma\Omega u} & \Sigma X \\
\| & & \| & & \downarrow{\scriptstyle \epsilon_Z} & & \| \\
X & \xrightarrow{f} & Y & \xrightarrow{g} & Z & \xrightarrow{u} & \Sigma X.
\end{array}$$

As the top row is a cofibre sequence with its boundary map, then by definition, so is the bottom because all vertical arrows are isomorphisms. (The boundary map holds the information about the coaction of ΣX on the cofibre, thus, having a commutative diagram as above also means that the coactions of the two cofibre sequences are isomorphic, see Remark 4.2.3.)

For the converse, assume that

$$X \xrightarrow{f} Y \xrightarrow{g} Z$$

is a cofibre sequence in $\mathrm{Ho}(\mathcal{C})$ with boundary map u. As in the previous step, naturality of the counit gives us the commutative diagram

$$\begin{array}{ccccccc}
\Sigma\Omega X & \xrightarrow{\Sigma\Omega f} & \Sigma\Omega Y & \xrightarrow{\Sigma\Omega g} & \Sigma\Omega Z & \xrightarrow{\Sigma\epsilon_X^{-1} \circ \Sigma\eta_X^{-1} \circ \Sigma\Omega u} & \Sigma^2\Omega X \\
\downarrow{\scriptstyle \epsilon_X} & & \downarrow{\scriptstyle \epsilon_Y} & & \downarrow{\scriptstyle \epsilon_Z} & & \downarrow{\scriptstyle \Sigma\epsilon_Z} \\
X & \xrightarrow{f} & Y & \xrightarrow{g} & Z & \xrightarrow{u} & \Sigma X.
\end{array}$$

As the bottom row is a cofibre sequence with its boundary map and the vertical maps are isomorphisms in $\mathrm{Ho}(\mathcal{C})$, this means that the top row is a cofibre sequence with its boundary map too.

Our claim is now that "desuspending" the top row still gives us a cofibre sequence

$$\Omega X \xrightarrow{\Omega f} \Omega Y \xrightarrow{\Omega g} \Omega Z \xrightarrow{\epsilon_X^{-1} \circ \eta_X^{-1} \circ \Omega u} \Sigma\Omega X.$$

Of course, Ωf is part of some cofibre sequence $\Omega X \xrightarrow{\Omega f} \Omega Y \xrightarrow{g'} W \xrightarrow{u'} \Sigma\Omega X$, to which we apply Σ and put it in the following commutative diagram.

$$\begin{array}{ccccccc}
\Sigma\Omega X & \xrightarrow{\Sigma\Omega f} & \Sigma\Omega Y & \xrightarrow{\Sigma\Omega g} & \Sigma\Omega Z & \xrightarrow{\Sigma\epsilon_X^{-1}\circ\Sigma\eta_X^{-1}\circ\Sigma\Omega u} & \Sigma^2\Omega X \\
\| & & \| & & & & \| \\
\Sigma\Omega X & \xrightarrow{\Sigma\Omega f} & \Sigma\Omega Y & \xrightarrow{\Sigma g'} & \Sigma W & \xrightarrow{\Sigma u'} & \Sigma^2\Omega X
\end{array}$$

We know that both rows are cofibre sequences, so Lemma 4.2.2 gives us a fill-in map $k\colon \Sigma\Omega Z \longrightarrow \Sigma W$. The long exact sequence of cofibre sequences (Theorem 3.6.4) together with the Five Lemma tell us that this map k is an isomorphism in $\mathrm{Ho}(\mathcal{C})$. Furthermore, we have that $k = \Sigma k'$ for some $k' \in \mathrm{Ho}(\mathcal{C})$ that is also an isomorphism (specifically, $k' = (\eta_W^{-1}) \circ \Omega k \circ \eta_{\Omega Z}$). This means that we have a commutative diagram

$$\begin{array}{ccccccc}
\Omega X & \xrightarrow{\Omega f} & \Omega Y & \xrightarrow{\Omega g} & \Omega Z & \xrightarrow{\epsilon_X^{-1}\circ\eta_X^{-1}\circ\Omega u} & \Sigma\Omega X \\
\| & & \| & & \downarrow{\scriptstyle k'} & & \| \\
\Omega X & \xrightarrow{\Omega f} & \Omega Y & \xrightarrow{g'} & W & \xrightarrow{u'} & \Sigma\Omega X,
\end{array}$$

where all vertical maps are isomorphisms and the bottom row is a cofibre sequence by assumption. Thus, the top row must be a cofibre sequence too. Lemma 3.6.2 and Corollary 3.6.3 allow us to shift the top row cofibre sequence two places to the right and obtain a cofibre sequence which is the top row of the next commutative diagram.

$$\begin{array}{ccccccc}
\Omega Z & \xrightarrow{\epsilon_X^{-1}\circ\eta_X^{-1}\circ\Omega u} & \Sigma\Omega X & \xrightarrow{-\Sigma\Omega f} & \Sigma\Omega Y & \xrightarrow{-\Sigma\Omega g} & \Sigma\Omega Z \\
\| & & \downarrow{\scriptstyle \epsilon_X} & & \downarrow{\scriptstyle \epsilon_Y} & & \| \\
\Omega Z & \xrightarrow{\eta_X^{-1}\circ\Omega u} & X & \xrightarrow{-f} & Y & \xrightarrow{-\epsilon_Z^{-1}\circ g} & \Sigma\Omega Z
\end{array}$$

All vertical arrows are isomorphisms, which means that the bottom row is a cofibre sequence and has the desired boundary map, which was the claim of the lemma. □

The proof of the octahedral axiom (T4) in fact only requires our model category to be pointed rather than stable.

Lemma 4.2.6 (Proof of (T4)) *Let $\mathcal{C}$ be a pointed model category and suppose we have cofibre sequences*

$$X \xrightarrow{f_1} Y \xrightarrow{f_2} Z$$

$$X \xrightarrow{g_1} U \xrightarrow{g_2} V$$

$$Y \xrightarrow{u_1} U \xrightarrow{u_2} W$$

in $\mathrm{Ho}(\mathcal{C})$ *with $g_1 = u_1 \circ f_1$. Then there are maps $v_1 \colon Z \longrightarrow V$, $v_2 \colon V \longrightarrow W$ and $v_3 \colon W \longrightarrow \Sigma Z$ making the following diagram commute*

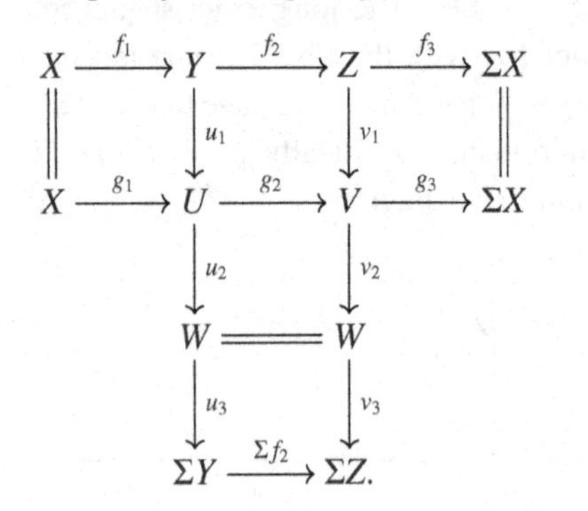

Furthermore,

- *$Z \xrightarrow{v_1} V \xrightarrow{v_2} W$ is a cofibre sequence with boundary map v_3,*
- *the coaction of ΣZ on W is given by*

$$W \xrightarrow{\odot} W \coprod \Sigma Y \xrightarrow{\mathrm{Id} \coprod \Sigma f_2} W \coprod \Sigma Z,$$

where the first map is the coaction of ΣY on W from the third cofibre sequence.

Proof Without loss of generality, let X, Y and U be fibrant and cofibrant in $\mathcal{C}$ and let f_1 and u_1 (and thus g_1) be cofibrations. We have a commutative diagram

$$\begin{array}{ccccc}
Y & \xleftarrow{f_1} & X & \longrightarrow & * \\
\downarrow{\scriptstyle u_1} & & \| & & \| \\
U & \xleftarrow{g_1} & X & \longrightarrow & * \\
\| & & \downarrow{\scriptstyle f_1} & & \| \\
U & \xleftarrow{u_1} & Y & \longrightarrow & *.
\end{array}$$

The pushout of the first row gives a map $f_2\colon Y \longrightarrow Z$, the pushout of the second row gives a map $g_2\colon U \longrightarrow V$ and the pushout of the third row gives a map $u_2\colon U \longrightarrow W$.

The commutative diagram also gives us induced maps of the pushouts of the three rows,

$$Z \xrightarrow{v_1} V \xrightarrow{v_2} W.$$

We see that $v_2 \circ v_1 = 0$, as it is induced by the maps $0 = u_2 \circ u_1 : Y \longrightarrow W$ and $0 = u_2 \circ u_1 \circ f_1 : X \longrightarrow W$.

By Lemma A.7.7 (the Patching Lemma), the map v_1 is a cofibration. The cofibre of v_1 is then given by $v_2\colon V \longrightarrow W$, as the cofibre of a map of pushouts is the pushout of a map of cofibres, see [Str11, Theorem 2.43].

We now claim that the coaction associated to the cofibre sequence

$$Z \xrightarrow{v_1} V \xrightarrow{v_2} W$$

is given by

$$W \xrightarrow{\odot} W \coprod \Sigma Y \xrightarrow{\mathrm{Id} \coprod \Sigma f_2} W \coprod \Sigma Z,$$

where $\odot$ is the coaction of ΣY on W from the third cofibre sequence. We denote this coaction by $\bullet$ to avoid confusion with the other coaction $\odot$. If our claim holds, the boundary map v_3 is given by

$$v_3 = (0 \coprod \mathrm{Id}) \circ \bullet = (0 \coprod \mathrm{Id}) \circ (\mathrm{Id} \coprod \Sigma f_2) \circ \odot \quad \text{while} \quad u_3 = (0 \coprod \mathrm{Id}) \circ \odot,$$

so $\Sigma f_2 \circ u_3 = v_3$. So we now have to check that this coaction $\bullet$ is the correct one. In other words, given $A \in \mathcal{C}$, we have to show that

$$[f] \bullet [h] = [f] \odot [h \circ f_2],$$

where $f\colon W \longrightarrow A$ and where $h\colon Z \longrightarrow PA$ represents an element of $[\Sigma Z, A]$. To obtain $[f] \bullet [h]$, we start with a commutative diagram

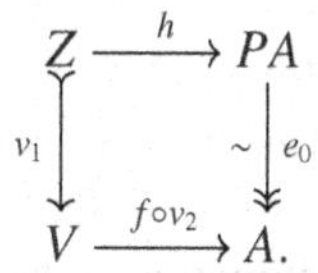

$$\begin{array}{ccc} Z & \xrightarrow{h} & PA \\ {\scriptstyle v_1}\downarrow & & \downarrow{\scriptstyle \sim}\, {\scriptstyle e_0} \\ V & \xrightarrow{f \circ v_2} & A. \end{array}$$

In this diagram, there is a lift $\alpha\colon V \longrightarrow PA$, and the map $\alpha \circ e_1$ factors over the cofibre of v_1

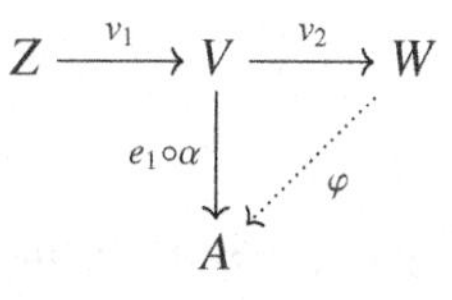

$$\begin{array}{ccccc} Z & \xrightarrow{v_1} & V & \xrightarrow{v_2} & W \\ & & {\scriptstyle e_1 \circ \alpha}\downarrow & \swarrow{\scriptstyle \varphi} & \\ & & A & & \end{array}$$

to give us $[\varphi] = [f] \bullet [h]$. In the same way, we get $[f] \odot [h \circ f_2]$ by starting with a diagram

$$\begin{array}{ccc} Y & \xrightarrow{h \circ f_2} & PA \\ {\scriptstyle u_1} \downarrow & & \downarrow {\scriptstyle \sim}\, {\scriptstyle e_0} \\ U & \xrightarrow{f \circ u_2} & A \end{array}$$

in which there is a lift $\beta \colon U \longrightarrow PA$. Then $e_1 \circ \beta$ factors over the cofibre of u_1

$$\begin{array}{ccccc} Y & \xrightarrow{u_1} & U & \xrightarrow{u_2} & W, \\ & & {\scriptstyle e_1 \circ \beta} \downarrow & \swarrow {\scriptstyle \psi} & \\ & & A & & \end{array}$$

which gives us $[\psi] = [f] \odot [h \circ f_2]$. But we can simply pick $\beta = \alpha \circ g_2$, because

$$\alpha \circ g_2 \circ u_1 = \alpha \circ v_1 \circ f_2 = h \circ f_2 \quad \text{and} \quad e_0 \circ \alpha \circ g_2 = f \circ v_2 \circ g_2 = f \circ u_2.$$

Then

$$\psi \circ u_2 = e_1 \circ \beta = e_1 \circ \alpha \circ g_2 = \varphi \circ v_2 \circ g_2 = \varphi \circ u_2,$$

so $\psi = \varphi$, which is what we wanted to prove. □

With the last lemma, we finally completed the proof that the cofibre sequences equip the homotopy category $\mathrm{Ho}(\mathcal{C})$ of a stable model category $\mathcal{C}$ with the structure of a triangulated category.

4.3 Comparison of Fibre and Cofibre Sequences

The fact that the homotopy category of a stable model category is triangulated means that we can use Proposition 4.1.5 to obtain a long exact sequence from a *cofibre sequence* using the functor $[A, -]$ for any $A \in \mathcal{C}$. This contrasts with Theorem 3.6.4, which converts cofibre sequences into long exact sequences via $[-, A]$. This is going to be important when comparing fibre and cofibre sequences in a stable model category. Our starting point is a lemma that holds for all pointed model categories.

Recall that

$$\epsilon_X \colon \Sigma\Omega X \longrightarrow X \quad \text{and} \quad \eta_X \colon X \longrightarrow \Omega\Sigma X$$

denote the counit and unit of the loop-suspension adjunction on $\mathrm{Ho}(\mathcal{C})$.

Lemma 4.3.1 *Let $\mathcal{C}$ be a pointed model category, and let*

$$\begin{array}{ccccccc} A & \xrightarrow{u} & B & \xrightarrow{v} & C & \xrightarrow{\partial'} & \Sigma A \\ \downarrow{\scriptstyle \alpha} & & \downarrow{\scriptstyle \beta} & & & & \downarrow{\scriptstyle -\epsilon_Z \circ \Sigma\alpha} \\ \Omega Z & \xrightarrow{\partial} & X & \xrightarrow{f} & Y & \xrightarrow{g} & Z \end{array}$$

be a commutative diagram in Ho($\mathcal{C}$), *where the top row is a cofibre sequence and its boundary map and the bottom row is a fibre sequence with its boundary map. Then there is a fill-in map $\gamma\colon C \longrightarrow Y$ making the resulting second and third square commute.*

As expected, there is a dual statement. We will only prove the second version.

Lemma 4.3.2 *Let $\mathcal{C}$ be a pointed model category, and let*

$$\begin{array}{ccccccc} A & \xrightarrow{u} & B & \xrightarrow{v} & C & \xrightarrow{\partial'} & \Sigma A \\ \downarrow{\scriptstyle -\Omega\gamma \circ \eta_A} & & & & \downarrow{\scriptstyle \beta} & & \downarrow{\scriptstyle \gamma} \\ \Omega Z & \xrightarrow{\partial} & X & \xrightarrow{f} & Y & \xrightarrow{g} & Z \end{array}$$

be a commutative diagram in Ho($\mathcal{C}$), *where the top row is a cofibre sequence and its boundary map and the bottom row is a fibre sequence with its boundary map. Then there is a fill-in map $\alpha\colon B \longrightarrow X$ making the resulting diagram commute.*

Proof Without loss of generality, let us assume that A, B and C are cofibrant and that u is a cofibration. Similarly, assume that X, Y and Z are fibrant and g is a fibration. Furthermore, we can assume C to be the mapping cone of u defined as the pushout of

$$B \xleftarrow{(u,0)} A \coprod A \xrightarrow{(i_0,i_1)} \mathrm{Cyl}(A)$$

because the cofibre sequence using the mapping cone is isomorphic to our original one.

In Ho($\mathcal{C}$) we have

$$g \circ \beta \circ v = \gamma \circ \partial' \circ v = 0,$$

so $\beta \circ v$ factors over the fibre of g, giving us a map $\alpha\colon B \longrightarrow X$ with $f \circ \alpha = \beta \circ v$. Recalling that C is the mapping cone of u, the map $\beta\colon C \longrightarrow Y$

is now the pushout of $f \circ \alpha$ and some other map $H\colon \mathrm{Cyl}(A) \longrightarrow Y$, resulting in the following commutative diagram

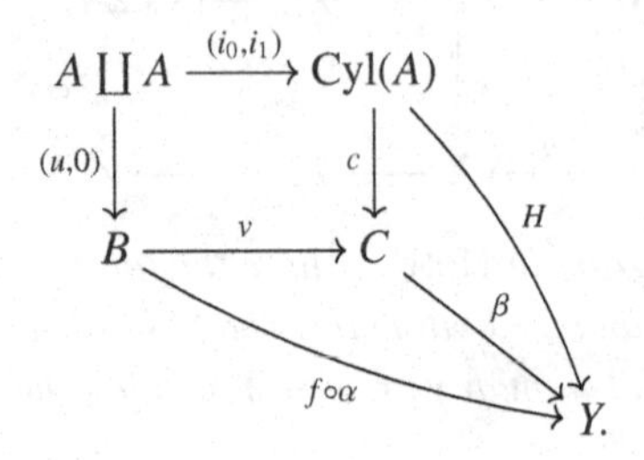

Now we have to show that with this α, the first square in our original diagram commutes. For convenience, let $\tilde{\gamma} = \Omega\gamma \circ \eta_A$. Thus, our claim is that

$$\alpha \circ u = \partial \circ (-\tilde{\gamma}).$$

We know that elements of $[\Sigma A, Z] \cong [A, \Omega Z]$ are represented by homotopies between the zero map $0\colon A \longrightarrow Z$ and itself. Furthermore, we can choose to have $[\gamma] \in [\Sigma A, Z]$ and $[\tilde{\gamma}] \in [A, \Omega Z]$ represented by the same homotopy $H'\colon \mathrm{Cyl}(A) \longrightarrow Z$ because $\tilde{\gamma}$ is the adjoint of γ.

With the definition of C as a pushout, $g \circ \beta = \gamma \circ \partial'\colon C \longrightarrow Z$ is given by the zero map on B and H' on $\mathrm{Cyl}(A)$. This means that we have the following commutative diagram

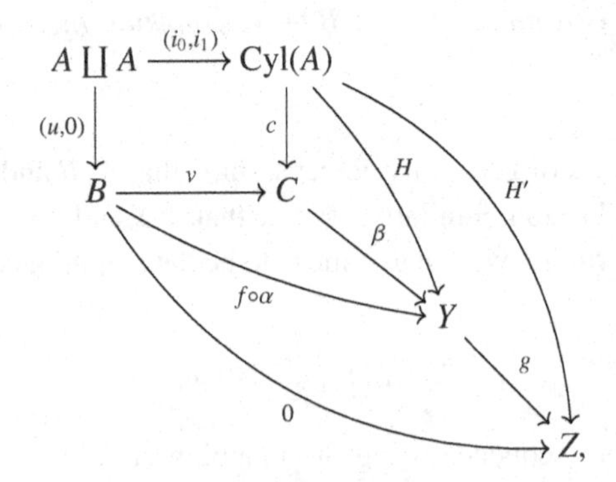

that is, $g \circ H = H'$.

Let us return to showing that $\alpha \circ u = \partial \circ (-\tilde{\gamma})$. By definition, $[\partial \circ (-\tilde{\gamma})]$ equals $0 \odot [-\tilde{\gamma}]$, thus, our claim amounts to showing that $[\alpha \circ u] = 0 \odot [-\tilde{\gamma}]$, or, equivalently,

$$[\alpha \circ u] \odot [H'] = 0.$$

To calculate this action, we look at the commutative square

$$\begin{array}{ccc} A & \xrightarrow{f\circ\alpha\circ u} & Y \\ {\scriptstyle i_0}\downarrow{\scriptstyle\sim} & & \downarrow{\scriptstyle g} \\ \mathrm{Cyl}(A) & \xrightarrow{H'} & Z \end{array}$$

in which there is a lift $K\colon \mathrm{Cyl}(A) \longrightarrow Y$. We then have $[k] = [\alpha \circ u] \odot [H']$, where $k\colon A \longrightarrow X$ is the unique map with $f \circ k = K \circ i_1$.

But we can choose $K = H$, where $H\colon \mathrm{Cyl}(A) \longrightarrow X$ was defined in the pushout diagram earlier, because

$$g \circ H = H' \quad \text{and} \quad H \circ i_0 = f \circ \alpha \circ u.$$

But also $H \circ i_1 = 0$, so $k = 0$, which is exactly what we wanted to prove. □

4.4 Consequences of Stability

We start with a result about the special relationship between cofibre and fibre sequences that holds when the pointed model category is also stable.

Corollary 4.4.1 *Let $\mathcal{C}$ be a stable model category. If*

$$X \xrightarrow{f} Y \xrightarrow{g} Z$$

is a fibre sequence in $\mathrm{Ho}(\mathcal{C})$ *with boundary map $\partial\colon \Omega Z \longrightarrow X$, then*

$$\Omega Z \xrightarrow{\partial} X \xrightarrow{f} Y$$

is a cofibre sequence with boundary map $-\epsilon_Z^{-1} \circ g\colon Y \longrightarrow \Sigma\Omega Z$.

Corollary 4.4.2 *Let $\mathcal{C}$ be a stable model category. If*

$$A \xrightarrow{u} B \xrightarrow{v} C$$

is a cofibre sequence in $\mathrm{Ho}(\mathcal{C})$ *with boundary map $w\colon C \longrightarrow \Sigma A$, then*

$$B \xrightarrow{v} C \xrightarrow{w} \Sigma A$$

is a fibre sequence with boundary map $u \circ (-\eta_A^{-1})\colon \Omega\Sigma A \longrightarrow B$.

We will only prove the first corollary as the second proof is very similar. People often suppress the η and ϵ (as well as $\Sigma\Omega$ and $\Omega\Sigma$) from the statement, as they are canonical isomorphisms, which leads to the slogan "in stable model categories, fibre and cofibre sequences agree".

Proof Let

$$X \xrightarrow{f} Y \xrightarrow{g} Z$$

be a fibre sequence with boundary map $\partial\colon \Omega Z \longrightarrow X$. Then ∂ is part of some cofibre sequence

$$\Omega Z \xrightarrow{\partial} X \xrightarrow{f'} Y' \xrightarrow{g'} \Sigma\Omega Z,$$

where g' is the boundary map. We can complete this to a commutative diagram

$$\begin{array}{ccccccc} \Omega Z & \xrightarrow{\partial} & X & \xrightarrow{f'} & Y' & \xrightarrow{g'} & \Sigma\Omega Z \\ \| & & \| & & & & \downarrow{\scriptstyle -\epsilon_Z} \\ \Omega Z & \xrightarrow{\partial} & X & \xrightarrow{f} & Y & \xrightarrow{g} & Z. \end{array}$$

By Lemma 4.3.1, there is a fill-in map $Y' \longrightarrow Y$ making the whole diagram commute. Because the bottom row is a fibre sequence, Theorem 3.6.1 tells us that applying $[A, -]$ to the bottom row results in a long exact sequence. Because $\mathrm{Ho}(\mathcal{C})$ is triangulated, applying $[A, -]$ to the top row also yields a long exact sequence by Proposition 4.1.5. Because our model category is stable, $-\epsilon_Z$ is a weak equivalence, so the Five Lemma implies that the fill-in map $Y' \longrightarrow Y$ is an isomorphism in $\mathrm{Ho}(\mathcal{C})$. Thus,

$$\begin{array}{ccccccc} \Omega Z & \xrightarrow{\partial} & X & \xrightarrow{f'} & Y' & \xrightarrow{g'} & \Sigma\Omega Z \\ \| & & \| & & \downarrow{\scriptstyle \sim} & & \| \\ \Omega Z & \xrightarrow{\partial} & X & \xrightarrow{f} & Y & \xrightarrow{-\epsilon_Z^{-1} \circ g} & \Sigma\Omega Z \end{array}$$

is a commutative diagram where the top row is a cofibre sequence and its boundary map and all the vertical maps are weak equivalences. Therefore, the bottom row must also be a cofibre sequence with its boundary map. □

We may now show that the homotopy cofibre of a map and the suspension of the homotopy fibre of a map are weakly equivalent.

Lemma 4.4.3 *Let $f\colon X \to Y$ be a map in a stable model category $\mathcal{C}$. There is a weak equivalence $Ff \longrightarrow \Omega Cf$ between the homotopy fibre of f and loops of the homotopy cofibre of f.*

Proof By Lemma 3.5.1 we have fibre sequences

$$\Omega Y \xrightarrow{\partial} Ff \xrightarrow{i} X \xrightarrow{f} Y \quad \text{and} \quad \Omega X \xrightarrow{-\Omega f} \Omega Y \xrightarrow{\partial} Ff \xrightarrow{i} X.$$

By Corollary 4.4.1 we have cofibre sequences

$$X \xrightarrow{f} Y \xrightarrow{q} Cf \xrightarrow{\partial'} \Sigma X \quad \text{and} \quad \Omega X \xrightarrow{-\Omega f} \Omega Y \xrightarrow{\partial} Ff \xrightarrow{-\epsilon_{Ff}^{-1} \circ i} \Sigma\Omega X.$$

Applying Σ to the last cofibre sequence gives the first row in the commutative diagram below.

$$\begin{array}{ccccccc}
\Sigma\Omega X & \xrightarrow{\Sigma\Omega f} & \Sigma\Omega Y & \xrightarrow{\Sigma\partial} & \Sigma Ff & \xrightarrow{\Sigma(\epsilon_{Ff}^{-1} \circ i)} & \Sigma^2\Omega X \\
\downarrow{\scriptstyle \epsilon_X} & & \downarrow{\scriptstyle \epsilon_Y} & & & & \downarrow{\scriptstyle \Sigma\epsilon_X} \\
X & \xrightarrow[f]{} & Y & \xrightarrow[q]{} & Cf & \xrightarrow[\partial']{} & \Sigma X
\end{array}$$

By (T3) (see Lemma 4.2.2), we can fill in the diagram to obtain a map

$$\Sigma Ff \longrightarrow Cf.$$

By Corollary 4.1.8, the map $\Sigma Ff \longrightarrow Cf$ is an isomorphism in $\mathrm{Ho}(\mathcal{C})$ and, hence, so is its adjoint $Ff \longrightarrow \Omega Cf$. □

We are now going to discuss some further useful results that hold in stable model categories.

A particularly handy consequence of stability is that finite products and coproducts are weakly equivalent. Hence, maps into a coproduct (or out of a product) are well understood.

Lemma 4.4.4 *Let $\mathcal{C}$ be a stable model category. Then for fibrant and cofibrant objects X and Y, the canonical map $\chi_{X,Y} \colon X \coprod Y \to X \prod Y$ is a weak equivalence.*

Proof Given such X and Y, we have two exact triangles

$$X = X \longrightarrow * \longrightarrow \Sigma X$$

$$* \longrightarrow Y = Y \longrightarrow *.$$

Hence, we can construct both their coproduct and product and obtain a map of exact triangles as below, where the middle map is the canonical map of a coproduct into a product.

$$\begin{array}{ccccccc}
X & \longrightarrow & X \coprod Y & \longrightarrow & Y & \longrightarrow & \Sigma X \\
\| & & \downarrow{\scriptstyle \chi_{X,Y}} & & \| & & \| \\
X & \longrightarrow & X \prod Y & \longrightarrow & Y & \longrightarrow & \Sigma X
\end{array}$$

By Corollary 4.1.8, the result follows. □

In most cases (including sequential spectra and chain complexes), one does not need to assume that X and Y are fibrant and cofibrant, as the functors $(-) \coprod A$ and $(-) \prod A$ preserve all weak equivalences.

Remark 4.4.5 The above lemma gives an equivalent definition of the addition in $\mathrm{Ho}(\mathcal{C})$. Given two maps $f, g \colon X \to Y$ between fibrant and cofibrant objects, one can model $f + g \in [X, Y]^{\mathcal{C}}$ by either of the composites

$$X \xrightarrow{\Delta} X \prod X \xrightarrow{f \prod g} Y \prod Y \xrightarrow{\chi_{Y,Y}^{-1}} Y \coprod Y \xrightarrow{\text{fold}} Y$$

$$X \xrightarrow{\Delta} X \prod X \xrightarrow{\chi_{X,X}^{-1}} X \coprod X \xrightarrow{f \coprod g} Y \coprod Y \xrightarrow{\text{fold}} Y.$$

4.5 Exact Functors and Quillen Functors

When dealing with categories with certain additional structure, we also have to consider functors that respect this additional structure. In the case of triangulated categories, this means that we have to study functors that send exact triangles to exact triangles. More precisely,

Definition 4.5.1 Let $\mathcal{T}$ and $\mathcal{T}'$ be triangulated categories. A functor

$$F \colon \mathcal{T} \longrightarrow \mathcal{T}'$$

is *exact* if there is a natural isomorphism

$$\tau \colon \Sigma \circ F(X) \xrightarrow{\cong} F \circ (\Sigma X)$$

such that for every exact triangle

$$X \xrightarrow{\alpha} Y \xrightarrow{\beta} Z \xrightarrow{\gamma} \Sigma X$$

in $\mathcal{T}$, the sequence

$$F(X) \xrightarrow{F(\alpha)} F(Y) \xrightarrow{F(\beta)} F(Z) \xrightarrow{\tau \circ F(\gamma)} \Sigma F(X)$$

is an exact triangle in $\mathcal{T}'$.

The key example of an exact functor is of course the following, which we may think of as an extension to Corollary 3.2.10.

Theorem 4.5.2 *Let $\mathcal{C}$ and $\mathcal{D}$ be stable model categories, and let*

$$F \colon \mathcal{C} \rightleftarrows \mathcal{D} \colon G$$

be a Quillen adjunction. Then the derived functors $\mathbb{L}F \colon \mathrm{Ho}(\mathcal{C}) \longrightarrow \mathrm{Ho}(\mathcal{D})$ and $\mathbb{R}G \colon \mathrm{Ho}(\mathcal{D}) \longrightarrow \mathrm{Ho}(\mathcal{C})$ are exact functors.

Proof We prove that the derived functor $\mathbb{L}F$ preserves cofibre sequences. The dual proof will show that the derived functor $\mathbb{R}G$ preserves fibre sequences. By Corollary 4.4.1, cofibre sequences and fibre sequences agree in a stable model category, hence, the second statement follows from the first.

We will write F instead of $\mathbb{L}F$ for convenience. Let

$$X \xrightarrow{\alpha} Y \xrightarrow{\beta} Z$$

be a cofibre sequence. Without loss of generality, let α be a cofibration between cofibrant objects and let Z be the cofibre of α. This means that Z is the pushout of α over a point. As F is a left adjoint and a left Quillen functor, it preserves cofibrations, cofibrant objects and pushouts, so $F(Z)$ is the cofibre of $F(\alpha)$.

Now let $\bullet$ denote the coaction associated to our original cofibre sequence, that is, for $A \in \mathrm{Ho}(\mathcal{C})$,

$$\bullet\colon [Z, A] \times [\Sigma X, A] \longrightarrow [Z, A], \quad ([f], [k]) \mapsto [f] \bullet [k].$$

Also, let $\odot$ denote the coaction associated to $F(X) \xrightarrow{F(\alpha)} F(Y) \xrightarrow{F(\beta)} F(Z)$. Our main claim amounts to showing that

$$[F(f)] \odot [F(k)] = F([f] \bullet [k]) \in [F(Z), F(A)].$$

Recall that to define $[f] \bullet [k]$, we begin with a lift in the commutative square

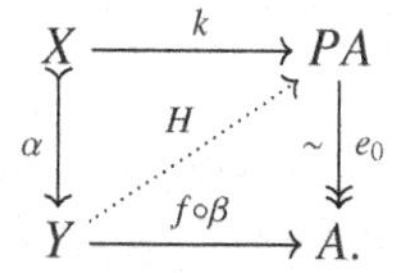

$$\begin{array}{ccc} X & \xrightarrow{k} & PA \\ {\scriptstyle\alpha}\downarrow & \overset{H}{\nearrow} & \downarrow{\scriptstyle e_0}\,\sim \\ Y & \xrightarrow{f \circ \beta} & A. \end{array} \tag{4.1}$$

Because k is a right homotopy from zero to zero, this satisfies

$$0 = e_1 \circ k = (e_1 \circ H) \circ \alpha.$$

This implies that $e_1 \circ H$ factors over the cofibre C of α, that is,

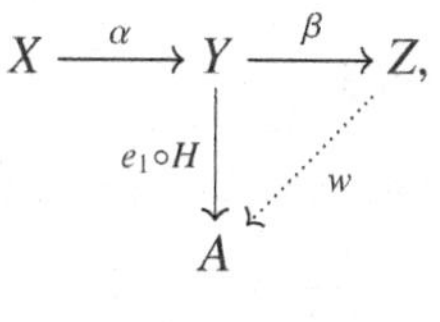

$$\begin{array}{ccccc} X & \xrightarrow{\alpha} & Y & \xrightarrow{\beta} & Z, \\ & & {\scriptstyle e_1 \circ H}\downarrow & \swarrow{\scriptstyle w} & \\ & & A & & \end{array}$$

and we define $[f] \bullet [k] = [w]$.

Applying F to the commutative square 4.1 gives a commutative diagram

$$\begin{array}{ccc} F(X) & \xrightarrow{F(k)} & F(PA) \\ {\scriptstyle F(\alpha)}\big\downarrow & \overset{F(H)}{\nearrow} & \big\downarrow{\scriptstyle F(e_0)} \\ F(Y) & \xrightarrow[F(f)\circ F(\beta)]{} & F(A). \end{array} \tag{4.2}$$

Since F is a left Quillen functor, the left-hand vertical is still a cofibration. The issue we now encounter is that $F(PA)$ is not necessarily a path object for $F(A)$. We will rectify this as follows. A path object for $F(A)$ is

$$FA \rightarrowtail^{s} PF(A) \xrightarrow{(e_0,e_1)} F(A) \times F(A).$$

Applying F to the projections $pr_1,\ pr_2 \colon A \times A \longrightarrow A$ induces a map

$$(q_1, q_2) \colon F(A \times A) \longrightarrow F(A) \times F(A).$$

Let $s' \colon A \longrightarrow PA$ be the acyclic cofibration that is part of the path object information for PA. We then get a diagram

$$\begin{array}{ccc} F(A) & \xrightarrow{s} & PF(A) \\ {\scriptstyle F(s')}\big\downarrow{\scriptstyle \simeq} & \overset{\lambda}{\nearrow} & \big\downarrow{\scriptstyle (e_0,e_1)} \\ F(PA) & \xrightarrow{(q_1,q_2)\circ(F(e_0),F(e_1))} & F(A) \times F(A) \end{array}$$

which commutes as each composite is the diagonal

$$\Delta \colon F(A) \longrightarrow F(A) \times F(A).$$

The lift exists as F sends the acyclic cofibration s' to an acyclic cofibration. It is a weak equivalence as both $F(s')$ and s are weak equivalences.

We can therefore extend diagram (4.2)

$$\begin{array}{ccccc} F(X) & \xrightarrow{F(k)} & F(PA) & \xrightarrow{\lambda} & PF(A) \\ {\scriptstyle F(\alpha)}\big\downarrow & \overset{F(H)}{\nearrow} & & & \big\downarrow{\scriptstyle e_0}\ {\scriptstyle \sim} \\ F(Y) & & \xrightarrow{F(f)\circ F(\beta)} & & F(A), \end{array}$$

so that $\lambda \circ F(H)$ is a lift. Note that

$$\lambda \circ F(k) \colon F(X) \longrightarrow PF(A)$$

is a representative of $[F(k)] \in [\Sigma F(X), F(A)]$. Hence, the diagram above is the first step to calculating $[F(f)] \odot [F(k)]$.

We now have

$$\begin{array}{ccc} F(X) \xrightarrow{F(\alpha)} & F(Y) & \xrightarrow{F(\beta)} F(Z), \\ & \Big\downarrow{\scriptstyle e_1\circ\lambda\circ F(H)=F(e_1\circ H)} & \swarrow \\ & F(A) & \end{array}$$

so we may take the dotted map, a representative for $[F(f)]\odot[F(k)]$, to be $F(w)$, where we defined $[w] = [f] \bullet [k]$. Hence, we have shown that

$$F([f] \bullet [k]) = F([w]) = [F(f)] \odot [F(k)]. \qquad \square$$

The above proof does not use the fact that F has an adjoint, only that it preserves finite colimits. Hence, we have the following slightly more general statement.

Corollary 4.5.3 *Let $F\colon \mathcal{C} \to \mathcal{D}$ be a left Quillen functor between stable model categories which commutes with finite colimits. Then $\mathbb{L}F$ is an exact functor. Dually, let $G\colon \mathcal{D} \to \mathcal{C}$ be a right Quillen functor between stable model categories which commutes with finite limits. Then $\mathbb{R}G$ is an exact functor.* $\square$

4.6 Toda Brackets

Toda brackets give additional structure to a triangulated category $\mathcal{T}$. We will introduce the definition and some basic properties, then discuss examples of how this set-up is used in the stable homotopy category.

Let

$$X \xrightarrow{\alpha} Y \xrightarrow{\beta} Z \quad \text{and} \quad Y \xrightarrow{\beta} Z \xrightarrow{\gamma} \Sigma W$$

be diagrams in a triangulated category $\mathcal{T}$. Assume that $\beta \circ \alpha = 0$ and $\gamma \circ \beta = 0$. Using (T1), α is part of some exact triangle

$$X \xrightarrow{\alpha} Y \xrightarrow{i} C\alpha \xrightarrow{j} \Sigma X.$$

Because $\beta \circ \alpha = 0$, we have a commutative diagram

$$\begin{array}{ccccccc} X & \xrightarrow{\alpha} & Y & \xrightarrow{i} & C\alpha & \xrightarrow{j} & \Sigma X \\ \downarrow & & \downarrow{\scriptstyle \beta} & & & & \downarrow \\ * & \longrightarrow & Z & = & Z & \longrightarrow & *, \end{array}$$

where the two rows are triangles by axiom (T1), so (T3) gives an arrow $\bar{\beta}\colon C\alpha \longrightarrow Z$ with $\bar{\beta} \circ i = \beta$.

We now have $0 = \gamma \circ \beta = (\gamma \circ \bar{\beta}) \circ i$, so using the same argument, we have a commutative diagram where the two rows are exact triangles

$$\begin{array}{ccccccc}
Y & \xrightarrow{i} & C\alpha & \xrightarrow{j} & \Sigma X & \xrightarrow{-\Sigma\alpha} & \Sigma Y \\
\downarrow & & \downarrow{\scriptstyle \gamma\circ\bar{\beta}} & & & & \downarrow \\
* & \longrightarrow & W & = & W & \longrightarrow & *.
\end{array}$$

By (T3), there is a $\tau \colon \Sigma X \longrightarrow W$ with $\tau \circ j = \gamma \circ \bar{\beta}$. Therefore, we arrive at the commutative diagram

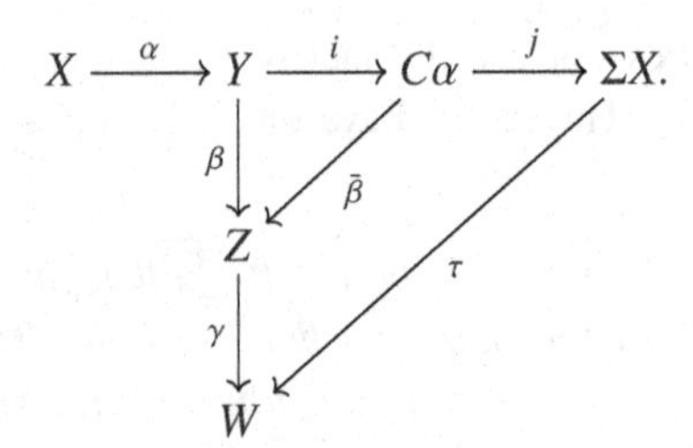

Definition 4.6.1 Let $\alpha \colon X \longrightarrow Y, \beta \colon Y \longrightarrow Z$ and $\gamma \colon Z \longrightarrow W$ be morphisms in a triangulated category $\mathcal{T}$ such that $\gamma \circ \beta = 0$ and $\beta \circ \alpha = 0$. Then the *Toda bracket* $\langle \alpha, \beta, \gamma \rangle$ is defined as the set of all possible elements $\tau \in \mathcal{T}(\Sigma X, W)$ arising in the manner described above.

It would be wonderfully simple if we could just set $\langle \alpha, \beta, \gamma \rangle$ to equal τ in the previous definition, but there are choices involved in the construction, meaning that the set in the definition can contain more than one element. We now give an identification of those choices with sets of maps in $\mathcal{T}$.

Lemma 4.6.2 *Let $\alpha \colon X \longrightarrow Y, \beta \colon Y \longrightarrow Z$ and $\gamma \colon Z \longrightarrow W$ be morphisms in a triangulated category $\mathcal{T}$ such that $\gamma \circ \beta = 0$ and $\beta \circ \alpha = 0$. Then $\langle \alpha, \beta, \gamma \rangle$ is a coset of*

$$\gamma_* \mathcal{T}(\Sigma X, Z) \oplus \Sigma\alpha^* \mathcal{T}(\Sigma Y, W)$$

in $\mathcal{T}(\Sigma X, W)$.

Proof By Proposition 4.1.5, using our previous notation, we have a diagram of exact sequences

$$\begin{array}{ccccccccc}
\mathcal{T}(\Sigma Y, Z) & \xrightarrow{\Sigma\alpha^*} & \mathcal{T}(\Sigma X, Z) & \xrightarrow{j^*} & \mathcal{T}(C\alpha, Z) & \xrightarrow{i^*} & \mathcal{T}(Y, Z) & \xrightarrow{\alpha^*} & \mathcal{T}(X, Z) \\
\downarrow{\scriptstyle \gamma_*} & & \downarrow{\scriptstyle \gamma_*} & & \downarrow{\scriptstyle \gamma_*} & & \downarrow{\scriptstyle \gamma_*} & & \downarrow{\scriptstyle \gamma_*} \\
\mathcal{T}(\Sigma Y, W) & \xrightarrow{\Sigma\alpha^*} & \mathcal{T}(\Sigma X, W) & \xrightarrow{j^*} & \mathcal{T}(C\alpha, W) & \xrightarrow{i^*} & \mathcal{T}(Y, W) & \xrightarrow{\alpha^*} & \mathcal{T}(X, W),
\end{array}$$

which is going to form the basis of our proof. Let $\bar{\beta}$ and $\bar{\beta}'$ be two lifts of β, so

$$i^*(\bar{\beta}) = i^*(\bar{\beta}') = \beta.$$

This means that

$$\bar{\beta} - \bar{\beta}' \in \ker(i^*) = \mathrm{Im}(j^*),$$

that is, there is a $\kappa \in \mathcal{T}(\Sigma X, Z)$ with $\bar{\beta}' = \bar{\beta} + \kappa \circ j$. Furthermore, for any such κ, $\bar{\beta} + \kappa \circ j$ is a lift of β.

Now let τ' and τ be two elements in the Toda bracket $\langle \alpha, \beta, \gamma \rangle$, that is,

$$\tau' \circ j = \gamma \circ \bar{\beta} = \tau \circ j.$$

As this also equals $\gamma \circ \bar{\beta}'$, we have

$$(\tau - \tau') \circ j = \gamma \circ (\bar{\beta}' - \bar{\beta}) = \gamma \circ \kappa \circ j.$$

This implies that

$$\tau' - \tau - \gamma \circ \kappa \in \ker(j^*) = \mathrm{Im}(\Sigma\alpha^*),$$

so there is an $\epsilon \in \Sigma\alpha^*\mathcal{T}(\Sigma Y, W)$ such that

$$\tau' = \tau + \gamma \circ \kappa + \epsilon \circ \Sigma\alpha.$$

Conversely, every $\epsilon \in \Sigma\alpha^*\mathcal{T}(\Sigma Y, W)$ creates another element in the Toda bracket as above. Therefore, we have shown that two elements τ and τ' in the Toda bracket can differ by elements in

$$\gamma_*\mathcal{T}(\Sigma X, Z) \oplus \Sigma\alpha^*(\Sigma Y, W). \qquad \square$$

Definition 4.6.3 The subgroup $\gamma_*\mathcal{T}(\Sigma X, Z) \oplus \Sigma\alpha^*\mathcal{T}(\Sigma Y, W)$ of $\mathcal{T}(\Sigma X, W)$ is called the *indeterminacy* of $\langle \alpha, \beta, \gamma \rangle$.

Remark 4.6.4 In [Qui67], Quillen gives a similar construction of the Toda bracket $\langle \alpha, \beta, \gamma \rangle$ in the homotopy category $\mathrm{Ho}(\mathcal{C})$ of a pointed rather than stable model category $\mathcal{C}$. Let

$$X \xrightarrow{\alpha} Y \xrightarrow{\beta} Z \xrightarrow{\gamma} W$$

be a diagram in $\mathrm{Ho}(\mathcal{C})$ with $\beta \circ \alpha = 0$ and $\gamma \circ \beta = 0$. Then we have a diagram

$$X \xrightarrow{\ \alpha\ } Y \xrightarrow{\ i\ } C\alpha \xrightarrow{\ \partial\ } \Sigma X,$$

where the first two arrows form a cofibre sequence and ∂ is the connecting map. Analogously, we have a fibre sequence with connecting map ∂',

$$\Omega W \xrightarrow{\ \partial'\ } F\gamma \xrightarrow{\ j\ } Z \xrightarrow{\ \gamma\ } W.$$

By our assumptions on α, β and γ, the vertical arrows in the following diagram exist and the Toda bracket $\langle \alpha, \beta, \gamma \rangle$ is the coset of possible choices for the rightmost arrow $\Sigma X \longrightarrow W$.

$$\begin{array}{ccccccc} X & \xrightarrow{\alpha} & Y & \xrightarrow{i} & C\alpha & \xrightarrow{\partial} & \Sigma X \\ \vdots & & \vdots & \searrow^{\beta} & \vdots & & \vdots \\ \downarrow & & \downarrow & & \downarrow & & \downarrow \\ \Omega W & \xrightarrow{\partial'} & F\gamma & \xrightarrow{j} & Z & \xrightarrow{\gamma} & W \end{array}$$

In the case of $\mathcal{C}$ being stable, this recovers Definition 4.6.1.

Remark 4.6.5 We could also start our construction of elements in $\langle \alpha, \beta, \gamma \rangle$ in a dual way by completing $\gamma \colon Z \longrightarrow W$ to an exact triangle

$$\Omega W \xrightarrow{i} F\gamma \xrightarrow{j} Z \xrightarrow{\gamma} W,$$

which leads us to a commutative diagram

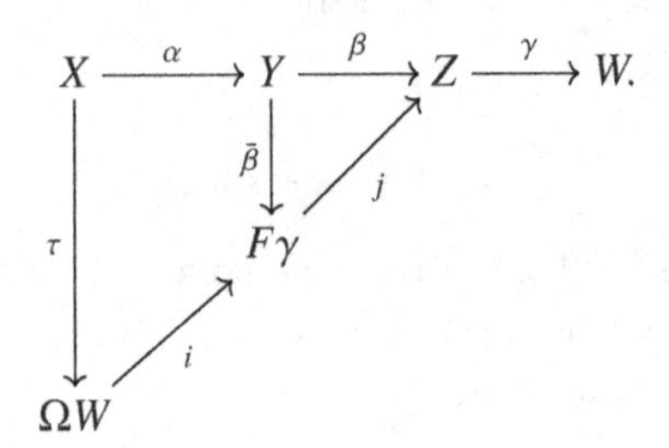

This construction gives rise to a definition of $\langle \alpha, \beta, \gamma \rangle$ which is equivalent to Definition 4.6.1.

Toda bracket relations have been extensively studied on the ring of the stable homotopy groups of spheres $\pi_*(\mathbb{S})$. For

$$\alpha \in \pi_i(\mathbb{S}), \quad \beta \in \pi_j(\mathbb{S}) \quad \text{and} \quad \gamma \in \pi_k(\mathbb{S}),$$

we get $\langle \alpha, \beta, \gamma \rangle \subseteq \pi_{i+j+k+1}(\mathbb{S})$. Some of the best-known relations are listed below, but many more can be found in, for example, [Tod62] or [Rav86]. Let

$$\eta \in \pi_1(\mathbb{S}) = \mathbb{Z}/2, \quad \nu \in \pi_3(\mathbb{S}), \quad \sigma \in \pi_7(\mathbb{S}), \quad \epsilon \in \pi_8(\mathbb{S}) \quad \text{and} \quad \mu \in \pi_9(\mathbb{S}).$$

Then

- $\eta^2 = \langle 2, \eta, 2 \rangle$,
- $8\sigma = \langle \nu, 8, \nu \rangle$,
- $\mu \in \langle 2, 8\sigma, \eta \rangle$, with indeterminacy generated by $\eta^2\sigma$ and $\eta\epsilon$.

We will show the first of the relations above. To do this, we look into the following phenomenon.

Let M denote the mod-2 Moore spectrum, that is, the cofibre of multiplication by 2 on $\mathbb{S}$. We have the following diagram

$$\mathbb{S} \xrightarrow{2} \mathbb{S} \xrightarrow{\text{incl}} M \xrightarrow{\text{pinch}} \mathbb{S}^1 = \Sigma\mathbb{S},$$

which is an exact triangle in the stable homotopy category. As a CW-spectrum, M has one cell in dimension 0 and one cell in dimension 1. The map

$$\text{incl}\colon \mathbb{S} \longrightarrow M$$

is the inclusion of the bottom cell into M, and the map

$$\text{pinch}\colon M \longrightarrow \mathbb{S}^1$$

denotes "killing" the bottom cell and thus only leaving the top cell $\mathbb{S}^1$. The mod-2 Moore spectrum M has the following remarkable anomaly, which the mod-p analogue for odd primes does not possess, in which case, $p \cdot \text{Id}_M = 0$.

Lemma 4.6.6 *The endomorphisms of the mod-2 Moore spectrum M in degree 0 are*

$$[M, M] = \mathbb{Z}/4\{\text{Id}_M\},$$

where

$$2\text{Id}_M = \text{incl} \circ \eta \circ \text{pinch} \neq 0.$$

(Recall that the curly brackets behind a module denote the generators of this module.)

Proof We calculate $[M, M]$ by repeated application of Theorem 3.6.1 to the exact triangle defining M. As the first map in the exact triangle $\mathbb{S} \to \mathbb{S}$ is multiplication by 2, the long exact Puppe sequence breaks up into short exact sequences

$$0 \longrightarrow \pi_0(\mathbb{S})/2 \xrightarrow{\text{incl}_*} \pi_0(M) \xrightarrow{\text{pinch}_*} {}_{(2)}\pi_{-1}(\mathbb{S}) \longrightarrow 0$$

and

$$0 \longrightarrow \pi_1(\mathbb{S})/2 \xrightarrow{\text{incl}_*} \pi_1(M) \xrightarrow{\text{pinch}_*} {}_{(2)}\pi_0(\mathbb{S}) \longrightarrow 0,$$

where

$${}_{(2)}\pi_n(\mathbb{S}) = \{x \in \pi_n(\mathbb{S}) \mid 2x = 0\}.$$

Thus, we obtain

$$\pi_0(M) = \mathbb{Z}/2\{\text{incl}\} \quad \text{and} \quad \pi_1(M) = \mathbb{Z}/2\{\text{incl} \circ \eta\}.$$

By Theorem 3.6.4, applying $[-, M]$ to the exact triangle defining M gives a long exact sequence which splits into short exact sequences as above. Hence, we have a short exact sequence

$$0 \longrightarrow \pi_1(M)/2 \xrightarrow{\text{pinch}^*} [M, M] \xrightarrow{\text{incl}^*} {}_{(2)}\pi_0(M) \longrightarrow 0.$$

At first glance, this means there are two possibilities for $[M, M]$ – it could either be

$$\mathbb{Z}/2\{\mathrm{Id}_M\} \oplus \mathbb{Z}/2\{\text{pinch} \circ \eta \circ \text{incl}\} \quad \text{or} \quad \mathbb{Z}/4\{\mathrm{Id}_M\}.$$

If the former was the case, we would have $2\mathrm{Id}_M = 0$. Applying $M \wedge -$ to the exact triangle

$$\mathbb{S} \xrightarrow{2} \mathbb{S} \xrightarrow{\text{incl}} M \xrightarrow{\text{pinch}} \mathbb{S}^1$$

yields another exact triangle

$$M \xrightarrow{2\mathrm{Id}_M} M \longrightarrow M \wedge M \longrightarrow \Sigma M.$$

If $2\mathrm{Id}_M = 0$, the above triangle splits, that is, $M \wedge M \cong M \vee \Sigma M$ by Lemma 4.1.12. However, we have seen in Lemma 2.5.7 that this is not the case. Therefore, we must have

$$2\mathrm{Id}_M = \text{incl} \circ \eta \circ \text{pinch} \neq 0$$

as claimed. □

A useful consequence is the following.

Corollary 4.6.7 *Let $x \in \pi_n(\mathbb{S})$ be an element with $2x = 0$. Then*

$$\eta x \in \langle 2, x, 2 \rangle.$$

In particular, $\eta^2 = \langle 2, \eta, 2 \rangle$.

Proof By Lemma 4.6.6, the following diagram commutes, where $\bar{x}$ is a lift of x.

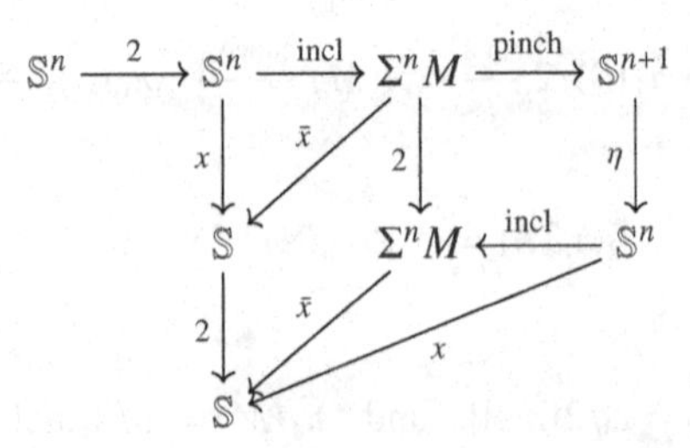

□

4.7 Muro's Exotic Triangulated Category

We saw in Section 4.2 that every stable model category gives rise to a triangulated category via its homotopy category. However, until relatively recently, it was not known whether every triangulated category arose as the homotopy category of a stable model category. Muro finally provided a counterexample in [MSS07], with technical additions by Schwede and Strickland. We will give a summary of this exotic triangulated category.

Let R be a commutative local ring with maximal ideal $\mathfrak{m} = (2) \neq 0$ satisfying $\mathfrak{m}^2 = 0$. For example, $R = \mathbb{Z}/4$ is such a ring. Let $\mathcal{F}(R)$ denote the category of finitely generated R–modules. Furthermore, we are going to use the following terminology.

Definition 4.7.1 A triangulated category $\mathcal{T}$ is *topological* if it is equivalent to a full triangulated subcategory of the homotopy category $\mathrm{Ho}(\mathcal{C})$ of some stable model category $\mathcal{C}$.

Muro's theorem is the following.

Theorem 4.7.2 *Let R and $\mathcal{F}(R)$ be as before. Then $\mathcal{F}(R)$ admits a triangulated structure such that the self-equivalence $\Sigma\colon \mathcal{F}(R) \longrightarrow \mathcal{F}(R)$ is the identity functor and*

$$R \xrightarrow{2} R \xrightarrow{2} R \xrightarrow{2} R$$

is an exact triangle. Furthermore, if $\mathcal{T}$ is a topological triangulated category, then every exact functor

$$\mathcal{F}(R) \longrightarrow \mathcal{T} \text{ or } \mathcal{T} \longrightarrow \mathcal{F}(R)$$

is trivial.

As a consequence, $\mathcal{F}(R)$ cannot be topological itself because then the identity functor would be trivial. We give an introduction to the key ingredients of the proof.

Definition 4.7.3 A sequence of morphisms in $\mathcal{F}(R)$

$$X \xrightarrow{f} Y \xrightarrow{g} Z \xrightarrow{h} X$$

is called a *candidate triangle* if $g \circ f = 0$, $h \circ g = 0$ and $f \circ h = 0$.

In particular,

$$X \xrightarrow{2} X \xrightarrow{2} X \xrightarrow{2} X$$

is always a candidate triangle for any $X \in \mathcal{F}(R)$.

We can also define the notion of a homotopy between morphisms of exact triangles in a similar fashion to the notion of chain homotopies.

Definition 4.7.4 Let (ϕ_1, ϕ_2, ϕ_3) and $(\phi'_1, \phi'_2, \phi'_3)$ be two morphisms of candidate triangles in $\mathcal{F}(R)$ (i.e. morphisms between the respective vertices of the candidate triangles making the obvious diagrams commute). Then a *homotopy* from (ϕ_1, ϕ_2, ϕ_3) to $(\phi'_1, \phi'_2, \phi'_3)$ consists of three morphisms (H_1, H_2, H_3) as follows

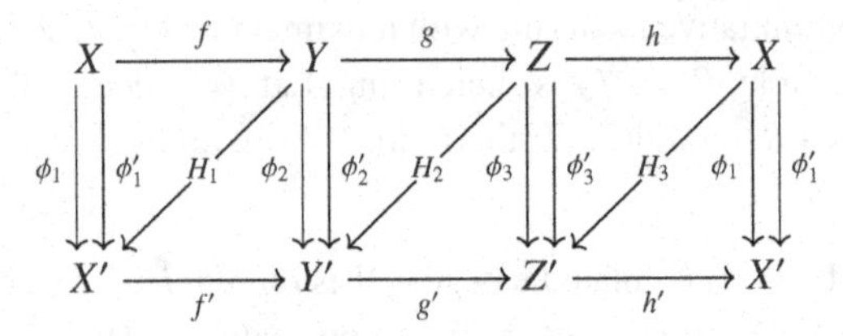

satisfying

$$\phi'_1 - \phi_1 = h' \circ H_3 + H_1 \circ f, \ \ \phi'_2 - \phi_2 = f' \circ H_1 + H_2 \circ g,$$
$$\text{and } \phi'_3 - \phi_3 = g' \circ H_2 + H_3 \circ h.$$

Definition 4.7.5 A candidate triangle in $\mathcal{F}(R)$ is *contractible* if the identity morphism is homotopic to the zero morphism.

We can now say what the triangulated structure on $\mathcal{F}(R)$ is.

Definition 4.7.6 A candidate triangle in $\mathcal{F}(R)$ is *exact* if it is isomorphic to the direct sum of a contractible candidate triangle and a candidate triangle of the form

$$X \xrightarrow{2} X \xrightarrow{2} X \xrightarrow{2} X$$

for some $X \in \mathcal{F}(R)$.

Muro showed that with this choice of self-equivalence and exact triangles, $\mathcal{F}(R)$ is indeed a triangulated category. The proof of this involves various results in commutative algebra, most prominently how over a ring satisfying our assumptions, morphisms can be decomposed into direct sums of much simpler morphisms.

The proof that $\mathcal{F}(R)$ is not topological depends on two different types of objects: “exotic” objects and “hopfian” objects.

Definition 4.7.7 Let $\mathcal{T}$ be a triangulated category and $A \in \mathcal{T}$. A *Hopf map* is a morphism

$$\eta \colon \Sigma A \longrightarrow A \ \text{ with } \ 2\eta = 0$$

such that for any exact triangle

$$A \xrightarrow{2} A \xrightarrow{i} C \xrightarrow{p} \Sigma A,$$

we have that $2\mathrm{Id}_C = i \circ \eta \circ p$. An object that admits a Hopf map is called *hopfian*.

Note that to obtain this condition for any exact triangle as above, it is enough to have it for one.

Of course, this terminology stems from the sphere spectrum: we have the classical Hopf map $\eta \in \pi_1(\mathbb{S}) \cong \mathbb{Z}/2$. The mapping cone of multiplication by 2 on the sphere is the mod-2 Moore spectrum M. Its identity map satisfies $2\mathrm{Id}_M = i \circ \eta \circ p$ (as we showed in Lemma 4.6.6), a phenomenon which is also crucial for Section 5.7.

Proposition 4.7.8 *In a topological triangulated category, every object is hopfian.*

Proof Let $\mathcal{C}$ be a stable model category. By Theorem 6.9.25, for every fibrant and cofibrant $A \in \mathcal{C}$, there exists a left Quillen functor (from sequential spectra in simplicial sets)

$$L\colon \mathcal{S} \longrightarrow \mathcal{C}$$

such that $L(\mathbb{S}) \simeq A$.

The left derived functor

$$\mathbb{L}\colon \mathrm{Ho}(\mathcal{S}) \longrightarrow \mathrm{Ho}(\mathcal{C})$$

is exact by Theorem 4.5.2. Thus, it sends the exact triangle

$$\mathbb{S} \xrightarrow{2} \mathbb{S} \xrightarrow{i} M \xrightarrow{p} \mathbb{S}^1$$

to an exact triangle

$$A \xrightarrow{2} A \xrightarrow{j} \mathbb{L}(M) \xrightarrow{q} \Sigma A$$

with $j = \mathbb{L}(i)$ and $q = \mathbb{L}(p)$. This also means that $\mathbb{L}$ sends the composite

$$2\mathrm{Id}_M\colon M \xrightarrow{p} \mathbb{S}^1 \xrightarrow{\eta} \mathbb{S} \xrightarrow{i} M$$

to the composite

$$2\mathrm{Id}_{\mathbb{L}(M)}\colon \mathbb{L}(M) \xrightarrow{q} \Sigma A \xrightarrow{\mathbb{L}(\eta)} A \xrightarrow{j} \mathbb{L}(M).$$

But this means that $\mathbb{L}(\eta)$ is a Hopf map for A, and A is hopfian. □

On the other hand, we have the following notion.

Definition 4.7.9 An object E in a triangulated category $\mathcal{T}$ is *exotic* if there is an exact triangle of the form

$$E \xrightarrow{2} E \xrightarrow{2} E \xrightarrow{q} \Sigma E.$$

In $\mathcal{F}(R)$, with the triangulation described earlier, every object is exotic. The relationship between exotic objects and hopfian objects boils down to the following.

Proposition 4.7.10 *Let $\mathcal{T}$ be a triangulated category, $E \in \mathcal{T}$ an exotic object and $A \in \mathcal{T}$ hopfian. Then the morphism groups $\mathcal{T}(E, A)$ and $\mathcal{T}(A, E)$ are trivial.*

Proof We only prove $\mathcal{T}(E, A) = 0$, as the other direction is very similar. First, we make the following observation. Given an exotic object E, we can compare the triangle

$$E \xrightarrow{2} E \xrightarrow{2} E \xrightarrow{q} \Sigma E$$

to its shifted version

$$E \xrightarrow{2} E \xrightarrow{q} \Sigma E \xrightarrow{-2} \Sigma E.$$

The axiom (T3) then gives us a fill-in morphism Ψ in the following diagram.

$$\begin{array}{ccccccc} E & \xrightarrow{2} & E & \xrightarrow{2} & E & \xrightarrow{q} & \Sigma E \\ \| & & \| & & \downarrow{\scriptstyle\Psi} & & \| \\ E & \xrightarrow{2} & E & \xrightarrow{q} & \Sigma E & \xrightarrow{-2} & \Sigma E \end{array}$$

Thus, $q = 2\Psi$, that is, the morphism q is always divisible by 2.

Now consider a morphism $f\colon E \longrightarrow A$ between an exotic element and a Hopf element. Then (T3) gives us a fill-in map g as follows.

$$\begin{array}{ccccccc} E & \xrightarrow{2} & E & \xrightarrow{2} & E & \xrightarrow{q} & \Sigma E \\ \downarrow{\scriptstyle f} & & \downarrow{\scriptstyle f} & & \downarrow{\scriptstyle g} & & \downarrow{\scriptstyle \Sigma f} \\ A & \xrightarrow{2} & A & \xrightarrow{i} & C & \xrightarrow{p} & \Sigma A \end{array}$$

This means that, using that A is hopfian,

$$i \circ f = 2g = (i \circ \eta \circ p) \circ g = i \circ \eta \circ \Sigma f \circ q.$$

We already showed that $q = 2\Psi$ for some Ψ and $2\eta = 0$ by assumption, which means that we arrive at $i \circ f = 0$. Hence, f factors over the cofibre of i, obtaining a map f' with $2f' = f$ as illustrated below.

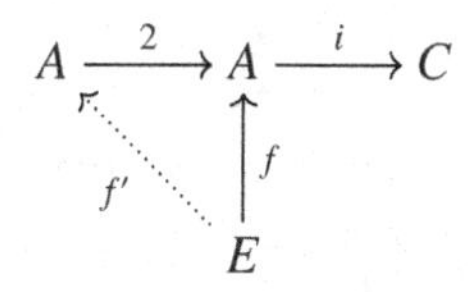

We can iterate this process to obtain a map $f'': E \longrightarrow A$ with $2f'' = f'$, so altogether $f = 4f''$ for some f''. But as E is an exotic object, $4\mathrm{Id}_E = 0$, which means that f is trivial. □

As a consequence, there are no non-trivial objects that can be hopfian and exotic at the same time. Therefore, $\mathcal{F}(R)$ does not contain any hopfian objects and thus cannot be topological, as claimed.

5

Modern Categories of Spectra

In this chapter, we introduce symmetric spectra and orthogonal spectra along with their associated stable model structures. These versions of spectra have various technical advantages over the sequential spectra of Chapter 2. Furthermore, they are Quillen equivalent to the category of sequential spectra (equipped with its stable model structure). Hence, one may choose between these models according to their relative strengths.

The primary advantage of symmetric and orthogonal spectra is that these model categories are *symmetric monoidal models* for the stable homotopy category. We will examine these monoidal structures further in Chapter 6 and show that symmetric spectra and orthogonal spectra are monoidally Quillen equivalent. Several other models of spectra also exist, and we will give short introductions to these later in this chapter. We end the chapter with a result that, roughly speaking, says that any model for the stable homotopy category will be Quillen equivalent to sequential spectra.

In each of the monoidal categories of spectra, there is some extra structure that accounts for "symmetries". This allows us to avoid the twist problem we saw earlier that prevents sequential spectra from being monoidal. In the case of symmetric spectra, this extra structure of symmetries comes from the symmetric groups Σ_n as the symmetry set of $\underline{n} = \{1, 2, \ldots, n\}$. This has the advantage that the extra structure compared to sequential spectra is somewhat minimal, but it will cause the weak equivalences of symmetric spectra to be quite different from the expected answer of the class of π_*-isomorphisms. In the case of orthogonal spectra, the extra symmetries come from the orthogonal groups $O(n)$ as the symmetry set of $\mathbb{R}^n$. This is a much larger group than Σ_n, and one that requires us to take more care over the continuity of the actions, but it allows orthogonal spectra to have the "correct" weak equivalences, the π_*-isomorphisms.

5.1 The Stable Homotopy Category – Revisited

In Subsection 1.1.4, we discussed key properties of the stable homotopy category. We then spent Chapter 2 introducing sequential spectra with the long-term aim of providing a good model for the stable homotopy category. We are finally in a good position to start proving this list of properties, as we have developed most of the necessary language in previous chapters.

Pre-Theorem 5.1.1 *The homotopy category of sequential spectra* $\mathrm{Ho}(\mathcal{S}^{\mathbb{N}})$ *satisfies Properties 1–12 from Subsection 1.1.4.*

We will prove these properties in separate lemmas and repeat the bullet points throughout the section. It will become clear in points 10, 11 and 12 why we are calling the above "Pre-Theorem" rather than "Theorem".

Let us begin with properties 1 and 2. Property 1 is

(1) There is an adjunction

$$\Sigma^\infty : \mathrm{Ho}(\mathrm{Top}_*) \rightleftarrows \mathcal{SHC} : \Omega^\infty.$$

We prove the following, where we abuse notation and do not distinguish between the functor Σ^∞ and its derived functor.

Lemma 5.1.1 *There is an adjunction*

$$\Sigma^\infty : \mathrm{Ho}(\mathrm{Top}_*) \rightleftarrows \mathrm{Ho}(\mathcal{S}^{\mathbb{N}}) : \Omega^\infty,$$

where Σ^∞ applied to a pointed CW-complex A is the suspension spectrum of A, and Ω^∞ sends an Ω-spectrum to its level zero space.

Proof By Lemma 2.3.16, the adjunction $(\Sigma^\infty, \mathrm{Ev}_0^{\mathbb{N}})$ is a Quillen adjunction. Hence, we have an adjunction on homotopy categories

$$\mathbb{L}\Sigma^\infty : \mathrm{Ho}(\mathrm{Top}_*) \rightleftarrows \mathrm{Ho}(\mathcal{S}^{\mathbb{N}}) : \mathbb{R}\,\mathrm{Ev}_0^{\mathbb{N}} = \Omega^\infty$$

with the equality on the right given by Section 2.4. Since Ω-spectra are the fibrant objects of the stable model structure, Ω^∞ sends an Ω-spectrum X to the space X_0. □

As a consequence, if X is an Ω-spectrum and A a pointed CW-complex, then we have natural isomorphisms

$$\pi(\Sigma^\infty A, X) \cong [\Sigma^\infty A, X] \cong [A, X_0],$$

where the left-hand side denotes maps of spectra up to homotopy and the right-hand side denotes maps in the homotopy category of pointed spaces. As $\Sigma^\infty A$

is cofibrant and X is fibrant, a homotopy between two maps $f, g\colon \Sigma^\infty A \longrightarrow X$ is a map

$$H\colon \Sigma^\infty A \wedge [0,1]_+ \longrightarrow X.$$

Because $(\Sigma^\infty A \wedge [0,1]_+)_n = \Sigma^n A \wedge [0,1]_+$, we see that a homotopy H is entirely determined by a map of spaces $H_0 : A \wedge [0,1]_+ \longrightarrow X_0$.

The next item on our list is the following.

(2) Let A and B be pointed CW-complexes. If A has only finitely many cells, there is a natural isomorphism

$$[\Sigma^\infty A, \Sigma^\infty B] \cong [A, B]^s,$$

where $[A, B]^s$ *is the set of stable homotopy classes of maps of spaces from Definition 1.1.16.*

Lemma 5.1.2 *Let A be a pointed CW-complex with finitely many cells and B a pointed CW-complex. Then there is a natural isomorphism*

$$[\Sigma^\infty A, \Sigma^\infty B] \cong [A, B]^s = \operatorname{colim}_n [\Sigma^n A, \Sigma^n B].$$

Proof We have that

$$[\Sigma^\infty A, \Sigma^\infty B] = [\Sigma^\infty A, R_\infty(\Sigma^\infty B)],$$

where R_∞ is the fibrant replacement functor constructed in Section 2.4. By Lemma 5.1.1,

$$[\Sigma^\infty A, R_\infty(\Sigma^\infty B)] = [A, (R_\infty(\Sigma^\infty B))_0] = [A, \operatorname{hocolim}_k(\Omega^k \Sigma^k B)].$$

Because A is finite, we have

$$[A, \operatorname{hocolim}_k(\Omega^k \Sigma^k B)] = \operatorname{colim}_k [A, \Omega^k \Sigma^k B] = \operatorname{colim}_k [\Sigma^k A, \Sigma^k B]. \qquad \square$$

Lemma 5.1.2 can be thought of as saying that the Spanier–Whitehead category $\mathcal{SW}$ is the full subcategory of the stable homotopy category $\mathcal{SHC}$ whose object class consists of the suspension spectra on finite CW-complexes. We can extend this result to the full subcategory of $\widehat{\mathcal{SW}}$ on finite CW-complexes as follows.

Lemma 5.1.3 *For finite CW-complexes A and B and non-negative integers d and e, there is a natural isomorphism*

$$[F_d^{\mathbb{N}} A, F_e^{\mathbb{N}} B] \cong \{(A, -d), (B, -e)\},$$

where the right hand side denotes morphisms in $\widehat{\mathcal{SW}}$.

Proof We suspend $d + e$ times so that we can use Lemma 5.1.2 to simplify the proof.

$$\begin{aligned}[F^{\mathbb{N}}_d A, F^{\mathbb{N}}_e B] &\cong [\Sigma^\infty \Sigma^e A, \Sigma^\infty \Sigma^d B] \\ &\cong \operatorname{colim}_a [\Sigma^{a+e} A, \Sigma^{a+d} B] \\ &\cong \operatorname{colim}_a [\Sigma^{a-d} A, \Sigma^{a-e} B] \\ &= \{(A, -d), (B, -e)\}\end{aligned}$$

□

We have to restrict the previous statements to finite complexes, as Example 1.2.7 and the isomorphism below show that $\widehat{\mathcal{SW}}$ can take incorrect values on infinite complexes

$$\begin{aligned}[\Sigma^\infty \textstyle\bigvee_{i\geqslant 0} S^i, \Sigma^\infty S^0] &\cong \textstyle\prod_{i\geqslant 0} [\Sigma^\infty S^i, \Sigma^\infty S^0] \\ &\cong \textstyle\prod_{i\geqslant 0} \left\{(S^i, 0), (S^0, 0)\right\} \\ &\ncong \left\{\left(\textstyle\bigvee_{i\geqslant 0} S^i, 0\right), (S^0, 0)\right\}.\end{aligned}$$

Theorem 2.3.14 showed that

$$\Sigma = (-) \wedge S^1 \colon \operatorname{Ho}(\mathcal{S}^{\mathbb{N}}) \longrightarrow \operatorname{Ho}(\mathcal{S}^{\mathbb{N}})$$

is an equivalence of categories, which shows Property 6:

(6) The functor

$$(-) \wedge S^1 \colon \mathcal{SHC} \longrightarrow \mathcal{SHC}$$

is an equivalence of categories.

As a consequence, $\mathcal{S}^{\mathbb{N}}$ is a stable model category, and thus $\operatorname{Ho}(\mathcal{S}^{\mathbb{N}})$ is additive (see Proposition 3.2.9) and has arbitrary small products and coproducts. By Lemma 4.4.4, finite products and coproducts agree. This gives us Properties 3 and 4:

(3) The set of maps in $\mathcal{SHC}$ can be equipped with the structure of graded abelian groups and composition is bilinear. We will use $[-,-]_*$ for the graded set of maps.

(4) The stable homotopy category $\mathcal{SHC}$ has arbitrary (small) products and coproducts. Finite products and coproducts coincide.

In Example 2.3.2, we defined $X \wedge A$ for a sequential spectrum X and a pointed CW-complex A via

$$(X \wedge A)_n = X_n \wedge A.$$

We also defined $F(A, Y)$ for a pointed CW-complex A and a spectrum Y as

$$F(A, Y)_n = \operatorname{Top}_*(A, Y_n).$$

This gives us the first part of Property 5.

(5) For a pointed CW-complex A and $X, Y \in \mathcal{SHC}$, there are objects $X \wedge A$ and $F(A, Y)$ in $\mathcal{SHC}$ such that

$$[X \wedge A, Y] \cong [X, F(A, Y)],$$

and for any pointed CW-complex B, there is an isomorphism in $\mathcal{SHC}$

$$(\Sigma^\infty B) \wedge A \cong \Sigma^\infty (B \wedge A).$$

Lemma 5.1.4 *Let X and Y be sequential spectra and A a pointed CW-complex. In* $\mathrm{Ho}(\mathcal{S}^{\mathbb{N}})$*, we have a natural isomorphism*

$$[X \wedge A, Y] \cong [X, F(A, Y)].$$

Proof We need to show that

$$- \wedge A \colon \mathcal{S}^{\mathbb{N}} \longrightarrow \mathcal{S}^{\mathbb{N}}$$

is a left Quillen functor. Since $(\mathrm{F}_d^{\mathbb{N}} B) \wedge A \cong \mathrm{F}_d^{\mathbb{N}}(B \wedge A)$ and $\mathrm{F}_d^{\mathbb{N}}$ is a left Quillen functor, it follows that $- \wedge A$ sends the generating q-cofibrations of sequential spectra to q-cofibrations. Hence, $- \wedge A$ preserves q-cofibrations of sequential spectra.

Smashing with a pointed CW-complex preserves π_*-isomorphisms of spectra by Proposition 2.2.14. These two properties together imply that $- \wedge A$ preserves cofibrations and acyclic cofibrations. □

The second part of Property 5 claims that for A and B pointed CW-complexes,

$$(\Sigma^\infty B) \wedge A \cong \Sigma^\infty (B \wedge A).$$

This follows from

$$(\Sigma^\infty B \wedge A)_n = \Sigma^n B \wedge A \cong \Sigma^n (B \wedge A).$$

Properties 7, 8 and 9 are concerned with the relation between sequential spectra and cohomology theories.

(7) Given a reduced cohomology theory $\widetilde{E}^*$, there is an object $E \in \mathcal{SHC}$ such that

$$\widetilde{E}^*(A) = [\Sigma^\infty A, E]_*$$

for any pointed CW-complex A. Moreover, the object E is unique up to isomorphism. We say that E represents *$\widetilde{E}^*$.*

(8) Every $E \in \mathcal{SHC}$ defines a reduced cohomology theory on pointed CW-complexes by

$$A \mapsto [\Sigma^\infty A, E]_* = \widetilde{E}^*(A).$$

(9) A map $E_1 \longrightarrow E_2$ in $\mathcal{SHC}$ induces a map of cohomology theories

$$\widetilde{E}_1^*(X) \longrightarrow \widetilde{E}_2^*(X).$$

The next proposition implies Property 7. Before we state the result, please note the following about notation. In Subsection 1.1.3, we denoted a cohomology theory of spaces by E^* and its reduced version by $\tilde{E}^*$. In the context of (co)homology theories of spectra, we will only consider reduced versions and therefore drop the tilde from the notation.

Proposition 5.1.5 *Let E be a spectrum. Then the functor*

$$E^* = [-, E]_{-*} \colon \operatorname{Ho}(\mathcal{S}^{\mathbb{N}}) \longrightarrow \mathrm{Ab}_*$$

is a cohomology theory on spectra, that is,

- *E^* sends an exact triangle $X \longrightarrow Y \longrightarrow Z \longrightarrow \Sigma X$ in $\mathcal{SHC}$ to a long exact sequence*

$$\cdots \longrightarrow E^{n-1}(X) \longrightarrow E^n(Z) \longrightarrow E^n(Y) \longrightarrow E^n(X) \longrightarrow E^{n+1}(Z) \longrightarrow \cdots,$$

- *for a set of spectra $\{X_\alpha\}_\alpha$, the maps $i_\alpha \colon X_\alpha \longrightarrow \bigvee_\alpha X_\alpha$ induce isomorphisms*

$$\prod_\alpha (i_\alpha)^* \colon E^*(\bigvee_\alpha X_\alpha) \longrightarrow \prod_\alpha E^*(X_\alpha),$$

- *for any spectrum X, there is a natural isomorphism $E^{n+1}(\Sigma X) \cong E^n(X)$.*

If $X = \Sigma^\infty A$ is a suspension spectrum, then the E^*–cohomology of X is the E^*–cohomology of A as defined in Proposition 1.1.25.

Proof Proposition 4.1.5 tells us that in a triangulated category, an exact triangle $X \longrightarrow Y \longrightarrow Z \longrightarrow \Sigma X$ gives us exact sequences

$$[\Sigma^{n+1} X, E] \to [\Sigma^n Z, E] \to [\Sigma^n Y, E] \to [\Sigma^n X, E]$$

for every E and $n \in \mathbb{Z}$. This proves the first point. The second point is a consequence of the fact that $\mathcal{S}^{\mathbb{N}}$ is a model category. The third point follows from the first point applied to the exact triangle

$$X \longrightarrow * \longrightarrow \Sigma X \longrightarrow \Sigma X.$$

The final claim is that for a CW-complex A, $E^n(\Sigma^\infty A) = [A, E_n]$. We have

$$E^n(\Sigma^\infty A) = [\Sigma^\infty A, E]_{-n} = [\Sigma^\infty A, \Sigma^n E] \cong [\mathrm{F}_n^{\mathbb{N}} \Sigma^n A, \Sigma^n E] \cong [\mathrm{F}_n^{\mathbb{N}} A, E] \cong [A, E_n].$$

□

The Brown Representability Theorem (Theorem 1.1.23 and Corollary 1.1.24) tells us that for a pointed CW-complex A and a cohomology theory $\tilde{E}^*$, there is a spectrum E with

$$\tilde{E}^n(A) = [A, E_n].$$

Proposition 1.1.25 and Corollary 1.1.26 tell us that the representing sequential spectrum E is unique up to isomorphism in $\mathrm{Ho}(\mathcal{S}^{\mathbb{N}})$, and that a morphism of spectra induces a map of cohomology theories. These are Properties 8 and 9. We summarise these results in the following.

Corollary 5.1.6 *The stable homotopy category $\mathcal{SHC}$ is equivalent to the category of cohomology theories on spectra.* □

Properties 10 and 11 discuss monoidal properties of $\mathcal{SHC}$.

(10) There is a monoidal product $\wedge^{\mathbb{L}}$ on $\mathcal{SHC}$ with

$$\Sigma^\infty A \wedge^{\mathbb{L}} \Sigma^\infty B \cong \Sigma^\infty (A \wedge B)$$

for all pointed CW-complexes A and B.

(11) There is an internal function object $\mathbb{R}\mathrm{Hom}(-,-)$ on $\mathcal{SHC}$ so that for $X, Y, Z \in \mathcal{SHC}$

$$[X \wedge^{\mathbb{L}} Y, Z] \cong [X, \mathbb{R}\mathrm{Hom}(Y, Z)].$$

Unfortunately, we cannot obtain those two properties from the model structure on sequential spectra. In order to obtain Properties 10 and 11 from a model category level, we will need the categories of spectra which we will define later in this chapter. There is a construction by Adams in [Ada74] of a monoidal product on $\mathcal{SHC}$, but it is not very practical, to say the least. For now, we will assume that $\mathcal{SHC}$ (not $\mathcal{S}^{\mathbb{N}}$ itself!) has a decent monoidal product which in particular satisfies

$$(E \wedge^{\mathbb{L}} \Sigma^\infty A)_k = E_k \wedge A.$$

This allows us to consider Property 12. We will encounter categories of spectra inducing Properties 10 and 11 later in this chapter and prove these properties in Chapter 6, see Lemma 6.3.21 and Theorem 6.4.8.

(12) Every $E \in \mathcal{SHC}$ defines a reduced homology theory via

$$E_n(X) = \pi_n(E \wedge^{\mathbb{L}} X).$$

More precisely, we get the following.

Proposition 5.1.7 *Let E be a spectrum. Then the functor*

$$E_* = \pi_*(E \wedge^{\mathbb{L}} -) = [\mathbb{S}, E \wedge^{\mathbb{L}} -]_* \colon \operatorname{Ho}(\mathcal{S}^{\mathbb{N}}) \longrightarrow \mathrm{Ab}_*$$

is a homology theory on spectra, that is,

- *an exact triangle $X \longrightarrow Y \longrightarrow Z \longrightarrow \Sigma X$ gives a long exact sequence*

$$\cdots \longrightarrow E_{n+1}(Z) \longrightarrow E_n(X) \longrightarrow E_n(Y) \longrightarrow E_n(Z) \longrightarrow E_{n-1}(X) \longrightarrow \cdots,$$

- *for a set of spectra $\{X_\alpha\}_\alpha$, the maps $i_\alpha \colon X_\alpha \longrightarrow \bigvee_\alpha X_\alpha$ induce isomorphisms*

$$\bigoplus_\alpha (i_\alpha)_* \colon \bigoplus_\alpha E_*(X_\alpha) \longrightarrow E_*(\bigvee_\alpha X_\alpha),$$

- *for any spectrum X, there is a natural isomorphism $E_{n+1}(\Sigma X) \cong E_n(X)$.*

If $X = \Sigma^\infty A$ is a suspension spectrum, then the homology of X is the homology of A as in

$$E_n(X) = E_n(A) = \operatorname{colim}_k \pi_{n+k}(E_n \wedge A).$$

Proof As in the proof of Proposition 5.1.5, the first point follows from having the long exact Puppe sequence

$$\cdots \longrightarrow \pi_{n+1}(E \wedge^{\mathbb{L}} Z) \to \pi_n(E \wedge^{\mathbb{L}} X) \to \pi_n(E \wedge^{\mathbb{L}} Y) \to \pi_n(E \wedge^{\mathbb{L}} Z) \to \cdots,$$

which is a consequence of $E \wedge^{\mathbb{L}} -$ being a triangulated functor. The second point holds as π_* sends coproducts to direct sums, and the third point follows from applying the first point to the exact triangle

$$X \longrightarrow * \longrightarrow \Sigma X \longrightarrow \Sigma X.$$

As for the last statement, Property 10 implies

$$E_n(\Sigma^\infty A) = \pi_n(E \wedge \Sigma^\infty A) = \operatorname{colim}_k \pi_{n+k}((E \wedge \Sigma^\infty A)_k) = \operatorname{colim}_k \pi_{n+k}(E_k \wedge A). \quad \square$$

Remark 5.1.8 The analogue of Corollary 5.1.6 for homology theories is false. The problem is the existence of *phantom maps*: maps of pointed CW-complexes $f \colon X \longrightarrow Y$ which are trivial on all (reduced) homology functors, but are not homotopic to the trivial map. Phantom maps are discussed in more detail in, for example, [HPS97] and [CS98].

5.2 Orthogonal Spectra

We define the category of orthogonal spectra and equip it with a stable model structure. We show that this model category is Quillen equivalent to the stable model structure on sequential spectra and, hence, is a model for the stable homotopy category. We leave the monoidal structure of orthogonal spectra for Chapter 6. The standard reference for orthogonal spectra is [MMSS01]. We take a different approach to that reference, using the results of Chapter 2 to create the stable model structure. In particular, there is a levelwise model structure on orthogonal spectra, but we will make little use of it. As well as these model structures, we will introduce h-cofibrations of spectra. These will be needed for the monoidal structures of Chapter 6.

Our starting place is to recall the notion of group actions on topological spaces. In the following, G will be a compact topological group, such as $O(n)$ or a finite group.

Definition 5.2.1 A *pointed G-space* X is a pointed topological space X with a continuous map $G \times X \to X$, $(g, x) \mapsto gx$ such that

$$(gh)x = g(hx), \; gx_0 = x_0 \text{ and } ex = x$$

for all $g, h \in G$ and $x \in X$.

A *map of pointed G-spaces* $f\colon X \longrightarrow Y$ is a continuous map of pointed spaces such that $f(gx) = gf(x)$ for all $x \in X$ and $g \in G$. We often describe such a map as a *G-equivariant map*.

Since the basepoint of the G-space X is G–fixed, the elements $g \in G$ give pointed maps $g\colon X \to X$.

The natural action of $O(n)$ on $\mathbb{R}^n$ induces an action on S^n (the one-point compactification of $\mathbb{R}^n$) which we call the *canonical action*.

Definition 5.2.2 An *orthogonal spectrum* X is a sequence of pointed topological spaces X_n, along with the following data and conditions for $n \in \mathbb{N}$.

1. The space X_n has a continuous action of $O(n)$ which fixes the basepoint.
2. There are maps of pointed spaces $\sigma_n\colon S^1 \wedge X_n \to X_{n+1}$.
3. The composite map σ_n^k given by

$$S^k \wedge X_n \xrightarrow{\mathrm{Id}\wedge\sigma_n} S^{k-1} \wedge X_{n+1} \xrightarrow{\mathrm{Id}\wedge\sigma_{n+1}} \cdots \xrightarrow{\sigma_{n+k-1}} X_{n+k}$$

is $(O(k) \times O(n))$-equivariant. Here, we treat $O(k) \times O(n)$ as a subgroup of $O(k + n)$, where $O(k)$ acts on the first k coordinates and $O(n)$ the last n coordinates.

A *morphism of orthogonal spectra* $f: X \to Y$ is a collection of $O(n)$-equivariant maps $f_n: X_n \to Y_n$ such that the square

$$\begin{array}{ccc} S^1 \wedge X_n & \xrightarrow{\mathrm{Id} \wedge f_n} & S^1 \wedge Y_n \\ \downarrow \sigma_n^X & & \downarrow \sigma_n^Y \\ X_{n+1} & \xrightarrow{f_{n+1}} & Y_{n+1} \end{array}$$

commutes for each $n \in \mathbb{N}$. We denote the category of orthogonal spectra as $\mathcal{S}^{\mathcal{O}}$.

Limits and colimits of orthogonal spectra are defined levelwise: the $O(n)$-action on level n of the (co)limit is induced by the $O(n)$-action on the components and the universal property of (co)limits. The structure maps for a colimit arise from the structure maps of the components and the fact that colimits of pointed spaces commute with smash products. The structure maps for a limit arise similarly. It follows that all small limits and colimits exist in $\mathcal{S}^{\mathcal{O}}$.

Examples 5.2.3 We can now give a series of simple examples of orthogonal spectra.

The trivial spectrum We define the spectrum $*$ to be the spectrum with level n given by a point. This object is the zero object of orthogonal spectra, as it is both initial and terminal.

The sphere spectrum We define the sphere spectrum $\mathbb{S}$ to be the orthogonal spectrum where level n is given by S^n with the canonical action of $O(n)$. The structure map

$$S^1 \wedge S^n \longrightarrow S^{n+1}$$

is the usual isomorphism, and we see that the maps σ_n^k are $O(k) \times O(n)$-equivariant.

Suspension spectra Generalising the above somewhat, we can make $\Sigma^\infty A$, the suspension spectrum of the pointed space A. Level n is given by $S^n \wedge A$, with $O(n)$ acting as $\alpha[s, a] = [\alpha s, a]$ for $\alpha \in O(n)$, $a \in A$ and $s \in S^n$. The structure maps are the isomorphisms as for the sphere spectrum

$$S^1 \wedge S^n \wedge A \longrightarrow S^{n+1} \wedge A,$$

and the composite maps σ_n^k are $O(k) \times O(n)$-equivariant as before.

Shifted suspension spectra The final generalisation is given by the shifted suspension functors. For A a pointed space and $d \in \mathbb{N}$, we define a spectrum $\mathrm{F}_d^{\mathcal{O}} A$ by

$$(\mathrm{F}_d^{\mathcal{O}} A)_n = \begin{cases} O(n)_+ \wedge_{O(n-d)} S^{n-d} \wedge A & n \geqslant d \\ * & n < d, \end{cases}$$

where $O(n-d)$ is the subgroup which acts on the first $n-d$ coordinates. The structure map is either trivial or the composite of the isomorphism

$$S^1 \wedge O(n)_+ \wedge_{O(n-d)} \left(S^{n-d} \wedge A\right) \longrightarrow O(n)_+ \wedge_{O(n-d)} \left(S^{1+n-d} \wedge A\right)$$

with $O(n) \to O(1+n)$ (and $O(n-d) \to O(1+n-d)$). Checking that the maps σ_n^k are $O(k) \times O(n)$-equivariant is an exercise in tracking the partition of $k+n$ into $[k|n-d|d]$ through the maps.

Another source of examples comes from an adjunction with sequential spectra. The left adjoint is difficult to define precisely at this stage, so we use the following lemma to see that it exists and leave the exact construction to Proposition 6.3.22.

Lemma 5.2.4 *There is an adjunction*

$$\mathbb{P}_{\mathbb{N}}^{\mathcal{O}} : \mathcal{S}^{\mathbb{N}} \rightleftarrows \mathcal{S}^{\mathcal{O}} : \mathbb{U}_{\mathbb{N}}^{\mathcal{O}},$$

where the functor $\mathbb{U}_{\mathbb{N}}^{\mathcal{O}}$ *preserves all colimits. We call* $\mathbb{U}_{\mathbb{N}}^{\mathcal{O}}$ *the* forgetful functor *and* $\mathbb{P}_{\mathbb{N}}^{\mathcal{O}}$ *the* prolongation functor.

Proof Let X be an orthogonal spectrum. Then we can define a sequential spectrum $\mathbb{U}_{\mathbb{N}}^{\mathcal{O}} X$ by forgetting structure as follows. Level n of $\mathbb{U}_{\mathbb{N}}^{\mathcal{O}} X$ is X_n (forgetting the $O(n)$-action) and the structure maps are those of X. As limits and colimits are constructed levelwise, it follows that $\mathbb{U}_{\mathbb{N}}^{\mathcal{O}}$ preserves them. It therefore follows that the left adjoint $\mathbb{P}_{\mathbb{N}}^{\mathcal{O}}$ exists. □

As well as forgetting to sequential spectra, we can also forget from orthogonal spectra to pointed spaces.

Definition 5.2.5 For each $d \in \mathbb{N}$, there is an adjunction

$$\mathrm{F}_d^{\mathcal{O}} : \mathrm{Top}_* \rightleftarrows \mathcal{S}^{\mathcal{O}} : \mathrm{Ev}_d^{\mathcal{O}}$$

between orthogonal spectra and pointed topological spaces. The left adjoint $\mathrm{F}_d^{\mathcal{O}}$ is the shifted suspension functor, and the right adjoint $\mathrm{Ev}_d^{\mathcal{O}}$ sends an orthogonal spectrum X to the space X_d. We refer to $\mathrm{Ev}_d^{\mathcal{O}}$ as an *evaluation functor*. When $d = 0$, we write Σ^∞ for $\mathrm{F}_0^{\mathcal{O}}$.

As with sequential spectra, we keep Ω^∞ for the derived functor of $\mathrm{Ev}_0^{\mathbb{N}}$, see Section 2.4.

Lemma 5.2.6 *The forgetful functors commute with the evaluation functors, and the shifted suspension functors commute with the prolongation functors, that is,*

$$\mathrm{Ev}_n^{\mathbb{N}}\,\mathbb{U}_{\mathbb{N}}^{\mathbb{O}} = \mathrm{Ev}_n^{\mathbb{O}} \qquad \mathrm{F}_n^{\mathbb{O}} \cong \mathbb{P}_{\mathbb{N}}^{\mathbb{O}}\,\mathrm{F}_n^{\mathbb{N}}.$$

Proof Let X be an orthogonal spectrum. Then level n of $\mathbb{U}_{\mathbb{N}}^{\mathbb{O}}X$ is precisely X_n. Hence, we have

$$\mathrm{Ev}_n^{\mathbb{N}}\,\mathbb{U}_{\mathbb{N}}^{\mathbb{O}} = \mathrm{Ev}_n^{\mathbb{O}}.$$

The functor $\mathbb{P}_{\mathbb{N}}^{\mathbb{O}}\,\mathrm{F}_n^{\mathbb{N}}$ is left adjoint to $\mathrm{Ev}_n^{\mathbb{N}}\,\mathbb{U}_{\mathbb{N}}^{\mathbb{O}}$, and $\mathrm{F}_n^{\mathbb{O}}$ is left adjoint to $\mathrm{Ev}_n^{\mathbb{O}}$. As the two right adjoints are equal, we must have a natural isomorphism

$$\mathrm{F}_n^{\mathbb{O}} \cong \mathbb{P}_{\mathbb{N}}^{\mathbb{O}}\,\mathrm{F}_n^{\mathbb{N}}$$

between the corresponding left adjoints. □

Given a spectrum X and a pointed space A, we can make a spectrum $X \wedge A$, which at level n is $X_n \wedge A$. The group $O(n)$ acts only on the X_n term and the structure map is

$$\sigma_n \wedge \mathrm{Id}_A \colon S^1 \wedge X_n \wedge A \longrightarrow X_{n+1} \wedge A.$$

We call this the *tensor* of orthogonal spectra with spaces. Similarly, we have a *cotensor* and can make a spectrum $\mathrm{Top}_*(A, X)$, which at level n is the space of maps $\mathrm{Top}_*(A, X_n)$. Again, the group $O(n)$ acts only on the X_n term, and the structure map is the composite

$$S^1 \wedge \mathrm{Top}_*(A, X_n) \to \mathrm{Top}_*(A, S^1 \wedge X_n) \to \mathrm{Top}_*(A, X_{n+1}).$$

The levelwise adjunction between smash products and function spectra gives an adjunction

$$\mathcal{S}^{\mathbb{O}}(X \wedge A, Y) \cong \mathcal{S}^{\mathbb{O}}(X, \mathrm{Top}_*(A, Y)).$$

We also have that $\mathbb{U}_{\mathbb{N}}^{\mathbb{O}}(\mathrm{Top}_*(A, X)) = \mathrm{Top}_*(A, \mathbb{U}_{\mathbb{N}}^{\mathbb{O}}X)$, hence, we have an isomorphism $\mathbb{P}_{\mathbb{N}}^{\mathbb{O}}(Z \wedge A) \cong \mathbb{P}_{\mathbb{N}}^{\mathbb{O}}Z \wedge A$ for any sequential spectrum Z.

Using these constructions we can define h-cofibrations of orthogonal spectra. These are similar to h-cofibrations of pointed spaces, see Section A.5 for a discussion of h-cofibrations in general model categories.

Definition 5.2.7 A map $i \colon A \longrightarrow X$ of orthogonal spectra is a *h-cofibration* if there is a map to the mapping cylinder of i

$$r \colon X \wedge [0,1]_+ \longrightarrow Mi$$

making the following diagram commute.

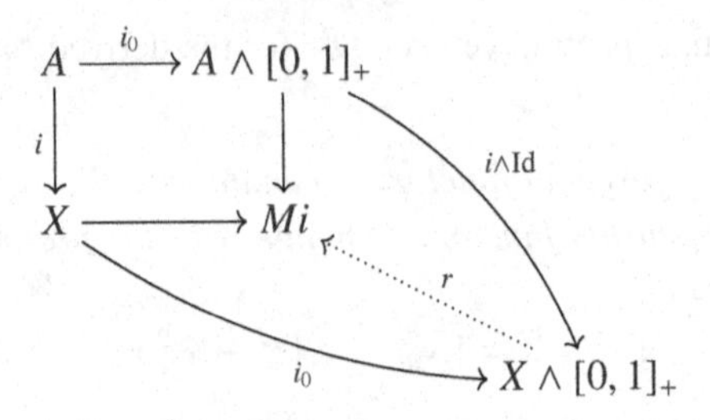

Lemma 5.2.8 *Each level of a h-cofibration of orthogonal spectra is a h-cofibration of pointed spaces.*

The tensor $- \wedge A$ with a pointed space A preserves h-cofibrations of orthogonal spectra.

Proof The tensor with spaces and mapping cylinders is constructed levelwise. □

The converse statement is false: a levelwise h-cofibration of sequential spectra is not necessarily a h-cofibration of sequential spectra. The levelwise retractions of Cyl(X) onto Mi of a levelwise h-cofibration do not have to give a map of spectra.

Recall from Section A.7 the notion of homotopy cofibres of maps of pointed spaces: the homotopy cofibre of $f: A \to X$ is the pushout of

$$A \wedge [0,1] \xleftarrow{i_0} A \xrightarrow{f} X,$$

where $[0,1]$ has basepoint 1. Recall further, that if f is a h-cofibration of pointed spaces, then the natural quotient map $Cf \to X/A$ to the cokernel of f is a homotopy equivalence by Lemma A.5.6.

We can define the homotopy cofibre of a map g of orthogonal spectra analogously, so level n is the homotopy cofibre of g_n. The proof of Lemma A.5.6 extends to orthogonal spectra, giving the next result.

Lemma 5.2.9 *Let $i: A \longrightarrow X$ be a h-cofibration of orthogonal spectra. Then the pushout X/A of i over a point is naturally homotopy equivalent to Ci, the homotopy cofibre of i.* □

Now we introduce the stable model structure on orthogonal spectra. The weak equivalences are our starting point. We choose the π_*-isomorphisms as we did for sequential spectra.

Definition 5.2.10 A map $f: X \to Y$ of orthogonal spectra is said to be a *weak equivalence* of the stable model structure if $\mathbb{U}_{\mathbb{N}}^{\mathcal{O}} f$ is a π_*-isomorphism of sequential spectra. That is, a map $f\colon X \to Y$ is a weak equivalence if for each $n \in \mathbb{Z}$, the map induced by f

$$\operatorname{colim}_k \pi_{n+k}(X_k) \longrightarrow \operatorname{colim}_k \pi_{n+k}(Y_k)$$

is an isomorphism.

It follows from the definition that (small) coproducts or products of π_*-isomorphisms are π_*-isomorphisms. We also see that $\mathbb{U}_{\mathbb{N}}^{\mathcal{O}}$ preserves and detects weak equivalences.

Lemma 5.2.11 *We have the following results on π_*-isomorphisms.*

1. *Let $f\colon X \to Y$ be a map of orthogonal spectra, and let Cf be the homotopy cofibre of f. Then f is a π_*-isomorphism if and only if Cf is π_*-isomorphic to $*$.*
2. *If f is a π_*-isomorphism and a levelwise h-cofibration, then the pushout of f along another map of spectra is a π_*-isomorphism and a levelwise h-cofibration.*
3. *If $(X^i)_{i\in\mathbb{N}}$ is a collection of orthogonal spectra and $f^i\colon X^i \to X^{i+1}$ a collection of levelwise h-cofibrations and π_*-isomorphisms, then the map from the initial object into the colimit*

$$X^0 \longrightarrow \operatorname{colim}_i X^i$$

is a π_*-isomorphism and a levelwise h-cofibration.

Proof The first statement follows from the long exact sequence of homotopy groups of sequential spectra that we have already seen.

For the second, Corollary A.5.4 states that h-cofibrations are preserved by pushouts, so we must show that the map g in the pushout square below is a π_*-isomorphism.

$$\begin{array}{ccc} X & \longrightarrow & P \\ \downarrow{\scriptstyle f} & & \downarrow{\scriptstyle g} \\ Y & \longrightarrow & Q. \end{array}$$

We know that f and g are levelwise h-cofibrations, so the homotopy cofibres of f and g are homotopy equivalent to the cokernels Y/X and Q/P by Lemma 5.2.9. The induced map $Y/X \to P/Q$ is a isomorphism since we have a pushout square. As f is a π_*-isomorphism, Y/X is π_*-isomorphic to a point, hence, so is P/Q. Thus, by the first part, g is a π_*-isomorphism as claimed.

For the third statement, we use Corollary A.7.10 and the Hurewicz model structure on spaces to see that the homotopy colimit of the system

$$\cdots \to X^i \xrightarrow{f^i} X^{i+1} \xrightarrow{f^{i+1}} X^{i+2} \to \cdots$$

is levelwise homotopy equivalent to the colimit of the sequence f^i. Hence, the natural map

$$\mathrm{colim}_i\, \pi_{n+k}(X^i_k) \longrightarrow \pi_{n+k}(\mathrm{colim}_i\, X^i_k)$$

is an isomorphism for all k and n. The result then follows by the definition of π_*-isomorphisms. □

We introduce orthogonal spectra versions of the maps λ_n that we used to make the stable model structure on sequential spectra, see Section 2.2. We abuse notation and use λ_n and k_n for both the sequential and orthogonal spectra versions. We find this acceptable as they agree under prolongation, because prolongation commutes with smash products of spaces and colimits and, hence, commutes with the mapping cone construction.

Definition 5.2.12 Let $\lambda_n \colon \mathrm{F}^{\mathcal{O}}_{n+1} S^1 \to \mathrm{F}^{\mathcal{O}}_n S^0$ be the adjoint of the map

$$S^1 \to \mathrm{Ev}^{\mathcal{O}}_{n+1} \mathrm{F}^{\mathcal{O}}_n S^0 = O(n+1)_+ \wedge_{O(1)} S^1,$$

which is $t \mapsto [\mathrm{Id}, t]$.

As previously, this map is not a q-cofibration. In fact, it is not even a levelwise cofibration of pointed spaces. We need to replace the λ_n by a different collection of maps before we can add them to the generating acyclic cofibrations.

Definition 5.2.13 Define $M\lambda_n$ to be the mapping cylinder of λ_n, that is, the pushout of

$$\begin{array}{ccc} \mathrm{F}^{\mathcal{O}}_{n+1} S^1 & \xrightarrow{\lambda_n} & \mathrm{F}^{\mathcal{O}}_n S^0 \\ \downarrow{\scriptstyle i_1} & & \downarrow{\scriptstyle s_n} \\ (\mathrm{F}^{\mathcal{O}}_{n+1} S^1) \wedge [0,1]_+ & \xrightarrow{t_n} & M\lambda_n, \end{array}$$

where i_1 is the inclusion into the one end of the interval. We then have a map

$$k_n = t_n \circ i_0 \colon \mathrm{F}^{\mathcal{O}}_{n+1} S^1 \to M\lambda_n$$

coming from the inclusion into the zero end of the interval. We also have a deformation retraction $r_n \colon M\lambda_n \to \mathrm{F}^{\mathcal{O}}_n S^0$ induced by the collapse map

$$(\mathrm{F}^{\mathcal{O}}_{n+1} S^1) \wedge [0,1]_+ \to \mathrm{F}^{\mathcal{O}}_{n+1} S^1.$$

The composition $r_n \circ k_n$ is the original map λ_n.

It remains to be shown that the maps λ_n and k_n are indeed π_*-isomorphisms.

Lemma 5.2.14 *For each $n \in \mathbb{N}$, the maps*

$$\lambda_n\colon F^{\mathcal{O}}_{n+1} S^1 \to F^{\mathcal{O}}_n S^0 \quad \textit{and} \quad k_n\colon F^{\mathcal{O}}_{n+1} S^1 \to M\lambda_n$$

are π_*-isomorphisms. Furthermore, the maps k_n are h-cofibrations of orthogonal spectra and levelwise q-cofibrations of pointed spaces.

Proof By our earlier work on π_*-isomorphisms of sequential spectra, we see that it suffices to show that $\Sigma^n \lambda_n$ is a π_*-isomorphism. When $a \geqslant n+1$, level a of $\Sigma^n \lambda_n$ is given by

$$\begin{aligned} S^n \wedge O(a)_+ \wedge_{O(a-n-1)} S^1 \wedge S^{a-n-1} &\xrightarrow{\cong} O(a)_+ \wedge_{O(a-n-1)} \left(S^{n+1} \wedge S^{a-n-1}\right) \\ &\xrightarrow{\cong} O(a)_+ \wedge_{O(a-n-1)} S^a \\ &\longrightarrow O(a)_+ \wedge_{O(a-n)} S^a \end{aligned}$$

induced by the map $O(a-n-1) \to O(a-n)$. We want to replace the domain and codomain with simpler spaces. The key is that we can regard S^a as space with the canonical action of $O(a)$ instead of an action of the smaller groups. We define a pair of isomorphisms

$$O(a)/O(a-n-1)_+ \wedge S^a \longrightarrow O(a)_+ \wedge_{O(a-n-1)} S^a$$

$$O(a)/O(a-n)_+ \wedge S^a \longrightarrow O(a)_+ \wedge_{O(a-n)} S^a.$$

We describe the first map, the second is similar. The domain is the smash product of the space of cosets of $O(a-n-1)$ (with a disjoint basepoint) and S^a, and $O(a)$ acts diagonally on the smash product. The codomain has $O(a)$ acting on the first factor only. The map is given by

$$[\sigma O(a-n-1), t] \mapsto [\sigma, \sigma^{-1} t],$$

and one can check this is well-defined with respect to the choice of coset representative, $O(a)$-equivariant, continuous, and an isomorphism between compact and Hausdorff spaces. The above isomorphisms, the map $\Sigma^n \lambda_n$ (at level a) and

$$\lambda'_{n,a}\colon O(a)/O(a-n-1)_+ \wedge S^a \to O(a)/O(a-n)_+ \wedge S^a$$

fit into a commutative square

$$\begin{array}{ccc}
S^n \wedge O(a)_+ \wedge_{O(a-n-1)} S^1 \wedge S^{a-n-1} & & \\
\downarrow \cong & & \\
O(a)_+ \wedge_{O(a-n-1)} S^a & \xleftarrow{\cong} & O(a)/O(a-n-1)_+ \wedge S^a \\
\downarrow & & \downarrow \lambda'_{n,a} \\
O(a)_+ \wedge_{O(a-n)} S^a & \xleftarrow{\cong} & O(a)/O(a-n)_+ \wedge S^a.
\end{array}$$

Thus, we now focus on the maps $\lambda'_{n,a}$ for varying a and fixed n.

We argue that $\lambda'_{n,a}$ is a $(2a-n-1)$-equivalence. To see this, note that we have a pair of fibre bundles

$$O(a-1) \longrightarrow O(a) \to S^{a-1} \qquad O(a-n) \longrightarrow O(n) \to O(a)/O(a-n).$$

Looking at the long exact sequence of homotopy groups of the first fibre sequence, it follows that

$$O(a-1)_+ \longrightarrow O(a)_+$$

induces an isomorphism on π_q for $q < a-2$ and a surjection on π_q when $q = a-2$. The long exact sequence of homotopy groups of the second fibre sequence shows that the space $O(a)/O(a-n)_+$ is $(a-n-1)$-connected. It follows that the map $\lambda'_{n,a}$ is a $(2a-n-1)$-equivalence, that is, $\pi_{m+1}\lambda'_{n,a}$ is an isomorphism for $m+a < 2a-n-1$ and a surjection for $m+a = 2a-n-1$. Now consider $\pi_{m+a}\lambda'_{n,a}$ for some $m \in \mathbb{Z}$ and take colimits over a. As a grows, the right-hand side of this inequality increases more quickly than the left-hand side, thus $\lambda'_{n,a}$ induces an isomorphism of stable homotopy groups. It follows that $\Sigma^n \lambda_n$ is a π_*-isomorphism as claimed.

To prove that k_n is a π_*-isomorphism, we use Lemma 5.2.11 to see that the map t_n of the definition is a π_*-isomorphism and hence, so is $k_n = t_n \circ i_0$.

The statements about cofibrations follow, as all the constructions are performed levelwise, and the spectrum $\mathrm{F}^{\mathbb{O}}_{n+1} S^1$ is levelwise cofibrant as a pointed space. □

We can now give the generating sets for the stable model structure on orthogonal spectra.

Definition 5.2.15 We say that a map $f\colon X \to Y$ of orthogonal spectra is a *stable fibration* if $\mathbb{U}^{\mathbb{O}}_{\mathbb{N}} f$ is a stable fibration of sequential spectra.

We then define sets of maps of orthogonal spectra as below, where $\square$ denotes the pushout product.

$$\begin{aligned} I^{\mathcal{O}}_{\text{stable}} &= I^{\mathcal{O}}_{\text{level}} = \{\mathrm{F}^{\mathcal{O}}_d(S^{a-1}_+ \longrightarrow D^a_+) \mid a, d \in \mathbb{N})\}, \\ J^{\mathcal{O}}_{\text{level}} &= \{\mathrm{F}^{\mathcal{O}}_d(D^a_+ \longrightarrow (D^a \times [0,1])_+) \mid a, d \in \mathbb{N})\}, \\ J^{\mathcal{O}}_{\text{stable}} &= J^{\mathcal{O}}_{\text{level}} \cup \{k_n \square (S^{a-1}_+ \to D^a_+) \mid a, n \in \mathbb{N}\}. \end{aligned}$$

In particular, we see that the domains of each set of maps are levelwise cofibrant as pointed spaces and each generating cofibration is a levelwise q-cofibration of pointed spaces.

Note that $I^{\mathcal{O}}_{\text{stable}}$ is precisely the prolongation functor $\mathbb{P}^{\mathcal{O}}_{\mathbb{N}}$ applied to the generating cofibrations for the stable (or levelwise) model structure on sequential spectra $I^{\mathbb{N}}_{\text{level}}$ by Lemma 5.2.6. The analogous statement is true for $J^{\mathcal{O}}_{\text{level}}$ and $J^{\mathcal{O}}_{\text{stable}}$ when compared to $J^{\mathbb{N}}_{\text{level}}$ and $J^{\mathbb{N}}_{\text{stable}}$.

Theorem 5.2.16 *The sets $I^{\mathcal{O}}_{\text{stable}}$ and $J^{\mathcal{O}}_{\text{stable}}$ and the class of π_*-isomorphisms form a cofibrantly generated model structure on $\mathcal{S}^{\mathcal{O}}$ called the* stable model structure. *The fibrations are called the* stable fibrations. *The cofibrations are called the* q-cofibrations, *and every q-cofibration is a h-cofibration of spectra, as well as a levelwise q-cofibration of pointed spaces.*

Proof We first prove that a q-cofibration is a h-cofibration. Since $\mathrm{F}^{\mathcal{O}}_d$ commutes with tensoring with pointed spaces and colimits, it follows that the generating cofibrations of the stable model structure are h-cofibrations of orthogonal spectra.

An arbitrary q-cofibration is a retract of a sequential colimit of pushouts of coproducts of the generating q-cofibrations. These constructions preserve h-cofibrations by Corollary A.5.4, so every q-cofibration is a h-cofibration. A similar argument shows that a q-cofibration is a levelwise q-cofibration of pointed spaces.

To prove the stable model structure exists, we apply the lifting lemma, Lemma A.6.12 to the adjunction

$$\mathbb{P}^{\mathcal{O}}_{\mathbb{N}} : \mathcal{S}^{\mathbb{N}} \rightleftarrows \mathcal{S}^{\mathcal{O}} : \mathbb{U}^{\mathcal{O}}_{\mathbb{N}}.$$

We must show that the sets $I^{\mathcal{O}}_{\text{stable}}$ and $J^{\mathcal{O}}_{\text{stable}}$ admit the small object argument, but this holds as $\mathbb{U}^{\mathcal{O}}_{\mathbb{N}}$ commutes with all colimits. We also note that the previously defined

$$k_n \colon \mathrm{F}^{\mathcal{O}}_{n+1} S^1 \longrightarrow M\lambda_n$$

is in fact equal to

$$\mathbb{P}^{\mathcal{O}}_{\mathbb{N}} k_n \colon \mathbb{P}^{\mathcal{O}}_{\mathbb{N}} \mathrm{F}^{\mathbb{N}}_{n+1} S^1 \longrightarrow \mathbb{P}^{\mathcal{O}}_{\mathbb{N}} M\lambda_n,$$

where the second k_n and λ_n are maps of sequential spectra.

We must also show that any element of $J^{\mathcal{O}}_{\text{stable}}$-cell is a π_*-isomorphism. Each map in $J^{\mathcal{O}}_{\text{level}}$ is a levelwise homotopy equivalence and hence, a π_*-isomorphism. The other maps in $J^{\mathcal{O}}_{\text{stable}}$ are of the form $k_n \square i$ for i a generating cofibration of spaces. We have seen that the maps k_n are π_*-isomorphisms and levelwise q-cofibrations. It follows that maps of the form $k_n \square i$ are levelwise q-cofibrations.

A q-cofibration of topological spaces is a h-cofibration, therefore, a levelwise q-cofibration is also a levelwise h-cofibration. This means we can apply Lemma 5.2.11, which shows that the maps of the form $k_n \square i$ are π_*-isomorphisms as well as levelwise h-cofibrations. Hence, we know that each map in $J^{\mathcal{O}}_{\text{stable}}$ is a π_*-isomorphism and a levelwise h-cofibration.

Such maps are preserved by pushouts, coproducts and countable colimits by Lemma 5.2.11, so it follows that the elements of $J^{\mathcal{O}}_{\text{stable}}$-cell are π_*-isomorphisms. We can therefore apply the lifting lemma, Lemma A.6.12 to obtain the final statement. □

Our characterisation of stable fibrations of sequential spectra lifts to orthogonal spectra.

Corollary 5.2.17 *A map of orthogonal spectra $f\colon X \to Y$ is a stable fibration if and only if f is a levelwise fibration of spaces and for each $n \in \mathbb{N}$, the map*

$$X_n \longrightarrow Y_n \underset{\Omega Y_{n+1}}{\times} \Omega X_{n+1}$$

induced by $\widetilde{\sigma}_n$ and f is a weak homotopy equivalence.

The class of stably fibrant orthogonal spectra is exactly the class of Ω-spectra, namely, those orthogonal spectra X whose adjoint structure maps

$$X_n \to \Omega X_{n+1}$$

are weak homotopy equivalences. □

It remains to be shown that this model category is stable and Quillen equivalent to the stable model structure on $\mathcal{S}^{\mathbb{N}}$. The first statement can be shown using the same approach as for sequential spectra, but it follows from the second.

Theorem 5.2.18 *The category of orthogonal spectra equipped with the stable model structure is Quillen equivalent to the stable model structure on sequential spectra. More precisely, the adjunction*

$$\mathbb{P}^{\mathcal{O}}_{\mathbb{N}} : \mathcal{S}^{\mathbb{N}} \rightleftarrows \mathcal{S}^{\mathcal{O}} : \mathbb{U}^{\mathcal{O}}_{\mathbb{N}}$$

is a Quillen equivalence. Hence, $\mathcal{S}^{\mathcal{O}}$ is a stable model category, and the derived functors of $\mathbb{P}_{\mathbb{N}}^{\mathcal{O}}$ and $\mathbb{U}_{\mathbb{N}}^{\mathcal{O}}$ induce an equivalence of categories

$$\mathrm{Ho}(\mathcal{S}^{\mathcal{O}}) \cong \mathrm{Ho}(\mathcal{S}^{\mathbb{N}}) = \mathcal{SHC}.$$

Proof The adjunction is a Quillen adjunction, as the fibrations and weak equivalences are defined by $\mathbb{U}_{\mathbb{N}}^{\mathcal{O}}$. Indeed, $\mathbb{U}_{\mathbb{N}}^{\mathcal{O}}$ preserves and detects all weak equivalences, so it suffices to check that for any cofibrant sequential spectrum X, the unit of the adjunction $X \to \mathbb{U}_{\mathbb{N}}^{\mathcal{O}}\mathbb{P}_{\mathbb{N}}^{\mathcal{O}}X$ is a π_*-isomorphism.

We claim that the unit of the adjunction is a π_*-isomorphism for spectra of the form $\mathrm{F}_n^{\mathbb{N}} S^0$.

Assuming this claim, the unit of the adjunction will be a π_*-isomorphism on spectra of the form $\mathrm{F}_n^{\mathbb{N}} A$ for a pointed CW-complex A, as $- \wedge A$ preserves π_*-isomorphisms and commutes with $\mathbb{U}_{\mathbb{N}}^{\mathcal{O}}$ and $\mathbb{P}_{\mathbb{N}}^{\mathcal{O}}$. Using Lemma 2.2.13, it follows that $X \to \mathbb{U}_{\mathbb{N}}^{\mathcal{O}}\mathbb{P}_{\mathbb{N}}^{\mathcal{O}}X$ is a π_*-isomorphism for any spectrum X that is a sequential colimit of pushouts of coproducts of the generating cofibrations. Since a cofibrant spectrum is a retract of a such a spectrum, the result will follow once we have proven the claim.

By Lemma 2.2.7, a map is a π_*-isomorphism if and only if its suspension is. So it suffices to consider $X = \mathrm{F}_n^{\mathbb{N}} S^n$ for $n \in \mathbb{N}$. The map $\mathrm{F}_n^{\mathbb{N}} S^n \to \mathrm{F}_0^{\mathbb{N}} S^0$ is the composite of suspensions of maps of the form λ_n and hence, is a π_*-isomorphism. Similarly, the prolongation of such a map is a π_*-isomorphism by Ken Brown's Lemma, see Lemma A.4.4. Now consider the commutative square

$$\begin{array}{ccc} \mathrm{F}_n^{\mathbb{N}} S^n & \longrightarrow & \mathbb{U}_{\mathbb{N}}^{\mathcal{O}}\mathbb{P}_{\mathbb{N}}^{\mathcal{O}} \mathrm{F}_n^{\mathbb{N}} S^n \\ \downarrow \simeq & & \downarrow \simeq \\ \mathrm{F}_0^{\mathbb{N}} S^0 & \longrightarrow & \mathbb{U}_{\mathbb{N}}^{\mathcal{O}}\mathbb{P}_{\mathbb{N}}^{\mathcal{O}} \mathrm{F}_0^{\mathbb{N}} S^0. \end{array}$$

The bottom map is the identity map of the sphere spectrum, hence, a stable equivalence, and so the result follows.

This implies that $\mathrm{Ho}(\mathcal{S}^{\mathcal{O}})$ is equivalent to the stable homotopy category. Since $\mathbb{P}_{\mathbb{N}}^{\mathcal{O}}$ and its derived functor commute with the suspension functor as this is smashing with a pointed CW-complex, it follows that $\mathrm{Ho}(\mathcal{S}^{\mathcal{O}})$ is stable. □

A formally identical proof to Proposition 2.1.6 shows that there is a levelwise model structure on orthogonal spectra with weak equivalences and fibrations given by the levelwise weak equivalences and fibrations of pointed spaces. The cofibrations of the levelwise model structure are the q-cofibrations. This model structure is cofibrantly generated by the sets $I_{\mathrm{level}}^{\mathcal{O}}$ and $J_{\mathrm{level}}^{\mathcal{O}}$.

We denote this model structure by $\mathcal{S}^{\mathcal{O}}_l$ and use $[-,-]^l$ for maps in the corresponding homotopy category. Since these generating sets are subsets of the generating sets for the stable model structure, we have the analogue of Lemma 2.3.15.

Lemma 5.2.19 *The identity functor from the levelwise structure on orthogonal spectra to the stable model structure is a left Quillen functor*

$$\mathrm{Id}\colon \mathcal{S}^{\mathcal{O}}_l \rightleftarrows \mathcal{S}^{\mathcal{O}} : \mathrm{Id}.$$

If E is an Ω-spectrum, then

$$[X, E] \cong [X, E]^l. \qquad \square$$

Inspecting the generating sets of the stable model structure gives the analogue of Lemma 2.3.16.

Lemma 5.2.20 *For $d \in \mathbb{N}$, the shifted suspension spectrum functor $\mathrm{F}^{\mathcal{O}}_d$ and the evaluation functor $\mathrm{Ev}^{\mathcal{O}}_d$ form a Quillen adjunction*

$$\mathrm{F}^{\mathcal{O}}_d : \mathrm{Top}_* \rightleftarrows \mathcal{S}^{\mathcal{O}} : \mathrm{Ev}^{\mathcal{O}}_d .$$

In particular, there is a Quillen adjunction

$$\Sigma^\infty : \mathrm{Top}_* \rightleftarrows \mathcal{S}^{\mathcal{O}} : \mathrm{Ev}^{\mathcal{O}}_0 . \qquad \square$$

5.3 Symmetric Spectra

We consider symmetric spectra on pointed topological spaces, with our account based on a mixture of [HSS00] (which uses simplicial sets in place of topological spaces) and [MMSS01]. We aim to give a short and direct account, showing that we can put a stable model structure on symmetric spectra that is Quillen equivalent to the sequential spectra of Chapter 2 and hence, is a model for the stable homotopy category. This requires substantially more effort than for orthogonal spectra, primarily as the weak equivalences for this stable model structure are not the π_*-isomorphisms.

The definition of symmetric spectra is particularly short and relatively simple.

Definition 5.3.1 A *symmetric spectrum* X is a sequence of pointed topological spaces X_n along with the following data and conditions for $n \in \mathbb{N}$.

1. X_n has a continuous action of the symmetric group Σ_n which fixes the basepoint.
2. There are maps of pointed spaces $\sigma_n : S^1 \wedge X_n \to X_{n+1}$.
3. The composite map σ_n^k given by

$$S^k \wedge X_n \xrightarrow{\mathrm{Id}\wedge\sigma_n} S^{k-1} \wedge X_{n+1} \xrightarrow{\mathrm{Id}\wedge\sigma_{n+1}} \cdots \xrightarrow{\sigma_{n+k-1}} X_{n+k}$$

is compatible with the $\Sigma_k \times \Sigma_n$-actions on domain and target. Here, we treat $\Sigma_k \times \Sigma_n$ as a subgroup of Σ_{k+n}, where Σ_k permutes the first k letters and Σ_n the last n letters.

A *morphism of symmetric spectra* $f : X \to Y$ is a collection of Σ_n-equivariant maps $f_n : X_n \to Y_n$ (so $f_n(\alpha x) = \alpha f_n(x)$ for each $x \in X$ and $\alpha \in \Sigma_n$) such that the square

$$\begin{array}{ccc} S^1 \wedge X_n & \xrightarrow{\mathrm{Id}\wedge f_n} & S^1 \wedge Y_n \\ \downarrow \sigma_n^X & & \downarrow \sigma_n^Y \\ X_{n+1} & \xrightarrow{f_{n+1}} & Y_{n+1} \end{array}$$

commutes for each $n \in \mathbb{N}$. We denote the category of symmetric spectra as $\mathcal{S}^\Sigma$.

In the above, the Σ_k-action on S^k is the one-point compactification of the permutation of coordinates action of Σ_k on $\mathbb{R}^k$. This will be referred to as the *canonical action* of Σ_k on S^k.

Note that every symmetric spectrum has an underlying sequential spectrum via forgetting the symmetric group actions: there is a forgetful functor

$$\mathbb{U}_{\mathbb{N}}^{\Sigma} : \mathcal{S}^\Sigma \longrightarrow \mathcal{S}^{\mathbb{N}},$$

where the level n of $\mathbb{U}_{\mathbb{N}}^{\Sigma} X$ is X_n (forgetting the Σ_n-action). As limits and colimits in $\mathcal{S}^\Sigma$ are defined levelwise, the functor $\mathbb{U}_{\mathbb{N}}^{\Sigma}$ preserves all limits and colimits and, thus, has a left and a right adjoint. We are only interested in the left adjoint,

$$\mathbb{P}_{\mathbb{N}}^{\Sigma} : \mathcal{S}^{\mathbb{N}} \longrightarrow \mathcal{S}^\Sigma.$$

A construction of this functor is given in Proposition 6.3.22.

Examples 5.3.2 With our knowledge of other types of spectra, it is relatively simple to give some examples of symmetric spectra.

Sphere spectrum We define the *sphere spectrum* $\mathbb{S}$ to be the symmetric spectrum with level n given by S^n with the canonical action of Σ_n. The structure map

$$S^1 \wedge S^n \longrightarrow S^{n+1}$$

is the usual isomorphism, and we see that the maps σ_n^k are $\Sigma_k \times \Sigma_n$-equivariant.

Suspension spectra Generalising the above somewhat, we can make $\Sigma^\infty A$, the *suspension spectrum* of the pointed space A. Level n is given by $S^n \wedge A$, with Σ_n acting as $\alpha[s, a] = [\alpha s, a]$ for $\alpha \in \Sigma_n$, $a \in A$ and $s \in S^n$. The structure maps are the isomorphisms as for the sphere spectrum

$$S^1 \wedge S^n \wedge A \longrightarrow S^{n+1} \wedge A,$$

and the composite maps σ_n^k are $\Sigma_k \times \Sigma_n$-equivariant.

Shifted suspension spectra The final generalisation along this path is the shifted suspension functors. For a pointed space A and $d \in \mathbb{N}$, we define a spectrum $\mathrm{F}_d^\Sigma A$, where

$$(\mathrm{F}_d^\Sigma A)_n = \begin{cases} (\Sigma_n)_+ \wedge_{\Sigma_{n-d}} S^{n-d} \wedge A & n \geqslant d \\ * & n < d, \end{cases}$$

and Σ_{n-d} is the subgroup which acts on the first $n-d$ letters. The structure map is either trivial or the composite of the inclusion

$$S^1 \wedge (\Sigma_n)_+ \wedge_{\Sigma_{n-d}} S^{n-d} \wedge A \longrightarrow (\Sigma_n)_+ \wedge_{\Sigma_{n-d}} S^{1+n-d} \wedge A$$

with $\Sigma_n \to \Sigma_{1+n}$. Checking that the maps σ_n^k are $\Sigma_k \times \Sigma_n$-equivariant is an exercise in tracking the partition of $k + n$ into $[k|n - d|d]$ through the maps.

Eilenberg–Mac Lane spectra Let R be a ring. We define a symmetric spectrum HR, the *Eilenberg–Mac Lane spectrum* of R, as follows. For K, a simplicial set, let $K \otimes R$ denote the simplicial abelian group whose set of k-simplices is the free R–module on the non-basepoint simplices of K. We identify the basepoint with 0. This can be turned into a topological space by geometric realisation $|K \otimes R|$. Choose a simplicial circle S_s^1 and an isomorphism $S^1 \cong |S_s^1|$. Level n of HR is given by

$$HR_n = |(S_s^1)^n \otimes R|,$$

where Σ_n permutes the factors of the n-fold smash product $(S_s^1)^n$. The structure map is induced by the chosen map $S^1 \cong |S_s^1|$ and the isomorphisms

$$|S_s^1| \wedge |(S_s^1)^n \otimes R| \cong |S_s^1 \wedge (S_s^1)^n \otimes R| \cong |(S_s^1)^{n+1} \otimes R|.$$

This is an Ω-spectrum whose underlying sequential spectrum is the Eilenberg–Mac Lane spectrum of Example 1.3.4. Note that the inclusion

$$(S_s^1)^n \longrightarrow (S_s^1)^n \otimes R$$

that sends a non-basepoint simplex x to $x \otimes 1$ induces a map $\mathbb{S} \longrightarrow HR$ called the Hurewicz map.

Tensor with spaces Given a symmetric spectrum X and a pointed space A, we can construct a new symmetric spectrum $X \wedge A$, where level n is simply $X_n \wedge A$. The structure maps and equivariance are inherited from X. Similarly, we have a cotensor with spaces, given by the symmetric spectrum $\mathrm{Top}_*(A, X)$, which at level n is $\mathrm{Top}_*(A, X_n)$.

As one should expect, suspension spectra are part of an adjunction between symmetric spectra and pointed topological spaces. We have already seen the left adjoints.

Definition 5.3.3 For each $d \in \mathbb{N}$, there is an adjunction

$$\mathrm{F}_d^\Sigma : \mathrm{Top}_* \rightleftarrows \mathcal{S}^\Sigma : \mathrm{Ev}_d^\Sigma$$

between symmetric spectra and pointed topological spaces. The left adjoint F_d^Σ is the shifted suspension functor, the right adjoint Ev_d^Σ sends a spectrum X to the space X_d. When $d = 0$, we write Σ^∞ for F_0^Σ.

As with other categories of spectra, we use Ω^∞ to denote the *derived functor* of Ev_0^Σ from the stable model structure on symmetric spectra to pointed topological spaces, see Section 2.4 for a discussion in the case of sequential spectra.

Now we show that we can equip symmetric spectra with a stable model structure. As with sequential spectra, it is useful to start with the levelwise model structure, then refine that into the stable model structure. This refinement will be an example of the left Bousfield localisations of Chapter 7, though we will not rely on the language of localisations in this chapter.

Definition 5.3.4 The *levelwise model structure* on symmetric spectra is a cofibrantly generated model structure with generating sets given by

$$\begin{aligned} I_{\text{level}}^\Sigma &= \{\mathrm{F}_d^\Sigma(S_+^{a-1} \to D_+^a) \mid a, d \in \mathbb{N})\}, \\ J_{\text{level}}^\Sigma &= \{\mathrm{F}_d^\Sigma(D_+^a \to (D^a \times [0,1])_+) \mid a, d \in \mathbb{N})\}. \end{aligned}$$

The cofibrations of this model structure are called the *q-cofibrations*, the fibrations are called the *levelwise fibrations* and the weak equivalences are called the *levelwise weak equivalences*.

The set of maps from X to Y in the levelwise homotopy category is denoted $[X, Y]^l$.

As in Theorem 2.1.7, the q-cofibrations are not simply the levelwise cofibrations of pointed spaces – a full description is given in [HSS00, Proposition 5.2.2].

Proposition 5.3.5 *The sets I^Σ_{level}, J^Σ_{level} and the class of levelwise weak homotopy equivalences (ignoring the Σ_n-action) define a cofibrantly generated model structure on symmetric spectra. A map f is a fibration if and only if each f_n is a fibration of pointed spaces (with the Σ_n-action ignored). This model structure is called the* levelwise model structure.

Proof Limits and colimits of spectra are defined levelwise, and the Σ_n-action on level n of the (co)limit is induced by the Σ_n-action on the components and the universal property of the (co)limit. The structure maps for a colimit arise from the structure maps of the components and the fact that colimits of pointed spaces commute with smash products. The structure maps for a limit arise similarly, using the adjoints of the structure maps. It follows that all small limits and colimits exist.

Following the Recognition Theorem for cofibrantly generated model categories, Theorem A.6.9, it suffices to show the following points. The proof is very similar to that for Proposition 2.1.6.

1. The levelwise equivalences satisfy the two-out-of-three condition.
2. The domains of the generating sets are small with respect to the class of levelwise cofibrations of spaces.
3. Maps in the class of J^Σ_{level}-cell are levelwise weak equivalences and are I^Σ_{level}-cofibrations.
4. The class of maps with the right lifting property with respect to I^Σ_{level} is exactly the class of levelwise weak equivalences that also have the right lifting property with respect to J^Σ_{level}.

The first is clear. For the second, we show that the spectrum $\mathrm{F}^\Sigma_d A$ is small with respect to the class of levelwise cofibrations of spaces for any compact pointed space A and $d \in \mathbb{N}$. This is sufficient, as every element of I^Σ_{level} is (in particular) a levelwise cofibration of pointed spaces. It follows that every map in I^Σ_{level}-cell is also such. Let

$$X^0 \longrightarrow X^1 \longrightarrow X^2 \longrightarrow \cdots$$

be a diagram of spectra, where each map is a levelwise cofibration of spectra and consider the map of sets

$$\operatorname{colim}_a \mathcal{S}^\Sigma(\mathrm{F}^\Sigma_d A, X^a) \longrightarrow \mathcal{S}^\Sigma(\mathrm{F}^\Sigma_d A, \operatorname{colim}_a X^a).$$

Since the functor Ev^{Σ}_d commutes with colimits, the above map is isomorphic to

$$\mathrm{colim}_a \,\mathrm{Top}_*(A, X^a_d) \longrightarrow \mathrm{Top}_*(A, \mathrm{colim}_a X^a_d).$$

This is an isomorphism for a compact space A: the image of a compact space can only intersect with finitely many cells of the target, see [MP12, Proposition 2.5.4] for details.

For the fourth point, let $i\colon A \longrightarrow B$ be a map of pointed spaces. Then, by adjointness, a map of symmetric spectra $f\colon X \longrightarrow Y$ has the right lifting property with respect to $\mathrm{F}^{\Sigma}_d\, i$ in the diagram

$$\begin{array}{ccc} \mathrm{F}^{\Sigma}_d A & \longrightarrow & X \\ {\scriptstyle \mathrm{F}^{\Sigma}_d i}\downarrow & \nearrow & \downarrow \\ \mathrm{F}^{\Sigma}_d B & \longrightarrow & Y \end{array}$$

if and only if f_d has the right lifting property with respect to i. Hence, f has the right lifting property with respect to $I^{\Sigma}_{\text{level}}$ if and only if it is a levelwise acyclic fibration of pointed spaces. Similarly, f has the right lifting property with respect to $J^{\Sigma}_{\text{level}}$ if and only if it is a levelwise fibration of pointed spaces. The statement then follows.

For the third point, the maps in $J^{\Sigma}_{\text{level}}$ have the left lifting property with respect to levelwise acyclic fibrations, hence, $J^{\Sigma}_{\text{level}}$ consists of $I^{\Sigma}_{\text{level}}$-cofibrations. The maps

$$D^a_+ \longrightarrow (D^a \times [0,1])_+$$

are weak homotopy equivalences of pointed spaces, and thus, by construction, the maps

$$\mathrm{F}^{\Sigma}_d(D^a_+ \longrightarrow (D^a \times [0,1])_+)$$

are levelwise weak homotopy equivalences.

We thus have a cofibrantly generated model category with the properties as claimed. □

We now want to investigate how the tensor product of symmetric spectra and pointed spaces interacts with this model structure.

Lemma 5.3.6 *If $f\colon X \to Y$ is a q-cofibration of symmetric spectra and $i\colon A \to B$ is a cofibration of pointed spaces, then the pushout product*

$$f \,\square\, i\colon X \wedge B \coprod_{X \wedge A} Y \wedge A \longrightarrow Y \wedge B$$

is a cofibration of symmetric spectra. If one of f or i is a weak equivalence, then so is $f \square i$. In particular,

$$Y \wedge -\colon \mathrm{Top}_* \longrightarrow \mathcal{S}^{\Sigma}$$

is a left Quillen functor from pointed spaces to symmetric spectra when Y is a cofibrant symmetric spectrum.

Proof Since $- \square -$ commutes with colimits in either factor, it suffices to show this for the generating cofibrations of the levelwise model structure of symmetric spectra and cofibrations of pointed spaces (see also Lemma 6.1.13). We have three facts: firstly, for cofibrant spaces A and B and $d \in \mathbb{N}$, there is an isomorphism

$$(\mathrm{F}^{\Sigma}_d A) \wedge B \cong \mathrm{F}^{\Sigma}_d(A \wedge B).$$

Secondly, if i and j are (acyclic) cofibrations of pointed spaces, then so is $i \square j$. Thirdly, F^{Σ}_d is a left Quillen functor from pointed spaces to symmetric spectra. These three together imply the statement for the generating (acyclic) cofibrations and thus all (acyclic) cofibrations. The last statement is the case $* \to Y$. □

We now want to develop the stable model structure. This will be a Bousfield localisation of the levelwise model structure at the class of *stable equivalences*. We start by defining the fibrant objects of the stable model structure, which will be the "local objects" in the terminology of Bousfield localisations from Chapter 7.

Definition 5.3.7 A symmetric spectrum X is said to be an *Ω-spectrum* if the adjoints of the structure maps $\widetilde{\sigma}_n\colon X_n \to \Omega X_{n+1}$ are weak equivalences for all $n \in \mathbb{N}$.

The Ω-spectra in symmetric spectra are exactly those spectra sent to an Ω-spectrum in sequential spectra via the forgetful functor $\mathbb{U}^{\Sigma}_{\mathbb{N}}$.

From these "local" objects, we can define the stable equivalences. This class is harder to understand than the π_*-isomorphisms of sequential spectra. One way to motivate this class is that we want the homotopy category of symmetric spectra to be the stable homotopy category, which is the representing category of cohomology theories.

We can relate cohomology theories to weak equivalences in the stable model structure of sequential spectra as follows. By Proposition 5.1.5, a sequential spectrum E defines a cohomology theory on sequential spectra by

$$E^*(-) = [-, E]_{-*}\colon \mathrm{Ho}(\mathcal{S}^{\mathbb{N}}) \longrightarrow \mathrm{Ab}_*.$$

By the Yoneda Lemma, $X \longrightarrow Y$ is a weak equivalence in the stable model structure on sequential spectra if and only if $E^*(Y) \to E^*(X)$ is an isomorphism for all spectra E. By Theorem 2.3.12 and Lemma 2.3.15, we may just as well use the homotopy category of the levelwise model structure and restrict E to the class of Ω-spectra. Hence, we can characterise the weak equivalences of the stable model structure using the levelwise homotopy category. For symmetric spectra, we will take this characterisation as a definition.

Definition 5.3.8 A *stable equivalence* is a map $f: X \longrightarrow Y$ of symmetric spectra such that for each Ω-spectrum E the induced morphism

$$f^*: [Y, E]^l \longrightarrow [X, E]^l$$

is an isomorphism.

Here, $[-,-]^l$ denotes morphisms in the homotopy category of symmetric spectra with the levelwise model structure.

The stable equivalences satisfy the two-out-of-three axiom. Furthermore, every levelwise weak equivalence is a stable equivalence, as weak equivalences induce isomorphisms in their corresponding homotopy category.

Similarly to sequential spectra, we want a model structure whose cofibrations are the q-cofibrations, whose weak equivalences are the stable equivalences and whose class of fibrant objects is precisely the class of Ω-spectra. We can follow the same pattern as for sequential spectra. The first step is to find a set of maps that will identify Ω-spectra.

Let $\lambda_n : \mathrm{F}^{\Sigma}_{n+1} S^1 \to \mathrm{F}^{\Sigma}_n S^0$ be the adjoint of the map

$$S^1 \to \mathrm{Ev}^{\Sigma}_{n+1} \mathrm{F}^{\Sigma}_n S^0 = (\Sigma_{n+1})_+ \wedge S^1 = \bigvee_{\alpha \in \Sigma_{n+1}} S^1,$$

which sends the domain to the $\alpha = \mathrm{Id}$ summand by the identity.

Define $M\lambda_n$ to be the mapping cylinder of λ_n, that is, the pushout of

$$\begin{array}{ccc} \mathrm{F}^{\Sigma}_{n+1} S^1 & \xrightarrow{\lambda_n} & \mathrm{F}^{\Sigma}_n S^0 \\ \downarrow{\scriptstyle i_1} & & \downarrow{\scriptstyle s_n} \\ (\mathrm{F}^{\Sigma}_{n+1} S^1) \wedge [0,1]_+ & \longrightarrow & M\lambda_n, \end{array}$$

where i_1 is the inclusion into the one end of the interval. We then have a map $k_n : \mathrm{F}^{\Sigma}_{n+1} S^1 \to M\lambda_n$ coming from the inclusion into the zero end of the interval and a deformation retraction $r_n : M\lambda_n \to F_n S^0$ induced by the collapse map

$$(\mathrm{F}^{\Sigma}_{n+1} S^1) \wedge [0,1]_+ \to \mathrm{F}^{\Sigma}_{n+1} S^1.$$

The composition $r_n \circ k_n$ is the original map λ_n.

The following lemma shows how the maps k_n are related to Ω-spectra and why they are stable equivalences.

Lemma 5.3.9 *The maps $\lambda_n \colon \mathrm{F}^{\Sigma}_{n+1} S^1 \to \mathrm{F}^{\Sigma}_n S^0$ for $n \in \mathbb{N}$ are stable equivalences. Hence, so are the maps $k_n \colon \mathrm{F}^{\Sigma}_{n+1} S^1 \to M\lambda_n$ for $n \in \mathbb{N}$.*

Proof Let E be an Ω-spectrum. As E is fibrant and $\mathrm{F}^{\Sigma}_{n+1} S^1$ is cofibrant in the levelwise model structure,

$$[\mathrm{F}^{\Sigma}_{n+1} S^1, E]^l \cong \pi(\mathrm{F}^{\Sigma}_{n+1} S^1, E),$$

where $\pi(-,-)$ denotes homotopy classes of maps between spectra. Since smashing with $[0,1]_+$ gives cylinder objects, we see that the model category definition of (left) homotopy agrees with levelwise homotopy of pointed topological spaces. It follows that we can use our adjunction $(\mathrm{F}^{\Sigma}_{n+1}, \mathrm{Ev}^{\Sigma}_{n+1})$ to obtain an isomorphism

$$\pi(\mathrm{F}^{\Sigma}_{n+1} S^1, E) \cong \pi(S^1, E_{n+1}) \cong \pi(S^0, \Omega E_{n+1}),$$

where the middle and right-hand terms refer to homotopy classes of maps of pointed spaces. Similarly, we have isomorphisms

$$[\mathrm{F}^{\Sigma}_n S^0, E]^l \cong \pi(\mathrm{F}^{\Sigma}_n S^0, E) \cong \pi(S^0, E_n).$$

Consider the square below, where the top map sends the homotopy class of a map f to the homotopy class of $\sigma_n \circ \Sigma f$ with σ_n being the structure map of E.

$$\begin{array}{ccc} \pi(S^0, E_n) & \longrightarrow & \pi(S^1, E_{n+1}) \\ \cong \downarrow & & \downarrow \cong \\ [\mathrm{F}^{\Sigma}_n S^0, E]^l & \xrightarrow{\lambda_n^*} & [\mathrm{F}^{\Sigma}_{n+1} S^1, E]^l \end{array}$$

This square commutes as the counit of the adjunction $(\mathrm{F}^{\Sigma}_n, \mathrm{Ev}^{\Sigma}_n)$ in level k is given by the composite structure map of E for $k \geqslant n$ and the trivial map for $k < n$. Therefore, composing with the adjunction isomorphism

$$\pi(S^1, E_{n+1}) \cong \pi(S^0, \Omega E_{n+1})$$

gives a commutative square, where $[f] \in \pi(S^0, E_n)$ is sent to $\widetilde{\sigma}_n \circ f$, where $\widetilde{\sigma}_n$ is the adjoint structure map of E.

$$\begin{array}{ccc} \pi(S^0, E_n) & \xrightarrow{(\widetilde{\sigma}_n)_*} & \pi(S^0, \Omega E_{n+1}) \\ \cong \downarrow & & \downarrow \cong \\ [\mathrm{F}^{\Sigma}_n S^0, E]^l & \xrightarrow{\lambda_n^*} & [\mathrm{F}^{\Sigma}_{n+1} S^1, E]^l \end{array}$$

As E is an Ω-spectrum, the map $\widetilde{\sigma}_n\colon E_n \longrightarrow \Omega E_{n+1}$ is a weak homotopy equivalence. Thus, $(\widetilde{\sigma}_n)_*$ is an isomorphism and λ_n is a stable equivalence.

We know that the retraction r_n is a levelwise homotopy equivalence, hence, a stable equivalence. It follows by the two-out-of-three property that so is k_n. □

We are ready to demonstrate that the stable model structure on symmetric spectra is more complicated to define than that of orthogonal spectra. As mentioned, the problem is that the π_*-isomorphisms are not the correct weak equivalences for this category.

Definition 5.3.10 A map $f\colon X \to Y$ of symmetric spectra is a π_*-isomorphism if the map induced by f

$$\pi_n(X) = \operatorname{colim}_k \pi_{k+n}(X_k) \xrightarrow{f_*} \operatorname{colim}_k \pi_{k+n}(Y_k) = \pi_n(Y)$$

is an isomorphism for all $n \in \mathbb{Z}$.

We see that for a symmetric spectrum X, $\pi_*(X) = \pi_*(\mathbb{U}^{\Sigma}_{\mathbb{N}} X)$.

Remark 5.3.11 The maps of symmetric spectra

$$\lambda_d\colon \mathrm{F}^{\Sigma}_{d+1} S^1 \longrightarrow \mathrm{F}^{\Sigma}_d S^0$$

for $d \in \mathbb{N}$ are not π_*-isomorphisms. Following the same argument as for orthogonal spectra, we look at a map of spectra which at level k is given by a quotient map smashed with the sphere S^k

$$(\Sigma_k/\Sigma_{k-d-1})_+ \wedge S^k \longrightarrow (\Sigma_k/\Sigma_{k-d})_+ \wedge S^k.$$

Taking homotopy groups π_{m+k} and a colimit over k, we obtain a non-injective map from a countable direct sum of $\pi_m^{\text{stable}}(S^0)$ to itself. It follows that λ_d is not a π_*–isomorphism. The homotopy groups of symmetric spectra are studied in more detail by Schwede [Sch08].

We do however have a containment in one direction: a π_*-isomorphism is a stable equivalence. We will make substantial use of this result in proving that the stable equivalences and q-cofibrations form a model structure on symmetric spectra. The basic idea is to construct a functor Q which sends π_*-isomorphisms to levelwise weak equivalences and a natural transformation $i\colon \mathrm{Id} \longrightarrow Q$ which is a levelwise weak equivalence on Ω-spectra.

Proposition 5.3.12 *Every π_*-isomorphism is a stable equivalence.*

Proof To avoid a much harder direct proof, we will use the closed monoidal structure on symmetric spectra. This is constructed in detail in Section 6.3. For now, we use the following properties.

- The category $\mathcal{S}^\Sigma$ is a closed symmetric monoidal category with product $\wedge$, unit $\mathbb{S}$ and internal function object $F_{\mathbb{S}}(-,-)$.
- There is a natural isomorphism $\mathcal{S}^\Sigma(A, F_{\mathbb{S}}(X,Y)) \cong \mathcal{S}^\Sigma(A \wedge X, Y)$.
- Furthermore, for pointed spaces A and B, we have

$$\mathrm{F}_n^\Sigma A \wedge \mathrm{F}_m^\Sigma B \cong \mathrm{F}_{n+m}^\Sigma (A \wedge B),$$

which is Lemma 6.3.21.

All the relevant constructions are entirely categorical, and they do not make use of the model structure we are still constructing at this stage. Therefore, we could have given Section 6.3, which contains the above constructions and properties, before the statement of this proposition.

We start by defining a functor $R\colon \mathcal{S}^\Sigma \longrightarrow \mathcal{S}^\Sigma$. The map

$$\lambda = \lambda_1 \colon \mathrm{F}_1^\Sigma S^1 \longrightarrow \mathrm{F}_0^\Sigma S^0 = \mathbb{S}$$

induces a map

$$\lambda^* \colon X = F_{\mathbb{S}}(\mathrm{F}_0^\Sigma S^0, X) \longrightarrow F_{\mathbb{S}}(\mathrm{F}_1^\Sigma S^1, X) =: RX.$$

Iterating this construction, we get a sequence of maps

$$X \xrightarrow{\lambda^*} RX \xrightarrow{R\lambda^*} R^2X \xrightarrow{R^2\lambda^*} R^3X \xrightarrow{R^3\lambda^*} \cdots,$$

and we define QX to be the homotopy colimit of this sequence in the levelwise model structure. The map from the first term to the homotopy colimit will be denoted $i\colon X \longrightarrow QX$. If we choose a functorial construction of homotopy colimits, then we have a functor and a natural transformation to that functor

$$Q\colon \mathcal{S}^\Sigma \longrightarrow \mathcal{S}^\Sigma \quad \text{and} \quad i\colon \mathrm{Id} \longrightarrow Q.$$

In practice though, we only need to be able to make a few commuting squares, which can be done with a non-functorial model for the homotopy colimit.

We have defined

$$\begin{aligned} R^nX = F_{\mathbb{S}}(\mathrm{F}_1^\Sigma S^1, R^{n-1}X) &= F_{\mathbb{S}}(\mathrm{F}_1^\Sigma S^1, F_{\mathbb{S}}(\mathrm{F}_1^\Sigma S^1, R^{n-2}X)) \\ &\cong F_{\mathbb{S}}(\mathrm{F}_1^\Sigma S^1 \wedge \mathrm{F}_1^\Sigma S^1, R^{n-2}X) \\ &\cong F_{\mathbb{S}}((\mathrm{F}_1^\Sigma S^1)^{\wedge n}, X). \end{aligned}$$

By Lemma 6.3.21, we have $(\mathrm{F}_1^\Sigma S^1)^{\wedge n} \cong \mathrm{F}_n^\Sigma S^n$, so there is a natural isomorphism

$$R^nX \cong F_{\mathbb{S}}(\mathrm{F}_n^\Sigma S^n, X).$$

We can use this to give a description of the levels of $R^n X$ via

$$\begin{aligned}\mathrm{Top}_*(A,(R^nX)_m) &\cong \mathrm{Top}_*(A,\mathrm{Ev}^\Sigma_m F_{\mathbb{S}}(\mathrm{F}^\Sigma_n S^n,X))\\ &\cong \mathcal{S}^\Sigma(\mathrm{F}^\Sigma_m A, F_{\mathbb{S}}(\mathrm{F}^\Sigma_n S^n,X))\\ &\cong \mathcal{S}^\Sigma(\mathrm{F}^\Sigma_m A\wedge \mathrm{F}^\Sigma_n S^n,X)\\ &\cong \mathcal{S}^\Sigma(\mathrm{F}^\Sigma_{n+m}(A\wedge S^n),X)\\ &\cong \mathrm{Top}_*(A\wedge S^n,X_{n+m})\\ &\cong \mathrm{Top}_*(A,\Omega^n X_{n+m}).\end{aligned}$$

The third isomorphism uses the closed symmetric monoidal structure of symmetric spectra as a category, and the fourth is another instance of Lemma 6.3.21. It follows that

$$(R^nX)_m \cong \mathrm{Ev}^\Sigma_m F_{\mathbb{S}}(\mathrm{F}^\Sigma_n S^n,X)\cong \Omega^n X_{n+m},$$

and that the map

$$(R^n\lambda^*)_m\colon \Omega^n X_{n+m}\longrightarrow \Omega^{n+1}X_{n+m+1}$$

is induced from the structure map of X. Thus, if X is an Ω-spectrum, the maps $R^n\lambda^*$ are levelwise weak homotopy equivalences. Furthermore, when X is an Ω-spectrum, the map $i\colon X\longrightarrow QX$ is a levelwise weak equivalence and QX is also an Ω-spectrum.

By the properties of homotopy colimits, given a map $f\colon X\longrightarrow Y$, we may construct a map $Qf\colon QX\longrightarrow QY$ giving a (homotopy) commutative square

$$\begin{array}{ccc} X & \xrightarrow{f} & Y\\ \downarrow{\scriptstyle i_X} & & \downarrow{\scriptstyle i_Y}\\ QX & \xrightarrow{Qf} & QY.\end{array}$$

If $f\colon X\longrightarrow Y$ is a π_*-isomorphism, we claim the map Qf is a levelwise weak equivalence. We calculate the homotopy groups of the levels of QX

$$\begin{aligned}\pi_a(QX_b) &= \mathrm{colim}_c\,\pi_a((R^cX)_b)\\ &\cong \mathrm{colim}_c\,\pi_a(\Omega^c X_{b+c})\\ &\cong \mathrm{colim}_c\,\pi_{a+c}(X_{b+c})\\ &= \pi_{a-b}(X),\end{aligned}$$

where the last term is the homotopy groups of the spectrum X. The claim now follows.

Now assume that $f\colon X\longrightarrow Y$ is a π_*-isomorphism and E is an Ω-spectrum. We have a diagram of sets of maps in the levelwise homotopy category of symmetric spectra.

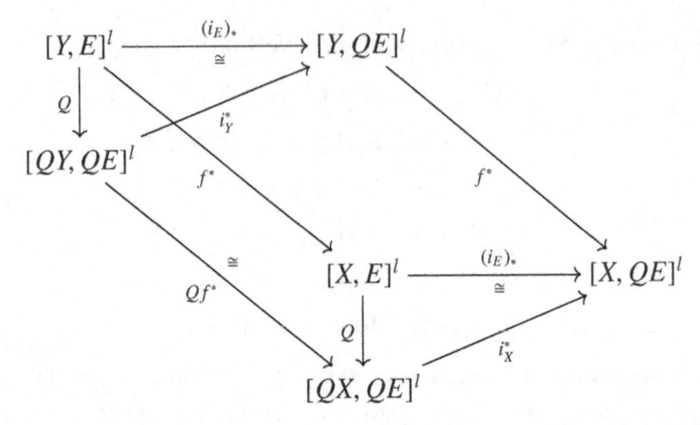

The diagram commutes by naturality of i and Q. Since i_E is a levelwise weak equivalence, the maps $(i_E)_*$ are isomorphisms. The maps labelled Q in the triangles are injective, and the maps i_X^* and i_Y^* are surjective.

The remainder of our proof that

$$f^*\colon [Y,E]^l \longrightarrow [X,E]^l$$

is an isomorphism is a diagram chase. Firstly, let us show that f^* is injective. Let α and β be elements of $[Y,E]^l$ such that $f^*\alpha = f^*\beta$. Then

$$Qf^*(Q\alpha) = Q(f^*\alpha) = Q(f^*\beta) = Qf^*(Q\beta).$$

Since Qf^* and Q are injective, this implies $\alpha = \beta$. Hence, f^* is injective.

Now take $\gamma \in [X,E]^l$. Then there is a $\delta \in [QY,QE]^l$ such that $Qf^*(\delta) = Q(\gamma)$. Similarly, there is an $\varepsilon \in [Y,E]^l$ such that $(i_E)_*(\varepsilon) = i_Y^*(\delta)$. Using the commutativity of the diagram, we see that

$$(i_E)_*f^*(\varepsilon) = f^*(i_E)_*(\varepsilon) = f^*i_Y^*(\delta) = i_X^*Qf^*(\delta) = i_X^*Q(\gamma) = (i_E)_*(\gamma).$$

Since $(i_E)_*$ is an isomorphism, $f^*(\varepsilon) = \gamma$, so f^* is surjective and therefore an isomorphism. □

Note that a direct proof of the existence of a symmetric spectrum with level m given by $\Omega^n X_{n+m}$ and the required properties is difficult to give without using the monoidal structure.

Lemma 5.3.13 *A stable equivalence between Ω-spectra is a levelwise weak equivalence of symmetric spectra.*

Proof Let $f\colon X \to Y$ be a stable equivalence between Ω-spectra. This means that

$$f^*\colon [Y,X]^l \longrightarrow [X,X]^l$$

is an isomorphism. Let g be the pre-image of Id_X. Similarly to the proof of Proposition 1.2.3, we see that g is an inverse isomorphism to f in the levelwise homotopy category. Hence, f and g are levelwise weak equivalences. □

Similarly to Section 2.2, we need to know that a number of operations will preserve stable equivalences. The first one requires some careful consideration of basepoints, as with Lemma 2.2.7. For convenience, it also makes some use of the monoidal structure of symmetric spectra similar to Proposition 5.3.12.

Lemma 5.3.14 *A map of symmetric spectra $f \colon X \to Y$ is a stable equivalence if and only if $\Sigma f \colon \Sigma X \to \Sigma Y$ is a stable equivalence.*

If A is a pointed CW-complex, then $- \wedge A$ preserves stable equivalences.

Proof The cofibrant replacement map in the levelwise model structure

$$X^{cof} \to X$$

is a levelwise weak equivalence and hence is a π_*-isomorphism of symmetric spectra. Suspension preserves π_*-isomorphisms of spectra by Lemma 2.2.7, hence, $\Sigma(X^{cof}) \to \Sigma X$ is a π_*-isomorphism and thus is a stable equivalence.

Let E be an Ω-spectrum. Then we have natural isomorphisms

$$[\Sigma X, E]^l \cong [\Sigma(X^{cof}), E]^l \cong [X^{cof}, \Omega E]^l \cong [X, \Omega E]^l.$$

Since $\Omega E = \mathrm{Top}_*(S^1, E)$ is also an Ω-spectrum, we see that Σ preserves stable equivalences (without assumptions on the basepoints of the spectra). Replacing S^1 by an arbitrary pointed CW-complex A gives the second statement.

For the converse, let E be an Ω-spectrum and consider the symmetric spectrum $F_{\mathbb{S}}(\mathrm{F}_1^\Sigma S^0, E)$. Similarly to the proof of Proposition 5.3.12, we see that

$$F_{\mathbb{S}}(\mathrm{F}_1^\Sigma S^0, E)_k \cong E_{k+1}$$

because

$$\begin{aligned}
\mathrm{Top}_*(A, \mathrm{Ev}_k^\Sigma F_{\mathbb{S}}(\mathrm{F}_1^\Sigma S^0, E)) &\cong \mathcal{S}^\Sigma(\mathrm{F}_k^\Sigma A, F_{\mathbb{S}}(\mathrm{F}_1^\Sigma S^0, E)) \\
&\cong \mathcal{S}^\Sigma(\mathrm{F}_k^\Sigma A \wedge \mathrm{F}_1^\Sigma S^0, E) \\
&\cong \mathcal{S}^\Sigma(\mathrm{F}_{k+1}^\Sigma A, E) \\
&\cong \mathrm{Top}_*(A, E_{k+1})
\end{aligned}$$

for any pointed space A. It follows that $F_{\mathbb{S}}(\mathrm{F}_1^\Sigma S^0, E)$ is an Ω-spectrum and the the structure map of E induces a levelwise weak equivalence

$$E \longrightarrow \Omega F_{\mathbb{S}}(\mathrm{F}_1^\Sigma S^0, E) \cong F_{\mathbb{S}}(\mathrm{F}_1^\Sigma S^1, E) = RE.$$

Assume that $f: X \longrightarrow Y$ is a map of symmetric spectra such that Σf is a stable equivalence. Since E is an Ω-spectrum, it follows from the isomorphisms

$$[X, E]^l \cong [X, \Omega F_{\mathbb{S}}(\mathrm{F}_1^\Sigma S^0, E)]^l \cong [\Sigma X, F_{\mathbb{S}}(\mathrm{F}_1^\Sigma S^0, E)]^l$$

that f induces an isomorphism on the third term and hence, f is a stable equivalence. □

Our next aim is to give the analogous result to Lemma 2.2.13, using stable equivalences in place of π_*-isomorphisms. We will use this list of results to prove the existence of the stable model structure on $\mathbb{S}^\Sigma$. The proof requires some long exact sequence arguments, similar to those for pointed spaces and the notion of h-cofibrations.

We can define h-cofibrations of symmetric spectra just as we did for orthogonal spectra in Definition 5.2.7.

Definition 5.3.15 A map $i: A \longrightarrow X$ of symmetric spectra is a *h-cofibration* if there is a map to the mapping cylinder of i

$$r: X \wedge [0,1]_+ \longrightarrow Mi$$

making the following diagram commute.

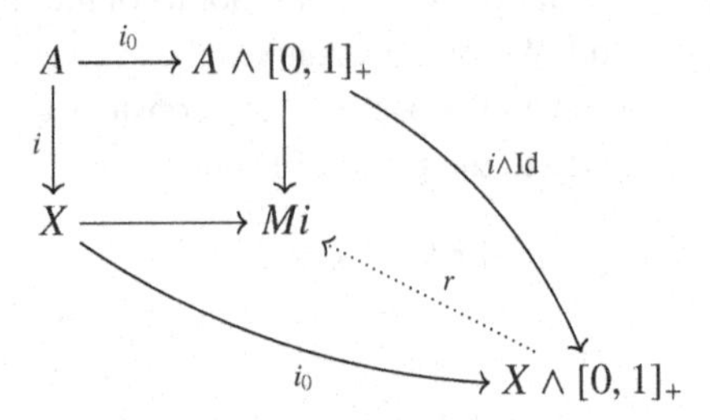

We can relate h-cofibrations to q-cofibrations and levelwise h-cofibrations as we did for orthogonal spectra.

Lemma 5.3.16 *Let $i: A \longrightarrow X$ be a h-cofibration of symmetric spectra. Then the pushout X/A of i over a point is naturally homotopy equivalent to Ci, the homotopy cofibre of i.*

Furthermore,

- *every q-cofibration of symmetric spectra is a h-cofibration,*
- *each level of a h-cofibration of symmetric spectra is a h-cofibration of pointed spaces,*
- *the tensor $- \wedge A$ with a pointed space A preserves h-cofibrations of symmetric spectra.*

Proof These follow from the same proofs as for Lemma 5.2.8, Lemma 5.2.9, Lemma A.5.6, and the part of Theorem 5.2.16 relating to h-cofibrations. □

Consider a q-cofibration $i\colon A \to X$ of cofibrant symmetric spectra. The cofibre of i is also cofibrant and is homotopy equivalent to X/A. The usual arguments for pointed spaces (which need cofibrancy assumptions) applied to each level of i gives a long exact Puppe sequence

$$\cdots \longrightarrow [\Sigma^{n+1}A, E]^l \longrightarrow [\Sigma^n X/A, E]^l \longrightarrow [\Sigma^n X, E]^l \longrightarrow [\Sigma^n A, E]^l \longrightarrow \cdots$$

for any spectrum E. Recall further, that this sequence ends as maps of sets

$$\cdots \longrightarrow [X/A, E]^l \longrightarrow [X, E]^l \longrightarrow [A, E]^l.$$

Lemma 5.3.17 *Several standard operations preserve stable equivalences.*

1. *If $g\colon X \longrightarrow Y$ is a h-cofibration of symmetric spectra and a stable equivalence, then the pushout of g along another map of spectra is also a stable equivalence.*
2. *Given a diagram as below, where i and i' are h-cofibrations and the vertical maps are all stable equivalences*

$$\begin{array}{ccccc} B & \xleftarrow{i} & A & \longrightarrow & C \\ \downarrow \simeq & & \downarrow \simeq & & \downarrow \simeq \\ B' & \xleftarrow{i'} & A' & \longrightarrow & C', \end{array}$$

then the induced map from the pushout P of the top row to the pushout of the second row P' is a stable equivalence.

3. *If $f^i\colon X^i \longrightarrow X^{i+1}$ for $i \in \mathbb{N}$ is a collection of maps which are both levelwise h-cofibrations and stable equivalences, then the map from the first object into the colimit*

$$X^0 \longrightarrow \operatorname{colim}_i X^i$$

is also a stable equivalence and a levelwise h-cofibration.

Proof Consider a diagram

$$Z \xleftarrow{f} X \xrightarrow{g} Y$$

with g a stable equivalence and f a h-cofibration of symmetric spectra. Let P be the pushout of this diagram. Since a h-cofibration of symmetric spectra is, in particular, a levelwise h-cofibration, we can use Part 2. of Lemma 2.2.13 to see that a cofibrant replacement of this diagram in the levelwise model structure (see Section A.7) will give a pushout that is π_*-isomorphic to P. Hence, we may assume that X, Y and Z are cofibrant and the maps f and g are q-cofibrations.

With this assumption, the horizontal cofibres are (up to homotopy equivalence) given by Y/X and P/Z, which are isomorphic. Choosing an Ω-spectrum E, we obtain a map of long exact sequences

$$\begin{array}{ccccccccc}
\cdots \longrightarrow & [\Sigma Y/X, E]^l & \longrightarrow & [\Sigma Y, E]^l & \xrightarrow{\cong} & [\Sigma X, E]^l & \longrightarrow & [Y/X, E]^l & \longrightarrow \cdots \\
& \downarrow \cong & & \downarrow & & \downarrow & & \downarrow \cong & \\
\cdots \longrightarrow & [\Sigma P/Z, E]^l & \longrightarrow & [\Sigma P, E]^l & \longrightarrow & [\Sigma Z, E]^l & \longrightarrow & [P/Z, E]^l & \longrightarrow \cdots
\end{array}$$

from which we see that $\Sigma Z \longrightarrow \Sigma P$ is a stable equivalence. By Lemma 5.3.14, the map $Z \longrightarrow P$ is a stable equivalence.

Now let us show 2. As with the first statement, we may assume without loss of generality that all spectra in the diagram are cofibrant. A long exact sequence argument just like the above shows that B/A is stably equivalent to B'/A'. The map $C \to C'$ is a stable equivalence, and we also have stable equivalences

$$P/C \cong B/A \xrightarrow{\simeq} B'/A' \cong P'/C'.$$

Another long exact sequence argument then shows that $P \longrightarrow P'$ is a stable equivalence.

As for the last point, once again, we begin by making every spectrum in the diagram cofibrant by taking a cofibrant replacement of X^0 and then factorising the map $Y^0 \to X^1$ into a q-cofibration $Y^0 \to Y^1$ followed by a levelwise acyclic fibration $Y^1 \to X^1$. Repeating inductively, we obtain a diagram

$$\begin{array}{ccccccc}
X^0 & \xrightarrow{f^0} & X^1 & \xrightarrow{f^1} & X^2 & \xrightarrow{f^2} & \cdots \\
\uparrow & & \uparrow & & \uparrow & & \\
Y^0 & \longrightarrow & Y^1 & \longrightarrow & Y^2 & \longrightarrow & \cdots
\end{array}$$

with each vertical map a levelwise acyclic fibration and each $Y^i \to Y^{i+1}$ a q-cofibration. Since the maps f^i in the original diagram were stable equivalences, so are the maps $Y^i \to Y^{i+1}$.

As the maps in each sequential diagram are levelwise h-cofibrations, Corollary A.7.10 implies that the colimit of each diagram is levelwise homotopy equivalent to the homotopy colimit. Moreover, the induced map

$$Y := \operatorname{colim} Y^i \simeq \operatorname{hocolim} Y^i \longrightarrow \operatorname{hocolim} X^i \simeq \operatorname{colim} X^i =: X$$

is a levelwise weak equivalence as the homotopy colimits are constructed levelwise.

By Corollary A.7.14, we obtain an exact sequence of sets

$$* \longrightarrow \lim^1[\Sigma Y^i, E]^l \longrightarrow [Y, E]^l \longrightarrow \lim_i[Y^i, E]^l \longrightarrow *.$$

As the maps $f^i \colon Y^i \longrightarrow Y^{i+1}$ in the sequential diagrams are stable equivalences, the maps

$$[Y^{i+1}, E]^l \xrightarrow{\cong} [Y^i, E]^l$$

are isomorphisms. Hence, the $\lim^1$ term is trivial and both maps below are isomorphisms

$$[Y, E]^l \longrightarrow \lim_i[Y^i, E]^l \longrightarrow [Y^0, E]^l.$$

It follows that $Y_0 \longrightarrow Y$ is a stable equivalence and hence, so is $X_0 \longrightarrow X$. □

We are ready to define a model structure on $\mathcal{S}^\Sigma$, where the weak equivalences are the stable equivalences and the cofibrations are the q-cofibrations. We will call this the stable model structure on symmetric spectra. We give the generating sets for this model structure below.

Definition 5.3.18 We define two sets of maps

$$\begin{aligned} I^\Sigma_{\text{stable}} &= I^\Sigma_{\text{level}} = \{\mathrm{F}^\Sigma_n(S^{a-1}_+ \to D^a_+) \mid a, n \in \mathbb{N})\} \\ J^\Sigma_{\text{stable}} &= J^\Sigma_{\text{level}} \cup \{k_n \square (S^{a-1}_+ \to D^a_+) \mid a, n \in \mathbb{N}\}, \end{aligned}$$

where k_n was defined just before Lemma 5.3.9.

We can now give our main theorem for this section, namely, that the desired stable model structure on symmetric spectra exists. We prove this as a series of lemmas which will also characterise the stable fibrations.

Theorem 5.3.19 *The sets I^Σ_{stable} and J^Σ_{stable} and the stable equivalences define a cofibrantly generated model structure on symmetric spectra called the* stable model structure.

*The cofibrations of this model structure are the class of q-*cofibrations, *and the fibrations are called the* stable fibrations. *An object is fibrant if and only if it is an Ω-spectrum.*

Proof We have a category with all small limits and colimits. Following the Recognition Theorem for cofibrantly generated model categories, Theorem A.6.9, it suffices to show the following points.

1. The stable equivalences satisfy the two-out-of-three condition.
2. The domains of the generating sets are small with respect to the class of levelwise cofibrations of spaces.

3. Maps in the class $J^{\Sigma}_{\text{stable}}$-cell are q-cofibrations that are stable equivalences, see Lemma 5.3.20.
4. The class of maps with the right lifting property with respect to $I^{\Sigma}_{\text{stable}}$ is exactly the class of stable equivalences that also have the right lifting property with respect to $J^{\Sigma}_{\text{stable}}$, see Lemma 5.3.22.

The first point is immediate. For the second, we are only concerned with the domains of the maps $k_n \square (S^{a-1}_+ \to D^a_+)$,

$$\mathrm{F}^{\Sigma}_{n+1} S^1 \wedge D^a_+ \coprod_{\mathrm{F}^{\Sigma}_{n+1} S^1 \wedge S^{a-1}_+} M\lambda_n \wedge S^{a-1}_+.$$

To deal with these terms, we use the two facts, namely, $(\mathrm{F}^{\Sigma}_d A) \wedge B \cong \mathrm{F}^{\Sigma}_d(A \wedge B)$ and that pushouts of small objects are small by Lemma A.6.5. This proves the second point.

The last two points will be proven in the following lemmas. □

The third point of the theorem is the following lemma.

Lemma 5.3.20 *The elements of* $J^{\Sigma}_{\text{stable}}$-cell *are stable equivalences and are* $I^{\Sigma}_{\text{stable}}$*-cofibrations.*

Proof The maps $k_n \colon \mathrm{F}^{\Sigma}_{n+1} S^1 \longrightarrow M\lambda_n$ are q-cofibrations of spectra, hence, so are the maps $k_n \square i$ by Lemma 5.3.6. It follows that $J^{\Sigma}_{\text{stable}}$ consists of q-cofibrations. In turn, any map formed by pushouts and sequential colimits of q-cofibrations is also a q-cofibration. Equally, we may use Lemma A.6.11.

The maps in $J^{\Sigma}_{\text{level}}$ are levelwise weak homotopy equivalences and therefore stable equivalences. The maps k_n are stable equivalences by Lemma 5.3.9. It remains to be checked that for a cofibration i of pointed spaces, the map $k_n \square i$ is a stable equivalence. By Lemmas 5.3.14 and 5.3.17, we know that stable equivalences that are q-cofibrations are preserved by smashing with cofibrant pointed spaces and taking pushouts.

The result then follows from Lemma 5.3.17. □

For now, let us say that a map is a *stable fibration* if it has the right lifting property with respect to $J^{\Sigma}_{\text{stable}}$. As with sequential spectra, we can characterise such maps. Indeed, the proof is formally identical to Proposition 2.3.10.

Lemma 5.3.21 *A map of symmetric spectra* $f \colon X \to Y$ *is a stable fibration if and only if* f *is a levelwise fibration and for each* $n \in \mathbb{N}$, *the map*

$$X_n \longrightarrow Y_n \underset{\Omega Y_{n+1}}{\times} \Omega X_{n+1}$$

induced by the structure map $\widetilde{\sigma}_n^X$ and f is a weak homotopy equivalence of pointed spaces. In particular, a spectrum X is stably fibrant if and only if it is an Ω-spectrum. □

The second condition of the lemma is often phrased as saying that a map f is a stable fibration if and only if it is a levelwise fibration and the square below is a homotopy pullback square.

$$\begin{array}{ccc} X_n & \xrightarrow{f_n} & Y_n \\ \downarrow{\scriptstyle \widetilde{\sigma}_n^X} & & \downarrow{\scriptstyle \widetilde{\sigma}_n^Y} \\ \Omega X_{n+1} & \xrightarrow{\Omega f_{n+1}} & \Omega Y_{n+1}. \end{array}$$

Recall from Proposition 5.3.5, that a map has the right lifting property with respect to $I^{\Sigma}_{\text{stable}} = I^{\Sigma}_{\text{level}}$ if and only if it is a levelwise acyclic fibration of pointed spaces.

Lemma 5.3.22 *A stable equivalence f has the right lifting property with respect to $J^{\Sigma}_{\text{stable}}$ if and only if it is a levelwise acyclic fibration of pointed spaces.*

Proof Let $f\colon X \to Y$ be a stable equivalence with the right lifting property with respect to $J^{\Sigma}_{\text{stable}}$. Then f is a levelwise fibration, and

$$\begin{array}{ccc} X_n & \xrightarrow{f_n} & Y_n \\ \downarrow{\scriptstyle \widetilde{\sigma}_n^X} & & \downarrow{\scriptstyle \widetilde{\sigma}_n^Y} \\ \Omega X_{n+1} & \xrightarrow{\Omega f_{n+1}} & \Omega Y_{n+1} \end{array}$$

is a homotopy pullback square for each $n \in \mathbb{N}$. We must show that each f_n is a weak homotopy equivalence.

Consider the kernel F of $f\colon X \to Y$, that is, the pullback of f over $*$. This spectrum is constructed levelwise, and the map $F \to *$ also has the right lifting property with respect to $J^{\Sigma}_{\text{stable}}$. This means that F is stably fibrant and therefore an Ω-spectrum by Lemma 5.3.21. Moreover, F is the homotopy fibre of f.

We can then construct Ci, the (levelwise) homotopy cofibre of the map

$$i\colon F \longrightarrow X.$$

There is a natural map $Ci \to Y$ and a map of cofibre sequences

$$\begin{array}{ccccc} X & \longrightarrow & Ci & \longrightarrow & \Sigma F \\ \downarrow{\scriptstyle =} & & \downarrow & & \downarrow \\ X & \xrightarrow[f]{} & Y & \longrightarrow & Cf. \end{array}$$

The map $\Sigma F \longrightarrow Cf$ is a π_*-isomorphism by Corollary 2.2.12, hence, the map $Ci \longrightarrow Y$ is a π_*-isomorphism by the Five Lemma applied to the long exact sequences of homotopy groups of underlying sequential spectra. It follows that $X \longrightarrow Ci$ is a stable equivalence.

The long exact sequence of the cofibre sequence $F \longrightarrow X \longrightarrow Ci \simeq Y$

$$[F,E]^l \leftarrow [X,E]^l \overset{\cong}{\leftarrow} [Ci,E]^l \leftarrow [\Sigma F,E]^l \leftarrow \cdots$$

shows that $[\Sigma F, E]^l$ is isomorphic to $[*, E]^l = 0$. Thus, $\Sigma F \to *$ is a stable equivalence and so is $F \to *$. By Lemma 5.3.13, the map $F \to *$ is a levelwise weak equivalence.

Hence, each level of F is weakly contractible, and

$$\pi_q(X_n) \longrightarrow \pi_q(Y_n)$$

is an isomorphism for each $q > 0$ and $n \in \mathbb{N}$ by the long exact sequences of homotopy groups of the spaces X_n, Y_n and F_n. We still need to show that the above is an isomorphism for $q = 0$. The map

$$\Omega f_n \colon \Omega X_n \longrightarrow \Omega Y_n$$

is a weak homotopy equivalence for all $n \in \mathbb{N}$. We can write f_n as the composite

$$X_n \longrightarrow Y_n \underset{\Omega Y_{n+1}}{\times} \Omega X_{n+1} \longrightarrow Y_n.$$

As the first map is a weak homotopy equivalence (f is a stable fibration) and the second map is the pullback of the acyclic fibration of spaces Ωf_{n+1}, we see that f_n is a weak equivalence.

For the converse, assume that $f \colon X \to Y$ is a levelwise acyclic fibration. Then f is a stable equivalence and $X_n \to Y_n$ and $\Omega X_{n+1} \to \Omega Y_{n+1}$ are weak homotopy equivalences (and fibrations). Consider the diagram

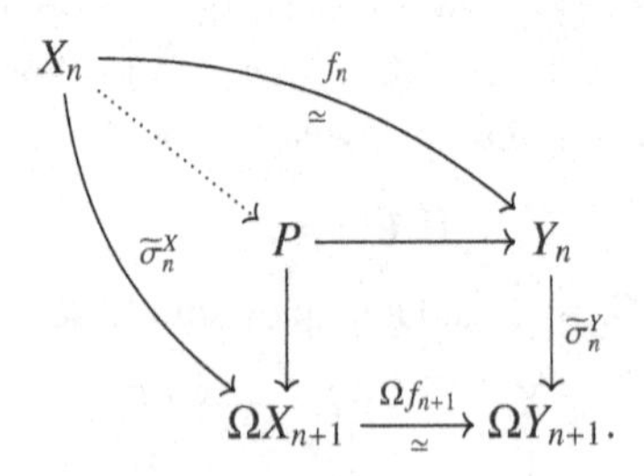

Since Ωf_{n+1} is an acyclic fibration of spaces, so is $P \to Y_n$, hence, it follows that $X_n \to P$ is a weak homotopy equivalence. Therefore, f has the right lifting property with respect to $J^{\Sigma}_{\text{stable}}$ by Lemma 5.3.21. □

This completes the proof of Theorem 5.3.19.

It remains to be shown that this model category is stable and is Quillen equivalent to the stable model structure on orthogonal spectra. We could prove the first point directly, but it will follow from the second.

Recall that at the beginning of this section, we defined a forgetful functor from symmetric spectra to sequential spectra $\mathbb{U}_{\mathbb{N}}^{\Sigma} X$ as follows. Level n of $\mathbb{U}_{\mathbb{N}}^{\Sigma} X$ is X_n (forgetting the Σ_n-action), and the structure maps are those of X. We may compare symmetric spectra and orthogonal spectra in a similar way.

Definition 5.3.23 Given an orthogonal spectrum Y, we can define a symmetric spectrum $\mathbb{U}_{\Sigma}^{\mathcal{O}} Y$ by forgetting structure as follows. Level n of $\mathbb{U}_{\Sigma}^{\mathcal{O}} Y$ is Y_n, with Σ_n acting through the standard inclusion $\Sigma_n \to O(n)$. The structure maps are those of X.

As limits and colimits are constructed levelwise, it follows that both forgetful functors preserve them. It therefore follows that the left adjoints called *prolongation functors* exist, and we have a diagram

$$\mathcal{S}^{\mathbb{N}} \underset{\mathbb{U}_{\mathbb{N}}^{\Sigma}}{\overset{\mathbb{P}_{\mathbb{N}}^{\Sigma}}{\rightleftarrows}} \mathcal{S}^{\Sigma} \underset{\mathbb{U}_{\Sigma}^{\mathcal{O}}}{\overset{\mathbb{P}_{\Sigma}^{\mathcal{O}}}{\rightleftarrows}} \mathcal{S}^{\mathcal{O}}.$$

The composite adjunction between $\mathcal{S}^{\mathbb{N}}$ and $\mathcal{S}^{\mathcal{O}}$ is the adjunction $(\mathbb{P}_{\mathbb{N}}^{\mathcal{O}}, \mathbb{U}_{\mathbb{N}}^{\mathcal{O}})$ of Theorem 5.2.18, which we have already shown to be a Quillen equivalence. We give a construction of the prolongation functors in Proposition 6.3.22.

Theorem 5.3.24 *The category of symmetric spectra equipped with the stable model structure is Quillen equivalent to the category of sequential spectra equipped with the stable model structure and to the category of orthogonal spectra equipped with the stable model structure.*

Hence, we have equivalences of triangulated categories

$$\mathrm{Ho}(\mathcal{S}^{\mathcal{O}}) \cong \mathrm{Ho}(\mathcal{S}^{\Sigma}) \cong \mathrm{Ho}(\mathcal{S}^{\mathbb{N}}) = \mathcal{SHC}.$$

Proof We prove that $(\mathbb{P}_{\mathbb{N}}^{\Sigma}, \mathbb{U}_{\mathbb{N}}^{\Sigma})$ is a Quillen adjunction and that $(\mathbb{P}_{\Sigma}^{\mathcal{O}}, \mathbb{U}_{\Sigma}^{\mathcal{O}})$ is a Quillen equivalence. Theorem 5.2.18 will then imply that the first adjunction is also a Quillen equivalence.

The forgetful functors $\mathbb{U}_{\mathbb{N}}^{\Sigma}$ and $\mathbb{U}_{\Sigma}^{\mathcal{O}}$ send stable fibrations to stable fibrations, as in all three model categories stable fibrations are defined as levelwise fibrations that satisfy a pullback diagram, see Proposition 2.3.10, Corollary 5.2.17 and Lemma 5.3.21. The forgetful functors also preserve acyclic stable fibrations, as these are just levelwise acyclic fibrations of pointed spaces in each case. Hence, we have Quillen adjunctions. Moreover, the forgetful functors preserve π_*-isomorphisms.

We prove that $(\mathbb{P}_\Sigma^{\mathcal{O}}, \mathbb{U}_\Sigma^{\mathcal{O}})$ is a Quillen equivalence by showing that the right adjoint reflects isomorphisms and that the derived unit of the adjunction is an isomorphism.

We claim that if $f \colon X \to Y$ is a map of orthogonal spectra such that $\mathbb{U}_\Sigma^{\mathcal{O}} f$ is a stable equivalence, then f is a π_*-isomorphism. Take a fibrant replacement Y' of Y in the stable model structure of orthogonal spectra and factor the composite

$$X \xrightarrow{f} Y \overset{\sim}{\rightarrowtail} Y'$$

as an acyclic cofibration followed by a stable fibration in the stable model structure

$$X \overset{\sim}{\rightarrowtail} X' \overset{f'}{\twoheadrightarrow} Y'.$$

Then $\mathbb{U}_\Sigma^{\mathcal{O}} f'$ is π_*-isomorphic to $\mathbb{U}_\Sigma^{\mathcal{O}} f$ and hence is also a stable equivalence of symmetric spectra. As $\mathbb{U}_\Sigma^{\mathcal{O}} X'$ and $\mathbb{U}_\Sigma^{\mathcal{O}} Y'$ are Ω-spectra, $\mathbb{U}_\Sigma^{\mathcal{O}} f'$ is a levelwise acyclic fibration, hence, so is f' itself. Thus, f is a π_*-isomorphism.

All that remains is to prove that for a cofibrant symmetric spectrum X, the derived unit map

$$X \longrightarrow \mathbb{U}_\Sigma^{\mathcal{O}} \mathbb{P}_\Sigma^{\mathcal{O}} X \longrightarrow \mathbb{U}_\Sigma^{\mathcal{O}} (\mathbb{P}_\Sigma^{\mathcal{O}} X)^{fib}$$

is a stable equivalence. In fact, it suffices to check this for the strict unit

$$X \longrightarrow \mathbb{U}_\Sigma^{\mathcal{O}} \mathbb{P}_\Sigma^{\mathcal{O}} X,$$

as the second map is $\mathbb{U}_\Sigma^{\mathcal{O}}$ applied to a π_*-isomorphism and hence is itself a π_*-isomorphism.

The rest of the proof follows the same argument as the proof of Theorem 5.2.18. Any cofibrant spectrum is a retract of a sequential colimit of pushouts of coproducts of the generating cofibrations. The functors $\mathbb{U}_\Sigma^{\mathcal{O}}$ and $\mathbb{P}_\Sigma^{\mathcal{O}}$ commute with sequential colimits, pushouts and smash products with pointed spaces. By Lemma 5.3.17, it suffices to prove that the (strict) unit of the adjunction is a stable equivalence for spectra of the form $F_n^\Sigma S^0$. As a map is a stable equivalence if and only if its suspension is, we consider the unit of the adjunction at $X = F_n^\Sigma S^n$ for $n \in \mathbb{N}$.

The map $F_n^\Sigma S^n \to F_0^\Sigma S^0$ is the composite of suspensions of maps of the form λ_n and hence is a stable equivalence. The prolongation of such a map is a π_*-isomorphism by Ken Brown's Lemma. Now consider the commutative square

$$\begin{array}{ccc} \mathbb{F}_n^{\Sigma} S^n & \longrightarrow & \mathbb{U}_{\Sigma}^{\mathbb{O}} \mathbb{P}_{\Sigma}^{\mathbb{O}} \mathbb{F}_n^{\Sigma} S^n \\ \downarrow \simeq & & \downarrow \simeq \\ \mathbb{F}_0^{\Sigma} S^0 & \longrightarrow & \mathbb{U}_{\Sigma}^{\mathbb{O}} \mathbb{P}_{\Sigma}^{\mathbb{O}} \mathbb{F}_0^{\Sigma} S^0. \end{array}$$

The bottom map is the identity map of the sphere spectrum, hence, a stable equivalence. The right-hand map is $\mathbb{U}_{\Sigma}^{\mathbb{O}}$ applied to the π_*-isomorphism

$$\mathbb{F}_n^{\mathbb{O}} S^n \longrightarrow \mathbb{F}_0^{\mathbb{O}} S^0$$

and hence is a π_*-isomorphism of symmetric spectra. Thus, the unit map is a stable equivalence of symmetric spectra. □

As for orthogonal spectra and sequential spectra, we can relate the stable model structure on symmetric spectra to the levelwise model structure and the Serre model structure on pointed spaces.

Lemma 5.3.25 *The identity functor from the levelwise model structure on symmetric spectra to the stable model structure is a left Quillen functor*

$$\mathrm{Id}\colon \mathcal{S}_l^{\Sigma} \rightleftarrows \mathcal{S}^{\Sigma} : \mathrm{Id}.$$

If E is an Ω*-spectrum, then*

$$[X, E] \cong [X, E]^l. \qquad \square$$

Lemma 5.3.26 *For* $d \in \mathbb{N}$*, the shifted suspension spectrum functor* $\mathbb{F}_d^{\Sigma}$ *and the evaluation functor* Ev_d^{Σ} *form a Quillen adjunction*

$$\mathbb{F}_d^{\Sigma}\colon \mathrm{Top}_* \rightleftarrows \mathcal{S}^{\Sigma} : \mathrm{Ev}_d^{\Sigma}.$$

In particular, there is a Quillen adjunction

$$\Sigma^{\infty}\colon \mathrm{Top}_* \rightleftarrows \mathcal{S}^{\Sigma} : \mathrm{Ev}_0^{\Sigma}. \qquad \square$$

One may ask if the levelwise model structure on symmetric spectra is Quillen equivalent to the levelwise model structure on sequential spectra, but this is irrelevant. We are only interested in the levelwise model structures as a stepping stone to the stable model structures, whose purpose is to model the stable homotopy category.

Since we can define stable equivalences for sequential spectra and orthogonal spectra using their respective levelwise model structures, we may ask how the stable equivalences are related to π_*-isomorphisms in those categories. The satisfactory answer is the following.

Lemma 5.3.27 *A map in sequential spectra or orthogonal spectra is a stable equivalence if and only if it is a π_*-isomorphism.*

Proof We work in the case of sequential spectra as the other case is similar. Let $f: X \to Y$ be a map of sequential spectra. Furthermore, let Z be a spectrum with fibrant replacement Z^{fib}, which is an Ω-spectrum. Then there is a commutative square

$$\begin{array}{ccc} [Y,Z] & \xrightarrow{f^*} & [X,Z] \\ \cong\downarrow & & \downarrow\cong \\ [Y,Z^{fib}]^l & \xrightarrow{f^*} & [X,Z^{fib}]^l. \end{array}$$

The map f is a weak equivalence of the stable model structure (i.e. a π_*-isomorphism) if and only if the top map is an isomorphism for all Z. That is equivalent to the bottom map being an isomorphism for all Z^{fib}. This occurs if and only if

$$[Y,E]^l \xrightarrow{f^*} [X,E]^l$$

is an isomorphism for all Ω-spectra E, that is, if and only if f is a stable equivalence. □

5.4 Properness of Spectra

There are a number of model-categorical results that apply to all three kinds of spectra that we have seen so far. For this section, we let $\mathcal{S}$ denote any of the three categories of spectra $\mathcal{S}^{\mathbb{N}}$, $\mathcal{S}^{\Sigma}$, or $\mathcal{S}^{\mathcal{O}}$. As we have shown that each of these model categories is stable, Theorem 4.2.1 tells us that their homotopy categories are triangulated. This implies all kinds of useful properties. In particular, Theorem 3.6.4, Theorem 3.6.1 and Corollaries 4.4.1 and 4.4.2 give us four long exact sequences involving fibre and cofibre sequences. Also, the natural map $X \vee Y \longrightarrow X \prod Y$ is a stable equivalence by Lemma 4.4.4.

We can use these properties to show that the categories of spectra $\mathcal{S}$ are proper.

Definition 5.4.1 Consider a commutative square in a model category $\mathcal{C}$

$$\begin{array}{ccc} A & \xrightarrow{f} & B \\ i\downarrow & & \downarrow j \\ C & \xrightarrow[g]{} & D. \end{array}$$

1. We say that $\mathcal{C}$ is *left proper* if whenever the square is a pushout square, i is a cofibration and f is a weak equivalence, the map g is also a weak equivalence.
2. We say that $\mathcal{C}$ is *right proper* if whenever the square is a pullback square, j is a fibration and g is a weak equivalence, the map f is also a weak equivalence.
3. We say that $\mathcal{C}$ is *proper* if it is both left and right proper.

Many familiar model structures are proper.

Example 5.4.2 The category of (pointed) spaces with either the Serre model structure or the Hurewicz model structure is proper. The category of (pointed) simplicial sets is proper, see [Hir03, Section 13.1].

The category of chain complexes with either the injective or projective model structure is proper [Hov01a].

Lemma 5.4.3 *The stable model structure on $\mathcal{S}$ is proper.*

Proof Consider a commutative square in $\mathcal{S}$

$$\begin{array}{ccc} A & \xrightarrow{f} & B \\ {\scriptstyle i}\downarrow & & \downarrow{\scriptstyle j} \\ C & \xrightarrow[g]{} & D. \end{array}$$

Assume that this is a pushout square with i a h-cofibration and f a stable equivalence (noting Lemma 5.3.27).

Corollary A.5.4 says that j is also a h-cofibration. We then use the same argument as for the first statement of Lemma 2.2.13 and the second statement of Lemma 5.2.11 to give the homotopy equivalences (see Lemma A.5.6) and the isomorphism below.

$$Ci \simeq C/A \cong D/B \simeq Cj.$$

Applying the functor $[-, E]$ to the cofibre sequences of i and j gives a long exact sequence. The Five Lemma shows that

$$g\colon C \longrightarrow D$$

is a stable equivalence. Every q-cofibration is a h-cofibration by Theorem 5.2.16 (whose proof also applies to sequential spectra) and Lemma 5.3.16. It follows that $\mathcal{S}$ is left proper.

Now assume that the square is a pullback, with j a levelwise fibration of pointed spaces and g a stable equivalence. Then i is also a levelwise fibration of pointed spaces. Since the Serre model structure on pointed spaces is right proper, a similar argument to the above gives levelwise weak equivalences and isomorphisms

$$Fi \simeq i^{-1}(*) \cong j^{-1}(*) \simeq Fj.$$

Applying the functor $[-, E]$ to the fibre sequences of i and j gives another long exact sequence. The Five Lemma shows that $f\colon A \to B$ is a stable equivalence. Since every fibration is a levelwise fibration, it follows that $\mathcal{S}$ is right proper. □

Note that the proof uses slightly weaker assumptions: we only require h-cofibrations for left properness and levelwise fibrations for right properness.

5.5 Other Categories of Spectra

There are many other categories of spectra. We discuss a few below and move on to mention the existence of other stable model categories which are related to spectra, but which are not a model for the stable homotopy category.

5.5.1 Spectra of Simplicial Sets

Chapter 2 and Section 5.3 could just as easily be written using pointed simplicial sets in place of pointed topological spaces. The primary reference is Hovey, Shipley and Smith [HSS00]. We give a summary of the main results as the proofs are similar to the techniques used in earlier sections.

Definition 5.5.1 A *sequential spectrum in simplicial sets* X is a sequence of pointed simplicial sets X_n, $n \in \mathbb{N}$, with *structure maps*

$$\sigma_n^X : \Sigma X_n \longrightarrow X_{n+1}.$$

A morphism $f\colon X \longrightarrow Y$ in sequential spectra in simplicial sets is a sequence of pointed maps of simplicial sets $f_n\colon X_n \longrightarrow Y_n$ such that for each $n \in \mathbb{N}$, the square below commutes.

$$\begin{array}{ccc} \Sigma X_n & \xrightarrow{\Sigma f_n} & \Sigma Y_n \\ \downarrow \sigma_n^X & & \downarrow \sigma_n^Y \\ X_{n+1} & \xrightarrow{f_{n+1}} & Y_{n+1} \end{array}$$

We denote this category $\mathcal{S}^{\mathbb{N}}(\mathrm{sSet}_*)$.

This category has all limits and colimits created levelwise in simplicial sets. The evaluation functors $\mathrm{Ev}_n^{\mathbb{N}}$ have left adjoints $\mathrm{F}_n^{\mathbb{N}}$ as in the topological case.

Proposition 5.5.2 *The category of sequential spectra in simplicial sets has a levelwise model structure, where the weak equivalences and fibrations are the levelwise weak equivalences and fibrations of simplicial sets. We call the cofibrations the q-cofibrations.*

A map $f\colon X \to Y$ is a q-cofibration if and only if each f_n is injective and the maps

$$X_{n+1} \coprod_{\Sigma X_n} \Sigma Y_n \longrightarrow Y_{n+1}$$

induced by f and the structure maps of X and Y are injective.

This category is cofibrantly generated, with generating sets analogous to those for the levelwise model structure in the topological case (Proposition 2.1.6), where the generating sets of topological spaces have been replaced with the generating sets of pointed simplicial sets. In other words,

$$\begin{aligned} I_{\text{level}} &= \{\mathrm{F}_d^{\mathbb{N}}\,\partial\Delta[n]_+ \longrightarrow \mathrm{F}_d^{\mathbb{N}}\,\Delta[n]_+ \mid n, d \in \mathbb{N}\} \\ J_{\text{level}} &= \{\mathrm{F}_d^{\mathbb{N}}\,\Lambda^r[n]_+ \longrightarrow \mathrm{F}_d^{\mathbb{N}}\,\Delta[n]_+ \mid d \in \mathbb{N},\ 0 < n,\ 0 \leqslant r \leqslant n\}. \end{aligned}$$

Definition 5.5.3 An Ω-spectrum in $\mathcal{S}^{\mathbb{N}}(\mathrm{sSet}_*)$ is a spectrum X such that each level X_n is a Kan complex and the adjoint structure maps

$$\widetilde{\sigma}_n^X\colon X_n \longrightarrow \Omega X_{n+1}$$

are weak equivalences of simplicial sets.

For the rest of this section we write $\mathcal{S}^{\mathbb{N}}(\mathrm{Top}_*)$ for the category of sequential spectra in topological spaces. Given such a spectrum Y, we can apply the singular complex functor levelwise and obtain a sequential spectrum $\operatorname{sing} Y$ in simplicial sets. The structure maps are defined by

$$\Omega(\operatorname{sing} Y)_n = \Omega(\operatorname{sing} Y_n) \cong \operatorname{sing}(\Omega Y_n) \xrightarrow{\widetilde{\sigma}_n^Y} \operatorname{sing}(Y_{n+1}) = (\operatorname{sing} Y)_{n+1}.$$

Equally, we may apply the geometric realisation functor $|-|$ levelwise to a sequential spectrum in simplicial sets X and obtain an object $|X|$ of $\mathcal{S}^{\mathbb{N}}(\mathrm{Top}_*)$. The structure maps are defined by

$$\Sigma(|X|)_n = \Sigma(|X|_n) = \Sigma|X_n| \cong |\Sigma X_n| \xrightarrow{\sigma_n^X} |X_{n+1}| = |X|_{n+1}.$$

Using geometric realisation, we may define a π_*-isomorphism of sequential spectra in simplicial sets as a map whose geometric realisation is a π_*-isomorphism of sequential spectra in topological spaces.

Theorem 5.5.4 *The category of sequential spectra in simplicial sets has a stable model structure with weak equivalences given by the π_*-isomorphisms and cofibrations given by the q-cofibrations.*

The fibrations are the levelwise Kan fibrations $f\colon X \longrightarrow Y$ such that the commutative square

$$\begin{array}{ccc} X_n & \xrightarrow{f} & Y_n \\ \downarrow{\scriptstyle \widetilde{\sigma}_n^X} & & \downarrow{\scriptstyle \widetilde{\sigma}_n^Y} \\ \Omega X_{n+1} & \xrightarrow{\Omega f} & \Omega Y_{n+1} \end{array}$$

is a homotopy pullback square of pointed simplicial sets for each $n \in \mathbb{N}$.

We may compare the stable model structure on sequential spectra in simplicial sets with the stable model structure on sequential spectra in topological spaces using the geometric realisation–singular complex adjunction.

Theorem 5.5.5 *Geometric realisation and the singular complex functor induce a Quillen equivalence*

$$|-|\colon \mathcal{S}^{\mathbb{N}}(\mathrm{sSet}_*) \rightleftarrows \mathcal{S}^{\mathbb{N}}(\mathrm{Top}_*) : \mathrm{sing}.$$

Proof Both functors preserve all levelwise weak equivalences, and the singular complex functor preserves levelwise fibrations and levelwise acyclic fibrations. The stable fibrations of $\mathcal{S}^{\mathbb{N}}(\mathrm{Top}_*)$ are defined as levelwise fibrations satisfying a homotopy pullback square property in pointed spaces analogous to Theorem 5.5.4. As sing is a right Quillen functor, it sends homotopy pullback squares in pointed spaces to homotopy pullback squares in pointed simplicial sets. Hence, we have a Quillen adjunction on the stable model structures.

We now prove it is a Quillen equivalence. Let X be a cofibrant spectrum in simplicial sets and Y a fibrant spectrum in topological spaces. The derived counit of the adjunction at Y is the composite of a pair of levelwise weak equivalences

$$|(\mathrm{sing}\, Y)^{cof}| \longrightarrow |\mathrm{sing}\, Y| \longrightarrow Y.$$

The derived unit of the adjunction at X is the composite

$$X \xrightarrow{\sim} \mathrm{sing}\,|X| \longrightarrow \mathrm{sing}\,|X^{fib}|.$$

The first map is a levelwise weak equivalence, and the second map is levelwise weakly equivalent to the π_*-isomorphism

$$X \longrightarrow X^{fib}.$$

□

We may repeat the above for symmetric spectra in simplicial sets. The starting point is simplicial sets with an action of the symmetric group on n letters, Σ_n. An action of Σ_n on a pointed simplicial set A is an action of Σ_n on each set A_k that preserves the basepoint, and these actions should commute with the face and degeneracy maps.

Definition 5.5.6 A *symmetric spectrum in simplicial sets* X is a sequence of pointed simplicial sets X_n, along with the following data and conditions, for $n \in \mathbb{N}$.

1. X_n has an action of Σ_n which fixes the basepoint.
2. There are maps of pointed simplicial sets $\sigma_n \colon S^1 \wedge X_n \to X_{n+1}$.
3. The composite map σ_n^k given by
$$S^k \wedge X_n \xrightarrow{\mathrm{Id}\wedge\sigma_n} S^{k-1} \wedge X_{n+1} \xrightarrow{\mathrm{Id}\wedge\sigma_{n+1}} \cdots \xrightarrow{\sigma_{n+k-1}} X_{n+k}$$
is compatible with the $\Sigma_k \times \Sigma_n$-actions on domain and target. Here, we treat $\Sigma_k \times \Sigma_n$ as a subgroup of Σ_{k+n}, where Σ_k permutes the first k letters and Σ_n the last n letters.

A *morphism of symmetric spectra* $f \colon X \to Y$ is a collection of Σ_n-equivariant maps
$$f_n \colon X_n \to Y_n$$
(so $f_n(\alpha x) = \alpha f_n(x)$ for each $x \in X$ and $\alpha \in \Sigma_n$) such that the square
$$\begin{array}{ccc} S^1 \wedge X_n & \xrightarrow{\mathrm{Id}\wedge f_n} & S^1 \wedge Y_n \\ \Big\downarrow{\scriptstyle \sigma_n^X} & & \Big\downarrow{\scriptstyle \sigma_n^Y} \\ X_{n+1} & \xrightarrow{f_{n+1}} & Y_{n+1} \end{array}$$
commutes for each $n \in \mathbb{N}$. We denote the category of symmetric spectra in simplicial sets as $\mathcal{S}^{\Sigma}(\mathrm{sSet}_*)$.

Just as for sequential spectra, we get levelwise and stable model structures on symmetric spectra in simplicial sets.

Proposition 5.5.7 *There is a levelwise model structure on symmetric spectra in simplicial sets, where the weak equivalences are the levelwise weak equivalences of simplicial sets and where the fibrations are the levelwise fibrations of simplicial sets. We call the cofibrations the q-cofibrations.*

This category is cofibrantly generated with generating sets analogous to those for the levelwise model structure in the topological case (see Definition 5.3.4), but using the generating sets of pointed simplicial sets, that is,

$$\begin{aligned} I_{\text{level}} &= \{\mathrm{F}_d^{\Sigma}(\partial\Delta[n]_+ \longrightarrow \mathrm{F}_d^{\Sigma}\Delta[n]_+) \mid n, d \in \mathbb{N}\} \\ J_{\text{level}} &= \{\mathrm{F}_d^{\Sigma}(\Lambda^r[n]_+ \longrightarrow \mathrm{F}_d^{\Sigma}\Delta[n]_+) \mid d \in \mathbb{N},\ 0 < n,\ 0 \leqslant r \leqslant n\}. \end{aligned}$$

Similarly, there is a stable model structure on $\mathcal{S}^{\Sigma}(\mathrm{sSet}_*)$. We omit the proofs that the levelwise and stable model structures exist, as they are similar in nature to those of Section 5.3.

Theorem 5.5.8 *There is a* stable model structure *on the category of symmetric spectra in simplicial sets $\mathcal{S}^{\Sigma}(\mathrm{sSet}_*)$ with weak equivalences given by the stable equivalences and cofibrations by the q-cofibrations.*

The fibrations are the levelwise Kan fibrations $f\colon X \longrightarrow Y$ such that the commutative square

$$\begin{array}{ccc} X_n & \xrightarrow{f} & Y_n \\ \downarrow{\scriptstyle \widetilde{\sigma}_n^X} & & \downarrow{\scriptstyle \widetilde{\sigma}_n^Y} \\ \Omega X_{n+1} & \xrightarrow{\Omega f} & \Omega Y_{n+1} \end{array}$$

is a homotopy pullback square of pointed simplicial sets for each $n \in \mathbb{N}$.

The primary reference for symmetric spectra in simplicial sets is [HSS00]. That reference takes a slightly different approach, using *injective* Ω-spectra to define stable equivalences, see [HSS00, Definition 3.1.1].

As with the topological case, there is a forgetful functor $\mathbb{U}_{\mathbb{N}}^{\Sigma}$ from symmetric spectra to sequential spectra. It has a left adjoint $\mathbb{P}_{\mathbb{N}}^{\Sigma}$.

Theorem 5.5.9 *There is a diagram of Quillen equivalences*

$$\begin{array}{ccc} \mathcal{S}^{\mathbb{N}}(\mathrm{sSet}_*) & \underset{\text{sing}}{\overset{|-|}{\rightleftarrows}} & \mathcal{S}^{\mathbb{N}}(\mathrm{Top}_*) \\ {\scriptstyle \mathbb{P}_{\mathbb{N}}^{\Sigma}}\downarrow\uparrow{\scriptstyle \mathbb{U}_{\mathbb{N}}^{\Sigma}} & & {\scriptstyle \mathbb{P}_{\mathbb{N}}^{\Sigma}}\downarrow\uparrow{\scriptstyle \mathbb{U}_{\mathbb{N}}^{\Sigma}} \\ \mathcal{S}^{\Sigma}(\mathrm{sSet}_*) & \underset{\text{sing}}{\overset{|-|}{\rightleftarrows}} & \mathcal{S}^{\Sigma}(\mathrm{Top}_*), \end{array}$$

such that the square of left adjoints commutes up to natural isomorphism (and hence, so does the square of right adjoints).

Proof The left-hand vertical pair is a Quillen adjunction, as the right adjoint $\mathbb{U}_{\mathbb{N}}^{\Sigma}$ preserves levelwise (acyclic) fibrations and preserves the homotopy pullback squares defining stable fibrations.

We have seen that the right-hand vertical pair is a Quillen equivalence by Theorem 5.3.24. The top horizontal adjunction is a Quillen equivalence by Theorem 5.5.5. A similar argument applies to show that the lower horizontal pair is also a Quillen equivalence.

The result follows from the two-out-of-three property for Quillen equivalences. □

A direct proof of the Quillen equivalence between symmetric spectra (in simplicial sets) and sequential spectra (in simplicial sets) is given as [HSS00, Theorem 4.2.5].

5.5.2 Diagram Spectra

We describe two other constructions from Mandell, May, Schwede and Shipley [MMSS01]. The first, $\mathcal{S}^{\mathcal{W}}$, is another model for the stable homotopy category. The category $\mathcal{S}^{\mathcal{W}}$, along with sequential spectra, symmetric spectra and orthogonal spectra, is an example of *diagram spectra*, as we discuss in more detail in Chapter 6.

The second construction, $\mathcal{S}^{\mathcal{F}}$, is a model for that part of the stable homotopy category whose objects have trivial homotopy groups in negative degrees, also known as connective spectra. A thorough treatment is given in [Sch99].

The first construction uses a very complicated category to define its category of spectra. While this is an obvious drawback, the suspension functor appears naturally from the definition.

Definition 5.5.10 We define $\mathcal{W} \subset \mathrm{Top}_*$ to be the full subcategory consisting of pointed spaces isomorphic to finite CW-complexes. A *$\mathcal{W}$-spectrum* is a functor from $\mathcal{W}$ to Top_* that is enriched over Top_*.

A *morphism of $\mathcal{W}$-spectra* is a natural transformation of Top_*-enriched functors. We denote the category of $\mathcal{W}$-spectra by $\mathcal{S}^{\mathcal{W}}$.

It is not immediately clear why $\mathcal{S}^{\mathcal{W}}$ deserve to be called "spectra", so we show how to define a sequential spectrum from a $\mathcal{W}$-spectrum. Let

$$F\colon \mathcal{W} \longrightarrow \mathrm{Top}_*$$

be a Top_*-enriched functor, and let A and B be finite CW-complexes. Since F is an enriched functor, there are maps of pointed spaces

$$F_{A,B}\colon \mathcal{W}(A,B) \longrightarrow \mathrm{Top}_*(F(A),F(B))$$
$$\widetilde{F}_{A,B}\colon F(A) \wedge \mathcal{W}(A,B) \longrightarrow F(B)$$

which correspond to each other under the adjunction of smash product and function spaces. The smash product functor of pointed spaces also induces a map

$$a_{A,B}\colon B \cong \mathcal{W}(S^0,B) \xrightarrow{A\wedge -} \mathcal{W}(A, A \wedge B).$$

Combining these maps gives the assembly map of F

$$F(A) \wedge B \xrightarrow{\mathrm{Id} \wedge a_{A,B}} F(A) \wedge \mathscr{W}(A, A \wedge B) \xrightarrow{\widetilde{F}_{A,A\wedge B}} F(A \wedge B).$$

We may now define a sequential spectrum $\mathbb{U}_{\mathbb{N}}^{\mathscr{W}} F$ by $(\mathbb{U}_{\mathbb{N}}^{\mathscr{W}} F)_n = F(S^n)$. Its structure maps are the assembly maps defined above with $A = S^n$ and $B = S^1$, giving a map

$$\Sigma F(S^n) \longrightarrow F(S^{n+1}).$$

We can think of the assembly map as allowing us to suspend with respect to any finite CW-complex.

Theorem 5.5.11 *There is a stable model structure on the category $\mathcal{S}^{\mathscr{W}}$ of $\mathscr{W}$-spectra, where the weak equivalences are the π_*-isomorphisms of underlying sequential spectra. The fibrations are defined termwise. The cofibrations are those maps with the left lifting property with respect to termwise acyclic fibrations.*

The functor $\mathbb{U}_{\mathbb{N}}^{\mathscr{W}} : \mathcal{S}^{\mathscr{W}} \longrightarrow \mathcal{S}^{\mathbb{N}}$ has a left adjoint, and this adjunction is a Quillen equivalence with respect to the stable model structures. □

This model structure is called the *absolute stable model structure* in [MMSS01]. In the other direction, we may take a very simple category and obtain a model for the category of *connective spectra*, that is, spectra whose homotopy groups are zero in negative degrees. Thus, we do not obtain a model for the stable homotopy category, but the resulting homotopy category is still interesting. Consequently, we do not use the term "spectra" for these objects.

Definition 5.5.12 We define $\mathscr{F}$ to be the category of finite pointed sets

$$\mathbf{n}_+ = \{0, 1, \ldots, n\}$$

with basepoint 0, for $n \in \mathbb{N}$. The morphisms are the pointed maps of sets.

An *$\mathscr{F}$-space* is a Top_*-enriched functor from $\mathscr{F}$ to Top_*. A *morphism of $\mathscr{F}$-spaces* is a Top_*-enriched natural transformation. The category of $\mathscr{F}$-spaces is denoted by $\mathcal{S}^{\mathscr{F}}$.

The category of $\mathscr{F}$-spaces can also be found in literature as the category of Γ-spaces.

Since a finite pointed set is a finite CW-complex, any $\mathscr{W}$-spectrum defines an $\mathscr{F}$-space. There is an adjunction

$$\mathbb{P}_{\mathscr{F}}^{\mathscr{W}} : \mathcal{S}^{\mathscr{F}} \rightleftarrows \mathcal{S}^{\mathscr{W}} : \mathbb{U}_{\mathscr{F}}^{\mathscr{W}}.$$

Given an $\mathscr{F}$-space X, we may define the homotopy groups of X to be the homotopy groups of the sequential spectrum underlying $\mathbb{P}_{\mathscr{F}}^{\mathscr{W}} X$. This gives us the notion of π_*-isomorphisms of $\mathscr{F}$-spaces.

Theorem 5.5.13 *The category of $\mathscr{F}$-spaces has a model structure with the weak equivalences being the π_*-isomorphisms. The cofibrations are those maps that have the left lifting property with respect to levelwise acyclic fibrations of pointed spaces.*

The prolongation–forgetful functor adjunction

$$\mathbb{P}_{\mathscr{F}}^{\mathscr{W}} : \mathcal{S}^{\mathscr{F}} \rightleftarrows \mathcal{S}^{\mathscr{W}} : \mathbb{U}_{\mathscr{F}}^{\mathscr{W}}$$

is a Quillen adjunction with this model structure.

If X is a cofibrant $\mathscr{F}$-space, then $\mathbb{P}_{\mathscr{F}}^{\mathscr{W}} X$ is a $\mathscr{W}$-spectrum with $\pi_k(\mathbb{P}_{\mathscr{F}}^{\mathscr{W}} X) = 0$ for all $k < 0$. If Y is a $\mathscr{W}$-spectrum with $\pi_k(Y) = 0$ for all $k < 0$, then the derived unit map at Y

$$\mathbb{P}_{\mathscr{F}}^{\mathscr{W}} (\mathbb{U}_{\mathscr{F}}^{\mathscr{W}} Y^{fib})^{cof} \longrightarrow \mathbb{P}_{\mathscr{F}}^{\mathscr{W}} (\mathbb{U}_{\mathscr{F}}^{\mathscr{W}} Y^{fib}) \longrightarrow Y^{fib} \longrightarrow Y$$

is a π_*-isomorphism.

We view the last two statements as saying that $\mathscr{F}$-spaces are a model for connective spectra. By Remark 2.4.6, connective spectra are strongly related to infinite loop spaces. In [Seg74], Segal uses $\mathscr{F}$-spaces (Γ-spaces) to examine this relation in detail.

5.5.3 More Spectra

S–Modules

We briefly mention the category of S–modules of Elmendorf, Kriz, Mandell and May [EKMM97]. We leave the (lengthy) definition to the reference and concentrate on the main theorem. Our reason for including S–modules on such a brief scale rather than omitting them altogether is that they provide a model for the stable homotopy category where every object is fibrant, which can sometimes provide a technical advantage.

Theorem 5.5.14 *There is a stable model structure on the category of S–modules which is Quillen equivalent to the category of orthogonal spectra equipped with the positive stable model structure of Proposition 6.7.1.*

Every object of the stable model structure on S–modules is fibrant.

The Quillen equivalence statement can be found in [MM02, Chapter 1], see also [Sch01a] for a Quillen equivalence between S–modules and symmetric spectra rather than orthogonal spectra. In fact, these Quillen equivalences are strong symmetric monoidal and pass to Quillen equivalences of rings, modules and commutative rings.

Spectra in General Model Categories

Hovey [Hov01b] provides a theory of how to generalise the notion of spectra and symmetric to spectra to more general model categories.

The base idea is that one takes a model category $\mathcal{C}$ and a left Quillen functor $T\colon \mathcal{C} \longrightarrow \mathcal{C}$ with right adjoint U and makes a category

$$\mathcal{S}^{\mathbb{N}}(\mathcal{C}, T),$$

where an object X is a collection $X_n \in \mathcal{C}$, $n \geqslant 0$, with maps $\sigma_n\colon TX_n \to X_{n+1}$.

A map of spectra $f\colon X \longrightarrow Y$ is then a sequence of maps $f_n\colon X_n \longrightarrow Y_n$ commuting with the structure maps.

After making a levelwise model structure, one constructs a stable model structure via a left Bousfield localisation, see Chapter 7. This requires $\mathcal{C}$ to be left proper and cellular (in the sense of [Hir03, Chapter 12]). The fibrant objects are the U-spectra, that is, those spectra which are levelwise fibrant and where the adjoints of the structure maps

$$X_n \longrightarrow UX_{n+1}$$

are weak equivalences.

Theorem 5.5.15 *Let $\mathcal{C}$ be a left proper and cellular model category, with a Quillen adjunction (T, U) between $\mathcal{C}$ and itself. Then, using the stable model structure, the functor T extends to a Quillen equivalence*

$$T\colon \mathcal{S}^{\mathbb{N}}(\mathcal{C}, T) \longrightarrow \mathcal{S}^{\mathbb{N}}(\mathcal{C}, T).$$

Similar statements are proven for symmetric spectra in the case that $\mathcal{C}$ is a symmetric monoidal model category and $T = K \otimes -$ for K a cofibrant object of $\mathcal{C}$.

Equivariant Spectra

Equivariant homotopy theory is a means of discussing homotopy theory "with symmetrics" via incorporating group actions. Rather than introducing group actions to the constructions that we already performed, one would have to start the whole book again from the category of pointed spaces with an action of a compact topological group G. The weak equivalences of G-spaces are maps $f\colon X \longrightarrow Y$ such that for each closed subgroup H of G, the induced map on fixed points $f^H\colon X^H \longrightarrow Y^H$ is a weak homotopy equivalence. In this category, one would look for stable phenomena, construct equivariant (co)homology theories and then develop categories of equivariant spectra. We will only give the starting points of equivariant stable homotopy theory and direct the interested reader to [LMSM86], [May96] and [MM02].

A G-spectrum is not indexed over $\mathbb{N}$ like a sequential or orthogonal spectrum, but is indexed over real G-representations V. We do not index over all representations, but rather a particular *set* of representations called a *G-universe*. A *representation sphere* S^V is the one-point compactification of a representation V. The maps $g\colon V \longrightarrow V$ induce maps $g\colon S^V \longrightarrow S^V$, and thus, S^V is a G-space. A *G-spectrum* consists of the following data

- a G-space $X(V)$ for every representation V in the G-universe,
- structure maps of G-spaces $S^{W-V} \wedge X(V) \longrightarrow X(W)$, with the domain equipped with the diagonal action of G.

Morphisms of G-spectra are then defined as a collection of G-maps between G-spaces that commute with the structure maps. We say that a morphism $f\colon X \longrightarrow Y$ of G-spectra is a weak equivalence if the maps

$$\pi_*^H(f)\colon \pi_*^H(X) \longrightarrow \pi_*^H(Y)$$

are isomorphisms for all subgroups H of G. Here,

$$\pi_q^H(X) = \begin{cases} \operatorname{colim}_V[G/H_+ \wedge S^{V\oplus\mathbb{R}^q}, X(V)]^G & \text{if } q \geqslant 0 \\ \operatorname*{colim}_{V\supset\mathbb{R}^q}[G/H_+ \wedge S^{V-\mathbb{R}^q}, X(V)]^G & \text{if } q \geqslant 0, \end{cases}$$

where $[-,-]^G$ denotes the homotopy category of G-spaces. This leads to a model structure on G-spectra which is very rich in structure, as it incorporates all concepts of the homotopy theory of spectra, as well as inheriting a wealth of constructions from representation theory such as induction, restriction, inflation, fixed points, transfer maps and norm functors.

Motivic Spectra

Motivic homotopy theory [MV99] is the study of schemes up to a notion of homotopy, with the role of the unit interval taken by the affine line $\mathbb{A}^1$. To help study these schemes and their homotopy theory, one develops cohomology theories. From here, it is logical to construct motivic spectra, representing objects for cohomology theories of schemes. Interestingly, there are two different suspension coordinates for these spectra. This gives an extra grading in homotopy groups.

From here, one may develop motivic stable homotopy theory. This is interesting from the point of algebraic geometry and has applications in "classical" stable homotopy theory. For example, the motivic Adams spectral sequence has been used to calculate stable homotopy groups of spheres far beyond the range that they have previously been known, see, for example, [IX15]. The extra grading on suspensions provides crucial extra algebraic structure on the E_2-term. This again shows how the general machinery of spectra and stable homotopy theory encompasses and brings together a huge variety of interesting mathematical fields.

5.6 Compact Objects

A very useful feature of triangulated categories is the notion of generators, that is, objects that "generate" a whole triangulated category using exact triangles and coproducts. When these generators are "compact", we can understand the whole of the triangulated category by looking at those generators and the maps between them. This then leads to vital structural results for the stable homotopy category.

Definition 5.6.1 Let $\mathcal{T}$ be a triangulated category with all small coproducts, and let

$$\mathcal{G} = \{X_i \mid i \in I\}$$

be a set of objects in $\mathcal{T}$, where I denotes an indexing set. Then $\mathcal{G}$ is a *set of generators* for $\mathcal{T}$ if the only full triangulated subcategory of $\mathcal{T}$ which is closed under coproducts and contains $\mathcal{G}$ is $\mathcal{T}$ itself.

Our prime example is of course the case of $\mathcal{T}$ being the homotopy category of a stable model category. In this case, the objects X_i are often called "homotopy generators" of $\mathcal{C}$. While we will not make this distinction, it is important to avoid confusion with the notion of a generator (separator) of a category as in [Mac71, Section V.7].

Definition 5.6.2 Let $\mathcal{T}$ be a triangulated category with all small coproducts. We say that an object $A \in \mathcal{T}$ is *compact* if the functor $\mathcal{T}(A, -)_*$ commutes with arbitrary coproducts.

The above definition relates to the definition of compactness of topological spaces, because if a space A is compact, the functor $[A, -]$ commutes with arbitrary coproducts.

Lemma 5.6.3 *Let $\mathcal{T}$ be a triangulated category with all small coproducts. The class of compact objects is closed under finite coproducts, suspension and desuspension.* □

Lemma 5.6.4 *For a set of compact objects $\{X_i\}$ in a triangulated category $\mathcal{T}$ with all small coproducts, the following two statements are equivalent.*

- *The set $\mathcal{G} = \{X_i \mid i \in I\}$ is a set of generators of $\mathcal{T}$.*
- *A morphism $f\colon A \longrightarrow B$ in $\mathcal{T}$ is an isomorphism if and only if*

$$\mathcal{T}(X_i, f)_*\colon \mathcal{T}(X_i, A)_* \longrightarrow \mathcal{T}(X_i, B)_*$$

is an isomorphism of groups for all $X_i \in \mathcal{G}$.

Proof Assume that $\mathcal{G}$ is a set of generators and that $\mathcal{T}(X_i, f)_*$ is an isomorphism for all $i \in I$. We want to show that f itself is an isomorphism. Let $\mathcal{T}'$ denote the full subcategory on those objects K such that $\mathcal{T}(K, f)_*$ is an isomorphism. By the Five Lemma, $\mathcal{T}'$ is a triangulated subcategory of $\mathcal{T}$. It is closed under coproducts and contains all of $\mathcal{G}$, hence, $\mathcal{T}'$ is equal to $\mathcal{T}$ by the definition of a set of generators. The Yoneda Lemma implies that f is an isomorphism.

Now assume the second condition. Note that this assumption is equivalent to assuming that $Z \cong 0$ for an object $Z \in \mathcal{T}$ is equivalent to $\mathcal{T}(X_i, Z) = 0$ for all $X_i \in \mathcal{G}$. Let $X \in \mathcal{T}$ and

$$\mathbb{W} = \{* \longrightarrow X_i \mid i \in I\}.$$

Then Example 7.1.5 (which only requires a triangulated category with all small coproducts) applied to this set gives an exact triangle

$$\widetilde{W}_\infty \longrightarrow X \longrightarrow X_\infty \longrightarrow \Sigma\widetilde{W}_\infty$$

with $\widetilde{W}_\infty$ built from $\mathcal{G}$ using coproducts, triangles and (de)suspensions, and

$$\mathcal{T}(X_i, X_\infty) = 0 \ \text{ for all } \ i \in I.$$

By our assumption, $X_\infty \cong 0$, and so $\widetilde{W}_\infty \longrightarrow X$ is an isomorphism in $\mathcal{T}$. Thus, X is in the full triangulated subcategory generated by $\mathcal{G}$. □

Remark 5.6.5 Having a set of compact generators is a non-trivial condition. For example, the triangulated categories $\mathrm{Ho}(L_{H\mathbb{Z}}\mathcal{S})$ and $\mathrm{Ho}(L_{H\mathbb{F}_p}\mathcal{S})$ which will be introduced in Chapter 7 have no non-zero compact objects by [HS99, Corollary B.13].

Let us look at the example of $\mathcal{T} = \mathcal{SHC} = \mathrm{Ho}(\mathcal{S}^{\mathbb{N}})$.

Proposition 5.6.6 *Let X be a sequential spectrum and $k \in \mathbb{N}$. There are natural isomorphisms of abelian groups*

$$\pi_k(X) \cong [\Sigma^\infty S^k, X] \quad \text{and} \quad \pi_{-k}(X) \cong [\mathrm{F}^{\mathbb{N}}_k S^0, X],$$

where $\mathrm{F}^{\mathbb{N}}_k S^0$ denotes the shifted suspension spectrum.

Proof We begin with the case of $[\Sigma^\infty S^k, X]$. Using the adjunction

$$\Sigma^\infty : \mathrm{Ho}(\mathrm{Top}_*) \rightleftarrows \mathcal{SHC} : \Omega^\infty$$

from Lemma 5.1.1, we can work in the homotopy category of pointed spaces. That adjunction and our explicit description of fibrant replacement in Section 2.4 show that $[\Sigma^\infty S^k, X]$ is isomorphic to the abelian groups

$$[S^k, \mathrm{hocolim}_a \Omega^a X_a] \cong \mathrm{colim}_a [S^{k+a}, X_a] \cong \pi_k(X).$$

The first isomorphism follows since S^k is a compact space. For the second isomorphism, we note that we only need to consider the basepoint components of the X_a, as the adjoint structure maps $X_a \longrightarrow \Omega X_{a+1}$ only depend on the basepoint components of the X_{a+1}. Equally, the structure map $\Sigma X_a \longrightarrow X_{a+1}$ has image in the basepoint component of X_{a+1}.

For the case of $[\mathrm{F}_k^{\mathbb{N}} S^0, X]$, we use the stability of the stable model structure on sequential spectra to suspend k times and get the first isomorphism below. The second isomorphism is the π_*-isomorphism $\mathrm{F}_k^{\mathbb{N}} S^K \longrightarrow \mathrm{F}_0^{\mathbb{N}} S^0$. The remainder are as for the previous case.

$$\begin{aligned} [\mathrm{F}_k^{\mathbb{N}} S^0, X] &\cong [\mathrm{F}_k^{\mathbb{N}} S^k, \Sigma^k X] \\ &\cong [\mathrm{F}_0^{\mathbb{N}} S^0, \Sigma^k X] \\ &\cong [S^0, \mathrm{hocolim}_a \Omega^a \Sigma^k X_a] \\ &\cong \mathrm{colim}_a [S^a, \Sigma^k X_a] \\ &\cong \mathrm{colim}_a \pi_a(\Sigma^k X_a) \\ &= \pi_0(\Sigma^k X) \end{aligned}$$

The result then follows from the formula

$$\pi_0(\Sigma^k X) \cong \pi_{-k}(X)$$

used in the proof of Lemma 2.2.7. □

Since π_*-isomorphisms are the weak equivalences of the stable model structure on sequential spectra, our definition of graded maps in the homotopy category gives the following.

Corollary 5.6.7 *Let X be a sequential spectrum. Then $[\mathbb{S}, X]_* = 0$ if and only if $X \simeq *$.* □

As a consequence of the above and Lemma 2.2.9, we have the following.

Corollary 5.6.8 *The sphere spectrum is a compact generator for $\mathcal{SHC}$.* □

We will encounter compact generators in various places in this book. In any case, there is a very useful comprehensive list of examples of compact generators in stable homotopy theory in [SS03b].

Having a set of compact generators makes it much easier to check if exact functors which preserve arbitrary coproducts are equivalences.

Lemma 5.6.9 *Let $\mathcal{T}$ be a triangulated category with infinite coproducts and a set of compact generators $\mathcal{G} = \{X_i \mid i \in I\}$. Furthermore, let $F\colon \mathcal{T} \longrightarrow \mathcal{T}$ be an exact endofunctor that commutes with arbitrary coproducts. If*

$$\{F(X_i) \mid i \in I\} = \mathcal{G}$$

and

$$F\colon \mathcal{T}(X_i, X_j) \xrightarrow{\cong} \mathcal{T}(F(X_i), F(X_j)) \text{ for all } i, j \in I,$$

then F is an equivalence of categories.

The idea is that all of $\mathcal{T}$ is "generated" from the X_i using exact triangles and coproducts, so a functor respecting exactly those must inevitably be an equivalence.

Proof We are going to show that F is fully faithful and essentially surjective.

Consider the full subcategories of $\mathcal{T}$ for each $i \in I$

$$\mathcal{T}_i = \{Y \in \mathcal{T} \mid F\colon \mathcal{T}(X_i, Y) \longrightarrow \mathcal{T}(F(X_i), F(Y)) \text{ is an isomorphism}\}.$$

As F is exact, $\mathcal{T}_i$ is triangulated. As X_i and $F(X_i)$ are compact for each $i \in I$, $\mathcal{T}_i$ is closed under coproducts. By assumption, $X_j \in \mathcal{T}_i$ for all $j \in I$. This means that $\mathcal{T}_i = \mathcal{T}$ for all i. Similarly, the full subcategory

$$\bar{\mathcal{T}} = \{X \in \mathcal{T} \mid F\colon \mathcal{T}(X, Y) \xrightarrow{\cong} \mathcal{T}(F(X), F(Y)) \text{ for all } Y \in \mathcal{T}\}$$

is triangulated and closed under coproducts, and we have just shown that it contains all the X_i. Thus, $\bar{\mathcal{T}} = \mathcal{T}$, so our functor is fully faithful.

It remains to be shown that F is essentially surjective. Let $\tilde{\mathcal{T}}$ denote the essential image of F, that is, all $X' \in \mathcal{T}$ for which there is an X with $F(X) \cong X'$. We claim that $\tilde{\mathcal{T}}$ is closed under exact triangles. For this, let $X', Y' \in \tilde{\mathcal{T}}$, and let $X, Y \in \mathcal{T}$ with $F(X) = X'$ and $F(Y) = Y'$. If X' and Y' are part of an exact triangle

$$X' \xrightarrow{f'} Y' \longrightarrow Z' \longrightarrow \Sigma X',$$

then we must show that $Z' \in \tilde{\mathcal{T}}$. As F is fully faithful, there is a morphism f with $F(f)$ equal to the composite

$$F(X) \xrightarrow{\cong} X' \xrightarrow{f'} Y' \xrightarrow{\cong} F(Y).$$

The map $F(f)$ is part of an exact triangle

$$F(X) \xrightarrow{F(f)} F(Y) \longrightarrow Z \longrightarrow \Sigma F(X)$$

for some Z. By the axioms of triangulated categories, we have a morphism of exact triangles

$$\begin{array}{ccccccc} X' & \xrightarrow{f'} & Y' & \longrightarrow & Z' & \longrightarrow & \Sigma X' \\ \downarrow \cong & & \downarrow \cong & & \downarrow \cong & & \downarrow \cong \\ F(X) & \xrightarrow{F(f)} & F(Y) & \longrightarrow & Z & \longrightarrow & \Sigma F(X), \end{array}$$

where the third vertical arrow $Z' \longrightarrow Z$ is an isomorphism because the other vertical arrows are.

We can complete f to an exact triangle

$$X \xrightarrow{f} Y \longrightarrow Cf \longrightarrow \Sigma X.$$

As F is exact, we have that

$$F(X) \xrightarrow{F(f)} F(Y) \longrightarrow F(Cf) \longrightarrow \Sigma F(X)$$

is an exact triangle. Comparing with the exact triangle

$$F(X) \xrightarrow{F(f)} F(Y) \longrightarrow Z \longrightarrow \Sigma F(X)$$

yields that $Z' \cong Z \cong F(Cf)$, which implies that Z' lies in the essential image of F. Thus, $\tilde{\mathcal{T}} = \mathcal{T}$, which was our original claim. □

Corollary 5.6.10 *Consider an adjunction between compactly generated triangulated categories with all small coproducts*

$$F : \mathcal{T} \rightleftarrows \mathcal{T}' : G.$$

If G commutes with small coproducts and the unit and counit

$$\eta_Y : Y \longrightarrow (G \circ F)(Y), \quad \epsilon_X : (F \circ G)(X) \longrightarrow X$$

are isomorphisms on the generators, the adjunction is an equivalence of triangulated categories. □

Definition 5.6.11 Let $\mathcal{T}$ be a triangulated category, and let $\mathcal{F}$ be a full triangulated subcategory of $\mathcal{T}$. Then $\mathcal{F}$ is *thick* if $\mathcal{F}$ is closed under retracts.

If in addition $\mathcal{F}$ is closed under arbitrary coproducts, then $\mathcal{F}$ is a *localising subcategory*.

Proposition 5.6.12 *Let $\mathcal{T}$ be a triangulated category with all small coproducts and a set of compact generators*

$$\mathcal{G} = \{X_i \mid i \in I\}.$$

Then an object of $\mathcal{T}$ is compact if and only if it lies in the thick subcategory generated by $\mathcal{G}$, that is, in the smallest thick subcategory of $\mathcal{T}$ containing $\mathcal{G}$.

Proof First, we show that the compact objects of $\mathcal{T}$ form a thick subcategory. Let

$$A_1 \longrightarrow A_2 \longrightarrow A_3 \longrightarrow \Sigma A_1$$

be an exact triangle with two of the objects being compact. Now let X_i be a set of objects in $\mathcal{T}$. Applying

$$\oplus_i \mathcal{T}(-, X_i)_* \longrightarrow \mathcal{T}(-, \amalg X_i)_*$$

gives us a morphism of two long exact sequences

$$\begin{array}{ccccccccc} \cdots \longrightarrow & \oplus_i \mathcal{T}(A_1, X_i)_{*-1} & \longrightarrow & \oplus_i \mathcal{T}(A_3, X_i)_* & \longrightarrow & \oplus_i \mathcal{T}(A_2, X_i)_* & \longrightarrow & \oplus_i \mathcal{T}(A_1, X_i)_* & \longrightarrow \cdots \\ & \downarrow & & \downarrow & & \downarrow & & \downarrow & \\ \cdots \longrightarrow & \mathcal{T}(A_1, \amalg X_i)_{*-1} & \longrightarrow & \mathcal{T}(A_3, \amalg X_i)_* & \longrightarrow & \mathcal{T}(A_2, \amalg X_i)_* & \longrightarrow & \mathcal{T}(A_1, \amalg X_i)_* & \longrightarrow \cdots . \end{array}$$

By the Five Lemma, all vertical maps are isomorphisms, hence, the class of compact objects is closed under exact triangles. Now let A be compact and B a retract of A. We have a morphism of two retracts of graded abelian groups,

$$\begin{array}{ccccc} \oplus_i \mathcal{T}(B, X_i)_* & \longrightarrow & \oplus_i \mathcal{T}(A, X_i)_* & \longrightarrow & \oplus_i \mathcal{T}(B, X_i)_* \\ \downarrow & & \downarrow & & \downarrow \\ \mathcal{T}(B, \amalg X_i)_* & \longrightarrow & \mathcal{T}(A, \amalg X_i)_* & \longrightarrow & \mathcal{T}(B, \amalg X_i)_*. \end{array}$$

The middle vertical map is an isomorphism. As retracts of isomorphisms are isomorphisms, the other two vertical maps are isomorphisms too, and we can conclude that retracts of compact objects are compact. Thus, the compact objects of $\mathcal{T}$ form a thick subcategory containing the generators $\mathcal{G}$, which are compact by assumption. Thus, we have

$$(\text{thick subcategory generated by } \mathcal{G}) \subseteq (\text{compact objects of } \mathcal{T}).$$

The converse is [HPS97, Corollary 2.3.12]. We will give a sketch proof. Let $\mathcal{F}$ denote the thick subcategory of $\mathcal{T}$ generated by the elements of $\mathcal{G}$. Let

$$H\colon \mathcal{F} \longrightarrow \mathrm{Ab}_*$$

be a homology functor, that is, a functor that takes exact triangles to long exact sequences, which also takes coproducts to direct sums. Then H can be extended to the entire category of $\mathcal{T}$ by

$$\hat{H}\colon \mathcal{T} \longrightarrow \mathrm{Ab}_*, \quad \hat{H}(X) = \operatorname{colim} H(X_\alpha),$$

where the colimit is taken over $\{X_\alpha \longrightarrow X \mid X_\alpha \in \mathcal{F}\}$. This $\hat{H}$ is again a homology functor, and every other extension of H onto $\mathcal{T}$ is canonically isomorphic to $\hat{H}$ [HPS97, Corollary 2.3.11].

Now let X be a compact object of $\mathcal{T}$. We would like to show that it is in $\mathcal{F}$. Because X is compact, the functor $H = \mathcal{T}(X, -)_*$ defines a homology functor on $\mathcal{F}$. As before, let

$$\hat{H}(X) = \operatorname{colim} H(X_\alpha) = \operatorname{colim} \mathcal{T}(X, X_\alpha)_*,$$

which extends

$$\mathcal{T}(X, -)_*\colon \mathcal{F} \longrightarrow \mathrm{Ab}_*$$

uniquely to $\mathcal{T}$. Since X is compact, the functor $\mathcal{T}(X, -)_*$ is already a well-defined homology functor on $\mathcal{T}$. So uniqueness of the extension implies

$$\mathcal{T}(X, X)_* = \hat{H}(X) = \operatorname{colim} \mathcal{T}(X, X_\alpha)_*.$$

This means that the identity of X is an element of this colimit and thus factors through some X_α. In other words, X is a retract of X_α. As X_α lies in $\mathcal{F}$, which is closed under retracts, X lies in $\mathcal{F}$ too. □

In Definition 2.1.8, we defined a CW-spectrum as a sequential spectrum X such that

- each space X_n is a pointed CW-complex,
- each structure map $\sigma\colon \Sigma X_n \longrightarrow X_{n+1}$ is a cellular map that is an isomorphism onto a subcomplex of X_{n+1}.

We also noted that if we apply the structure map to a cell in X_n, we obtain a cell in X_{n+1}. The colimit of such a sequence of cells is called a stable cell.

Now Proposition 5.6.12 helps us identify the compact objects of the stable homotopy category.

Theorem 5.6.13 *For an object X in the stable homotopy category $\mathcal{SHC}$, the following are equivalent.*

- *X is compact.*
- *X is in the thick subcategory generated by the sphere spectrum.*
- *X is isomorphic in $\mathcal{SHC}$ to a CW-spectrum with finitely many stable cells.*

Proof The first two points are equivalent by Proposition 5.6.12.

The class of spectra isomorphic to finite CW-spectra is a thick subcategory of $\mathcal{SHC}$. Furthermore, it contains the sphere spectrum. Therefore, it contains the thick subcategory generated by the sphere spectrum, that is, the class of compact objects.

Conversely, let us show that a spectrum that is isomorphic to a finite CW-spectrum is compact. Let X be a CW-spectrum with finitely many stable cells. Without loss of generality, we assume that the smallest dimension of a stable cell is 0. We construct a CW-approximation to X. The 0–skeleton $X^{(0)}$ consists of $\mathrm{F}_0^{\mathbb{N}}(*)$ (the basepoint) and a finite wedge of spectra of the form $\mathrm{F}_d^{\mathbb{N}} D_+^d$, one for each stable 0–cell. We choose the indices d so that there is a map $D_+^d \longrightarrow X_d$ representing each stable 0–cell. By adjunction, these representatives give a map $X^{(0)} \longrightarrow X$.

A stable 1–cell of X is represented by a map of pointed spaces $D_+^{n+1} \longrightarrow X_n$ (a cell of the CW-complex X_n). The attaching map of this cell $S_+^n \longrightarrow X_n$ lands in the n–skeleton of X_n. By increasing n, we may assume that the image of the attaching map lands in the image of the stable 0–cells. By adjunction, we have a map $\mathrm{F}_n^{\mathbb{N}} S_+^n \longrightarrow X$. After repeating this for each stable 1–cell, we may form the pushout diagram

$$\begin{array}{ccc} \bigvee_\alpha \mathrm{F}_{d_\alpha}^{\mathbb{N}} S_+^{d_\alpha} & \longrightarrow & X^{(0)} \\ \downarrow & & \downarrow \\ \bigvee_\alpha \mathrm{F}_{d_\alpha}^{\mathbb{N}} D_+^{d_\alpha+1} & \longrightarrow & X^{(1)}, \end{array}$$

where α runs over the finite set of stable 1–cells of X. This pushout defines an exact triangle

$$\bigvee_\alpha \mathrm{F}_{d_\alpha}^{\mathbb{N}} S_+^{d_\alpha} \longrightarrow X^{(0)} \longrightarrow X^{(1)} \longrightarrow \Sigma\left(\bigvee_\alpha \mathrm{F}_{d_\alpha}^{\mathbb{N}} S_+^{d_\alpha}\right).$$

We continue in this way until all stable cells are added. As X only has finitely many stable cells, this process will terminate at $X^{(m)}$ for some m. As with spaces, there is a map $X^{(m)} \longrightarrow X$ which is a π_*-isomorphism.

Thus, up to isomorphism in $\mathcal{SHC}$, X can be constructed using finitely many coproducts and cofibre sequences of objects of the form $\mathrm{F}_d^{\mathbb{N}} S_+^n$. Applying $\mathrm{F}_d^{\mathbb{N}}$ to the cofibre sequence of spaces

$$S_+^n \longrightarrow D_+^{n+1} \longrightarrow S^{n+1}$$

gives an exact triangle

$$\mathrm{F}_d^{\mathbb{N}} S_+^n \longrightarrow \mathrm{F}_d^{\mathbb{N}} S^0 \longrightarrow \mathrm{F}_d^{\mathbb{N}} S^{n+1} \longrightarrow \Sigma \mathrm{F}_d^{\mathbb{N}} S_+^n.$$

Since spectra of the form $F_d^{\mathbb{N}} S^{n+1}$ are shifts of the sphere spectrum, every finite CW-spectrum is in the thick subcategory generated by the sphere. □

Corollary 5.6.14 *The homotopy groups of a compact spectrum X are finitely generated over $\mathbb{Z}$ in each degree, and only finitely many negative homotopy groups are non-zero.*

Proof This is true for the sphere spectrum, see, for example, [Ser53] or [Spa81, Chapter 9.7], so it holds for any CW-spectrum with finitely many cells. □

The following result is phrased for sequential spectra, but similar results hold for any orthogonal spectra and, where appropriate, symmetric spectra.

Lemma 5.6.15 *Let*

$$X_1 \xrightarrow{f_1} X_2 \xrightarrow{f_2} X_3 \xrightarrow{f_3} \cdots$$

be a sequential diagram of cofibrant sequential spectra. If each X_i is an Ω-spectrum, then so is the homotopy colimit of the diagram.

Furthermore, there is an exact triangle

$$\bigvee_n X_n \xrightarrow{\mathrm{Id}-\bigvee_n f_n} \bigvee_n X_n \longrightarrow \operatorname{hocolim} X_n \longrightarrow \Sigma\left(\bigvee_n X_n\right).$$

For any compact spectrum A there is a natural isomorphism

$$\operatorname{colim}[A, X_n] \cong [A, \operatorname{hocolim} X_n].$$

Proof The first statement follows from Corollary A.7.10: Ω commutes with homotopy colimits as loops and homotopy colimits are constructed levelwise.

Using Example A.7.12 and Lemma A.7.13, we may write the homotopy colimit as the homotopy cofibre of

$$\mathrm{Id} - \bigvee_n f_n \colon \bigvee_n X_n \longrightarrow \bigvee_n X_n.$$

Applying $[A, -]$ for a compact spectrum A to this exact triangle gives a short exact sequence

$$0 \longrightarrow \bigoplus_n [A, X_n] \xrightarrow{\mathrm{Id}-\bigoplus_n (f_n)_*} \bigoplus_n [A, X_n] \longrightarrow [A, \operatorname{hocolim}_n X_n] \longrightarrow 0.$$

Note that this is in fact a short exact sequence because $\mathrm{Id} - \bigoplus_n (f_n)_*$ is injective. The cokernel of the map $\mathrm{Id} - \bigoplus_n (f_n)_*$ is $\operatorname{colim}_n [A, X_n]$. □

We may think of this as understanding maps from a compact object of the stable homotopy category into a sequential homotopy colimit. Maps out of a sequential homotopy colimit are examined in Corollary A.7.14, with the case of maps into a sequential homotopy limit appearing as Corollary A.7.23.

5.7 Rigidity of Spectra

Previously in this chapter, we got to know several different model categories of spectra, which all have their individual technical advantages and disadvantages. What they all have in common is that their homotopy category is the stable homotopy category $\mathcal{SHC}$. We have also seen that each individual category of spectra that we discussed is Quillen equivalent to sequential spectra, and thus, those categories of spectra are also Quillen equivalent to each other.

However, there is a wealth of different categories of spectra all modelling $\mathcal{SHC}$, and it would be tedious to compare them individually. The following result by Schwede [Sch07a, Sch01b] tells us that, in fact, all such categories of spectra are Quillen equivalent. This means that all higher homotopy structure of spectra is captured by the triangulated structure of $\mathcal{SHC}$ alone.

Theorem 5.7.1 (Schwede) *Let $\mathcal{C}$ be a stable model category. If there is an equivalence of triangulated categories*

$$\Psi\colon \mathcal{SHC} \longrightarrow \mathrm{Ho}(\mathcal{C}),$$

then $\mathcal{C}$ is Quillen equivalent to sequential spectra $\mathcal{S}^{\mathbb{N}}$.

This phenomenon is known as *rigidity*, that is, the stable homotopy category is rigid.

In particular, we can replace sequential spectra in the statement with any other model category of spectra. The idea of the theorem stems from a time when lots of categories of highly structured spectra were being developed, the construction of which usually involved a manually constructed Quillen equivalence with a previously known category of spectra. The Rigidity Theorem removes the necessity of this step, as it says that all stable model categories modelling the stable homotopy category are automatically Quillen equivalent on a model category level.

Providing a complete proof of Schwede's Rigidity Theorem would be beyond the scope of this book, but we are going to present some of the key ideas, as they will demonstrate the use of many tools in triangulated categories and indeed spectra.

The model category $\mathcal{C}$ of the theorem could be any stable model category for now, which makes it awkward to work with. So the first step will be to reformulate the problem in order to remove $\mathcal{C}$. The construction of stable framings from Subsection 6.9.2 tells us that for any (fibrant and cofibrant) object X in a stable model category $\mathcal{C}$, there is a Quillen adjunction

$$X \wedge -\colon \mathcal{S}^{\mathbb{N}} \rightleftarrows \mathcal{C} : \mathrm{Map}(X, -),$$

which sends the sphere to X. (A weaker version had been previously devised by Schwede and Shipley [SS02] as the "universal property of spectra".)

In the situation of the theorem, let X be a fibrant and cofibrant replacement of $\Psi(S^0)$. Denote by F the composite

$$F\colon \mathrm{Ho}(\mathcal{S}^{\mathbb{N}}) \xrightarrow{X \wedge^{\mathbb{L}} -} \mathrm{Ho}(\mathcal{C}) \xrightarrow{\Psi^{-1}} \mathrm{Ho}(\mathcal{S}^{\mathbb{N}}).$$

Note that

$$F(\mathbb{S}) = \Psi^{-1}(X \wedge^{\mathbb{L}} \mathbb{S}) = \Psi^{-1}(\Psi(\mathbb{S})) = \mathbb{S}.$$

We see immediately that $X \wedge^{\mathbb{L}} -$ is an equivalence of categories if and only if F is, and as $X \wedge^{\mathbb{L}} -$ is the left derived functor of a Quillen functor, this would prove the theorem. Thus, we have reduced the Rigidity Theorem to the following statement:

Theorem 5.7.2 *Let $F\colon \mathcal{SHC} \longrightarrow \mathcal{SHC}$ be an exact endofunctor which sends the sphere spectrum to itself. Then F is an equivalence of categories.*

The rest of this section will outline the proof of this result.

Because the sphere spectrum is a compact generator for the stable homotopy category, a consequence of Lemma 5.6.9 is the following.

Corollary 5.7.3 *Let $F\colon \mathcal{SHC} \longrightarrow \mathcal{SHC}$ be an exact endofunctor such that $F(\mathbb{S}) \cong \mathbb{S}$. If*

$$F\colon [\mathbb{S}, \mathbb{S}]_n \longrightarrow [\mathbb{S}, \mathbb{S}]_n$$

is an isomorphism for all n, then F is an equivalence of categories. □

To prove the Rigidity Theorem it is sufficient to show that for each prime p,

$$F\colon [\mathbb{S}, \mathbb{S}]_n \otimes \mathbb{Z}_{(p)} \longrightarrow [\mathbb{S}, \mathbb{S}]_n \otimes \mathbb{Z}_{(p)}$$

is an isomorphism for all n, where $\mathbb{Z}_{(p)}$ denotes the p-local integers. We are going to give an extremely brief outline of the steps towards this.

The statement is proved by induction on the Adams filtration of the elements of $\pi_*(\mathbb{S}) \otimes \mathbb{Z}_{(p)}$. This is based on the idea that the exact endofunctor preserves not just composition, but also Toda brackets, see Section 4.6, and each element of $\pi_*(\mathbb{S})$ is "built" from elements of Adams filtration 1 using higher Toda brackets [Coh68].

For $p = 2$, the elements of $\pi_*(\mathbb{S}) \otimes \mathbb{Z}_{(p)}$ of Adams filtration 1 are the Hopf elements η, ν and σ, and for each odd prime the only element in Adams filtration 1 is $\alpha_1 \in \pi_{2p-3}(\mathbb{S})$. Thus, the statement is reduced to checking that, up to a p-local unit,

$$F(\eta) = \eta, F(\nu) = \nu, F(\sigma) = \sigma \text{ and } F(\alpha_1) = \alpha_1 \text{ for all } p.$$

Again, as ν and σ satisfy strong Toda bracket relations with η, the statement for those last two elements can be deduced from the statement for η with some brief calculations. It remains to be shown that

$$F(\eta) = \eta \quad \text{and} \quad F(\alpha_1) = \alpha_1.$$

The proof that $F(\alpha_1) = \alpha_1$ is very lengthy and technical, and it occupies the majority of [Sch07a]. However, the proof that $F(\eta) = \eta$ is a very neat argument. Consider the mod-2 Moore spectrum M as part of the exact triangle

$$\mathbb{S} \xrightarrow{2} \mathbb{S} \xrightarrow{\text{incl}} M \xrightarrow{\text{pinch}} \Sigma\mathbb{S}.$$

Its identity map Id_M satisfies

$$2\mathrm{Id}_M = \text{incl} \circ \eta \circ \text{pinch},$$

see Lemma 4.6.6. As an exact endofunctor F has to send $2\mathrm{Id}_M$ to $2\mathrm{Id}_M$ again (because it is, in particular, additive), it cannot send η to 0, otherwise $F(2\mathrm{Id}_M) = 2\mathrm{Id}_M$ would also be trivial. Hence,

$$F(\eta) = \eta \in \pi_1(\mathbb{S}) = \mathbb{Z}/2.$$

Therefore, the key aspect of the Rigidity Theorem is the various reduction steps in the proof:

- from checking that $F\colon [X, Y] \xrightarrow{\cong} [F(X), F(Y)]$ for all $X, Y \in \mathcal{T}$ to only checking that F is an isomorphism on $[\mathbb{S}, \mathbb{S}] \otimes \mathbb{Z}_{(p)}$ using compact generators,
- from asking that F is an isomorphism on all of $[\mathbb{S}, \mathbb{S}] \otimes \mathbb{Z}_{(p)}$ to checking this only on the Hopf elements and on α_1 using induction on the Adams filtration,
- finally reducing the previous statement to just one element for each p, namely, η for $p = 2$ and α_1 for p odd.

We return to the question of rigidity in Subsection 7.4.2.

6

Monoidal Structures

The aim of this chapter is to investigate symmetric monoidal products on our categories of spectra and the stable homotopy category. After motivating this monoidal product in terms of the smash product on spaces and the Spanier–Whitehead category $\mathcal{SW}$, we show that symmetric spectra and orthogonal spectra are symmetric monoidal model categories. As a consequence, the stable homotopy category is a closed symmetric monoidal category, and this monoidal structure is compatible with the triangulated structure in the sense of Theorem 6.1.14.

The remainder of the chapter is investigating the consequences of this monoidal structure at both the homotopy category and model category level. At the homotopy level, we can give a modern interpretation of Spanier–Whitehead duality. At the model category level, we discuss model categories of ring spectra, modules over ring spectra and commutative ring spectra.

We end the chapter with an overview of some of the fundamental properties of spectra and the stable homotopy category, demonstrating that they are central to the study of stable homotopy theory. At the model category level, Theorem 6.8.1 states that the positive stable model structure on symmetric spectra (see Proposition 6.7.1) is initial amongst stable simplicial monoidal model categories. At the homotopy level, we show that the homotopy category of any stable model category has an "action" of $\mathcal{SHC}$ in Theorem 6.9.28.

6.1 Monoidal Model Categories

A particularly useful piece of structure for a category to have is a monoidal structure. In this section, we give a categorical definition, examples and necessary conditions for the monoidal product to interact with the model structure and pass to the homotopy category. We focus on the case of *symmetric* monoidal structures, as this is the structure we aim to give to the stable homotopy category.

Further references for this material are Hovey [Hov99], along with Borceaux [Bor94] and Kelly [Kel05] for the categorical underpinnings.

Definition 6.1.1 A *symmetric monoidal category* is a category $\mathcal{C}$ with a functor

$$\otimes \colon \mathcal{C} \times \mathcal{C} \longrightarrow \mathcal{C}$$

called a *monoidal product*, an object $\mathcal{I}$ called the *monoidal unit* and isomorphisms

- $a \colon (X \otimes Y) \otimes Z \longrightarrow X \otimes (Y \otimes Z)$ (associativity)
- $u \colon \mathcal{I} \otimes X \longrightarrow X$ (unit)
- $\tau \colon X \otimes Y \longrightarrow Y \otimes X$ (symmetry)

which are natural in X, Y and Z and satisfy the coherence diagrams given below. We often omit the maps from our notation and refer to $(\mathcal{C}, \otimes, \mathcal{I})$ as a symmetric monoidal category. We also refer to the product $\otimes$ as being commutative and call τ the twist functor.

The four coherence diagrams are as follows. The first says that four-fold associativity is coherent.

$$\begin{array}{ccccc}
((W \otimes X) \otimes Y) \otimes Z & \xrightarrow{a} & (W \otimes X) \otimes (Y \otimes Z) & \xrightarrow{a} & W \otimes (X \otimes (Y \otimes Z)) \\
\downarrow{\scriptstyle a \otimes \mathrm{Id}} & & & & \uparrow{\scriptstyle \mathrm{Id} \otimes a} \\
((W \otimes (X \otimes Y)) \otimes Z & & \xrightarrow{\quad\quad a \quad\quad} & & W \otimes ((X \otimes Y) \otimes Z)
\end{array}$$

The second says that the twist is self-inverse.

$$X \otimes Y \xrightarrow{\tau} Y \otimes X \xrightarrow{\tau} X \otimes Y, \qquad \tau \circ \tau = \mathrm{Id}$$

The third gives a coherence between the twist map and associativity.

$$\begin{array}{ccc}
(X \otimes Y) \otimes Z & \xrightarrow{a} & X \otimes (Y \otimes Z) \\
\downarrow{\scriptstyle \tau} & & \downarrow{\scriptstyle \tau} \\
Z \otimes (X \otimes Y) & & (Y \otimes Z) \otimes X \\
\downarrow{\scriptstyle \mathrm{Id} \otimes \tau} & & \downarrow{\scriptstyle \tau \otimes \mathrm{Id}} \\
Z \otimes (Y \otimes X) & \xleftarrow{a} & (Z \otimes Y) \otimes X
\end{array}$$

The fourth gives a coherence between the unit, the twist and associativity.

$$\begin{array}{ccc} (X \otimes \mathcal{I}) \otimes Y & \xrightarrow{a} & X \otimes (\mathcal{I} \otimes Y) \\ {\scriptstyle \tau\otimes \mathrm{Id}} \downarrow & & \downarrow {\scriptstyle \mathrm{Id}\otimes u} \\ (\mathcal{I} \otimes X) \otimes Y & \xrightarrow[u\otimes \mathrm{Id}]{} & X \otimes Y \end{array}$$

Many other coherence diagrams will follow from these, such as n-fold associativity, and that τ is the identity on $\mathcal{I} \otimes \mathcal{I}$. A discussion on coherence is given in [Mac71, Sections VII.1 and VII.7].

Many of standard categories have a symmetric monoidal structure.

Examples 6.1.2 The category of sets is a symmetric monoidal category with the Cartesian product as the monoidal operation. The one-point set is the monoidal unit. This product extends to simplicial sets, with the product acting levelwise

$$(A \times B)_n = A_n \times B_n,$$

and face and degeneracy maps given by the products of those for A and B. Similarly, the category of pointed simplicial sets is a symmetric monoidal category under the smash product. Its unit is the simplicial set S^0.

Topological spaces (that is, compactly generated weak Hausdorff spaces) are a symmetric monoidal category under the Cartesian product, using the Kelly product topology with the one-point space as the unit. Pointed spaces are also symmetric monoidal categories with the smash product as the monoidal product. The unit is S^0.

Chain complexes over a commutative ring R are a symmetric monoidal category with monoidal product $- \otimes -$ given by

$$(X \otimes Y)_n = \bigoplus_{a+b=n} X_a \otimes Y_b, \qquad \partial(x_a \otimes y_b) = (\partial x_a) \otimes y_b + (-1)^a x_a \otimes \partial y_b.$$

We will see later that each of these examples is a *closed* symmetric monoidal category, that is, there is a function object related to the monoidal product. Another source of examples is the category of maps in a monoidal category. We will introduce two monoidal products on maps.

Definition 6.1.3 Let $(\mathcal{C}, \otimes, \mathcal{I})$ be a symmetric monoidal category. For maps $f\colon A \longrightarrow B$ and $g\colon X \longrightarrow Y$ in $\mathcal{C}$, we define their *pushout product* to be the natural map

$$f \square g\colon B \otimes X \coprod_{A\otimes X} A \otimes Y \longrightarrow B \otimes Y.$$

Let $g\colon X \longrightarrow Y$ be a map in $\mathcal{C}$ and A be an object of $\mathcal{C}$. Then,

$$\mathrm{Id}_A \mathbin{\square} g = \mathrm{Id}_{A\otimes Y}\colon A\otimes Y \longrightarrow A\otimes Y.$$

In particular, $\mathrm{Id}_{\mathcal{J}} \mathbin{\square} g = \mathrm{Id}_Y$. Assume that $\mathcal{C}$ has an initial object $\emptyset$ and that $\emptyset\otimes X = \emptyset$ for all X in $\mathcal{C}$. (This will hold when $\mathcal{C}$ is a closed symmetric monoidal category.) Let $i_A\colon \emptyset \longrightarrow A$ be the unique map from the initial object. Then

$$i_A \mathbin{\square} g = A\otimes g\colon A\otimes X \longrightarrow A\otimes Y.$$

In particular, $i_{\mathcal{J}} \mathbin{\square} g = g$.

Examples 6.1.4 Let $(\mathcal{C},\otimes,\mathcal{J})$ be a symmetric monoidal category. The category of maps in $\mathcal{C}$ with commuting squares as the morphisms has a symmetric monoidal product given by the termwise product. Given maps $f\colon A \longrightarrow B$ and $g\colon X \longrightarrow Y$, we define their termwise product as

$$f\otimes g\colon A\otimes X \longrightarrow B\otimes Y.$$

The monoidal unit is the identity map of the unit.

Now assume that $\mathcal{C}$ has an initial object $\emptyset$ and that $\emptyset\otimes X = \emptyset$ for all X in $\mathcal{C}$. The category of maps in $\mathcal{C}$ is a symmetric monoidal category with the monoidal product given by the pushout product. The monoidal unit is the map $\emptyset \longrightarrow \mathcal{J}$.

Definition 6.1.5 Let $(\mathcal{C},\otimes,\mathcal{J})$ be a symmetric monoidal category. We say that $\mathcal{C}$ is *closed symmetric monoidal* if there is a functor

$$\mathrm{Hom}\colon \mathcal{C}^{op}\times\mathcal{C} \longrightarrow \mathcal{C}$$

and a natural isomorphism

$$\phi\colon \mathcal{C}(A\otimes B, C) \longrightarrow \mathcal{C}(A, \mathrm{Hom}(B,C)).$$

We call Hom the *internal function object*. Omitting the natural isomorphisms as previously, we refer to $(\mathcal{C},\otimes,\mathcal{J},\mathrm{Hom})$ as a closed symmetric monoidal category.

The addition of an internal function object adds a great deal of structure to the category. In particular, we see that for any A in $\mathcal{C}$ there is an adjunction

$$A\otimes -\colon \mathcal{C} \rightleftarrows \mathcal{C} : \mathrm{Hom}(A,-),$$

so that $A\otimes -$ will preserve colimits and $\mathrm{Hom}(A,-)$ will preserve limits. Moreover, the functor $\mathrm{Hom}(\mathcal{J},-)$ is adjoint to $\mathrm{Id}_{\mathcal{C}} \cong \mathcal{J}\otimes -$ and hence is (naturally isomorphic to) the identity functor of $\mathcal{C}$.

For A, B and C in $\mathcal{C}$, the sequence of natural isomorphisms

$$\mathcal{C}^{op}(\mathrm{Hom}(A, B), C) = \mathcal{C}(C, \mathrm{Hom}(A, B)) \cong \mathcal{C}(C \otimes A, B) \cong \mathcal{C}(A, \mathrm{Hom}(C, B))$$

implies that there is an adjunction

$$\mathrm{Hom}(-, B)\colon \mathcal{C} \rightleftarrows \mathcal{C}^{op} : \mathrm{Hom}(-, B).$$

It follows that for any B in $\mathcal{C}$, the functor $\mathrm{Hom}(-, B)$ sends colimits to limits.

For any A and B in $\mathcal{C}$ the isomorphism

$$\mathcal{C}(\emptyset \otimes A, B) \cong \mathcal{C}(\emptyset, \mathrm{Hom}(A, B))$$

says that there is a unique map $\emptyset \otimes A \longrightarrow B$. Hence, we have a unique map $\emptyset \otimes A \longrightarrow \emptyset$ and the two composites

$$\emptyset \otimes A \longrightarrow \emptyset \longrightarrow \emptyset \otimes A \qquad \emptyset \longrightarrow \emptyset \otimes A \longrightarrow \emptyset$$

are the identity maps. Thus, $\emptyset \longrightarrow \emptyset \otimes A$ is an isomorphism.

Examples 6.1.6 Sets, simplicial sets and topological spaces are all closed symmetric monoidal categories. For sets, the function object is given by the set of maps. For simplicial sets, the function object $\mathrm{Hom}(K, L)$ has n-simplices given by

$$\mathrm{Hom}(K, L)_n = \mathrm{sSet}(K \times \Delta[n], L)$$

with face and degeneracy maps coming from those for $\Delta[n]$. For topological spaces, we use the modified compact–open topology on the set of continuous maps from a space X to a space Y, as introduced in Subsection 1.1.1.

Pointed spaces and pointed simplicial sets are also closed symmetric monoidal categories, using smash products and function spaces defined via pointed maps.

We now investigate symmetric monoidal categories that are also model categories. Given a model category $\mathcal{C}$ that is a symmetric monoidal category, we want conditions on the model structure so that the monoidal product and internal function object pass to derived versions on the homotopy category. A first condition is that the adjunction

$$(A \otimes -, \mathrm{Hom}(A, -))$$

should be a Quillen adjunction when A is cofibrant. We may also like to have that the adjunction

$$(\mathrm{Hom}(-, B), \mathrm{Hom}(-, B))$$

is a Quillen adjunction (using the opposite model structure on $\mathcal{C}^{op}$) when B is fibrant. The definition of a monoidal model category gives a much stronger requirement that will imply both of these conditions.

Definition 6.1.7 Let $(\mathcal{C}, \otimes, \mathcal{I}, \mathrm{Hom})$ be a closed symmetric monoidal category such that $\mathcal{C}$ is also a model category. We say that the *pushout product axiom* holds for $\mathcal{C}$ if the following conditions are satisfied.

1. For some cofibrant replacement of the unit $\mathcal{I}^{cof} \longrightarrow \mathcal{I}$ and any cofibrant A in $\mathcal{C}$, the map
$$\mathcal{I}^{cof} \otimes A \to \mathcal{I} \otimes A \cong A$$
is a weak equivalence.
2. For cofibrations $f\colon A \longrightarrow B$ and $g\colon X \longrightarrow Y$ in $\mathcal{C}$, the map
$$f \,\square\, g\colon B \otimes X \coprod_{A \otimes X} A \otimes Y \longrightarrow B \otimes Y$$
is a cofibration.
3. For cofibrations f and g in $\mathcal{C}$ with one of f or g a weak equivalence, the map $f \,\square\, g$ is an acyclic cofibration.

By Lemma 6.1.10, if the first condition on the unit holds for some cofibrant replacement of $\mathcal{I}$, then it holds for every cofibrant replacement of $\mathcal{I}$.

There are several adjoint versions of the above conditions involving fibrations as well as cofibrations. These may be easier to check in some specific examples where the fibrations are well understood.

Lemma 6.1.8 *Let $(\mathcal{C}, \otimes, \mathcal{I}, \mathrm{Hom})$ be a closed symmetric monoidal category with a model structure. The following two conditions are equivalent.*

1. *Given cofibrations $f\colon A \longrightarrow B$ and $g\colon X \longrightarrow Y$ in $\mathcal{C}$, the map*
$$f \,\square\, g\colon B \otimes X \coprod_{A \otimes X} A \otimes Y \longrightarrow B \otimes Y$$
is a cofibration that is acyclic if one of f or g is as well.
2. *Given a cofibration $f\colon A \longrightarrow B$ and a fibration $h\colon P \longrightarrow Q$ in $\mathcal{C}$, the map*
$$\mathrm{Hom}_\square(f, h)\colon \mathrm{Hom}(B, P) \longrightarrow \mathrm{Hom}(B, Q) \underset{\mathrm{Hom}(A,Q)}{\times} \mathrm{Hom}(A, P)$$
is a fibration that is acyclic if one of f or g is as well.

Proof This follows from how the adjunction relation between $- \otimes -$ and $\mathrm{Hom}(-,-)$ interacts with pushouts and pullbacks. □

Definition 6.1.9 A *symmetric monoidal model category* is a closed symmetric monoidal category $(\mathcal{C}, \otimes, \mathcal{I}, \mathrm{Hom})$ with a model structure on $\mathcal{C}$ which satisfies the pushout product axiom.

We will see that this definition gives us the desired Quillen adjunctions and a symmetric monoidal structure on the homotopy category.

Lemma 6.1.10 *Let* $(\mathcal{C}, \otimes, \mathcal{I}, \mathrm{Hom})$ *be a symmetric monoidal model category. Then when A is cofibrant and B is fibrant, the adjunctions*

$$(A \otimes -, \mathrm{Hom}(A, -)) \quad \textit{and} \quad (\mathrm{Hom}(-, B), \mathrm{Hom}(-, B))$$

are Quillen adjunctions.

Proof Consider the cofibration $f\colon \emptyset \longrightarrow A$ from the initial object to A and an (acyclic) cofibration $g\colon X \to Y$. The pushout product axiom states that

$$f \,\square\, g\colon A \otimes X \cong A \otimes X \coprod_{\emptyset \otimes X} \emptyset \otimes Y \longrightarrow A \otimes Y$$

is an (acyclic) cofibration. Hence, $A \otimes -$ is a left Quillen functor. The other case follows using the adjoint descriptions of the pushout product axiom from Lemma 6.1.8 with the fibration $B \longrightarrow *$ from B to the terminal object. □

Theorem 6.1.11 *Let* $(\mathcal{C}, \otimes, \mathcal{I}, \mathrm{Hom})$ *be a symmetric monoidal model category. Then* $(\mathrm{Ho}(\mathcal{C}), \otimes^{\mathbb{L}}, \mathcal{I}, \mathbb{R}\mathrm{Hom})$ *is a closed symmetric monoidal category.*

Proof Let $f\colon A \longrightarrow B$ and $g\colon X \longrightarrow Y$ be weak equivalences between cofibrant objects. Then, as $A \otimes -$ and $- \otimes Y$ are Quillen functors, the maps

$$A \otimes g\colon A \otimes X \longrightarrow A \otimes Y \quad \text{and} \quad f \otimes Y\colon A \otimes Y \longrightarrow B \otimes Y$$

are weak equivalences, hence, so is $f \otimes g$. It follows that $\otimes$ has a derived functor. A similar argument shows that Hom also has a derived functor. We must show that the functors

$$- \otimes^{\mathbb{L}} -\colon \quad \mathrm{Ho}(\mathcal{C}) \times \mathrm{Ho}(\mathcal{C}) \longrightarrow \mathrm{Ho}(\mathcal{C})$$
$$\mathbb{R}\mathrm{Hom}(-,-)\colon \quad \mathrm{Ho}(\mathcal{C})^{op} \times \mathrm{Ho}(\mathcal{C}) \longrightarrow \mathrm{Ho}(\mathcal{C})$$

induce a closed symmetric monoidal structure on the homotopy category.

We begin by constructing the unit, symmetry and associativity isomorphisms. We use the model category versions of these statements, but we must check that they pass to homotopy categories. This will follow from two facts. Firstly, when A is cofibrant, the functor $A \otimes -$ preserves cylinder objects on cofibrant objects, therefore, it preserves homotopies. Secondly, $- \otimes -$ preserves all weak equivalences between cofibrant objects. It follows that a, τ and u induce maps at the level of homotopy categories.

We have assumed that for any A in $\mathcal{C}$, the map

$$\mathcal{I} \otimes^{\mathbb{L}} A = \mathcal{I}^{cof} \otimes A^{cof} \longrightarrow \mathcal{I} \otimes A^{cof} \xrightarrow{u} A^{cof} \longrightarrow A$$

is a weak equivalence. Hence, $\mathcal{I}$ is a unit for $\otimes^{\mathbb{L}}$. The twist isomorphism on $\mathcal{C}$ induces one on the homotopy category

$$\tau\colon A \otimes^{\mathbb{L}} B = A^{cof} \otimes B^{cof} \longrightarrow B^{cof} \otimes A^{cof} = B \otimes^{\mathbb{L}} A.$$

We may define an associativity isomorphism via the sequence of maps

$$\begin{aligned} (A \otimes^{\mathbb{L}} B) \otimes^{\mathbb{L}} C &= (A^{cof} \otimes B^{cof})^{cof} \otimes C^{cof} \\ &\xrightarrow{\sim} (A^{cof} \otimes B^{cof}) \otimes C^{cof} \\ &\xrightarrow{a} A^{cof} \otimes (B^{cof} \otimes C^{cof}) \\ &\xleftarrow{\sim} A^{cof} \otimes (B^{cof} \otimes C^{cof})^{cof} \\ &= A \otimes^{\mathbb{L}} (B \otimes^{\mathbb{L}} C). \end{aligned}$$

The two maps labelled "$\sim$" come from choices of cofibrant replacements and are weak equivalences. Hence, the composite is an isomorphism in the homotopy category.

We next show that the adjunction isomorphism

$$\phi\colon \mathcal{C}(A \otimes B, C) \longrightarrow \mathcal{C}(A, \operatorname{Hom}(B, C))$$

passes to the level of homotopy categories. As already mentioned, when A is cofibrant, the functor $A \otimes -$ preserves cylinder objects on cofibrant objects. Similarly, when B is cofibrant, $\operatorname{Hom}(B, -)$ preserves path objects on fibrant objects, and for C fibrant, $\operatorname{Hom}(-, C)$ sends cylinder objects on cofibrant objects to path objects on fibrant objects. It follows that for A and B cofibrant and C fibrant we obtain an isomorphism

$$[A \otimes B, C] \longrightarrow [A, \operatorname{Hom}(B, C)]$$

after one takes quotients by the homotopy relation. Since the functors involved preserve weak equivalences between suitably cofibrant and fibrant objects, we have an adjunction at the level of homotopy categories.

The coherence diagrams all follow from their model category versions, using the explicitly described associativity, unit and twist maps given above. □

Examples 6.1.12 Simplicial sets and topological spaces (with the Serre or Hurewicz model structure) are all symmetric monoidal model categories. The same is true of the pointed versions.

The projective model structure on chain complexes is a symmetric monoidal model category. However, the category of chain complexes with the injective model structure is not symmetric monoidal: the pushout product axiom fails. Recall that the cofibrations of this model structure are the monomorphisms. Consider the monomorphisms $0 \longrightarrow \mathbb{Z}/2$ and $\mathbb{Z} \longrightarrow \mathbb{Q}$, viewed as chain complexes in degree 0. Their pushout product is $\mathbb{Z}/2 \longrightarrow 0$, which is not a monomorphism.

When $\mathcal{C}$ is cofibrantly generated, it suffices to check conditions (2.) and (3.) of the pushout product axiom on the generating sets.

Lemma 6.1.13 *Let $(\mathcal{C}, \otimes, \mathcal{I}, \mathrm{Hom})$ be a closed symmetric monoidal category with a cofibrantly generated model structure with generating sets I and J. Assume that for some cofibrant replacement of the unit $\mathcal{I}^{cof} \longrightarrow \mathcal{I}$ and any cofibrant A in $\mathcal{C}$, the map*

$$\mathcal{I}^{cof} \otimes A \to \mathcal{I} \otimes A \cong A$$

is a weak equivalence.

If every map in $I \,\square\, I$ is a cofibration and every map in $I \,\square\, J$ is an acyclic cofibration, then $\mathcal{C}$ is a symmetric monoidal model category.

Proof By Lemma 6.1.8, the maps in I have the left lifting property with respect to maps of the form $\mathrm{Hom}_\square(i, h)$, where $i \in I$ and h is an acyclic fibration. Hence, maps of the form $\mathrm{Hom}_\square(i, h)$ are acyclic fibrations. As every cofibration has the left lifting property with respect to such maps, Lemma 6.1.8 implies that any map of the form $f \,\square\, i$ is a cofibration for f a cofibration and $i \in I$. Using symmetry and repeating this argument shows that $f \,\square\, g$ is a cofibration whenever f and g are cofibrations.

The case of an acyclic cofibration paired with a cofibration is similar. □

We note there is an adjoint version of the unit condition for a monoidal model category: the map

$$X \cong \mathrm{Hom}(\mathcal{I}, X) \longrightarrow \mathrm{Hom}(\mathcal{I}^{cof}, X)$$

must be a weak equivalence for any fibrant X.

Now that we have defined stable model categories (whose homotopy categories are triangulated) and closed symmetric monoidal model categories (whose homotopy categories are closed symmetric monoidal), it is logical to ask how these two definitions interact.

A proposed list of axioms for a triangulated category with a compatible tensor product can be found in [HPS97]. The following theorem is based on that list. Further axioms for compatibility of a tensor product with a triangulated category are given in [May01]. In particular, that reference considers axioms for a tensor products of an exact triangle with another exact triangle.

Theorem 6.1.14 *Let* $(\mathcal{C}, \otimes, \mathcal{I}, \mathrm{Hom})$ *be a symmetric monoidal stable model category. Then* $(\mathrm{Ho}(\mathcal{C}), \otimes^{\mathbb{L}}, \mathcal{I}, \mathbb{R}\,\mathrm{Hom})$ *is a triangulated category with a closed symmetric monoidal structure satisfying the following list of properties for all* A*,* X*,* Y *and* Z *in* $\mathrm{Ho}(\mathcal{C})$.

1. *There is a natural isomorphism*

$$e_{X,Y} \colon (\Sigma X) \otimes^{\mathbb{L}} Y \longrightarrow \Sigma(X \otimes^{\mathbb{L}} Y).$$

2. *The functor* $- \otimes^{\mathbb{L}} A$ *is an exact functor.*
3. *The functor* $\mathbb{R}\,\mathrm{Hom}(A, -)$ *is an exact functor.*
4. *The functor* $\mathbb{R}\,\mathrm{Hom}(-, A)$ *sends an exact triangle of the form*

$$X \xrightarrow{f} Y \xrightarrow{g} Z \xrightarrow{h} \Sigma X$$

to the exact triangle

$$\mathbb{R}\,\mathrm{Hom}(\Sigma X, A) \xrightarrow{\mathbb{R}\,\mathrm{Hom}(h, \mathrm{Id}_A)} \mathbb{R}\,\mathrm{Hom}(Z, A) \xrightarrow{\mathbb{R}\,\mathrm{Hom}(g, \mathrm{Id}_A)} \mathbb{R}\,\mathrm{Hom}(Y, A) \xrightarrow{\mathbb{R}\,\mathrm{Hom}(f, \mathrm{Id}_A)} \mathbb{R}\,\mathrm{Hom}(X, A).$$

5. *Let* a *and* b *be integers, and let* (-1) *denote the additive inverse of the identity map in the ring* $[\mathcal{I}, \mathcal{I}]$*. Then the diagram below commutes, where the horizontal isomorphisms are induced by* e *and* τ.

$$\begin{array}{ccc} \Sigma^a \mathcal{I} \otimes^{\mathbb{L}} \Sigma^b \mathcal{I} & \xrightarrow{\cong} & \Sigma^{a+b} \mathcal{I} \\ \downarrow \tau & & \downarrow (-1)^{ab} \\ \Sigma^b \mathcal{I} \otimes^{\mathbb{L}} \Sigma^a \mathcal{I} & \xrightarrow{\cong} & \Sigma^{a+b} \mathcal{I} \end{array}$$

6. *The following diagram commutes*

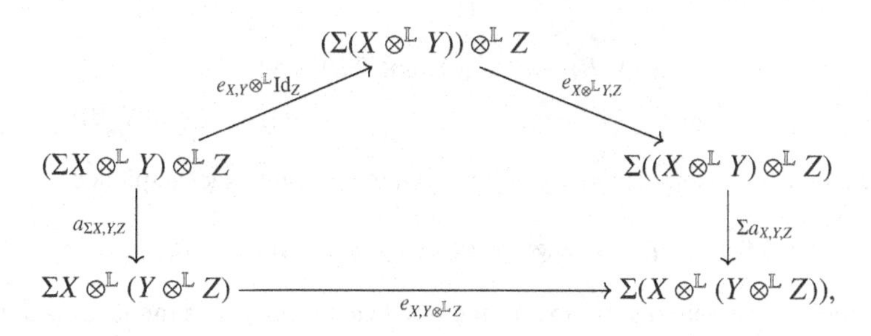

where $a_{X,Y,Z}$ *denotes the associativity isomorphism of* $\otimes^{\mathbb{L}}$.

7. *The following diagram commutes, where u denotes the unit isomorphism of the monoidal structure.*

$$\begin{array}{ccc} \Sigma X \otimes^{\mathbb{L}} \mathcal{I} & \xrightarrow{u} & \Sigma X \\ {\scriptstyle e_{X,\mathcal{I}}} \downarrow & \nearrow {\scriptstyle \Sigma u} & \\ \Sigma(X \otimes^{\mathbb{L}} \mathcal{I}) & & \end{array}$$

Proof When Y is cofibrant, the functor $- \otimes Y$ is a left adjoint which preserves cylinder objects between cofibrant objects. It follows that for cofibrant X, there is an isomorphism

$$e_{X,Y} \colon \Sigma X \otimes Y \longrightarrow \Sigma(X \otimes Y).$$

This induces the natural map $e_{X,Y}$ in $\mathrm{Ho}(\mathcal{C})$, see Definition 3.1.2. The last two coherence diagrams in the theorem follow as the suspensions are defined using pushouts, and the natural transformation e is defined using the universal property of pushouts.

Let $X \longrightarrow Y \longrightarrow Z \longrightarrow \Sigma X$ be an exact triangle in $\mathrm{Ho}(\mathcal{C})$, and let $A \in \mathrm{Ho}(\mathcal{C})$. Since $- \otimes^{\mathbb{L}} A$, is the left derived functor of a Quillen functor, it is exact by Theorem 4.5.2. Hence, the sequence

$$X \otimes^{\mathbb{L}} A \longrightarrow Y \otimes^{\mathbb{L}} A \longrightarrow Z \otimes^{\mathbb{L}} A \longrightarrow \Sigma X \otimes^{\mathbb{L}} A$$

is also an exact triangle, and the natural isomorphism $e_{X,A}$ shows that there is no ambiguity in the last term.

Similarly, $\mathbb{R}\,\mathrm{Hom}(A, -)$ is exact. We may use the isomorphisms induced by the natural transformation e

$$\mathbb{R}\,\mathrm{Hom}(\Sigma X, A) \cong \Sigma^{-1}\mathbb{R}\,\mathrm{Hom}(X, A) \quad \text{and} \quad \mathbb{R}\,\mathrm{Hom}(X, \Sigma^{-1}A) \cong \Sigma^{-1}\mathbb{R}\,\mathrm{Hom}(X, A)$$

to see how Σ interacts with this functor.

The functor $\mathbb{R}\,\mathrm{Hom}(-, B)$ takes cofibre sequences to fibre sequences, which we can see as follows. The sequence

$$\cdots \to [A, \mathbb{R}\,\mathrm{Hom}(\Sigma X, B)] \xrightarrow{\partial} [A, \mathbb{R}\,\mathrm{Hom}(Z, B)] \to [A, \mathbb{R}\,\mathrm{Hom}(Y, B)] \\ \to [A, \mathbb{R}\,\mathrm{Hom}(X, B)] \to \cdots$$

is exact as it is isomorphic, by adjunction, to the long exact sequence

$$\cdots \to [\Sigma X \otimes^{\mathbb{L}} A, B] \to [Z \otimes^{\mathbb{L}} A, B] \to [Y \otimes^{\mathbb{L}} A, B] \to [X \otimes^{\mathbb{L}} A, B] \to \cdots.$$

Moreover, the boundary map ∂ in the first exact sequence is induced by a map

$$\mathbb{R}\,\mathrm{Hom}(\Sigma X, B) \longrightarrow \mathbb{R}\,\mathrm{Hom}(Z, B)$$

(which in turn is induced by the map $Z \longrightarrow \Sigma X$). By Remark 4.2.3, such a boundary map in a long exact sequence is equivalent to the action of a fibre sequence. Therefore,

$$\mathbb{R}\operatorname{Hom}(Z,B) \xrightarrow{\mathbb{R}\operatorname{Hom}(g,\mathrm{Id}_B)} \mathbb{R}\operatorname{Hom}(Y,B) \xrightarrow{\mathbb{R}\operatorname{Hom}(f,\mathrm{Id}_B)} \mathbb{R}\operatorname{Hom}(X,B)$$

is a fibre sequence whose action is induced by the coaction of the cofibre sequence

$$X \xrightarrow{f} Y \xrightarrow{g} Z.$$

Therefore, we have the exact triangle of point 4.

The remaining point is the sign convention 5. Given two exact triangles

$$X \xrightarrow{f} Y \xrightarrow{g} Z \xrightarrow{h} \Sigma X \text{ and } A \xrightarrow{a} B \xrightarrow{b} C \xrightarrow{c} \Sigma A,$$

we can apply the 3×3 Lemma, Lemma 4.1.13, to the square

$$\begin{array}{ccc} A \otimes^{\mathbb{L}} X & \xrightarrow{\mathrm{Id} \otimes^{\mathbb{L}} f} & A \otimes^{\mathbb{L}} Y \\ {\scriptstyle a \otimes^{\mathbb{L}} \mathrm{Id}} \downarrow & & \downarrow {\scriptstyle a \otimes^{\mathbb{L}} \mathrm{Id}} \\ B \otimes^{\mathbb{L}} X & \xrightarrow{\mathrm{Id} \otimes^{\mathbb{L}} f} & B \otimes^{\mathbb{L}} Y. \end{array}$$

In the resulting large grid, the lower right-hand corner commutes up to a sign of -1. This corner is

$$\begin{array}{ccc} C \otimes^{\mathbb{L}} Z & \xrightarrow{e \circ \mathrm{Id} \otimes^{\mathbb{L}} h} & \Sigma(C \otimes^{\mathbb{L}} X) \\ {\scriptstyle e' \circ c \otimes^{\mathbb{L}} \mathrm{Id}} \downarrow & \ominus & \downarrow {\scriptstyle e' \circ c \otimes^{\mathbb{L}} \mathrm{Id}} \\ \Sigma(A \otimes^{\mathbb{L}} Z) & \xrightarrow{e \circ \mathrm{Id} \otimes^{\mathbb{L}} h} & \Sigma^2(A \otimes^{\mathbb{L}} Z), \end{array}$$

where e' is the natural transformation $(- \otimes^{\mathbb{L}} \Sigma -) \longrightarrow \Sigma(- \otimes^{\mathbb{L}} -)$ induced by e and τ. Taking both triangles to be

$$\mathbb{J} \longrightarrow * \longrightarrow \Sigma\mathbb{J} \xrightarrow{\mathrm{Id}} \Sigma\mathbb{J}$$

gives the result. □

Remark 6.1.15 We will refer to the homotopy category of a symmetric monoidal stable model category as a *tensor-triangulated category*.

We now consider functors between symmetric monoidal (model) categories that interact well with the monoidal structures. We focus on the case of most interest, where we consider adjunctions between closed symmetric monoidal categories. Our approach is based on [SS03a].

Definition 6.1.16 Let $(\mathcal{C}, \otimes_{\mathcal{C}}, \mathcal{I}_{\mathcal{C}}, \mathrm{Hom}_{\mathcal{C}})$ and $(\mathcal{D}, \otimes_{\mathcal{D}}, \mathcal{I}_{\mathcal{D}}, \mathrm{Hom}_{\mathcal{D}})$ be symmetric monoidal categories. A functor $G \colon \mathcal{C} \longrightarrow \mathcal{D}$ is said to be *lax symmetric monoidal* (also known as *weak symmetric monoidal*) if there is a natural transformation

$$\phi_{c,c'} \colon G(c) \otimes_{\mathcal{D}} G(c') \longrightarrow G(c \otimes_{\mathcal{C}} c')$$

and a map

$$\nu \colon \mathcal{I}_{\mathcal{D}} \longrightarrow G(\mathcal{I}_{\mathcal{C}})$$

satisfying three coherence diagrams describing the interaction with the unit, twist map and associativity isomorphism.

We say that F is *strong symmetric monoidal* if ϕ is a natural isomorphism and ν an isomorphism.

A functor $F \colon \mathcal{C} \longrightarrow \mathcal{D}$ is said to be *op-lax symmetric monoidal* if there is a natural transformation

$$\widetilde{\phi}_{c,c'} \colon F(c \otimes_{\mathcal{C}} c') \longrightarrow F(c) \otimes_{\mathcal{D}} F(c')$$

and a map

$$\widetilde{\nu} \colon F(\mathcal{I}_{\mathcal{C}}) \longrightarrow \mathcal{I}_{\mathcal{D}}$$

satisfying three coherence diagrams describing the interaction with the unit, twist map and associativity isomorphism.

As the coherence diagrams in the above definition are very similar to those of Definition 6.1.1, we do not spell them out here.

Remark 6.1.17 A lax symmetric monoidal functor F lifts to a functor from commutative monoids in $\mathcal{C}$ to commutative monoids in $\mathcal{D}$. An op-lax symmetric monoidal functor lifts to the level of commutative comonoids.

If the morphisms $\widetilde{\phi}$ and $\widetilde{\nu}$ of an op-lax symmetric monoidal functor F are isomorphisms, then the functor F is strong symmetric monoidal. The notation $\widetilde{\phi}$ and $\widetilde{\nu}$ has been chosen to reflect the standard example: op-lax symmetric monoidal functors tend to be left adjoints to lax symmetric monoidal functors.

Lemma 6.1.18 *Let* $(\mathcal{C}, \otimes_{\mathcal{C}}, \mathcal{I}_{\mathcal{C}}, \mathrm{Hom}_{\mathcal{C}})$ *and* $(\mathcal{D}, \otimes_{\mathcal{D}}, \mathcal{I}_{\mathcal{D}}, \mathrm{Hom}_{\mathcal{D}})$ *be symmetric monoidal categories with an adjunction*

$$F \colon \mathcal{C} \rightleftarrows \mathcal{D} \colon G.$$

Then the right adjoint G is lax symmetric monoidal if and only the left adjoint F is op-lax symmetric monoidal.

Proof We only show one direction of the proof as the other one is dual. We first consider the maps relating the monoidal units. As (F, G) is an adjunction, a map

$$\nu\colon \mathcal{I}_{\mathcal{C}} \longrightarrow G(\mathcal{I}_{\mathcal{D}})$$

corresponds to a map

$$\widetilde{\nu}\colon F(\mathcal{I}_{\mathcal{C}}) \longrightarrow \mathcal{I}_{\mathcal{D}}.$$

Assume that G is lax monoidal, so there is a natural map

$$\phi_{d,d'}\colon G(d) \otimes_{\mathcal{C}} G(d') \longrightarrow G(d \otimes_{\mathcal{D}} d')$$

satisfying certain coherence conditions. The adjoint ϕ' of ϕ is the composite

$$F(G(d) \otimes_{\mathcal{C}} G(d')) \xrightarrow{F\phi_{d,d'}} F(G(d \otimes_{\mathcal{D}} d')) \xrightarrow{\varepsilon} d \otimes_{\mathcal{D}} d'.$$

We define $\widetilde{\phi}_{d,d'}$ to be the composite below.

$$F(c \otimes_{\mathcal{C}} c') \xrightarrow{F(\eta_c \otimes_{\mathcal{C}} \eta_{c'})} F(GF(c) \otimes_{\mathcal{C}} GF(c')) \xrightarrow{\phi'_{F(c),F(c')}} F(c) \otimes_{\mathcal{D}} F(c')$$

The coherence conditions for G imply the coherence conditions for F. □

Definition 6.1.19 Let $(\mathcal{C}, \otimes_{\mathcal{C}}, \mathcal{I}_{\mathcal{C}}, \mathrm{Hom}_{\mathcal{C}})$ and $(\mathcal{D}, \otimes_{\mathcal{D}}, \mathcal{I}_{\mathcal{D}}, \mathrm{Hom}_{\mathcal{D}})$ be symmetric monoidal model categories. Let

$$F\colon \mathcal{C} \rightleftarrows \mathcal{D} : G$$

be an adjunction such that G is lax symmetric monoidal with structure maps ϕ and ν. The adjunction is a *weak symmetric monoidal Quillen adjunction* if the adjoint structure maps $\widetilde{\phi}$ and $\widetilde{\nu}$ satisfy the following conditions.

1. For a cofibrant replacement $\mathcal{I}_{\mathcal{C}}{}^{cof} \to \mathcal{I}_{\mathcal{C}}$ of the unit of $\mathcal{C}$, the composite
$$F(\mathcal{I}_{\mathcal{C}}{}^{cof}) \longrightarrow F(\mathcal{I}_{\mathcal{C}}) \xrightarrow{\widetilde{\nu}} \mathcal{I}_{\mathcal{D}}$$
is a weak equivalence in $\mathcal{D}$.
2. For cofibrant c and c' in $\mathcal{C}$, the map
$$\widetilde{\phi}_{c,c'}\colon F(c \otimes_{\mathcal{C}} c') \longrightarrow F(c) \otimes_{\mathcal{D}} F(c')$$
is a weak equivalence in $\mathcal{D}$.

We say that (F, G) is a *strong symmetric monoidal Quillen adjunction* if, in addition, $\widetilde{\nu}$ is an isomorphism and the adjoint structure $\widetilde{\phi}$ is an isomorphism for all c and c'.

If the unit condition of a weak symmetric monoidal Quillen adjunction holds for one cofibrant replacement of the unit $\mathcal{I}_{\mathcal{C}}$, then it holds for any cofibrant replacement.

Examples 6.1.20 The primary example of a weak symmetric monoidal Quillen adjunction is the rational Dold–Kan adjunction, as explained in detail in [SS03a].

Given a map of commutative rings $f\colon R \longrightarrow S$, the adjunction

$$S \otimes_R -\colon \mathrm{Ch}(R) \rightleftarrows \mathrm{Ch}(S) : f^*$$

on chain complexes with the respective projective model structures is a strong symmetric monoidal Quillen adjunction.

The Quillen adjunction given by geometric realisation and the singular functor

$$|-|\colon \mathrm{sSet}_* \rightleftarrows \mathrm{Top}_* : \mathrm{sing}$$

is strong symmetric monoidal.

Theorem 6.1.21 *Let* $(\mathcal{C}, \otimes_{\mathcal{C}}, \mathcal{I}_{\mathcal{C}}, \mathrm{Hom}_{\mathcal{C}})$ *and* $(\mathcal{D}, \otimes_{\mathcal{D}}, \mathcal{I}_{\mathcal{D}}, \mathrm{Hom}_{\mathcal{D}})$ *be symmetric monoidal model categories. Let*

$$F\colon \mathcal{C} \rightleftarrows \mathcal{D} : G$$

be a weak symmetric monoidal Quillen adjunction. Then $\mathbb{L}F$ *is a strong symmetric monoidal functor and* $\mathbb{R}G$ *is a lax symmetric monoidal functor.*

Proof We prove that $\mathbb{L}F$ is a strong symmetric monoidal functor. Lemma 6.1.18 will then give the statement for $\mathbb{R}G$.

Similar arguments to those in the proof of Theorem 6.1.11 show that the natural transformations $\widetilde{\nu}$ and $\widetilde{\phi}$ pass to the homotopy category. The unit map on homotopy categories is induced by the inverse of the composite

$$\mathbb{L}F(\mathcal{I}) = F(\mathcal{I}_{\mathcal{C}}{}^{cof}) \longrightarrow F(\mathcal{I}_{\mathcal{C}}) \xrightarrow{\widetilde{\nu}} \mathcal{I}_{\mathcal{D}},$$

which is a weak equivalence by assumption.

The monoidal map on homotopy categories is the composite

$$\begin{array}{rcl} \mathbb{L}F(c) \otimes^{\mathbb{L}}_{\mathcal{D}} \mathbb{L}F(c') & = & F(c^{cof})^{cof} \otimes_{\mathcal{D}} F(c'^{cof})^{cof} \\ & \longrightarrow & F(c^{cof}) \otimes_{\mathcal{D}} F(c'^{cof}) \\ & \xleftarrow{\sim} & F(c^{cof} \otimes_{\mathcal{C}} c'^{cof}) \\ & \longrightarrow & F((c^{cof} \otimes_{\mathcal{C}} c'^{cof})^{cof}) \\ & = & \mathbb{L}F(c \otimes^{\mathbb{L}}_{\mathcal{C}} c'), \end{array}$$

where the middle map is $\widetilde{\phi}$, which is a weak equivalence by assumption.

The coherence conditions follow from the model category versions, using the explicit maps above. □

Remark 6.1.22 It seems to be more common that the left adjoint of an adjunction is strong symmetric monoidal, rather than the right adjoint. Of course, when one has an equivalence of categories, the left adjoint is strong symmetric monoidal if and only if the right adjoint is strong symmetric monoidal. This follows from the relation between ϕ and $\widetilde{\phi}$ of Lemma 6.1.18.

This applies to show that the derived functors of a weak symmetric monoidal Quillen equivalence are strong monoidal.

We now review the concept of an enriched category and how an enrichment can interact usefully with model structures. The ideas are similar to those we have already seen for monoidal model categories.

Definition 6.1.23 Let $(\mathcal{C}, \otimes, \mathcal{I}, \mathrm{Hom})$ be a closed symmetric monoidal category. An *enrichment* of a category $\mathcal{D}$ in $\mathcal{C}$ is a functor

$$\mathrm{map}_{\mathcal{D}}(-,-)\colon \mathcal{D}^{op} \times \mathcal{D} \longrightarrow \mathcal{C}$$

satisfying the following conditions and properties.

1. For each $A \in \mathcal{D}$, there is a map $\mathcal{I} \longrightarrow \mathrm{map}_{\mathcal{D}}(A, A)$ called the identity element.
2. For each triple $A,\ B,\ C \in \mathcal{D}$, there is a composition map
$$\mathrm{map}_{\mathcal{D}}(B, C) \otimes \mathrm{map}_{\mathcal{D}}(A, B) \longrightarrow \mathrm{map}_{\mathcal{D}}(A, C).$$
3. The composition is associative and unital.

We have encountered this concept before.

Example 6.1.24 An additive category is a category with an enrichment in abelian groups.

A closed symmetric monoidal category is enriched in itself, with the function object providing the enrichment.

In the cases of interest to us, we have more structure, namely, a tensor and cotensor that are compatible with the enrichment.

Definition 6.1.25 Let $(\mathcal{C}, \otimes, \mathcal{I}, \mathrm{Hom})$ be a closed symmetric monoidal category. A category $\mathcal{D}$ is a *closed module over* $\mathcal{C}$ (or a $\mathcal{C}$*–module*) if there are functors $-\bar{\otimes}-$ called the *tensor* and $(-)^{(-)}$ called the *cotensor*

$$\begin{aligned} -\bar{\otimes}-\colon \mathcal{C} \times \mathcal{D} &\longrightarrow \mathcal{D} \\ (-)^{(-)}\colon \mathcal{C} \times \mathcal{D}^{op} &\longrightarrow \mathcal{D} \\ \mathrm{map}_{\mathcal{D}}(-,-)\colon \mathcal{D}^{op} \times \mathcal{D} &\longrightarrow \mathcal{C} \end{aligned}$$

together with the following isomorphisms and coherence conditions.

1. An associativity isomorphism $a\colon (C \otimes C') \bar{\otimes} D \longrightarrow C \bar{\otimes} (C' \bar{\otimes} D)$.
2. A unit isomorphism $u\colon \mathcal{I} \bar{\otimes} D \longrightarrow D$.
3. A coherence diagram for associativity.
4. A coherence diagram relating the unit isomorphism of the tensor and the unit isomorphism of $\mathcal{C}$.
5. A coherence diagram relating the two unit isomorphisms and the symmetry of $\mathcal{C}$.
6. Compatibility isomorphisms

$$\mathcal{D}(D, E^C) \cong \mathcal{D}(C \bar{\otimes} D, E) \cong \mathcal{C}(C, \mathrm{map}_{\mathcal{D}}(D, E))$$

   for $C \in \mathcal{C}$ and $D, E \in \mathcal{D}$.

We can think of the tensor as giving an action of $\mathcal{C}$ on $\mathcal{D}$. The functor $\bar{\otimes}$ and its adjoints are known as an *adjunction of two variables*.

Examples 6.1.26 For a ring R, the category of R–modules is a closed module over the category of abelian groups.

A closed symmetric monoidal category is a closed module over itself. The function object gives the enrichment and cotensor, and the tensor is the monoidal product.

The category of sequential spectra is a closed module over the category of pointed spaces, see Definition 2.3.1.

Just as we passed from closed symmetric monoidal categories to symmetric monoidal model categories with Definition 6.1.7, we can define a model category version of closed modules.

Definition 6.1.27 Let $(\mathcal{C}, \otimes, \mathcal{I}, \mathrm{Hom})$ be a symmetric monoidal model category. A model category $\mathcal{D}$ is called a $\mathcal{C}$*–model category* if it is a closed module over $\mathcal{C}$ and the following conditions hold.

1. For a cofibration i in $\mathcal{C}$ and a cofibration j in $\mathcal{D}$, the pushout product $i \square j$ is a cofibration of $\mathcal{D}$ that is a weak equivalence if one of i or j is as well.
2. For a cofibrant replacement $\mathcal{I}^{cof} \longrightarrow \mathcal{I}$ of the unit of $\mathcal{C}$, the map

$$\mathcal{I}^{cof} \bar{\otimes} D \longrightarrow \mathcal{I} \bar{\otimes} D \cong D$$

   is a weak equivalence for any cofibrant $D \in \mathcal{D}$.

One can use an analogue of Lemma 6.1.8 to give equivalent characterisations of the above in terms of the enrichment or cotensor.

Definition 6.1.28 When $\mathcal{C}$ is the category of pointed simplicial sets, a $\mathcal{C}$–model category is often called a *simplicial model category*.

When $\mathcal{C}$ is the category of pointed topological sets, a $\mathcal{C}$–model category is often called a *topological model category*.

Analogously to Theorem 6.1.14, the homotopy category of a $\mathcal{C}$–model category has an action of $\mathrm{Ho}(\mathcal{C})$. We do not include a proof, as it would be very similar.

Theorem 6.1.29 *Let* $(\mathcal{C}, \otimes, \mathcal{I}, \mathrm{Hom})$ *be a symmetric monoidal model category and* $\mathcal{D}$ *a* $\mathcal{C}$*–model category. The* $\mathrm{Ho}(\mathcal{D})$ *is a closed module over* $\mathrm{Ho}(\mathcal{C})$. □

6.2 A Smash Product on the Stable Homotopy Category

The category of pointed topological spaces is a symmetric monoidal model category with the smash product providing the monoidal product. Its unit is the two point space S^0, and the internal function object is given by the space of continuous maps (with a suitable topology).

It makes sense to extend this smash product to the Spanier–Whitehead category $\mathcal{SW}$ and further to the stable homotopy category $\mathcal{SHC}$, where we denote it $- \wedge^{\mathbb{L}} -$. We want the functor

$$\Sigma^\infty \colon \mathrm{Ho}(\mathrm{Top}_*) \longrightarrow \mathcal{SHC}$$

to be strong symmetric monoidal, so that

$$\Sigma^\infty A \wedge^{\mathbb{L}} \Sigma^\infty B \cong \Sigma^\infty (A \wedge B)$$

for any pointed CW-complexes A and B. This will also imply that the unit of the smash product of spectra will be the sphere spectrum $\mathbb{S} = \Sigma^\infty S^0$. Furthermore, for any spectrum X, we want the smash product of spectra to be related to the smash product of a pointed CW-complex with a spectrum

$$X \wedge^{\mathbb{L}} \Sigma^\infty A \cong X \wedge A,$$

where the second term has

$$(X \wedge A)_n = X_n \wedge A.$$

Finally, we want an internal function object making the stable homotopy category into a closed symmetric monoidal category.

This extension (and defining the internal function object for spectra) took a long time to construct, as it does not come from a symmetric monoidal product on the model category of sequential spectra. Indeed, the primary reason for the development of symmetric spectra or orthogonal spectra is that these are symmetric monoidal model categories, and hence, their monoidal structures pass to the level of homotopy categories.

We can define an operation on sequential spectra that behaves like a symmetric monoidal product, up to (coherent) homotopy. We will call this the "handi-crafted smash product". Keeping track of the homotopies and coherence is a substantial task, see Adams [Ada74, Section III.4]. The starting idea is that the smash product should use the spaces and structure maps from both inputs. Let X and Y be sequential spectra. We can define a new spectrum by

$$(X \underset{\text{hand}}{\wedge} Y)_k = \begin{cases} X_n \wedge Y_n & k = 2n \\ X_{n+1} \wedge Y_n & k = 2n+1 \end{cases}$$

with structure maps given by

$$\Sigma(X_n \wedge Y_n) \cong (\Sigma X_n) \wedge Y_n \xrightarrow{\sigma_n^X \wedge Y_n} X_{n+1} \wedge Y_n$$

$$\Sigma(X_{n+1} \wedge Y_n) = S^1 \wedge X_{n+1} \wedge Y_n \xrightarrow{\tau_{S^1, X_{n+1}} \wedge \mathrm{Id}} X_{n+1} \wedge S^1 \wedge Y_n \xrightarrow{\mathrm{Id} \wedge \sigma_n^Y} X_{n+1} \wedge Y_{n+1}.$$

We leave the proof that this product gives a symmetric monoidal product on $\mathcal{SHC}$ with the sphere spectrum as the unit to [Ada74]. We will instead show that $\mathcal{S}^\Sigma$ and $\mathcal{S}^\mathcal{O}$ are symmetric monoidal model categories, and hence, their homotopy categories have closed symmetric monoidal products. For now, we continue to investigate this handi-crafted smash product in order to motivate the constructions of the symmetric monoidal products on symmetric and orthogonal spectra.

One can see that the definition of the handi-crafted smash product is not going to be associative. There are many other similar choices one could make, but none of these will be strictly associative. One solution often employed by mathematicians is to take *all* choices. We define an operation $- \otimes -$ as

$$(X \otimes Y)_n = \bigvee_{a+b=n} X_a \wedge Y_b.$$

We make no claim that this object is a spectrum, as it is not clear how to define the suspension in a way that makes use of the structure maps of both X and Y. While this operation is associative, the unit is the spectrum with S^0 in degree zero and $*$ elsewhere.

Interestingly, we see that a sequential spectrum X has an "action" of the sphere spectrum in the sense that the structure maps of X induce maps of each level

$$\mu_{X,n} \colon (\mathbb{S} \otimes X)_n = \bigvee_{a+b=n} S^a \wedge X_b \longrightarrow X_n.$$

Again, the category in which we are working is left undefined. If we pretend that a spectrum is a "module" over $\mathbb{S}$, then we can make $X \otimes Y$ into a module over $\mathbb{S}$: we follow algebra and take the monoidal product over $\mathbb{S}$

$$X \otimes_{\mathbb{S}} Y := \operatorname{coeq}(X \otimes \mathbb{S} \otimes Y \rightrightarrows X \otimes Y).$$

For this to make sense, we need the sphere spectrum to be a "commutative ring object" with regards to the operation $- \otimes -$. We can see that this is not true for sequential spectra, even before we attempt to improve the categorical setting for $- \otimes -$. Consider the map of spaces

$$\mu_{\mathbb{S},2} \colon (\mathbb{S} \otimes \mathbb{S})_2 = (S^0 \wedge S^2) \vee (S^1 \wedge S^1) \vee (S^2 \wedge S^0) \longrightarrow S^2$$

induced by the structure map of the sphere spectrum. The component map $a \colon S^1 \wedge S^1 \longrightarrow S^2$ is precisely the standard isomorphism. If $\mathbb{S}$ were a commutative object with respect to $\otimes$, then we would have a commutative diagram

$$\begin{array}{ccc} S^1 \wedge S^1 & \xrightarrow{a} & S^2. \\ {\scriptstyle \tau_{S^1,S^1}} \downarrow & \nearrow_{a} & \\ S^1 \wedge S^1 & & \end{array}$$

In particular, the two maps labelled a have to be the *same*. Taking singular homology, we obtain the diagram

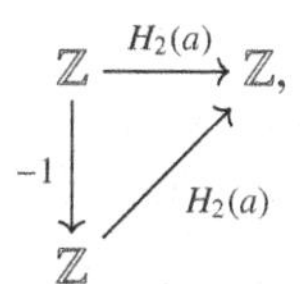

which cannot commute, regardless of whether the isomorphism $H_2(a)$ is represented by 1 or -1. We can view the problem as coming from a lack of symmetries on the sphere spectrum itself. This can be resolved in symmetric and orthogonal spectra.

6.3 Closed Monoidal Structures on Spectra

The solution to the difficulties we encountered previously is to recast the definition of spectra in a very categorical manner. We will see that this approach to constructing a commutative smash product will work for symmetric and orthogonal spectra. Our approach is based on that of Mandell et al. [MMSS01].

We will need the notion of enriched categories, enriched functors, enriched ends and coends, enriched Kan extensions and the enriched Yoneda Lemma. Our primary sources are [Kel05] and [Bor94]. In all cases, we will use pointed topological spaces Top_* for our enrichment.

We define three enriched categories, each with the same object set, but different spaces of morphisms.

Definition 6.3.1 We define three enriched categories with objects given by the non-negative integers

$$\mathbb{N} = \{0, 1, 2, 3, \dots\}.$$

($\mathcal{N}$) The Top_*–enriched category $\mathcal{N}$ has morphism spaces given by

$$\mathcal{N}(a,b) = \begin{cases} S^0 & \text{if } a = b \\ * & \text{if } a \neq b. \end{cases}$$

(Σ) The Top_*–enriched category Σ has morphism spaces given by

$$\Sigma(a,b) = \begin{cases} (\Sigma_a)_+ & \text{if } a = b \\ * & \text{if } a \neq b. \end{cases}$$

($\mathcal{O}$) The Top_*–enriched category $\mathcal{O}$ has morphism spaces given by

$$\mathcal{O}(a,b) = \begin{cases} O(a)_+ & \text{if } a = b \\ * & \text{if } a \neq b. \end{cases}$$

In the case of Σ, it may be helpful to think of n as the set $\{1, 2, \dots, n\}$ and the morphism spaces given by (discrete) spaces of isomorphisms, with a disjoint basepoint. For $\mathcal{O}$, it may be helpful to think of n as the inner product space $\mathbb{R}^n$ and the morphisms spaces given by spaces of linear isometries, with a disjoint basepoint.

There are maps of enriched categories

$$\mathcal{N} \longrightarrow \Sigma \longrightarrow \mathcal{O},$$

which send n to n. On morphism spaces, these maps are given by the identity when $a \neq b$, and for $a = b$, they are induced by the inclusion maps

$$* \longrightarrow \Sigma_a \longrightarrow O(a),$$

where the first sends the point to the identity of Σ_a and the second is the standard inclusion of Σ_a into $O(a)$, which sends a permutation to the corresponding permutation of the axes.

We now define some categories of enriched functors and natural transformations. We will use these to define spectra in terms of modules over a monoid.

Recall that a Top_*–enriched functor $F\colon \mathcal{E} \longrightarrow \mathrm{Top}_*$ from an enriched category $\mathcal{E}$ to Top_* is a collection of maps of pointed spaces

$$F(a,b)\colon \mathcal{E}(a,b) \longrightarrow \mathrm{Top}_*(F_a, F_b)$$

for each $a, b \in \mathcal{E}$. These maps are required to be compatible with composition and satisfy associativity and unital coherence conditions. We may rewrite the map $F(a, b)$ as a map

$$\mathcal{E}(a, b) \wedge F_a \longrightarrow F_b$$

and think of $\mathcal{E}$ acting of F.

Definition 6.3.2 We define three categories of enriched functors.

($\mathcal{N}$) A *sequential space* is a Top_*–enriched functor from $\mathcal{N}$ to pointed topological spaces. The category of sequential spaces and enriched natural transformations is denoted $\mathcal{N}\mathrm{Top}_*$.
(Σ) A *symmetric space* is a Top_*–enriched functor from Σ to pointed topological spaces. The category of symmetric spaces and enriched natural transformations is denoted $\Sigma\mathrm{Top}_*$.
($\mathcal{O}$) An *orthogonal space* is a Top_*–enriched functor from $\mathcal{O}$ to pointed topological spaces. The category of orthogonal spaces and enriched natural transformations is denoted $\mathcal{O}\mathrm{Top}_*$.

In other words, a sequential space is a sequence of pointed topological spaces. A symmetric space X is a sequence of spaces X_n with maps

$$(\Sigma_n)_+ \wedge X_n \longrightarrow X_n,$$

which are associative and unital. These maps are precisely the data of an action of Σ_n on the pointed space X_n. A symmetric space is also called a *symmetric sequence* in [HSS00].

An orthogonal space X is a sequence of spaces X_n with maps

$$O(n)_+ \wedge X_n \longrightarrow X_n,$$

which are associative and unital. These maps are precisely the data of an action of $O(n)$ on the pointed space X_n. One may also call an orthogonal space an *orthogonal sequence.*

Lemma 6.3.3 *The enriched categories $\mathcal{N}$, Σ and $\mathcal{O}$ have symmetric monoidal products. We use $+$ to denote all three of them, as the three operations are very similar. On objects, it sends (a, b) to $a + b$.*

($\mathcal{N}$) *There is a symmetric monoidal product $+ \colon \mathcal{N} \times \mathcal{N} \longrightarrow \mathcal{N}$, which on mapping spaces*

$$+ \colon \mathcal{N}(a, b) \wedge \mathcal{N}(c, d) \longrightarrow \mathcal{N}(a + c, b + d)$$

is either the identity $* \to *$, the identity $S^0 \wedge S^0 \to S^0$, or the initial map $* \to S^0$.

(Σ) *There is a symmetric monoidal product* $+\colon \Sigma \times \Sigma \longrightarrow \Sigma$*, which on mapping spaces*

$$+\colon \Sigma(a,b) \wedge \Sigma(c,d) \longrightarrow \Sigma(a+c,b+d)$$

is only non-trivial when $a = b$ *and* $c = d$*. In this case, it is induced by the block inclusion*

$$\Sigma_a \times \Sigma_c \longrightarrow \Sigma_{a+c},$$

where (σ, τ) *is sent to the permutation which acts as* σ *on the first* a *letters and* τ *on the last* c *letters.*

($\mathcal{O}$) *There is a symmetric monoidal product* $+\colon \mathcal{O} \times \mathcal{O} \longrightarrow \mathcal{O}$*, which on mapping spaces*

$$+\colon \mathcal{O}(a,b) \wedge \mathcal{O}(c,d) \longrightarrow \mathcal{O}(a+c,b+d)$$

is only non-trivial when $a = b$ *and* $c = d$*. In this case, it is induced by the block sum of matrices*

$$\begin{array}{rcl} O(a) \times O(c) & \longrightarrow & O(a+c) \\ (A,B) & \longmapsto & \left(\begin{array}{c|c} A & 0 \\ \hline 0 & B \end{array}\right). \end{array}$$

Proof We see that the products are unital and associative.

For commutativity, we need to give an enriched natural transformation from $+$ to $+$ pre-composed with the swap map. We give the components in each case.

($\mathcal{N}$) The map $S^0 \longrightarrow \mathcal{N}(a+b,b+a) = S^0$ is the identity map.

(Σ) The map $S^0 \longrightarrow \Sigma(a+b,b+a) = (\Sigma_{a+b})_+$ sends the non-basepoint to the block permutation map, that is, the permutation which swaps the first a letters with the last b letters.

($\mathcal{O}$) The map $S^0 \longrightarrow \mathcal{O}(a+b,b+a) = O(a+b)_+$ sends the non-basepoint to the block permutation matrix

$$\left(\begin{array}{c|c} 0 & \mathrm{Id}_a \\ \hline \mathrm{Id}_b & 0 \end{array}\right).$$

We then need to see that these maps induce a commutative square as below. We start with $\mathcal{O}$. The only non-trivial case is where $a = b$ and $c = d$.

$$\begin{array}{ccc} \mathcal{O}(a,b) \wedge \mathcal{O}(c,d) & \xrightarrow{+} & \mathcal{O}(a+c,b+d) \\ \Big\downarrow \tau & & \Big\downarrow \\ \mathcal{O}(c,d) \wedge \mathcal{O}(a,b) & \xrightarrow{+} & \mathcal{O}(c+a,d+b) \end{array}$$

The right-hand vertical map is given by conjugation by the block permutations. Commutativity follows from the equation

$$\left(\begin{array}{c|c} B & 0 \\ \hline 0 & A \end{array}\right) = \left(\begin{array}{c|c} 0 & \mathrm{Id}_d \\ \hline \mathrm{Id}_b & 0 \end{array}\right)\left(\begin{array}{c|c} A & 0 \\ \hline 0 & B \end{array}\right)\left(\begin{array}{c|c} 0 & \mathrm{Id}_a \\ \hline \mathrm{Id}_c & 0 \end{array}\right).$$

The symmetric case follows by restriction. The sequential case follows as the twist map is the identity. □

Note that for $\mathcal{N}$, the elements of the natural transformation in the proof have domain and codomain S^0, so the only sensible choice is the identity map.

We can use these symmetric monoidal products on our enriched categories to produce a symmetric monoidal product on the categories of enriched functors to pointed spaces. The original reference is [Day70].

Before we construct our smash products, we will give the definitions of enriched ends and coends and recap how they relate to the Yoneda Lemma and left Kan extensions. Let $\mathcal{V}$ denote a closed symmetric monoidal category with all small limits and colimits. We use $\otimes$ for the monoidal product of $\mathcal{V}$ and hom for the internal function object.

Definition 6.3.4 Let $\mathcal{C}$ be a small $\mathcal{V}$–enriched category and $\mathcal{D}$ a $\mathcal{V}$–enriched category with all small limits and colimits. Let

$$M\colon \mathcal{C}^{op} \times \mathcal{C} \longrightarrow \mathcal{D}$$

be an enriched functor. We define the *enriched coend* of M as

$$\int^{c\in\mathcal{C}} M(c,c) = \mathrm{coeq}\left(\coprod_{a,b\in\mathcal{C}} M(a,b)\otimes \mathcal{C}(b,a) \rightrightarrows \coprod_{c\in\mathcal{C}} M(c,c)\right).$$

The two maps are the two different actions of $\mathcal{C}$ on M.

We define the *enriched end* of F to be

$$\int_{c\in\mathcal{C}} M(c,c) = \mathrm{eq}\left(\prod_{c\in\mathcal{C}} M(c,c) \rightrightarrows \prod_{a,b\in\mathcal{C}} \mathrm{hom}(\mathcal{C}(a,b), M(a,b))\right).$$

The two maps are adjoints to the action maps.

Let $\mathcal{C}$ be a small $\mathcal{V}$–enriched category. Given two $\mathcal{V}$–enriched functors

$$F,\ G\colon \mathcal{C} \longrightarrow \mathcal{D},$$

we can define the $\mathcal{V}$ object $\mathrm{Nat}(-,-)$ of enriched natural transformations from F to G in terms of an enriched end

$$\mathrm{Nat}(F,G) = \int_{c\in\mathcal{C}} \mathrm{hom}(F_c, G_c).$$

Lemma 6.3.5 (Yoneda) *Let $\mathcal{C}$ be a small $\mathcal{V}$–enriched category. For any $\mathcal{V}$–enriched functor $F\colon \mathcal{C} \longrightarrow \mathcal{V}$ and any $c \in \mathcal{C}$, there is a natural isomorphism in $\mathcal{V}$*

$$\mathrm{Nat}(\mathcal{C}(c,-),F) \cong F_c.$$

We may write the object $\mathrm{Nat}(\mathcal{C}(c,-),F) \in \mathcal{V}$ in terms of an enriched end so that the Yoneda Lemma becomes

$$F_c \xrightarrow{\cong} \int_{d\in\mathcal{C}} \mathrm{hom}(\mathcal{C}(c,d),F_d).$$

This description of the Yoneda Lemma suggests the following adjoint form.

Lemma 6.3.6 *Let $\mathcal{C}$ be a small $\mathcal{V}$–enriched category. For any $\mathcal{V}$–enriched functor $F\colon \mathcal{C} \longrightarrow \mathcal{V}$ and any $d \in \mathcal{C}$, there is a natural isomorphism*

$$\int^{c\in\mathcal{C}} \mathcal{C}(c,d)\otimes F_c \xrightarrow{\cong} F_d$$

given on term c by using the action of $\mathcal{C}(c,d)$ on F_c.

We can describe enriched left Kan extensions in terms of an enriched coend.

Lemma 6.3.7 *Let $\mathcal{C}$, $\mathcal{D}$ and $\mathcal{E}$ be $\mathcal{V}$–enriched categories. Further, assume that $\mathcal{C}$ and $\mathcal{E}$ are small categories. Given enriched functors*

$$F\colon \mathcal{C} \longrightarrow \mathcal{D} \text{ and } G\colon \mathcal{C} \longrightarrow \mathcal{E},$$

the left Kan extension $\mathrm{Lan}_G F$ of F along G is given by the enriched coend

$$(\mathrm{Lan}_G F)_e = \int^{c\in\mathcal{C}} \mathcal{E}(G(c),e)\otimes F_c.$$

Let us return to sequential, orthogonal and symmetric spaces.

Definition 6.3.8 Let $\mathcal{E}$ be some Top_*–enriched symmetric monoidal category. The *convolution product* $F \otimes G$ of two enriched functors F and G from $\mathcal{E}$ to Top_* is defined to be the left Kan extension of $\wedge \circ (F,G)$ along $+$.

$$\begin{array}{ccccc} \mathcal{E}\times\mathcal{E} & \xrightarrow{(F,G)} & \mathrm{Top}_*\times\mathrm{Top}_* & \xrightarrow{\wedge} & \mathrm{Top}_* \\ \big\downarrow{\scriptstyle +} & & & \nearrow{\scriptstyle F\otimes G} & \\ \mathcal{E} & & & & \end{array}$$

Note that we do not ask for the diagram to commute, instead, we require the universal property

$$\mathcal{E}\mathrm{Top}_*(F\otimes G,H) \cong (\mathcal{E}\times\mathcal{E})\mathrm{Top}_*(\wedge\circ(F,G),H\circ +),$$

which in terms of ends is given by

$$\int_{a\in\mathcal{E}} \mathrm{Top}_*((F\otimes G)_a, H_a) \cong \int_{b,c\in\mathcal{E}} \mathrm{Top}_*(F_b \wedge G_c, H_{b+c}).$$

Using Lemma 6.3.7, we can write the Kan extension as a coend

$$(F\otimes G)_a = \int^{b,c\in\mathcal{E}} \mathcal{E}(b+c,a)\wedge F_b \wedge G_c.$$

In our case, the neatness of enriched categories leads to the following alternative descriptions of the monoidal products.

Lemma 6.3.9 *The categories* $\mathcal{N}\mathrm{Top}_*$, $\Sigma\mathrm{Top}_*$ *and* $\mathcal{O}\mathrm{Top}_*$ *are symmetric monoidal. The unit in each case is the functor which sends* 0 *to* S^0 *and everything else to a point.*

($\mathcal{N}$) If F *and* G *are in* $\mathcal{N}\mathrm{Top}_*$, *then*

$$(F\otimes G)_a = \int^{b,c\in\mathcal{N}} \mathcal{N}(b+c,a)\wedge F_b\wedge G_c \cong \bigvee_{b+c=a} F_b\wedge G_c.$$

(Σ) If F *and* G *are in* $\Sigma\mathrm{Top}_*$, *then*

$$(F\otimes G)_a = \int^{b,c\in\Sigma} \Sigma(b+c,a)\wedge F_b\wedge G_c \cong \bigvee_{b+c=a} (\Sigma_a)_+ \underset{\Sigma_b\times\Sigma_c}{\wedge} F_b\wedge G_c.$$

($\mathcal{O}$) If F *and* G *are in* $\mathcal{O}\mathrm{Top}_*$, *then*

$$(F\otimes G)_a = \int^{b,c\in\mathcal{O}} \mathcal{O}(b+c,a)\wedge F_b\wedge G_c \cong \bigvee_{b+c=a} O(a)_+ \underset{O(b)\times O(c)}{\wedge} F_b\wedge G_c. \quad \square$$

Definition 6.3.10 We can define a *sphere spectrum* for each of these categories.

($\mathcal{N}$) The sphere spectrum in $\mathcal{N}\mathrm{Top}_*$ sends n to S^n.
(Σ) The sphere spectrum in $\Sigma\mathrm{Top}_*$ sends n to S^n, which we may think of as $(S^1)^{\wedge n}$. The group Σ_n acts on S^n by permuting the factors.
($\mathcal{O}$) The sphere spectrum in $\mathcal{O}\mathrm{Top}_*$ sends n to S^n, which we may think of as the one-point compactification of $\mathbb{R}^n$. The group $O(n)$ acts on S^n by the standard action of $O(n)$ on $\mathbb{R}^n$.

Note that the action of Σ_n on $S^n \cong (S^1)^{\wedge n}$ is equal to the action of Σ_n on the one-point compactification of $\mathbb{R}^n$ via the standard inclusion $\Sigma_n \longrightarrow O(n)$.

Lemma 6.3.11 *In* $\Sigma\mathrm{Top}_*$ *and* $\mathcal{O}\mathrm{Top}_*$, *the sphere spectrum is a commutative monoid with respect to* $\otimes$.

Proof We give the multiplication map for each of our two cases.

(Σ) The multiplication map of the sphere spectrum is given by

$$\begin{aligned}(\mathbb{S}\otimes\mathbb{S})_a &= \bigvee\nolimits_{b+c=a}(\Sigma_a)_+ \underset{\Sigma_b\times\Sigma_c}{\wedge} S^b\wedge S^c \\ &\cong \bigvee\nolimits_{b+c=a}(\Sigma_a)_+ \underset{\Sigma_b\times\Sigma_c}{\wedge} S^{b+c} \\ &\longrightarrow S^a,\end{aligned}$$

where the last map is evaluation using the Σ_a-action on $S^a = S^{b+c}$.

($\mathcal{O}$) The multiplication map of the sphere spectrum is given by

$$\begin{aligned}(\mathbb{S}\otimes\mathbb{S})_a &= \bigvee\nolimits_{b+c=a} O(a)_+ \underset{O(b)\times O(c)}{\wedge} S^b\wedge S^c \\ &\cong \bigvee\nolimits_{b+c=a} O(a)_+ \underset{O(b)\times O(c)}{\wedge} S^{b+c} \\ &\longrightarrow S^a,\end{aligned}$$

where the last map is evaluation using the $O(a)$-action on $S^a = S^{b+c}$.

We leave the unit and associativity conditions to the reader.

The coend description is the easiest way to see that these are commutative monoids. It suffices to check that the following diagram commutes when $a = b + c$.

$$\begin{array}{ccc}\mathcal{O}(b+c,a)\wedge S^b\wedge S^c & \longrightarrow & S^a \\ \downarrow & \nearrow & \\ \mathcal{O}(c+b,a)\wedge S^c\wedge S^b & & \end{array}$$

The vertical map is defined to be pre-multiplication by the block permutation matrix for b and c on $\mathcal{O}(b+c,a)$ and the twist map on the spheres. The other maps are then evaluation maps and the natural isomorphisms $S^c\wedge S^b \cong S^a$. The diagram commutes as the block permutation cancels out the twist map.

The case of symmetric spectra holds by restriction. □

Lemma 6.3.12 *In* $\mathcal{N}\mathrm{Top}_*$, *the sphere spectrum is a monoid with respect to* $\otimes$, *but it is* not *commutative.*

Proof We focus on the case where $a = b + c$. Recall that

$$\mathcal{N}(c+b,a) = \mathcal{N}(b+c,a) = S^0.$$

The action map is given by the natural isomorphism of smash products of spheres

$$\mathcal{N}(b+c,a)\wedge S^b\wedge S^c = S^b\wedge S^c \longrightarrow S^a.$$

One can check this is associative and unital. However, this monoid is not commutative. Consider the following diagram.

$$\begin{array}{ccccc} \mathcal{N}(b+c,a) \wedge S^b \wedge S^c & \xrightarrow{=} & S^b \wedge S^c & \longrightarrow & S^a \\ \downarrow & & \downarrow & \nearrow & \\ \mathcal{N}(c+b,a) \wedge S^c \wedge S^b & \xrightarrow[=]{} & S^c \wedge S^b & & \end{array}$$

The vertical maps are the identity on $S^0 = \mathcal{N}(b+c,a)$ and the twist maps on $S^b \wedge S^c$. This diagram does not commute when $b = c = 1$ and does not even commute up to homotopy, as the twist map has degree -1 and the action maps have degree 1, as we saw in Section 6.2. □

We may view the above result as saying that the sphere spectrum in sequential spectra lacks the symmetries to make it a commutative monoid. This is resolved for symmetric spectra by using the symmetric groups and for orthogonal spectra by using the orthogonal groups.

We are now able to relate our constructions to the original categories of spectra.

Theorem 6.3.13 *In each case, the category of $\mathbb{S}$–modules is a category of spectra.*

($\mathcal{N}$) *The category of $\mathbb{S}$–modules in $\mathcal{N}\mathrm{Top}_*$ is isomorphic to the category of sequential spectra, $\mathcal{S}^{\mathbb{N}}$.*

(Σ) *The category of $\mathbb{S}$–modules in $\Sigma\mathrm{Top}_*$ is isomorphic to the category of symmetric spectra $\mathcal{S}^{\Sigma}$.*

($\mathcal{O}$) *The category of $\mathbb{S}$–modules in $\mathcal{O}\mathrm{Top}_*$ is isomorphic to the category of orthogonal spectra $\mathcal{S}^{\mathcal{O}}$.*

Proof We start with $\mathcal{N}\mathrm{Top}_*$, then the other cases build on this by considering the additional information of the symmetric and orthogonal group actions.

($\mathcal{N}$) An enriched functor F from $\mathcal{N}$ to Top_* is an $\mathbb{S}$–module if and only if there is an associative and unital map $\mu\colon \mathbb{S} \otimes F \longrightarrow F$. At level a, this map takes the form

$$\mu_a\colon \bigvee_{b+c=a} S^b \wedge F_c \longrightarrow F_{b+c} = F_a.$$

Associativity implies that the component $S^b \wedge F_c \longrightarrow F_{b+c}$ is the composite of maps $S^1 \wedge F_c \longrightarrow F_{c+1}$. The unit condition implies that $S^0 \wedge F_c \longrightarrow F_c$

acts as the identity. Therefore, such a map is equivalent to having structure maps

$$\sigma_n^F \colon S^1 \wedge F_n = \Sigma F_n \longrightarrow F_{n+1}$$

for each n.

(Σ) Following the same proof as for $\mathcal{N}$, we look at the components of the $\mathbb{S}$-action on a module $F \in \Sigma\mathrm{Top}_*$

$$\mu_a \colon \bigvee_{b+c=a} (\Sigma_a)_+ \underset{\Sigma_b \times \Sigma_c}{\wedge} S^b \wedge F_c \longrightarrow F_{b+c} = F_a.$$

As before, associativity implies that each map $S^b \wedge F_c \longrightarrow F_{b+c}$ is a composite of structure maps $\sigma_n^F \colon \Sigma F_n \longrightarrow F_{n+1}$. The (b, c) component of μ_a is required to be Σ_a-equivariant. This precisely says that

$$S^b \wedge F_c \longrightarrow F_{b+c}$$

is $\Sigma_b \times \Sigma_c$-equivariant, regarding the right-hand side as a $\Sigma_b \times \Sigma_c$–module via block sum of permutations.

($\mathcal{O}$) This is similar to the case of Σ. □

Following the original plan of Section 6.2, we can define a monoidal product on $\mathbb{S}$–modules via the tensor product over $\mathbb{S}$.

$$X \wedge Y = X \otimes_{\mathbb{S}} Y := \mathrm{coeq}\,(X \otimes \mathbb{S} \otimes Y \rightrightarrows X \otimes Y)$$

The first map of the coequaliser is induced by the action $\mathbb{S} \otimes Y \to Y$ of the sphere spectrum on Y. The second map is induced by the action of the sphere spectrum on X, but pre-composed with a twist map

$$X \otimes \mathbb{S} \longrightarrow \mathbb{S} \otimes X \longrightarrow X.$$

Just as in the world of algebra, the smash product is symmetric monoidal.

Corollary 6.3.14 *The category of symmetric spectra $\mathcal{S}^\Sigma$ has a symmetric monoidal smash product $\wedge$ with the sphere spectrum as the unit. For X and Y in $\mathcal{S}^\Sigma$, the smash product is defined to be*

$$X \wedge Y = X \otimes_{\mathbb{S}} Y.$$

The category of orthogonal spectra $\mathcal{S}^{\mathcal{O}}$ has a symmetric monoidal smash product $\wedge$ with the sphere spectrum as the unit. For X and Y in $\mathcal{S}^{\mathcal{O}}$, the smash product is defined to be

$$X \wedge Y = X \otimes_{\mathbb{S}} Y.$$

□

We can incorporate the $\mathbb{S}$–module structure of our categories of spectra into the enriched categories and obtain a description of spectra as enriched functors from some enriched category to pointed topological spaces. This description will give very compact formulae for the shifted suspension functors and the smash product of symmetric spectra and orthogonal spectra.

If $\mathcal{E}$ is a Top_*–enriched category, then any element a of $\mathcal{E}$ defines a functor

$$\begin{aligned} \mathcal{E}(a,-)\colon \mathcal{E} &\longrightarrow \mathrm{Top}_* \\ b &\longmapsto \mathcal{E}(a,b). \end{aligned}$$

Similarly, one has a functor $\mathcal{E}(a,-) \wedge A$ given by the termwise smash with a pointed topological space A.

In the case of $\mathcal{N}$ (and Σ and $\mathcal{O}$), we can use the functor $\mathcal{N}(a,-)$ to define an $\mathbb{S}$–module $\mathbb{S} \otimes \mathcal{N}(a,-)$ in $\mathcal{N}\mathrm{Top}_*$. We have seen these functors before.

Lemma 6.3.15 *Let A be a pointed topological space.*

($\mathcal{N}$) For $a \in \mathcal{N}$, there is a natural isomorphism

$$(\mathbb{S} \otimes \mathcal{N}(a,-)) \wedge A \cong \mathrm{F}_a^{\mathbb{N}} A$$

of sequential spectra.

(Σ) For $a \in \Sigma$, there is a natural isomorphism

$$(\mathbb{S} \otimes \Sigma(a,-)) \wedge A \cong \mathrm{F}_a^{\Sigma} A$$

of symmetric spectra.

($\mathcal{O}$) For the $a \in \mathcal{O}$, there is a natural isomorphism

$$(\mathbb{S} \otimes \mathcal{O}(a,-)) \wedge A \cong \mathrm{F}_a^{\mathcal{O}} A$$

of orthogonal spectra.

Proof For symmetric spectra,

$$((\mathbb{S} \otimes \Sigma(a,-)) \wedge A)_b = (\mathbb{S} \otimes \Sigma(a,-))_b \wedge A = \bigvee_{b=c+d} (\Sigma_b)_+ \underset{\Sigma_c \times \Sigma_d}{\wedge} S^c \wedge \Sigma(a,d) \wedge A.$$

There is only one factor which is non-trivial, namely, $d = a$, where it takes value

$$(\Sigma_b)_+ \wedge_{\Sigma_{b-a}} S^{b-a} \wedge A,$$

if $b \geqslant a$ and $*$ otherwise. This is precisely the definition of the shifted suspension.

For sequential and orthogonal spectra similar calculations apply. □

Using ends and coends, we show directly that the adjoint to

$$\begin{array}{rcl}(\mathbb{S}\otimes\mathcal{N}(a,-))\wedge(-)\colon \mathrm{Top}_* & \longrightarrow & \mathcal{S}^{\mathbb{N}} \\ A & \longmapsto & (\mathbb{S}\otimes\mathcal{N}(a,-))\wedge A\end{array}$$

is evaluation of a sequential spectrum at level a. A map from this functor to an $\mathbb{S}$–module X in $\mathcal{S}^{\mathbb{N}}$ is precisely a map from the functor $\mathcal{N}(a,-)\wedge A$ to the object of $\mathcal{N}\mathrm{Top}_*$ that underlies the sequential spectrum X. Now consider the following sequence of isomorphisms.

$$\begin{array}{rcl}\mathcal{N}\mathrm{Top}_*(\mathcal{N}(a,-)\wedge A, X) & = & \int_{b\in\mathcal{N}}\mathrm{Top}_*(\mathcal{N}(a,b)\wedge A, X_b) \\ & \cong & \int_{b\in\mathcal{N}}\mathrm{Top}_*(A,\mathrm{Top}_*(\mathcal{N}(a,b),X_b)) \\ & \cong & \mathrm{Top}_*(A,\int_{b\in\mathcal{N}}\mathrm{Top}_*(\mathcal{N}(a,b),X_b)) \\ & \cong & \mathrm{Top}_*(A,X_a)\end{array}$$

The first is the definition of the set of maps, the second is the standard smash product–function space adjunction, and the third is moving a limit into the second variable. The final isomorphism is an instance of the enriched Yoneda Lemma. Formally similar arguments can be applied to symmetric and orthogonal spectra.

We can use these functors $\mathbb{S}\otimes\mathcal{N}(a,-)$ to make new enriched categories, giving a description of spectra as enriched functors without needing to mention $\mathbb{S}$–modules.

Definition 6.3.16 We define three enriched categories, one for each of our three cases. The object set in each case remains the non-negative integers.

($\mathcal{N}$) Define a Top_*–enriched category $\mathcal{N}_{\mathbb{S}}$ by setting $\mathcal{N}_{\mathbb{S}}(a,b)=*$ for $a>b$ and the following for $a\leqslant b$.

$$\begin{array}{rcl}\mathcal{N}_{\mathbb{S}}(a,b) & = & \mathcal{S}^{\mathbb{N}}(\mathbb{S}\otimes\mathcal{N}(b,-),\mathbb{S}\otimes\mathcal{N}(a,-)) \\ & \cong & (\mathbb{S}\otimes\mathcal{N}(a,-))_b \\ & \cong & \int^{c\in\mathcal{N}}\mathcal{N}(a+c,b)\wedge S^c\cong S^{b-a}\end{array}$$

The composition is given by composition in $\mathcal{S}^{\mathbb{N}}$.

(Σ) Define a Top_*–enriched category $\Sigma_{\mathbb{S}}$ by setting $\Sigma_{\mathbb{S}}(a,b)=*$ for $a>b$ and the following for $a\leqslant b$.

$$\begin{array}{rcl}\Sigma_{\mathbb{S}}(a,b) & = & \mathcal{S}^{\Sigma}(\mathbb{S}\otimes\Sigma(b,-),\mathbb{S}\otimes\Sigma(a,-)) \\ & \cong & (\mathbb{S}\otimes\Sigma(a,-))_b \\ & \cong & \int^{c\in\Sigma}\Sigma(a+c,b)\wedge S^c\cong(\Sigma_b)_+\wedge_{\Sigma_{b-a}}S^{b-a}\end{array}$$

The composition is given by composition in $\mathcal{S}^{\Sigma}$.

($\mathcal{O}$) Define a Top$_*$–enriched category $\mathcal{O}_{\mathbb{S}}$ by setting $\mathcal{O}_{\mathbb{S}}(a, b) = *$ for $a > b$ and the following for $a \leqslant b$.

$$\begin{aligned} \mathcal{O}_{\mathbb{S}}(a, b) &= \mathcal{S}^{\mathcal{O}}(\mathbb{S} \otimes \mathcal{O}(b, -), \mathbb{S} \otimes \mathcal{O}(a, -)) \\ &\cong \mathbb{S} \otimes \mathcal{O}(a, -)_b \\ &\cong \int^{c \in \mathcal{O}} \mathcal{O}(a + c, b) \wedge S^c \cong O(b)_+ \wedge_{O(b-a)} S^{b-a}. \end{aligned}$$

The composition is given by composition in $\mathcal{S}^{\mathcal{O}}$.

We may consider Top$_*$–enriched functors from these categories to Top$_*$. This gives three categories

$$\mathcal{N}_{\mathbb{S}}\mathrm{Top}_*, \quad \Sigma_{\mathbb{S}}\mathrm{Top}_* \quad \text{and} \quad \mathcal{O}_{\mathbb{S}}\mathrm{Top}_*.$$

Theorem 6.3.17 *There are equivalences of categories*

$$\mathcal{S}^{\mathbb{N}} \cong \mathcal{N}_{\mathbb{S}}\mathrm{Top}_*, \quad \mathcal{S}^{\Sigma} \cong \Sigma_{\mathbb{S}}\mathrm{Top}_* \quad \textit{and} \quad \mathcal{S}^{\mathcal{O}} \cong \mathcal{O}_{\mathbb{S}}\mathrm{Top}_*.$$

Proof The proof is the same in each case. We use the notation of $\mathcal{S}^{\mathcal{O}}$ for definiteness.

For one direction, we use Theorem 6.3.13, namely, that the category of orthogonal spectra is isomorphic to the category of $\mathbb{S}$–modules in $\mathcal{O}$Top$_*$. Let $X\colon \mathcal{O}_{\mathbb{S}} \longrightarrow \mathrm{Top}_*$ be a Top$_*$–enriched functor. We want to show that X is an $\mathbb{S}$–module in $\mathcal{O}$Top$_*$.

The natural isomorphism

$$\mathcal{O}(a, a) \cong \mathcal{O}_{\mathbb{S}}(a, a)$$

shows that X_a has an action of $\mathcal{O}(a, a)$. Since $\mathcal{O}(a, b) = *$ when $a \neq b$, this shows that X has an underlying enriched functor from $\mathcal{O}$ to Top$_*$, so $X \in \mathcal{O}\mathrm{Top}_*$.

For any $b \in \mathbb{N}$ there is a map

$$S^b \longrightarrow O(a + b)_+ \wedge_{O(b)} S^b = \mathcal{O}_{\mathbb{S}}(a, a + b),$$

which sends x to the class of (e, x), where e is the identity of $O(a+b)$. Smashing this map with $X(a)$ gives a map

$$S^b \wedge X(a) \longrightarrow \mathcal{O}_{\mathbb{S}}(a, a + b) \wedge X(a) \longrightarrow X(a + b).$$

It follows that X defines a $\mathbb{S}$–module in $\mathcal{O}$Top$_*$ and thus an orthogonal spectrum.

For the converse, let X be an orthogonal spectrum and $a \leqslant b$. The structure maps of X and the action of the orthogonal groups induce a map

$$O(b)_+ \wedge_{O(b-a)} S^{b-a} \wedge X(a) \longrightarrow O(b)_+ \wedge_{O(b-a)} X(b) \longrightarrow X(b).$$

This map is unital and associative, so this gives a Top_*–enriched functor

$$\mathcal{O}_{\mathbb{S}} \longrightarrow \mathrm{Top}_*. \qquad \square$$

Using this new description of the various kinds of spectra, we obtain a compact description of the symmetric monoidal product via Lemma 6.3.7.

Lemma 6.3.18 *The smash product of symmetric spectra is given by the formula*

$$(X \wedge Y)_a = \int^{b,c \in \Sigma_{\mathbb{S}}} \Sigma_{\mathbb{S}}(b + c, a) \wedge X_b \wedge Y_c.$$

The smash product of orthogonal spectra is given by the formula

$$(X \wedge Y)_a = \int^{b,c \in \mathcal{O}_{\mathbb{S}}} \mathcal{O}_{\mathbb{S}}(b + c, a) \wedge X_b \wedge Y_c.$$

Proof Let X and Y be symmetric spectra. Recall that

$$X \wedge Y = X \otimes_{\mathbb{S}} Y$$

by Corollary 6.3.14. Lemma 6.3.6 gives isomorphisms

$$X \cong \int^{b \in \Sigma_{\mathbb{S}}} \Sigma_{\mathbb{S}}(b, -) \wedge X_b \qquad Y \cong \int^{c \in \Sigma_{\mathbb{S}}} \Sigma_{\mathbb{S}}(c, -) \wedge Y_c.$$

Using these and the operation $- \otimes_{\mathbb{S}} -$ gives an isomorphism

$$X \otimes_{\mathbb{S}} Y = \int^{b,c \in \Sigma_{\mathbb{S}}} \Sigma_{\mathbb{S}}(b, -) \otimes_{\mathbb{S}} \Sigma_{\mathbb{S}}(c, -) \wedge X_b \wedge Y_c.$$

The result then follows from the isomorphisms

$$\begin{aligned} \Sigma_{\mathbb{S}}(b, -) \otimes_{\mathbb{S}} \Sigma_{\mathbb{S}}(c, -) &= (\mathbb{S} \otimes \Sigma(b, -)) \otimes_{\mathbb{S}} (\mathbb{S} \otimes \Sigma(c, -)) \\ &\cong \mathbb{S} \otimes (\Sigma(b, -) \otimes \Sigma(c, -)) \\ &\cong \mathbb{S} \otimes \Sigma(b + c, -) \\ &= \Sigma_{\mathbb{S}}(b + c, -). \end{aligned}$$

An analogous proof gives the case of orthogonal spectra. $\square$

Our next aim is to show that these symmetric monoidal structures are closed and related well to the smash product of spaces with spectra and the enrichment of spaces in spectra.

Definition 6.3.19 For G and H symmetric spectra, the *internal function object* from G to H is a symmetric spectrum $F_{\mathbb{S}}(G, H)$ defined by

$$F_{\mathbb{S}}(G, H)_b = \int_{c \in \Sigma_{\mathbb{S}}} \mathrm{Top}_*(G_c, H_{c+b}).$$

For G and H orthogonal spectra, the *internal function object* from G to H is an orthogonal spectrum $F_{\mathbb{S}}(G,H)$ defined by

$$F_{\mathbb{S}}(G,H)_b = \int_{c \in \mathcal{O}_{\mathbb{S}}} \mathrm{Top}_*(G_c, H_{c+b}).$$

The structure maps are given by the sequence of maps below. We use symmetric spectra for definiteness, the case of orthogonal spectra is formally identical.

$$\begin{array}{rcl} F_{\mathbb{S}}(G,H)_b \wedge \Sigma_{\mathbb{S}}(b,d) & = & \left(\int_{c \in \Sigma_{\mathbb{S}}} \mathrm{Top}_*(G_c, H_{c+b})\right) \wedge \Sigma_{\mathbb{S}}(b,d) \\ & \longrightarrow & \int_{c \in \Sigma_{\mathbb{S}}} (\mathrm{Top}_*(G_c, H_{c+b}) \wedge \Sigma_{\mathbb{S}}(b,d)) \\ & \longrightarrow & \int_{c \in \Sigma_{\mathbb{S}}} \mathrm{Top}_*(G_c, H_{c+d}) \\ & = & F_{\mathbb{S}}(G,H)_d. \end{array}$$

The first brings the smash product (a left adjoint) inside an enriched end (a limit) and the second map is induced by the maps

$$H_{c+b} \wedge \Sigma_{\mathbb{S}}(b,d) \longrightarrow H_{c+b} \wedge \Sigma_{\mathbb{S}}(c+b, c+d) \longrightarrow H_{c+d}.$$

The first map comes from the monoidal structure on $\Sigma_{\mathbb{S}}$. For any a, b and c in $\Sigma_{\mathbb{S}}$, we may combine the unit $S^0 \to \Sigma_{\mathbb{S}}(c,c)$ with the monoidal product to obtain the map

$$\Sigma_{\mathbb{S}}(a,b) \longrightarrow \Sigma_{\mathbb{S}}(a,b) \wedge \Sigma_{\mathbb{S}}(c,c) \longrightarrow \Sigma_{\mathbb{S}}(a+c, b+c).$$

We can also recognise an internal function object $F_{\otimes}$ on the symmetric monoidal categories $\Sigma\mathrm{Top}_*$, $\mathcal{O}\mathrm{Top}_*$. Then we can define the internal function object $F_{\mathbb{S}}$ of spectra as one would do for algebra, namely, as an equaliser of

$$F_{\otimes}(X,Y) \rightrightarrows F_{\otimes}(X \otimes \mathbb{S}, Y).$$

As we have already shown, the various descriptions of the smash product agree, and so too do the descriptions of the function objects.

Lemma 6.3.20 *The categories of symmetric spectra $\mathcal{S}^{\Sigma}$ and orthogonal spectra $\mathcal{S}^{\mathcal{O}}$ are closed symmetric monoidal.*

Proof We need to prove that for symmetric spectra X, Y and Z, there is a natural isomorphism

$$\mathcal{S}^{\Sigma}(X \wedge Y, Z) \cong \mathcal{S}^{\Sigma}(X, F_{\mathbb{S}}(Y,Z)).$$

This is done by manipulating enriched ends and coends. We start by writing the smash product in terms of the universal property

$$\begin{aligned} \mathcal{S}^{\Sigma}(X \wedge Y, Z) &\cong \int_{b,c \in \Sigma_{\mathbb{S}}} \mathrm{Top}_*(X_b \wedge Y_c, Z_{b+c}) \\ &\cong \int_{b,c \in \Sigma_{\mathbb{S}}} \mathrm{Top}_*(X_b, \mathrm{Top}_*(Y_c, Z_{b+c})) \\ &\cong \int_{b \in \Sigma_{\mathbb{S}}} \mathrm{Top}_*\Big(X_b, \int_{c \in \Sigma_{\mathbb{S}}} \mathrm{Top}_*(Y_c, Z_{b+c})\Big) \\ &\cong \int_{b \in \Sigma_{\mathbb{S}}} \mathrm{Top}_*(X_b, F_{\mathbb{S}}(Y, Z)_b) \\ &= \mathcal{S}^{\Sigma}(X, F_{\mathbb{S}}(Y, Z)). \end{aligned}$$

□

We can use our understanding of the smash product to see how the adjoints to the evaluation functors interact with the smash product.

Lemma 6.3.21 *In the categories of symmetric spectra and orthogonal spectra, there are natural isomorphisms of spectra*

$$\mathrm{F}^{\Sigma}_n A \wedge \mathrm{F}^{\Sigma}_m B \cong \mathrm{F}^{\Sigma}_{n+m}(A \wedge B) \text{ and } \mathrm{F}^{\mathcal{O}}_n A \wedge \mathrm{F}^{\mathcal{O}}_m B \cong \mathrm{F}^{\mathcal{O}}_{n+m}(A \wedge B)$$

for all $m, n \in \mathbb{N}$ and all pointed spaces A and B.

Proof From the original description of the shifted suspension functors of Chapter 5, and the smash product over the sphere spectrum of Corollary 6.3.14, this is difficult to see. Instead, we argue using the description in terms of representable functors of $\Sigma_{\mathbb{S}}$ and $\mathcal{O}_{\mathbb{S}}$ of Lemma 6.3.15. We choose $\Sigma_{\mathbb{S}}$ for definiteness. One then has

$$\begin{aligned} \mathrm{F}^{\Sigma}_n A \wedge \mathrm{F}^{\Sigma}_m B &= \int^{b,c \in \Sigma_{\mathbb{S}}} \Sigma_{\mathbb{S}}(b+c, -) \wedge \Sigma_{\mathbb{S}}(n, b) \wedge A \wedge \Sigma_{\mathbb{S}}(m, c) \wedge B \\ &\cong \Sigma_{\mathbb{S}}(n+m, -) \wedge A \wedge B \\ &= \mathrm{F}^{\Sigma}_{n+m}(A \wedge B). \end{aligned}$$

The isomorphism in the middle is two applications of Lemma 6.3.6. □

We also obtain a precise description of the prolongation functors originally introduced in Chapter 5. We will see that $\mathbb{P}^{\mathcal{O}}_{\Sigma}$ is strong symmetric monoidal in Theorem 6.4.9.

Proposition 6.3.22 *The prolongation functors from sequential spectra to symmetric spectra and symmetric spectra to orthogonal spectra are given by the formulas*

$$(\mathbb{P}^{\Sigma}_{\mathbb{N}} X)_a = \int^{c \in \mathcal{N}_{\mathbb{S}}} \Sigma_{\mathbb{S}}(c, a) \wedge X_c \text{ and } (\mathbb{P}^{\mathcal{O}}_{\Sigma} Y)_a = \int^{c \in \Sigma_{\mathbb{S}}} \mathcal{O}_{\mathbb{S}}(c, a) \wedge Y_c.$$

Proof Using Theorem 6.3.17 to describe spectra as categories of enriched functors, we can identify the adjoint forgetful functor $\mathbb{U}_{\Sigma}^{\mathcal{O}}$ as a pullback along a map of enriched categories

$$\begin{array}{rcl} \Sigma_{\mathbb{S}} & \longrightarrow & \mathcal{O}_{\mathbb{S}} \\ a & \longmapsto & a \\ \Sigma_{\mathbb{S}}(a,b) = (\Sigma_b)_+ \wedge_{\Sigma_{b-a}} S^{b-a} & \longrightarrow & O(b)_+ \wedge_{O(b-a)} S^{b-a} = \mathcal{O}_{\mathbb{S}}(a,b). \end{array}$$

Here, the map on spaces is induced by the standard inclusion of the symmetric group into the orthogonal group. This map $\Sigma_{\mathbb{S}} \longrightarrow \mathcal{O}_{\mathbb{S}}$ is a morphism of symmetric monoidal categories. The key fact is that the block permutations of Σ_n are sent to block permutations matrices of $O(n)$ via the standard inclusion.

Since left adjoints are unique (up to natural isomorphism), we recognise the left adjoint $\mathbb{P}_{\Sigma}^{\mathcal{O}}$ on a symmetric spectrum X as the left Kan extension of X along $\Sigma_{\mathbb{S}} \longrightarrow \mathcal{O}_{\mathbb{S}}$. The result follows from Lemma 6.3.7. □

Remark 6.3.23 The categories Σ and $\mathcal{O}$ have larger versions that can also be used to construct equivalent categories of spectra. Instead of Σ, one could use the category of finite subsets of $\mathbb{N}$ and bijections, and one could use finite-dimensional subspaces of $\mathbb{R}^{\infty}$ instead of $\mathcal{O}$. This is useful when constructing equivariant spectra.

6.4 Monoidal Model Categories of Spectra

6.4.1 Homotopical Properties of the Smash Product

We are going to prove that orthogonal spectra and symmetric spectra are symmetric monoidal model categories. The key results concern the interaction of the smash product with stable equivalences and π_*-isomorphisms in symmetric spectra and orthogonal spectra.

Lemma 6.4.1 *The levelwise model structures on orthogonal spectra and symmetric spectra are symmetric monoidal model structures.*

Proof We prove that these model categories satisfy the pushout product axiom from Definition 6.1.7 by showing that the assumptions of Lemma 6.1.13 hold.

We work in orthogonal spectra for definiteness, the case of symmetric spectra is the same. Let $\mathrm{F}_n^{\mathcal{O}}\, i$ and $\mathrm{F}_m^{\mathcal{O}}\, j$ be two generating cofibrations of orthogonal spectra, where i and j are generating cofibrations of pointed topological spaces. Then

$$\mathrm{F}_n^{\mathcal{O}}\, i \,\square\, \mathrm{F}_m^{\mathcal{O}}\, j \cong \mathrm{F}_{n+m}^{\mathcal{O}}(i \,\square\, j)$$

by Lemma 6.3.21. The model category Top_* with the Serre model structure is symmetric monoidal, and $\mathrm{F}^{\mathcal{O}}_{n+m}$ is a left Quillen functor from pointed spaces to the levelwise model structure. Therefore, we see that $\mathrm{F}^{\mathcal{O}}_{n+m}(i \square j)$ is a q-cofibration.

The case of a generating cofibration and generating acyclic cofibration follows by the same argument. The unit part of the pushout product axiom holds as the unit is already cofibrant. □

Theorem 6.4.2 *If $f\colon X \to Y$ is a stable equivalence (or π_*-isomorphism) of symmetric spectra, then $X \wedge Z \longrightarrow Y \wedge Z$ is also a stable equivalence (or π_*-isomorphism) for any cofibrant symmetric spectrum Z.*

Proof Assume that we know the result for $Z = \mathrm{F}^{\Sigma}_n S^n$. Lemma 5.3.14 gives the result for the cases $Z = \mathrm{F}^{\Sigma}_n S^0$ and $Z = \mathrm{F}^{\Sigma}_n A$.

Now let Z be a $I^{\Sigma}_{\mathrm{stable}}$-cell complex, so $Z = \mathrm{colim}_i(Z_i)$ with $Z_0 = *$, and each map $Z_i \longrightarrow Z_{i+1}$ is a pushout of coproducts of maps in $I^{\Sigma}_{\mathrm{stable}}$. By Lemma 6.4.6, $X \wedge Z$ is the colimit of a sequence of h-cofibrations and hence is a homotopy colimit. By Lemmas 2.2.13 and 5.3.17, each map

$$X \wedge Z_i \longrightarrow Y \wedge Z_i$$

is a stable equivalence (or π_*-isomorphism). Hence, the induced map on homotopy colimits

$$X \wedge Z \longrightarrow Y \wedge Z$$

is a stable equivalence (or π_*-isomorphism). The general case follows by passing to retracts.

We now return to the case of $- \wedge \mathrm{F}^{\Sigma}_n S^n$ and π_*-isomorphisms of symmetric spectra. The smash product with a shifted suspension spectrum is given by tensoring with a representable functor by Lemma 6.3.15

$$\mathrm{F}^{\Sigma}_n S^n \wedge X = (\mathbb{S} \otimes \Sigma(n,-) \wedge S^n) \otimes_{\mathbb{S}} X \cong (\Sigma(n,-) \wedge S^n) \otimes X.$$

We then calculate the value of this tensor product at $a \in \mathbb{N}$ to be

$$(\Sigma_a)_+ \wedge_{\Sigma_{a-n}} X_{a-n} \wedge S^n.$$

Ignoring the group actions, this space is homeomorphic to

$$(\Sigma_a/\Sigma_{a-n})_+ \wedge X_{a-n} \wedge S^n,$$

where the isomorphism is given by choosing coset representations. This isomorphism requires Σ_a to be discrete. We may then choose the cosets inductively, leading to a commutative square

$$\begin{array}{ccc} (\Sigma_a)_+ \wedge_{\Sigma_{a-n}} \Sigma X_{a-n} \wedge S^n & \xrightarrow{\mathrm{Id} \wedge \sigma_{a-n} \wedge \mathrm{Id}} & (\Sigma_{a+1})_+ \wedge_{\Sigma_{a+1-n}} X_{a+1-n} \wedge S^n \\ \downarrow & & \downarrow \\ (\Sigma_a/\Sigma_{a-n})_+ \wedge \Sigma X_{a-n} \wedge S^n & \xrightarrow{\mathrm{Id} \wedge \sigma_{a-n} \wedge \mathrm{Id}} & (\Sigma_{a+1}/\Sigma_{a+1-n})_+ \wedge X_{a+1-n} \wedge S^n. \end{array}$$

Taking homotopy groups shows that $\pi_*(\mathrm{F}_n^\Sigma S^n \wedge X)$ is a countable direct sum of copies of $\pi_*(X)$, and that this isomorphism is natural in X. It follows that $- \wedge \mathrm{F}_n^\Sigma S^n$ preserves π_*-isomorphisms.

It remains to be shown that $- \wedge \mathrm{F}_n^\Sigma S^n$ preserves stable equivalences of symmetric spectra. Since the cofibrant replacement $X^{cof} \longrightarrow X$ is a levelwise weak equivalence, it is also a π_*-isomorphism. Hence,

$$X^{cof} \wedge Z \longrightarrow X \wedge Z$$

is a π_*-isomorphism and thus a stable equivalence. We conclude that it suffices to prove that $- \wedge \mathrm{F}_n^\Sigma S^n$ preserves stable equivalences between cofibrant spectra.

Let E be an Ω-spectrum. Recall from the proof of Proposition 5.3.12 that

$$F_{\mathbb{S}}(\mathrm{F}_n^\Sigma S^n, E) \cong R^n E$$

is an Ω-spectrum, which at level m is the space $\Omega^n E_{n+m}$.

By Lemma 6.4.1, there is a natural isomorphism

$$[X \wedge \mathrm{F}_n^\Sigma S^n, E]^l \cong [X, F_{\mathbb{S}}(\mathrm{F}_n^\Sigma S^n, E)]^l,$$

which shows that $X \wedge \mathrm{F}_n^\Sigma S^n \longrightarrow Y \wedge \mathrm{F}_n^\Sigma S^n$ is a stable equivalence. □

We have the analogous result for orthogonal spectra, recalling that for orthogonal spectra, the class of stable equivalences is equal to the class of π_*-isomorphisms by Lemma 5.3.27.

Theorem 6.4.3 *If $f: X \to Y$ is a π_*-isomorphism of orthogonal spectra, then $X \wedge Z \longrightarrow Y \wedge Z$ is a π_*-isomorphism for any cofibrant orthogonal spectrum Z.*

Proof We follow the same plan as for Theorem 6.4.2. Assume that $- \wedge \mathrm{F}_n^{\mathcal{O}} S^n$ preserves π_*-isomorphisms, then Lemmas 2.2.7, 2.2.13 and 5.2.11 give the general case.

Let Cf be the homotopy cofibre of f. Then $(Cf) \wedge \mathrm{F}_n^{\mathcal{O}} S^n$ is the homotopy cofibre of

$$f \wedge \mathrm{F}_n^{\mathcal{O}} S^n : X \wedge \mathrm{F}_n^{\mathcal{O}} S^n \longrightarrow Y \wedge \mathrm{F}_n^{\mathcal{O}} S^n.$$

Thus, it suffices to prove that $W \wedge \mathrm{F}_n^{\mathcal{O}} S^n$ is π_*-isomorphic to $*$ for any W with $\pi_*(W) = 0$.

We prove this directly, arguing via elements of homotopy groups. Recall that the map $\lambda_n\colon \mathrm{F}^{\mathcal{O}}_{n+1} S^1 \to \mathrm{F}^{\mathcal{O}}_n S^0$ is the adjoint of the map

$$S^1 \to \mathrm{Ev}^{\mathcal{O}}_{n+1} \mathrm{F}^{\mathcal{O}}_n S^0 = O(n+1)_+ \wedge_{O(1)} S^1,$$

which is $t \mapsto [\mathrm{Id}, t]$, see Definition 5.2.12. The central point is that the two maps

$$\mathrm{F}^{\mathcal{O}}_{2n} S^{2n} \cong \mathrm{F}^{\mathcal{O}}_n S^n \wedge \mathrm{F}^{\mathcal{O}}_n S^n \underset{\mathrm{Id}\wedge\lambda_n}{\overset{\lambda_n\wedge\mathrm{Id}}{\rightrightarrows}} \mathrm{F}^{\mathcal{O}}_n S^n$$

are homotopic. These maps are determined by their adjoints

$$S^{2n} \rightrightarrows O(2n)_+ \wedge_{O(n)} S^{2n} \cong (O(2n)/O(n))_+ \wedge S^{2n}.$$

The top map sends $t \in S^{2n}$ to the equivalence class $[\mathrm{Id}, t]$, and the lower map sends it to the equivalence class $[\tau_{n,n}, t]$, where $\tau_{n,n}$ is the block permutation matrix exchanging the first n coordinates with the last n. As $O(2n)/O(n)$ is path-connected, these two maps are homotopic. Hence, $\lambda_n \wedge \mathrm{Id}$ and $\mathrm{Id} \wedge \lambda_n$ are homotopic.

Let $\alpha \in \pi_q(\mathrm{F}^{\mathcal{O}}_n S^n \wedge W)$ be represented by $f\colon \mathrm{F}^{\mathcal{O}}_r S^{q+r} \longrightarrow \mathrm{F}^{\mathcal{O}}_n S^n \wedge W$. Our calculation on λ_n shows that the two composites

$$\mathrm{F}^{\mathcal{O}}_n S^n \wedge \mathrm{F}^{\mathcal{O}}_r S^{q+r} \xrightarrow{\mathrm{Id}\wedge f} \mathrm{F}^{\mathcal{O}}_n S^n \wedge \mathrm{F}^{\mathcal{O}}_n S^n \wedge W \underset{\mathrm{Id}\wedge\lambda_n\wedge\mathrm{Id}}{\overset{\lambda_n\wedge\mathrm{Id}\wedge\mathrm{Id}}{\rightrightarrows}} \mathrm{F}^{\mathcal{O}}_n S^n \wedge W$$

are homotopic. The lower composite is $\mathrm{Id} \wedge g$, where

$$g = (\lambda_n \wedge \mathrm{Id}) \circ f\colon \mathrm{F}^{\mathcal{O}}_r S^{q+r} \xrightarrow{f} \mathrm{F}^{\mathcal{O}}_n S^n \wedge W \xrightarrow{\lambda_n\wedge\mathrm{Id}} W.$$

We may choose r large enough so that g, and hence $\mathrm{Id} \wedge g$, is homotopic to zero. The upper composite of our pair of maps is equal to

$$\mathrm{F}^{\mathcal{O}}_{n+r} S^{n+q+r} \cong \mathrm{F}^{\mathcal{O}}_n S^n \wedge \mathrm{F}^{\mathcal{O}}_r S^{q+r} \xrightarrow{\lambda_n\wedge\mathrm{Id}} \mathrm{F}^{\mathcal{O}}_r S^{q+r} \xrightarrow{f} \mathrm{F}^{\mathcal{O}}_n S^n \wedge W,$$

which is another representative for α, as

$$\mathrm{F}^{\mathcal{O}}_{n+r} S^{n+q+r} \longrightarrow \mathrm{F}^{\mathcal{O}}_r S^{q+r}$$

induced by λ_n is a π_*-isomorphism. Putting this together, we have shown that $\alpha = 0$ and thus,

$$\pi_q(\mathrm{F}^{\mathcal{O}}_n S^n \wedge W) = 0$$

as claimed. □

Corollary 6.4.4 *Working in either orthogonal spectra or symmetric spectra, let $X \longrightarrow Y$ be a stable equivalence of cofibrant spectra. Then*

$$X \wedge Z \longrightarrow Y \wedge Z$$

is a stable equivalence for any spectrum Z. Furthermore, let W be an orthogonal spectrum and $n, m \in \mathbb{N}$. Then

$$\pi_m(\mathrm{F}_n^{\mathcal{O}} S^n \wedge W) \cong \pi_{m+n}(W).$$

Proof For the first statement, let $Z^{cof} \to Z$ be a cofibrant replacement of Z, which is a levelwise weak equivalence and hence a stable equivalence. There is a commutative square

$$\begin{array}{ccc} X \wedge Z & \longrightarrow & Y \wedge Z \\ \uparrow & & \uparrow \\ X \wedge Z^{cof} & \longrightarrow & Y \wedge Z^{cof}. \end{array}$$

We have seen that the vertical maps and the lower horizontal map are stable equivalences, so the result follows by the two-out-of-three axiom.

By the first part, the map $\lambda_n \colon \mathrm{F}_n^{\mathcal{O}} S^n \longrightarrow \mathrm{F}_0^{\mathcal{O}} S^0$ induces a stable equivalence

$$\mathrm{F}_n^{\mathcal{O}} S^n \wedge W \longrightarrow \mathrm{F}_0^{\mathcal{O}} S^0 \wedge W \cong W.$$

Therefore, one has isomorphisms of homotopy groups of orthogonal spectra

$$\pi_{m+n}(W) \cong \pi_{m+n}(\mathrm{F}_n^{\mathcal{O}} S^n \wedge W) \cong \pi_{m+n}(\mathrm{F}_n^{\mathcal{O}} S^0 \wedge W \wedge S^n) \cong \pi_m(\mathrm{F}_n^{\mathcal{O}} S^0 \wedge W). \quad \square$$

Remark 6.4.5 In algebra, a module M over a ring R is said to be flat if $M \otimes -$ is exact. Hence, tensoring with a flat module preserves homology isomorphisms.

Therefore, we may describe the previous theorems as saying that a cofibrant spectrum is "flat". Or more simply, "cofibrant implies flat" in the stable model structures of symmetric and orthogonal spectra.

Before we prove the pushout product axiom for spectra, we show that smashing an acyclic cofibration with a spectrum gives a h-cofibration that is a stable equivalence. This result is needed to prove the existence of model structures of modules and rings. The statement is known as the *monoid axiom* in [SS00].

Lemma 6.4.6 *Let $\mathcal{S}$ denote either symmetric spectra or orthogonal spectra. If $f \colon X \longrightarrow Y$ is a h-cofibration in $\mathcal{S}$ and Z is a spectrum in $\mathcal{S}$, then the map*

$$f \wedge \mathrm{Id} \colon X \wedge Z \longrightarrow Y \wedge Z$$

is a h-cofibration.

Proof A h-cofibration is defined using a pushout and smashing with spaces (see Definition 5.2.7 and Section A.5). Both these operations are preserved by $- \wedge Z$. It follows that for a h-cofibration f, the map $f \wedge Z$ will be a h-cofibration too. □

Corollary 6.4.7 (The monoid axiom) *Let $\mathcal{S}$ denote either symmetric or orthogonal spectra. Let $X \longrightarrow Y$ be a stable equivalence and q-cofibration of spectra and Z a spectrum. Then*

$$X \wedge Z \longrightarrow Y \wedge Z$$

is a stable equivalence and a h-cofibration.

Pushouts and sequential colimits of such maps are also stable equivalences and h-cofibrations.

Proof Let $X \longrightarrow Y$ be a stable equivalence and q-cofibration. Since every q-cofibration is a h-cofibration,

$$X \wedge Z \longrightarrow Y \wedge Z$$

is a h-cofibration. Let $C = Y/X$ be the cofibre of $X \longrightarrow Y$. By Lemmas 5.2.9 and 5.3.16, C is homotopy equivalent to the homotopy cofibre of f.

The spectrum C is cofibrant and stably equivalent to $*$ because f is a stable equivalence. We know that $- \wedge Z^{cof}$ preserves stable equivalences, so the homotopy cofibre of

$$f \wedge Z^{cof} : X \wedge Z^{cof} \longrightarrow Y \wedge Z^{cof}$$

is also stably equivalent to $*$ as well as stably equivalent to $C \wedge Z^{cof}$. Now $C \wedge -$ preserves stable equivalences, so $C \wedge Z^{cof} \simeq *$ in turn is stably equivalent to $C \wedge Z$. This implies that

$$X \wedge Z \longrightarrow Y \wedge Z$$

is a stable equivalence too, because it is a h-cofibration whose homotopy cofibre is stably equivalent to $*$.

The last statement is a consequence of Lemmas 2.2.13 and 5.3.16. □

Theorem 6.4.8 *The stable model structures on orthogonal spectra and symmetric spectra are symmetric monoidal model structures.*

Proof We work in orthogonal spectra for definiteness. The case of symmetric spectra is the same. We have seen that the levelwise model structure is monoidal in Lemma 6.4.1. The levelwise model structure and the stable model structure have the same cofibrations. Therefore, all that remains is to prove the pushout product axiom for a q-cofibration and an acyclic q-cofibration $j : X \longrightarrow Y$.

By Lemma 6.1.13, we may assume that the q-cofibration is a generating cofibration

$$\mathrm{F}_n^{\mathcal{O}} i\colon \mathrm{F}_n^{\mathcal{O}} A \longrightarrow \mathrm{F}_n^{\mathcal{O}} B.$$

The pushout product is now

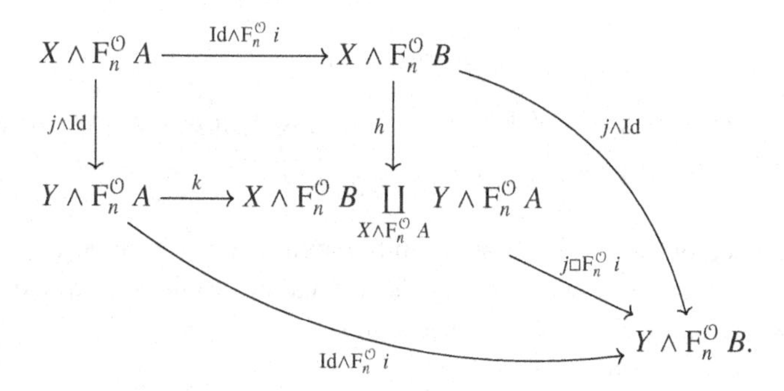

Since $\mathrm{F}_n^{\mathcal{O}} A$ and $\mathrm{F}_n^{\mathcal{O}} B$ are cofibrant, Theorem 6.4.3 implies that both maps labelled $j \wedge \mathrm{Id}$ are stable equivalences and q -cofibrations. Therefore, the pushout h is a stable equivalence (and a q-cofibration). It follows that $j \square \mathrm{F}_n^{\mathcal{O}} i$ is a stable equivalence.

As the unit is cofibrant, the other point of the pushout product axiom holds. □

Theorem 6.4.9 *The Quillen equivalence between symmetric spectra and orthogonal spectra*

$$\mathbb{P}_\Sigma^{\mathcal{O}} : \mathcal{S}^\Sigma \rightleftarrows \mathcal{S}^{\mathcal{O}} : \mathbb{U}_\Sigma^{\mathcal{O}}$$

is a strong symmetric monoidal Quillen equivalence.

Proof We want to give a natural isomorphism

$$\mathbb{P}_\Sigma^{\mathcal{O}}(X \wedge Y) \cong \mathbb{P}_\Sigma^{\mathcal{O}} X \wedge \mathbb{P}_\Sigma^{\mathcal{O}} Y$$

for symmetric spectra X and Y. We use the formulae of Lemma 6.3.18 and Proposition 6.3.22 to obtain this isomorphism. Starting at one side we have

$$\begin{aligned}
\mathbb{P}_\Sigma^{\mathcal{O}}(X \wedge Y) &= \int^{c\in\Sigma_{\mathbb{S}}} \mathcal{O}_{\mathbb{S}}(c,-) \wedge (X \wedge Y)_c \\
&\cong \int^{a,b,c\in\Sigma_{\mathbb{S}}} \mathcal{O}_{\mathbb{S}}(c,-) \wedge \Sigma_{\mathbb{S}}(a+b,c) \wedge X_a \wedge Y_b \\
&\cong \int^{a,b\in\Sigma_{\mathbb{S}}} \mathcal{O}_{\mathbb{S}}(a+b,-) \wedge X_a \wedge Y_b,
\end{aligned}$$

where the last isomorphism is (a contravariant version of) Lemma 6.3.6.

Starting at the other end, we use the same formulae to obtain the first two comparisons.

$$\begin{aligned}
\mathbb{P}_\Sigma^{\mathcal{O}} X \wedge \mathbb{P}_\Sigma^{\mathcal{O}} Y &= \int^{d,e\in\Sigma_\mathbb{S}} \mathcal{O}_\mathbb{S}(d+e,-) \wedge (\mathbb{P}_\Sigma^{\mathcal{O}} X)_d \wedge (\mathbb{P}_\Sigma^{\mathcal{O}} X)_e \\
&\cong \int^{d,e\in\mathcal{O}_\mathbb{S}} \mathcal{O}_\mathbb{S}(d+e,-) \wedge \left(\int^{a\in\Sigma_\mathbb{S}} \mathcal{O}_\mathbb{S}(a,d) \wedge X_a\right) \\
&\qquad \wedge \left(\int^{b\in\Sigma_\mathbb{S}} \mathcal{O}_\mathbb{S}(b,e) \wedge Y_b\right) \\
&\cong \int^{a,b\in\Sigma_\mathbb{S}} \int^{d,e\in\mathcal{O}_\mathbb{S}} \mathcal{O}_\mathbb{S}(d+e,-) \wedge \mathcal{O}_\mathbb{S}(a,d) \wedge X_a \wedge \mathcal{O}_\mathbb{S}(b,e) \wedge Y_b \\
&\cong \int^{a,b\in\Sigma_\mathbb{S}} \mathcal{O}_\mathbb{S}(a+b,-) \wedge X_a \wedge Y_b
\end{aligned}$$

The third isomorphism follows as colimits commute with colimits, and the last is Lemma 6.3.6. Thus, we have our desired natural isomorphism and $\mathbb{P}_\Sigma^{\mathcal{O}}$ is strong monoidal. It is strong symmetric monoidal as the map $\Sigma_\mathbb{S} \longrightarrow \mathcal{O}_\mathbb{S}$ is a morphism of symmetric monoidal categories. □

Hence, at the level of derived categories, we have strong symmetric monoidal equivalences of triangulated categories

$$\mathbb{L}\mathbb{P}_\Sigma^{\mathcal{O}} \colon \operatorname{Ho}(\mathcal{S}^\Sigma) \rightleftarrows \operatorname{Ho}(\mathcal{S}^{\mathcal{O}}) \colon \mathbb{R}\mathbb{U}_\Sigma^{\mathcal{O}}.$$

Since the monoidal structures on $\operatorname{Ho}(\mathcal{S}^\Sigma)$ and $\operatorname{Ho}(\mathcal{S}^{\mathcal{O}})$ agree, we have constructed a single closed symmetric monoidal structure on $\mathcal{SHC}$. Moreover, this monoidal structure has the properties listed in Theorem 6.1.14. Summarising, we have the following.

Corollary 6.4.10 *The stable homotopy category is a tensor-triangulated category.* □

Similar arguments to the above apply to symmetric spectra in simplicial sets, see Hovey et al. [HSS00] for a complete treatment. Therefore, the stable model structure on symmetric spectra in simplicial sets, $\mathcal{S}^\Sigma(\mathrm{sSet}_*)$ is a symmetric monoidal model category whose homotopy category is strong symmetric monoidally equivalent to $\mathcal{SHC}$.

6.4.2 Homotopy Groups of a Smash Product

In this section, we prove that the handi-crafted smash product of sequential spectra is related to the smash products of orthogonal and symmetric spectra, and we give a formula for the homotopy groups of a smash product.

Since symmetric spectra and orthogonal spectra are symmetric monoidally Quillen equivalent, it suffices to compare the handi-crafted smash product on sequential spectra with the smash product of orthogonal spectra. Note that the implied relationship to symmetric spectra will have to phrased carefully as not every stable equivalence of symmetric spectra is a π_*-isomorphism.

We start by proving that for cofibrant orthogonal spectra X and Y, there are natural isomorphisms

$$\pi_n(X \wedge Y) \cong \operatorname{colim}_a \pi_{n+a}(X_a \wedge Y) \cong \operatorname{colim}_{a,b} \pi_{n+a+b}(X_a \wedge Y_b).$$

In the first and second terms, π_* indicates homotopy groups of spectra, in the last term, π_* denotes homotopy groups of spaces. Note that the second isomorphism follows automatically from the definition of homotopy groups of spectra, and note that we need to move S^1 past X_a (a twist map) before we may use the structure maps of Y. We will then relate this formula to the homotopy groups of a handi-crafted smash product.

Proposition 6.4.11 *For cofibrant orthogonal spectra X and Y, there is a natural isomorphism of abelian groups*

$$\operatorname{colim}_{a,b} \pi_{n+a+b}(X_a \wedge Y_b) \longrightarrow \pi_n(X \wedge Y).$$

Proof The map of orthogonal spectra

$$\lambda\colon \mathrm{F}^{\mathcal{O}}_{a+1} X_a \wedge S^1 \longrightarrow \mathrm{F}^{\mathcal{O}}_a X_a$$

of Definition 5.2.12 is a π_*-isomorphism. Since Y is cofibrant, $- \wedge Y$ preserves π_*-isomorphisms by Theorem 6.4.3. Consequently, there is a natural isomorphism

$$\pi_n(\mathrm{F}^{\mathcal{O}}_{a+1} X_a \wedge S^1 \wedge Y) \xrightarrow{\cong} \pi_n(\mathrm{F}^{\mathcal{O}}_a X_a \wedge Y).$$

We have a commutative diagram

$$\begin{array}{ccc}
\mathrm{F}^{\mathcal{O}}_{a+1} X_a \wedge S^1 & \xrightarrow{\mathrm{F}^{\mathcal{O}}_{a+1} \sigma^X_a} & \mathrm{F}^{\mathcal{O}}_{a+1} X_{a+1} \\
\downarrow \lambda & & \downarrow \\
\mathrm{F}^{\mathcal{O}}_a X_a & \longrightarrow & X,
\end{array}$$

where the unlabelled maps come from the counit maps $\mathrm{F}^{\mathcal{O}}_{a+1} \mathrm{Ev}^{\mathcal{O}}_{a+1} \to \mathrm{Id}$ and $\mathrm{F}^{\mathcal{O}}_a \mathrm{Ev}^{\mathcal{O}}_a \to \mathrm{Id}$. The top map is induced by the structure map on X.

As the map λ is a π_*-isomorphism, there are natural maps

$$\pi_n(\mathrm{F}_a^{\mathcal{O}} X_a) \xrightarrow{\lambda_*^{-1}} \pi_n(\mathrm{F}_{a+1}^{\mathcal{O}} X_a \wedge S^1) \longrightarrow \pi_n(\mathrm{F}_{a+1}^{\mathcal{O}} X_{a+1})$$

and thus,

$$\pi_n(\mathrm{F}_a^{\mathcal{O}} X_a \wedge Y) \longrightarrow \pi_n(\mathrm{F}_{a+1}^{\mathcal{O}} X_a \wedge S^1 \wedge Y) \longrightarrow \pi_n(\mathrm{F}_{a+1}^{\mathcal{O}} X_{a+1} \wedge Y).$$

By Corollary 6.4.4, these induce a natural map on colimits

$$\psi_n(X, Y)\colon\ \operatorname{colim}_a \pi_{n+a}(X_a \wedge Y) \cong \operatorname{colim}_a \pi_n(\mathrm{F}_a^{\mathcal{O}} X_a \wedge Y) \longrightarrow \pi_n(X \wedge Y).$$

We want to prove that this map is an isomorphism. It is true for $Y = \mathbb{S} = \mathrm{F}_0^{\mathcal{O}} S^0$ by the definition of homotopy groups of an orthogonal spectrum.

Consider the class of those Y in the homotopy category of orthogonal spectra where $\psi_n(X, Y)$ is an isomorphism. Both $\operatorname{colim}_a \pi_{n+a}(X_a \wedge -)$ and $\pi_n(X \wedge -)$ commute with coproducts and send exact triangles to long exact sequences. (For the second, we note that $X \wedge -$ and $\mathrm{F}_a^{\mathcal{O}} X_a \wedge -$ preserve cofibre sequences, that π_n sends cofibre sequences to long exact sequences and that exact sequences are preserved by sequential colimits.) Therefore, this class defines a full triangulated subcategory of $\mathcal{SHC}$ which is closed under coproducts and contains the sphere spectrum.

By Corollary 5.6.8, it follows that the class of Y such that $\psi_n(X, Y)$ is an isomorphism is the entire stable homotopy category. In other words, $\psi_n(X, -)$ is an isomorphism on every object in the homotopy category of orthogonal spectra. Therefore, for all cofibrant X and Y,

$$\operatorname{colim}_a \pi_{n+a}(X_a \wedge Y) \cong \operatorname{colim}_{a,b} \pi_{n+a+b}(X_a \wedge Y_b) \cong \pi_n(X \wedge Y). \qquad \square$$

Using the symmetric monoidal model structure on orthogonal spectra, we can follow Proposition 5.1.7 and define a homology theory on the homotopy category of spectra from any cofibrant orthogonal spectrum E by

$$E_* = \pi_*(E \wedge^{\mathbb{L}} -).$$

Similarly, we can define a reduced homology theory on the category of pointed spaces by

$$\widetilde{E}_* = \pi_*(E \wedge^{\mathbb{L}} \Sigma^\infty -),$$

see Proposition 5.1.7. Using the formula for the homotopy groups of a smash product, we can see a relation between the values of $\widetilde{E}_*$ on the levels of a spectrum X and $E_*(X)$.

Corollary 6.4.12 *Let E and X be cofibrant orthogonal spectra. Then*

$$E_n(X) \cong \operatorname{colim}_b \widetilde{E}_n(X_b).$$

Proof The result follows from the isomorphisms

$$\begin{aligned} E_n(X) = \pi_n(E \wedge^{\mathbb{L}} X) &= \pi_n(E \wedge X) \\ &= \operatorname{colim}_{a,b} \pi_{n+a+b}(E_a \wedge X_b) \\ &\cong \operatorname{colim}_b \pi_{n+b}(E \wedge X_b) \\ &\cong \operatorname{colim}_b \widetilde{E}_{n+b}(X_b). \end{aligned}$$

Here, $\wedge^{\mathbb{L}}$ denotes the derived smash product in $\mathcal{SHC}$. We have the second equality, as the spectra are all cofibrant. □

Since the smash product preserves colimits, the derived smash product interacts well with homotopy colimits. This allows us to give a neat formula for the homology of a sequential homotopy colimit. We start by looking again at sequential homotopy colimits.

Let $A \in \mathcal{S}^{\mathcal{O}}$ be cofibrant. Then, for a sequential diagram of orthogonal spectra

$$X_1 \xrightarrow{f_1} X_2 \xrightarrow{f_2} X_3 \xrightarrow{f_3} \cdots,$$

there is a weak equivalence

$$A \wedge \operatorname{hocolim} X_n \simeq \operatorname{hocolim}(A \wedge X_n).$$

This follows by replacing the sequential diagram with a cofibrant diagram (see Example A.7.9) and from the fact that $A \wedge -$ preserves cofibrant diagrams, as it is a left Quillen functor.

If the maps of the diagram were h-cofibrations, then Corollary A.7.10 says that it suffices to take the colimit without performing any cofibrant replacements. In this case, the above weak equivalence becomes an equality, as $A \wedge -$ preserves h-cofibrations by Lemma 6.4.6.

Lemma 6.4.13 *Let E be a cofibrant orthogonal spectrum representing the homology theory E_*. Then*

$$E_*(\operatorname{hocolim}(X_n)) = \operatorname{colim} E_*(X_n).$$

Proof This follows from the above discussion and Lemma 5.6.15. □

Let X and Y be cofibrant orthogonal spectra. If we forget down to sequential spectra, we may take the homotopy groups of the handi-crafted smash product

$$\pi_n(\mathbb{U}_{\mathbb{N}}^{\mathcal{O}} X \underset{\text{hand}}{\wedge} \mathbb{U}_{\mathbb{N}}^{\mathcal{O}} Y) \cong \operatorname{colim}_a \pi_{n+2a}(X_a \wedge Y_a) \cong \operatorname{colim}_{a,b} \pi_{n+a+b}(X_a \wedge Y_b).$$

Hence, we see that the homotopy groups of the underlying handi-crafted smash product agree with those of the smash product of orthogonal spectra.

We could further ask for a map

$$\mathbb{U}_{\mathbb{N}}^{\mathcal{O}} X \underset{\text{hand}}{\wedge} \mathbb{U}_{\mathbb{N}}^{\mathcal{O}} Y \longrightarrow \mathbb{U}_{\mathbb{N}}^{\mathcal{O}}(X \wedge Y).$$

It is possible to make a map on each level. For example on the even levels, the map $S^0 \to \mathcal{O}_{\mathbb{S}}(2n, 2n) = O(2n)_+$ sends the non-basepoint to the identity matrix. This induces a natural map

$$X_n \wedge Y_n \longrightarrow (X \wedge Y)_{2n} = \int^{a,b \in \mathcal{O}_{\mathbb{S}}} \mathcal{O}_{\mathbb{S}}(a + b, 2n) \wedge X_a \wedge Y_b,$$

as any term of a colimit admits a preferred map to the colimit itself. Similarly on odd indices, we have $S^0 \to \mathcal{O}_{\mathbb{S}}(2n + 1, 2n + 1) = O(2n + 1)_+$, which sends the non-basepoint to the identity matrix. This induces a natural map

$$X_{n+1} \wedge Y_n \longrightarrow (X \wedge Y)_{2n+1} = \int^{a,b \in \mathcal{O}_{\mathbb{S}}} \mathcal{O}_{\mathbb{S}}(a + b, 2n + 1) \wedge X_a \wedge Y_b.$$

However, the square involving these maps and the structure maps of the domain and codomain is commutative only up to homotopy. Further details are given in [MMSS01, Section 11]. The relation between the handi-crafted smash product of sequential spectra and the smash product of symmetric spectra is discussed in [HSS00, Remark 4.2.16].

6.5 Spanier–Whitehead Duality

Spanier–Whitehead duality is a relation between the homology and cohomology of finite CW-complexes. It was the original motivation for constructing the Spanier–Whitehead category $\mathcal{SW}$, see [SW55]. Now that we have a closed symmetric monoidal structure on the stable homotopy category, we can place those early results in a more formal framework which shows that compact spectra interact exceptionally well with the closed monoidal structure on $\mathcal{SHC}$. In this section, we give a brief account of the concepts. A discussion of historical versions can be found in [Ada74, Section III.5] and further details can be found in [LMSM86, Chapter III] and [DP80].

We work in the stable homotopy category and use "spectrum" to mean an object of $\mathcal{SHC}$ and "map" to mean a map in $\mathcal{SHC}$. While one can translate the following to statements about symmetric or orthogonal spectra, one would then need to keep track of fibrancy and cofibrancy conditions.

Definition 6.5.1 The *Spanier–Whitehead dual* of a spectrum X is

$$DX = \mathbb{R}\operatorname{Hom}(X, \mathbb{S}),$$

where $\mathbb{R}\operatorname{Hom}$ is the internal function object of $\mathcal{SHC}$. This is often shortened to just the *dual* of X.

Given a spectrum X, the adjoint of the identity map of DX gives an evaluation map

$$\varepsilon\colon DX \wedge^{\mathbb{L}} X \longrightarrow \mathbb{S}.$$

A spectrum X is *strongly dualisable* if there is a map

$$\eta\colon \mathbb{S} \longrightarrow X \wedge^{\mathbb{L}} DX$$

such that the two composites below are the identity maps.

$$X \cong \mathbb{S} \wedge^{\mathbb{L}} X \xrightarrow{\eta \wedge^{\mathbb{L}} \mathrm{Id}} X \wedge^{\mathbb{L}} DX \wedge^{\mathbb{L}} X \xrightarrow{\mathrm{Id} \otimes \varepsilon} X \wedge^{\mathbb{L}} \mathbb{S} \cong X$$
$$X \cong X \wedge^{\mathbb{L}} \mathbb{S} \xrightarrow{\mathrm{Id} \wedge^{\mathbb{L}} \eta} DX \wedge^{\mathbb{L}} X \wedge^{\mathbb{L}} DX \xrightarrow{\varepsilon \otimes \mathrm{Id}} \mathbb{S} \wedge^{\mathbb{L}} X \cong X$$

Example 6.5.2 The dual of $\mathbb{S}^n$ is $\mathbb{S}^{-n}$ for $n \in \mathbb{Z}$. This follows from the more general case $D\mathrm{F}_k^{\mathcal{O}} S^l \cong \mathrm{F}_l^{\mathcal{O}} S^k$ for $k, l \in \mathbb{N}$. We have isomorphisms

$$\begin{aligned}
[A, \mathbb{R}\operatorname{Hom}(\mathrm{F}_k^{\mathcal{O}} S^l, \mathbb{S})] &\cong [A \wedge \mathrm{F}_k^{\mathcal{O}} S^l, \mathbb{S}] \\
&\cong [A \wedge \mathrm{F}_k^{\mathcal{O}} S^{l+k}, \Sigma^k \mathbb{S}] \\
&\cong [A \wedge S^l, \Sigma^k \mathbb{S}] \\
&\cong [A \wedge S^l, \mathrm{F}_0^{\mathcal{O}} S^k] \\
&\cong [A \wedge S^l, \mathrm{F}_l^{\mathcal{O}} S^{k+l}] \\
&\cong [A, \mathrm{F}_l^{\mathcal{O}} S^k].
\end{aligned}$$

In particular, we can conclude that the sphere $\mathbb{S}$ and all of its (de)suspensions are strongly dualisable.

Let M denote the mod-p^n Moore spectrum. We have an exact triangle in $\mathcal{SHC}$

$$\mathbb{S} \xrightarrow{p^n} \mathbb{S} \longrightarrow M \longrightarrow \mathbb{S}^1.$$

We know that $\mathbb{R}\operatorname{Hom}(-, \mathbb{S})$ is an exact functor by Theorem 6.1.14, so we have an exact triangle

$$D\mathbb{S}^1 \xrightarrow{p^n} D\mathbb{S}^1 \longrightarrow DM \longrightarrow D\mathbb{S}.$$

Using that $D\mathbb{S}^1 = \mathbb{S}^{-1}$, the above becomes

$$\mathbb{S}^{-1} \xrightarrow{p^n} \mathbb{S}^{-1} \longrightarrow DM \longrightarrow \mathbb{S}$$

and thus, $DM = \Sigma^{-1} M$.

We can give an alternative characterisation of strongly dualisable spectra. For any spectra B and X there is a natural map

$$\nu_{B,X} \colon B \wedge^{\mathbb{L}} X \longrightarrow \mathbb{R}\mathrm{Hom}(DX, B)$$

induced by

$$B \wedge^{\mathbb{L}} X \wedge^{\mathbb{L}} DX \xrightarrow{\mathrm{Id} \wedge^{\mathbb{L}} \tau} B \wedge^{\mathbb{L}} DX \wedge^{\mathbb{L}} X \xrightarrow{\varepsilon} B \wedge^{\mathbb{L}} \mathbb{S} \cong B.$$

Lemma 6.5.3 *When the spectrum X is strongly dualisable, there is an adjunction*

$$- \wedge^{\mathbb{L}} DX \colon \mathcal{SHC} \rightleftarrows \mathcal{SHC} : - \wedge^{\mathbb{L}} X.$$

A spectrum X is strongly dualisable if and only if the natural map

$$\nu_{B,X} \colon B \wedge^{\mathbb{L}} X \longrightarrow \mathbb{R}\mathrm{Hom}(DX, B)$$

is an isomorphism for all B.

Proof After tensoring with a spectrum A, the maps η and ε from Definition 6.5.1 form the counit and unit of the adjunction. The triangle identities of an adjunction then follow from the two conditions relating η and ε.

Assume that X is strongly dualisable. Then we have natural isomorphisms

$$[A, B \wedge^{\mathbb{L}} X] \cong [A \wedge^{\mathbb{L}} DX, B] \cong [A, \mathbb{R}\mathrm{Hom}(DX, B)]$$

for every A, which show that the map $\nu_{B,X}$ is an isomorphism.

Conversely, the natural isomorphisms

$$[A, B \wedge^{\mathbb{L}} X] \cong [A, \mathbb{R}\mathrm{Hom}(DX, B)] \cong [A \wedge^{\mathbb{L}} DX, B]$$

give an adjunction. The unit η comes from the case of $A = \mathbb{S}$ and $B = DX$ and corresponds to the identity of DX under the given isomorphisms. □

Corollary 6.5.4 *Let X be a strongly dualisable spectrum. Then*

$$\nu_{\mathbb{S},X} \colon X \longrightarrow DDX$$

is an isomorphism. Furthermore, for strongly dualisable X and for any spectra A and B, the map

$$\mathbb{R}\mathrm{Hom}(A, B) \wedge^{\mathbb{L}} X \longrightarrow \mathbb{R}\mathrm{Hom}(A, B \wedge^{\mathbb{L}} X)$$

induced by the adjoint to the evaluation $\mathbb{R}\mathrm{Hom}(A, B) \wedge^{\mathbb{L}} A \longrightarrow B$ is an isomorphism. □

Theorem 6.5.5 *A spectrum is strongly dualisable if and only if it is compact.*

Proof Consider the class of strongly dualisable spectra, thought of as those spectra X such that

$$\nu_{B,X}\colon B \wedge X \longrightarrow \mathbb{R}\operatorname{Hom}(DX, B)$$

is an isomorphism for all B. This class forms a thick subcategory of $\mathcal{SHC}$, as it is closed under retracts and because the functors

$$B \otimes - \quad \text{and} \quad \mathbb{R}\operatorname{Hom}(D(-), B)$$

are exact by Corollary 6.4.10. This thick subcategory contains $\mathbb{S}$ and hence contains all compact objects by Theorem 5.6.13.

Conversely, assume X is strongly dualisable. We have a natural isomorphism

$$[DX, -] \cong [\mathbb{S}, - \wedge^{\mathbb{L}} X],$$

and the right-hand side commutes with coproducts. Therefore, DX is compact. By the previous step, as DX is compact, it is also strongly dualisable. Thus, $DDX \cong X$ is compact. □

As a consequence, we obtain the following.

Theorem 6.5.6 (Spanier–Whitehead duality) *Let E and X be spectra, and let X be compact. Then there is a natural isomorphism*

$$E_n(X) = [\mathbb{S}, E \wedge X]_n \cong [DX, E]_n = E^{-n}(DX). \qquad \square$$

Remark 6.5.7 If we let $X = \Sigma^\infty A$, we can try to use this to apply Brown representability to homology theories. This works for finite CW-complexes, but we must be cautious when using infinite complexes due to phantom maps, see Remark 5.1.8.

6.6 Ring Spectra and Modules

Using our symmetric monoidal product on orthogonal and symmetric spectra, we can define ring spectra (monoids), commutative ring spectra (commutative monoids) and modules over a ring spectrum. We prove that there are model categories of modules over a ring spectrum in symmetric spectra and orthogonal spectra and that these categories are Quillen equivalent (in a suitable sense).

We then turn to ring spectra and discuss model categories of ring spectra in symmetric spectra and orthogonal spectra and compare them via Quillen equivalences.

Let $\mathcal{S}$ denote either orthogonal spectra or symmetric spectra.

Definition 6.6.1 A *ring spectrum* is a spectrum $R \in \mathcal{S}$ with maps in $\mathcal{S}$

$$\mu \colon R \wedge R \longrightarrow R \text{ and } \eta \colon \mathbb{S} \longrightarrow R$$

such that μ is associative and unital with respect to η. A *map of ring spectra* $f \colon R \longrightarrow S$ is a map of spectra that commutes with the multiplication and unit maps.

A spectrum R is called a *commutative ring spectrum* if it is a ring spectrum and the following diagram commutes

$$\begin{array}{ccc} R \wedge R & \xrightarrow{\mu} & R, \\ \downarrow{\scriptstyle \tau} & \nearrow{\scriptstyle \mu} & \\ R \wedge R & & \end{array}$$

where τ is the twist map. A *map of commutative ring spectra* $f \colon R \longrightarrow T$ is a map of ring spectra.

Ring spectra are also often called *algebras*.

Examples 6.6.2 We have already seen some examples of ring spectra. We recap these examples here.

Sphere spectrum The sphere spectrum is a commutative ring spectrum in symmetric spectra and orthogonal spectra.

Eilenberg–Mac Lane spectra For R a (commutative) ring, the Eilenberg–Mac Lane spectrum HR from Examples 5.3.2 is a (commutative) ring spectrum in symmetric spectra. Level n of HR is given by

$$HR_n = |(S^1_s)^n \otimes R|.$$

The (commutative) ring structure comes from the natural map

$$(K \otimes R) \wedge (L \otimes R) \longrightarrow (K \wedge L) \otimes R$$

induced by sending the simplex $(k \cdot r, l \cdot s)$ to $(k \wedge l) \cdot rs$.

Endomorphism spectra For any spectrum X, the spectrum $F_{\mathbb{S}}(X,X)$ is a ring spectrum. The composition map

$$F_{\mathbb{S}}(X,X) \wedge F_{\mathbb{S}}(X,X) \longrightarrow F_{\mathbb{S}}(X,X)$$

is defined via its adjoint map

$$F_{\mathbb{S}}(X,X) \wedge F_{\mathbb{S}}(X,X) \wedge X \longrightarrow X,$$

which is two instances of the evaluation map $F_{\mathbb{S}}(X,X) \wedge X \longrightarrow X$, which is adjoint to the identity of $F_{\mathbb{S}}(X,X)$. The unit map $\mathbb{S} \longrightarrow F_{\mathbb{S}}(X,X)$ is defined as the adjoint of the identity map on X. In general, there is no reason for this ring spectrum to be commutative.

Dual spectra For an unpointed space A, the spectrum $DA_+ = F_{\mathbb{S}}(A_+,\mathbb{S})$ is a commutative ring spectrum. The unit map is induced by the terminal map $A \longrightarrow *$ in unpointed spaces. The multiplication map is given by the composite

$$F_{\mathbb{S}}(A_+,\mathbb{S}) \wedge F_{\mathbb{S}}(A_+,\mathbb{S}) \longrightarrow F_{\mathbb{S}}(A_+ \wedge A_+, \mathbb{S} \wedge \mathbb{S}) \cong F_{\mathbb{S}}((A \times A)_+,\mathbb{S}) \longrightarrow F_{\mathbb{S}}(A_+,\mathbb{S}),$$

where the last map is induced by the diagonal $A \longrightarrow A \times A$. Since the diagonal on unpointed spaces is coassociative and cocommutative, the spectrum DA_+ has an associative and commutative multiplication.

More examples of ring spectra can be found in [Sch07b, Chapter I.2]. Constructing point-set models for commutative ring spectra can be quite involved. As an example of some of the complexities, a point-set model for the spectrum of real K-theory KO (see Subsection 7.4.2) can be found in work of Joachim [Joa01].

Definition 6.6.3 A spectrum M is a *left module over a ring spectrum* R if there is a map

$$\nu \colon R \wedge M \longrightarrow M,$$

which is associative and unital. A *map of module spectra* $f \colon M \longrightarrow N$ is a map of spectra that commutes with the module action maps. A *right module over* R is defined similarly, but with R on the right of the module.

By default, a module will mean a left module.

Examples 6.6.4 We already know several examples of module spectra.

- Every spectrum is a module over the sphere spectrum.
- If R is a ring spectrum, then it is a module over itself.
- For any spectrum X, the *free R–module on* X, $R \wedge X$, is a left R–module, with R acting on itself.
- If M is a module over a ring R, then the Eilenberg–Mac Lane spectrum HM is a module over HR.
- The spectrum X is a module over $F_{\mathbb{S}}(X,X)$.

One motivation for the study of ring spectra is that they induce cohomology theories with cup products.

Lemma 6.6.5 *If E is a (commutative) ring spectrum and A is an unpointed space, then $\widetilde{E}^*(A_+)$ is a (commutative) graded ring.*

If M is a module over E, then $\widetilde{M}^*(A_+)$ is a graded module over $\widetilde{E}^*(A_+)$.

Proof Similarly to the multiplication on dual spectra, we define the multiplication using the multiplication μ of E and the diagonal map $\Delta \colon A \longrightarrow A \times A$ as below.

$$\widetilde{E}^*(A_+) \otimes \widetilde{E}^*(A_+) = [A_+, E]_{-*} \otimes [A_+, E]_{-*} \longrightarrow [A_+ \wedge A_+, E \wedge E]_{-*} \longrightarrow [A_+, E]_{-*}$$

This is commutative when E has a commutative multiplication. The second statement is similar. □

We may now consider the categories of ring spectra and the category of modules over a fixed ring spectrum. We leave commutative ring spectra to Section 6.7 and turn to modules over a ring spectrum.

Definition 6.6.6 For a ring spectrum $R \in \mathcal{S}$, the category of (left) R–modules is denoted R–mod.

As with algebra, there is an adjunction

$$R \wedge - \colon \mathcal{S} \rightleftarrows R\text{–mod} : U,$$

where the right adjoint U is the forgetful functor and the left adjoint sends a spectrum X to the free R–module $R \wedge X$. There is a right adjoint to U denoted $F_{\mathcal{S}}(R, -)$. The action map

$$R \wedge F_{\mathcal{S}}(R, M) \longrightarrow F_{\mathcal{S}}(R, M)$$

is defined to be the adjoint of multiplication followed by evaluation

$$R \wedge R \wedge F_{\mathcal{S}}(R, M) \xrightarrow{\mu \wedge \mathrm{Id}} R \wedge F_{\mathcal{S}}(R, M) \longrightarrow M.$$

The category of R–modules has all small limits and colimits. Colimits are defined as the colimit of the diagram of underlying spectra, with action map given by commuting $R \wedge -$ past the colimit (which is an isomorphism). The case for limits is similar, and it follows that the functor U preserves all limits and colimits.

The category of R–modules is tensored, cotensored and enriched over pointed spaces. The tensor $M \wedge A$ is given by smashing a module M with a space A and by letting R act on M as before. The cotensor $\mathrm{Top}_*(A, M)$ has the same underlying spectrum as for the cotensor of spectra, with action map adjoint to

$$A \wedge R \wedge \mathrm{Top}_*(A, M) \xrightarrow{\tau_{A,R} \wedge \mathrm{Id}} R \wedge A \wedge \mathrm{Top}_*(A, M) \longrightarrow R \wedge M \xrightarrow{\nu_M} M.$$

The enrichment over pointed spaces is given by an equaliser

$$\mathrm{Top}_*^R(M, N) = \mathrm{eq}\,(\mathrm{Top}_*(M, N) \rightrightarrows \mathrm{Top}_*(R \wedge M, N)).$$

The two maps are defined by their adjoints, with the unmarked maps being the evaluations.

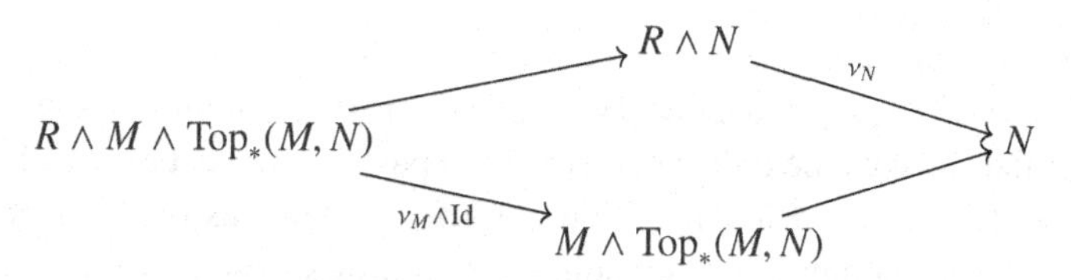

We give the main result on modules over ring spectra. This is similar to the part of [SS00, Theorem 4.1] which relates to modules, but we do not make the strict smallness assumptions of that result.

Theorem 6.6.7 *Let $R \in \mathcal{S}$ be a ring spectrum. There is a model category of modules over R where f is a weak equivalence or fibration if $Uf \in \mathcal{S}$ is a weak equivalence or fibration.*

This model category on R–mod *is cofibrantly generated, proper and stable. If R is commutative,* R–mod *is a closed monoidal model category.*

Furthermore, if R is cofibrant, then a cofibration of R–modules forgets to a cofibration of $\mathcal{S}$.

Proof We use the lifting lemma, Lemma A.6.12 to obtain the model structure. The most complicated part is verifying the smallness conditions. The generating sets of R–mod are given by applying $R \wedge -$ to the generating sets for $\mathcal{S}$, see Definitions 5.2.15 and 5.3.18. Lemma 6.4.6 says that the generating cofibrations of R–mod are h-cofibrations. As U preserves all colimits, the smallness conditions will hold if the domains of the generating sets of $\mathcal{S}$ are small with respect to the class of h-cofibrations in the category $\mathcal{S}$.

By Lemma A.5.7, shifted suspension spectra $\mathrm{F}_d^{\Sigma} A$ and $\mathrm{F}_d^{\mathcal{O}} A$ are small in $\mathcal{S}$ with respect to the class of h-cofibrations when A is a pointed compact topological space. The domains of the generating sets of $\mathcal{S}$ are defined in terms of pushouts of shifted suspension spectra on compact spaces, hence, they are small with respect to the class of h-cofibrations by Lemma A.6.5.

The final condition to verify for the lifting lemma is the condition on acyclic cofibrations. Lemma 6.4.6 and Corollary 6.4.4 show that $R \wedge -$ takes the generating acyclic cofibrations of $\mathcal{S}$ to stable equivalences that are h-cofibrations. As these are preserved by pushouts and sequential colimits, the results follow.

Properness follows as colimits, limits, fibrations and weak equivalences are given in the underlying category of spectra and every cofibration of R–modules is a h-cofibration by Lemma 6.4.6. Stability also follows from considering underlying spectra.

When R is commutative, the tensor product over R given by

$$M \wedge_R N = \mathrm{coeq}\,(M \wedge R \wedge N \underset{\mathrm{Id} \wedge \nu_N}{\overset{\nu_M \wedge \mathrm{Id}}{\rightrightarrows}} M \wedge N)$$

defines a monoidal product just as with rings and modules in algebra. The internal function object is given by an equaliser. Proposition 6.6.8 will show that R–mod satisfies the unit statement of the pushout product axiom. The rest of the pushout product axiom follows as the (co)domains of the generating sets are free R–modules. The monoid axiom follows from that for spectra as the generating acyclic cofibrations are free R–modules.

When R is cofibrant, the functor $R \wedge -$ preserves cofibrations. Hence, the generating cofibrations of R–mod, and thus all cofibrations, forget to cofibrations of $\mathcal{S}$. □

We have the module equivalent of "cofibrant implies flat".

Proposition 6.6.8 *Let M be a cofibrant R–module in $\mathcal{S}$, where R is a commutative ring spectrum. Then $- \wedge_R M$ preserves stable equivalences and π_*-isomorphisms.*

Proof When $M = R \wedge X$ for a cofibrant spectrum X, this follows from Theorems 6.4.2 and 6.4.3. The same method as the proof of those theorems gives the general case. □

Now that we have categories of modules over a ring, it is natural to ask about change of rings functors. Let $f \colon R \longrightarrow R'$ be a map of rings and M an R'–module. Let

$$\mu_R \colon R \wedge R \longrightarrow R \text{ and } \mu_{R'} \colon R' \wedge R' \longrightarrow R'$$

denote the multiplication maps of R and R', with the action map of M denoted

$$\nu_M \colon R \wedge M \longrightarrow M.$$

We may define an R–module f^*M by taking the same underlying spectrum as M with action map given by pre-composition with f

$$R \wedge M \xrightarrow{f \wedge \mathrm{Id}} R' \wedge M \xrightarrow{\nu_M} M.$$

This functor $f^* \colon R'\text{–mod} \longrightarrow R\text{–mod}$ has both left and right adjoints and hence preserves all limits and colimits. Let N be an R–module with action map

$$\nu_N \colon R \wedge N \longrightarrow N.$$

Then the left adjoint is given by

$$R' \wedge_R N = \operatorname{coeq}(R' \wedge R \wedge N \underset{\mathrm{Id} \wedge \nu_N}{\overset{\mu_R \circ \mathrm{Id} \wedge f \wedge \mathrm{Id}}{\rightrightarrows}} R' \wedge N).$$

The right adjoint is given by a similar formula, namely,

$$F_R(R', N) = \operatorname{eq}(F_{\mathbb{S}}(R', N) \rightrightarrows F_{\mathbb{S}}(R' \wedge R, N)).$$

Theorem 6.6.9 *Let $\mathcal{S}$ denote either symmetric spectra or orthogonal spectra and let $f\colon R \longrightarrow R'$ be a map of ring spectra. Then the change of rings adjunction is a Quillen adjunction*

$$R' \wedge_R -\colon R\text{–mod} \rightleftarrows R'\text{–mod} : f^*.$$

Furthermore, if f is a stable equivalence, this adjunction is a Quillen equivalence.

Proof The right adjoint f^* preserves and detects fibrations and weak equivalences, so $(R' \wedge_R -, f^*)$ is a Quillen adjunction. When f is a stable equivalence, the derived unit on a cofibrant R–module M

$$M = R \wedge_R M \longrightarrow R' \wedge_R M$$

is a weak equivalence by Proposition 6.6.8. □

We now show that the homotopy category of R–modules does not depend on whether we work in orthogonal or symmetric spectra. In the following, we use the fact that the functor $\mathbb{P}_\Sigma^{\mathcal{O}}$ from symmetric spectra to orthogonal spectra is strong monoidal to produce the displayed functors at the level of module categories.

Let R be a cofibrant ring spectrum in symmetric spectra, and let R' be a ring spectrum in orthogonal spectra. If M is an R–module in symmetric spectra, then $\mathbb{P}_\Sigma^{\mathcal{O}} M$ is a $\mathbb{P}_\Sigma^{\mathcal{O}} R$–module with action map

$$\mathbb{P}_\Sigma^{\mathcal{O}} R \wedge \mathbb{P}_\Sigma^{\mathcal{O}} M \cong \mathbb{P}_\Sigma^{\mathcal{O}}(R \wedge M) \longrightarrow \mathbb{P}_\Sigma^{\mathcal{O}} M.$$

If N is a $\mathbb{U}_\Sigma^{\mathcal{O}} R'$–module in symmetric spectra, then $\mathbb{P}_\Sigma^{\mathcal{O}} N$ is a module over $\mathbb{P}_\Sigma^{\mathcal{O}} \mathbb{U}_\Sigma^{\mathcal{O}} R'$, hence, we define an R'–module LN in orthogonal spectra by

$$LN = R' \wedge_{\mathbb{P}_\Sigma^{\mathcal{O}} \mathbb{U}_\Sigma^{\mathcal{O}} R'} \mathbb{P}_\Sigma^{\mathcal{O}} N.$$

The right adjoint $\mathbb{U}_\Sigma^{\mathcal{O}}$ passes to the various module categories in a similar fashion, using the monoidal structure on the functor and the unit map. More details on these adjoints can be found in [SS03a].

Theorem 6.6.10 *Let R be a cofibrant ring spectrum in symmetric spectra and R' a ring spectrum in orthogonal spectra. Then there are commutative diagrams of Quillen adjunctions in which the horizontal adjunctions are Quillen equivalences.*

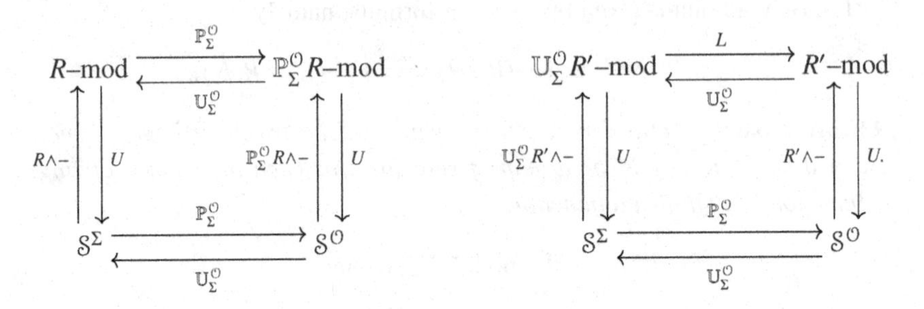

Proof The squares of right adjoints commute, hence, so do the squares of left adjoints. The functors labelled $\mathbb{U}^{\mathcal{O}}_{\Sigma}$ are right Quillen functors at the level of spectra. They are also right Quillen functors at the level of modules, as the fibrations and acyclic fibrations of modules are defined in terms the functors labelled U.

To see that the upper adjunctions are Quillen equivalences, we only need consider the derived units. For the left-hand square, this follows from the fact that

$$\mathbb{P}^{\mathcal{O}}_{\Sigma} : \mathcal{S}^{\Sigma} \rightleftarrows \mathcal{S}^{\mathcal{O}} : \mathbb{U}^{\mathcal{O}}_{\Sigma}$$

is a Quillen equivalence: the derived unit and counit of the top adjunction are weak equivalences of module spectra because the underlying maps of spectra are weak equivalences.

For the right-hand square, we use the same argument to see that

$$\mathbb{P}^{\mathcal{O}}_{\Sigma} : \mathbb{U}^{\mathcal{O}}_{\Sigma} R'\text{–mod} \rightleftarrows \mathbb{P}^{\mathcal{O}}_{\Sigma}\mathbb{U}^{\mathcal{O}}_{\Sigma} R'\text{–mod} : \mathbb{U}^{\mathcal{O}}_{\Sigma}$$

is a Quillen equivalence, and then we compose it with the change of rings Quillen equivalence of Theorem 6.6.9. □

We will now turn our attention to the category of ring spectra in $\mathcal{S}$. As with commutative ring spectra later, we sketch some of the key constructions and results. We leave details to our references.

The category of rings has all small limits, which are constructed by forgetting to the underlying category $\mathcal{S}$ as follows. Take a diagram of ring spectra

$$\phi_{i,j} : R_i \longrightarrow R_j$$

for i, j in some indexing set. Then the smash products of the canonical maps

$$\lim_i R_i \wedge \lim_j R_j \longrightarrow R_k \wedge R_k \longrightarrow R_k$$

induce a map to the limit $\lim_k R_k$. This defines a multiplication map on the limit. The construction of colimits is more complicated. One can either show that filtered colimits of ring spectra can be constructed in $\mathcal{S}$ and use Borceux [Bor94, Section 4.3] or take the approach of Elmendorf et al. [EKMM97, Propositions II.7.2 and VII.2.10].

The initial object in the category of ring spectra is the sphere spectrum $\mathbb{S}$, the terminal object is $*$. Hence, the category of ring spectra is not pointed, so, in particular, the model structures we will put on these categories are not stable.

Definition 6.6.11 The category of ring spectra in $\mathcal{S}$ is denoted ring–$\mathcal{S}$. The *free ring spectrum* $\mathbb{T}X$ *on* $X \in \mathcal{S}$ is defined by

$$\mathbb{T}X = \bigvee_{n \geqslant 0} X^{\wedge n}$$

with $X^{\wedge 0} := \mathbb{S}$.

Let $\mathbb{U}$ denote the forgetful functor from ring spectra to spectra. This is the right adjoint to the free functor

$$\mathbb{T} \colon \mathcal{S} \rightleftarrows \text{ring–}\mathcal{S} : \mathbb{U}.$$

The main theorems on ring spectra are the existence of the model structures, as well as the Quillen equivalences between ring spectra in symmetric spectra and ring spectra in orthogonal spectra.

Theorem 6.6.12 *There is a model structure on the category of ring spectra where f is a weak equivalence or fibration if $\mathbb{U}f \in \mathcal{S}$ is a weak equivalence or fibration.*

If $f \colon X \longrightarrow Y$ is a cofibration of ring spectra then $\mathbb{U}f$ is a h-cofibration of spectra. If X is a cofibrant ring spectrum, then $\mathbb{U}f$ is a q-cofibration of spectra.

In particular, a cofibrant ring spectrum is cofibrant as a spectrum. For the proof of the theorem, one applies the lifting lemma to the adjunction

$$\mathbb{T} \colon \mathcal{S} \rightleftarrows \text{ring–}\mathcal{S} : \mathbb{U}$$

and follows the strategy of [SS00, Theorem 4.1], see also [MMSS01, Theorem 12.1]. One ingredient for this is [SS00, Lemma 6.2], which can be modified to drop the smallness assumption of the original result by using h-cofibrations similarly to Theorem 6.6.7.

Since $\mathbb{P}_{\Sigma}^{\mathcal{O}} : \mathcal{S}^{\Sigma} \rightleftarrows \mathcal{S}^{\mathcal{O}} : \mathbb{U}_{\Sigma}^{\mathcal{O}}$ is a strong symmetric monoidal Quillen adjunction, it passes to the level of ring spectra.

Theorem 6.6.13 *There are commutative diagrams of Quillen adjunctions in which the horizontal adjunctions are Quillen equivalences.*

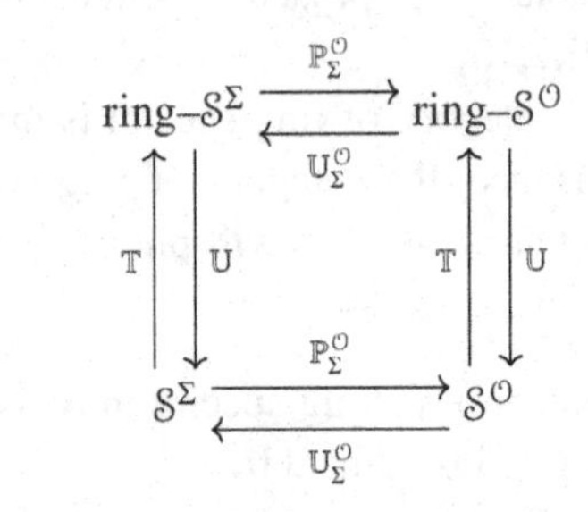

Proof Since the fibrations and weak equivalences in our model structures on ring spectra are defined in terms of the underlying model categories of spectra, we have a Quillen adjunction. Since a cofibrant ring symmetric spectrum is a cofibrant symmetric spectrum, the derived unit and counit are stable equivalences at the level of ring spectra. □

Given a commutative ring spectrum R, one can define the category of R–algebras to be the category of ring spectra under R. This has a model structure by [DS95, Remark 3.10]. As with Theorem 6.6.10, one can then compare R–algebras in symmetric spectra with $\mathbb{P}_{\Sigma}^{\mathcal{O}} R$–algebras in orthogonal spectra. The adjoints in this result become more complicated, so we refer to Schwede and Shipley [SS03a] for details.

6.7 Commutative Ring Spectra

There are model structures on the categories of commutative ring symmetric spectra and commutative ring orthogonal spectra. These model structures are even harder to construct than those for ring spectra. The problem arises from the Σ_n-action of the n-fold smash of a spectrum. As an example of the extra difficulties caused by symmetry, consider the case of chain complexes over the integers. The chain complex D^k is given by

$$D^k = (\cdots \longleftarrow 0 \leftarrow \mathbb{Z} \overset{\mathrm{Id}}{\longleftarrow} \mathbb{Z} \longleftarrow 0 \longleftarrow \cdots)$$

with the copies of $\mathbb{Z}$ in degrees k and $k-1$. It is therefore acyclic. However,

$$\mathbb{P}D^k = \bigoplus_{n \geqslant 0} (D^k)^{\otimes n}/\Sigma_n$$

is not acyclic. This implies that there is no model structure on integral commutative differential graded algebras with the homology isomorphisms as weak equivalences and with surjections as fibrations.

For spectra, this problem is resolved by Lemma 6.7.4, which implies that the analogous free commutative algebra functor $\mathbb{P}$ preserves weak equivalences between cofibrant spectra. A complete account of the existence of the model structure of commutative ring spectra would take us too far afield and would be best placed alongside a comprehensive account of E_∞–ring spectra and operads. We leave these to the numerous references on the subject, such as Elmendorf et al. [EKMM97], Harper [Har09, Har15] and May [May77]. Instead, we give an overview of the properties of the model structure on commutative ring spectra, which is lifted from the "positive stable model structure" on symmetric spectra and orthogonal spectra.

The positive stable model structures are required to avoid a problem originally identified by Lewis in [Lew91]. The problem is that if $\mathbb{S}$ is a cofibrant–fibrant spectrum that is also a commutative ring spectrum, then level zero of this spectrum would be a strictly commutative model X for QS^0, see Definition 1.1.14. By [Moo58, Theorem 3.19], the basepoint component of X is a product of Eilenberg–Mac Lane spaces, but this is not true of the basepoint component of QS^0.

The following result summarises Mandell et al. [MMSS01, Section 14].

Proposition 6.7.1 *There are* positive stable model structures *on symmetric spectra and orthogonal spectra. The weak equivalences are given by the stable equivalences. The cofibrations are called the* positive q-cofibrations, *which are precisely the class of q-cofibrations that are homeomorphisms in degree zero. The fibrations are called the* positive fibrations. *We denote the positive stable model structures by $\mathcal{S}^\Sigma_+$ and $\mathcal{S}^\mathcal{O}_+$.*

These model structures are cofibrantly generated proper model structures. Furthermore, there is a commutative square of Quillen equivalences

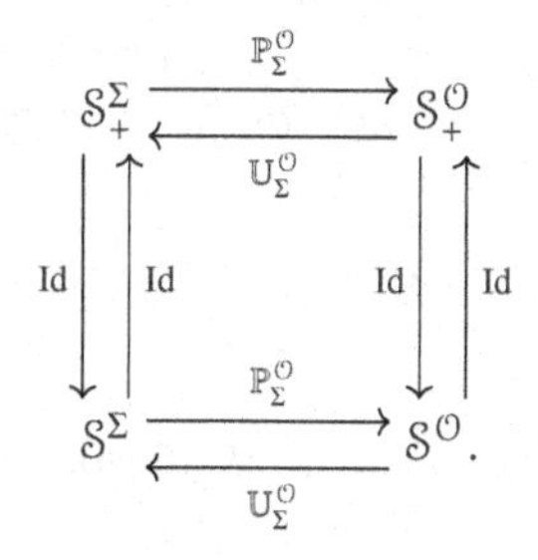

We give the generating sets for the positive stable model structures. They are the generating sets of the stable model structures, but with maps in the image of Σ^∞ removed.

$$\begin{aligned}
I^{\mathcal{O}}_{\text{stable+}} &= I^{\mathcal{O}}_{\text{level+}} = \{F^{\mathcal{O}}_d(S^{a-1}_+ \longrightarrow D^a_+) \mid a \in \mathbb{N},\ d > 0\} \\
J^{\mathcal{O}}_{\text{level+}} &= \{F^{\mathcal{O}}_d(D^a_+ \longrightarrow (D^a \times [0,1])_+) \mid a \in \mathbb{N},\ d > 0\} \\
J^{\mathcal{O}}_{\text{stable+}} &= J^{\mathcal{O}}_{\text{level+}} \cup \{k_d \square (S^{a-1}_+ \to D^a_+) \mid a \in \mathbb{N},\ d > 0\}
\end{aligned}$$

$$\begin{aligned}
I^{\Sigma}_{\text{stable+}} &= I^{\Sigma}_{\text{level+}} = \{F^{\Sigma}_d(S^{a-1}_+ \to D^a_+) \mid a \in \mathbb{N},\ d > 0\} \\
J^{\Sigma}_{\text{level+}} &= \{F^{\Sigma}_d(D^a_+ \to (D^a \times [0,1])_+) \mid a \in \mathbb{N},\ d > 0\} \\
J^{\Sigma}_{\text{stable+}} &= J^{\Sigma}_{\text{level+}} \cup \{k_d \square (S^{a-1}_+ \to D^a_+) \mid a \in \mathbb{N},\ d > 0\}
\end{aligned}$$

From these sets, we see that a positive cofibration is a q-cofibration, so every fibration of the stable model structure is a positive fibration. We also see that the sphere spectrum is not cofibrant in either positive model structure.

As with ring spectra, a limit of a diagram of commutative ring spectra may be constructed in the underlying category of $\mathcal{S}$. The proof that all small colimits exist can be found in Elmendorf et al. [EKMM97, Propositions II.7.2 and VII.2.10] or Borceux [Bor94, Section 4.3]. While an arbitrary colimit in the category of commutative ring spectra can be very complicated, filtered colimits are given by colimits in the underlying category of spectra. The pushout of a diagram

$$B \longleftarrow A \longrightarrow C$$

is given by the smash product $B \wedge_A C$ over the initial term. The initial object of commutative ring spectra is the sphere spectrum $\mathbb{S}$, and the terminal object is $*$, so again, this category is not pointed and hence cannot support a stable model structure.

Lemma 6.7.2 *There is an adjunction*

$$\mathbb{P}\colon \mathcal{S} \rightleftarrows \text{comm–ring–}\mathcal{S} : \mathbb{U},$$

where $\mathbb{U}$ *denotes the forgetful functor from commutative ring spectra to spectra and*

$$\mathbb{P}X = \bigvee_{n \geqslant 0} X^{\wedge n}/\Sigma_n$$

with $X^{\wedge 0} := \mathbb{S}$. *The* Σ_n*-action commutes the factors of the smash product.*

We may now give the first main result for this section, namely, the existence of model structures on commutative ring symmetric spectra and commutative ring orthogonal spectra. The proof is particularly difficult and is left to the references given at the start of the section and [MMSS01, Section 15].

Theorem 6.7.3 *There is a model structure on the category of commutative ring spectra where f is a weak equivalence or fibration if $\mathbb{U}f$ is a weak equivalence or positive fibration.*

The model structure is cofibrantly generated, with generating sets given by applying $\mathbb{P}$ to the generating sets for the positive stable model structures.

Work of Shipley [Shi04] gives an alternative model structure on symmetric spectra and orthogonal spectra that has better compatibility properties between commutative ring spectra and their underlying spectra.

The key fact that allows the model structure to exist is the following lemma. It says that the n-fold smash product of a cofibrant spectrum is Σ_n–free up to stable equivalence. The proof uses a small amount of equivariant homotopy theory, see [May96, Chapter 1] for an introduction.

Lemma 6.7.4 *For X, a positive cofibrant symmetric spectrum or a cofibrant orthogonal spectrum, the natural map*

$$\rho_{X,n} \colon (E\Sigma_n)_+ \wedge_{\Sigma_n} X^{\wedge n} \longrightarrow X^{\wedge n}/\Sigma_n$$

is a stable equivalence for any cofibrant X.

The functors $\mathbb{P}_{\Sigma}^{\mathcal{O}}$ and $\mathbb{U}_{\Sigma}^{\mathcal{O}}$ pass to the level of commutative ring spectra and form a Quillen equivalence by [MMSS01, Theorem 0.7].

Theorem 6.7.5 *There are commutative diagrams of Quillen adjunctions in which the horizontal adjunctions are Quillen equivalences.*

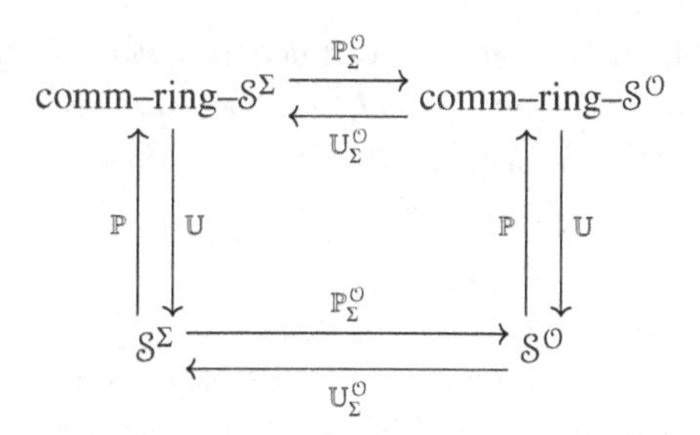

This statement is the commutative analogue of Theorem 6.6.13. It says that we can work with either symmetric spectra or orthogonal spectra. However, there are sometimes small differences in the behaviour of these model categories. An example is given by Kro [Kro07] which gives a *symmetric monoidal* fibrant replacement functor for the positive stable model structure on orthogonal spectra. The corresponding construction for symmetric spectra does not give a fibrant replacement functor.

We can repeat the above with symmetric spectra in simplicial sets, obtaining categories of rings, modules and commutative rings. Geometric realisation and the singular complex functor give Quillen equivalences between the various model categories of rings, modules and commutative rings. Details can be found in Mandell et al. [MMSS01, Section 19].

6.8 Applications of Monoidality

We now have commutative smash products on model categories of spectra and the stable homotopy category, as well as model categories of rings, modules and commutative rings. This is a major improvement on $\mathcal{S}^{\mathbb{N}}$, and these structures have major consequences and applications.

The primary place for such results is Elmendorf et al. [EKMM97], which has chapters on the algebraic K-theory of ring spectra and topological Hochschild homology and cohomology. We leave that material to the reference (see also Baker and Richter [BR04]) to focus on a number of results on the structure of the stable homotopy category. They make fundamental use of the monoidal structure of spectra.

The following result of Shipley [Shi01] is related to Theorem 6.9.31. This theorem makes use of the monoidal structure on the model category rather than on the triangulated homotopy category. It states that the positive stable model category of symmetric spectra in simplicial sets is initial among stable simplicial symmetric monoidal model categories.

Theorem 6.8.1 (Shipley) *Let $\mathcal{S}^{\Sigma}_{+}(\mathrm{sSet}_*)$ denote symmetric spectra in simplicial sets with the positive stable model structure, and let $\mathcal{C}$ be any stable simplicial symmetric monoidal model category. Then there is a simplicial strong symmetric monoidal left Quillen functor*

$$\mathcal{S}^{\Sigma}_{+}(\mathrm{sSet}_*) \longrightarrow \mathcal{C}.$$

Much of this book is focused on categories of spectra rather than on stable model categories in general. The following pair of theorems of Schwede and Shipley from [SS03b] shows that this is in fact no restriction at all: under some reasonable assumptions, any other stable model category can be considered as a category of spectra.

Theorem 6.8.2 (Schwede–Shipley) *Let $\mathcal{C}$ be a simplicial cofibrantly generated proper stable model category with a compact generator P. Then there exists a chain of simplicial Quillen equivalences between $\mathcal{C}$ and the model category of* End(P)*–modules. Here,* End(P) *is a ring spectrum satisfying*

$$\pi_*(\mathrm{End}(P)) \cong [P, P]_*^{\mathcal{C}}$$

as graded rings.

The ring spectrum can be constructed using the framings of Section 6.9. With that technology, one can consider the case where $\mathcal{C}$ has a *set* of compact generators P rather than a single generator. In this case, there is a category $\mathcal{E}(P)$ with objects given by the elements of P. This category is enriched over $\mathcal{S}^{\Sigma}(\mathrm{sSet}_*)$. There are natural isomorphisms

$$\pi_*(\mathcal{E}(P)(A, B)) \cong [A, B]_*^{\mathcal{C}}$$

in a manner that is associative and unital. One can then define an $\mathcal{E}(P)$–module to be a functor, enriched over symmetric spectra, from $\mathcal{E}(P)$ to $\mathcal{S}^{\Sigma}(\mathrm{sSet}_*)$.

Theorem 6.8.3 (Schwede–Shipley) *Let $\mathcal{C}$ be a simplicial cofibrantly generated proper stable model category with a set P of compact generators. Then there exists a chain of simplicial Quillen equivalences between $\mathcal{C}$ and the model category of $\mathcal{E}(P)$–modules.*

The next results of Shipley [Shi07] and Richter and Shipley [RS17] can be thought of as extensions of Theorem 6.8.2 in the case where $\mathcal{C}$ is the category Ch(R) of chain complexes of R–modules for a commutative ring R. These results are extremely significant, as they allow us to view the world of algebra as a special case of spectra.

Theorem 6.8.4 (Shipley) *There is a chain of weak symmetric monoidal Quillen equivalences between* Ch(R) *with the projective model structure and HR–module spectra in symmetric spectra, where HR denotes the Eilenberg–Mac Lane spectrum of R.*

The ring version also holds.

Theorem 6.8.5 (Shipley) *There is a chain of Quillen equivalences between differential graded R–algebras (ring objects in* Ch(R)*) and ring spectra under HR in symmetric spectra.*

The analogue of the above for commutative algebras is more complicated due to the difficulty of constructing a model structure on that category. However, when $R = \mathbb{Q}$, the projective model structure on rational chain complexes does lift to the level of commutative differential graded algebras.

Theorem 6.8.6 (Richter–Shipley) *There is a chain of Quillen equivalences between E_∞–monoids in* $\mathrm{Ch}(R)$ *and commutative ring spectra under HR in symmetric spectra.*

There is a chain of Quillen equivalences between commutative differential graded $\mathbb{Q}$–algebras (commutative ring objects in $\mathrm{Ch}(\mathbb{Q})$*) and commutative ring spectra under $H\mathbb{Q}$ in symmetric spectra.*

Versions of these last two results were known for some time at the homotopical level, see Elmendorf et al. [EKMM97, Section IV.2] and Robinson [Rob87].

6.9 Homotopy Mapping Objects and Framings

6.9.1 Unstable Framings

Let $\mathcal{C}$ be a pointed simplicial model category (see Definition 6.1.28). Then the homotopy category of $\mathcal{C}$ is tensored, cotensored and enriched over the homotopy category of pointed simplicial sets by Theorem 6.1.29. In particular, let $X \in \mathcal{C}$ be cofibrant and $Y \in \mathcal{C}$ be fibrant. Then the enrichment gives a Kan complex $\mathrm{map}_{\mathcal{C}}(X, Y)$ such that

$$\pi_n(\mathrm{map}_{\mathcal{C}}(X, Y)) = [S^n, \mathrm{map}_{\mathcal{C}}(X, Y)]^{\mathrm{sSet}_*} \cong [X \otimes S^n, Y]^{\mathcal{C}}.$$

We can think of $\mathrm{map}_{\mathcal{C}}(X, Y)$ as a "mapping space" as it is a space (simplicial set) consisting of maps from X to Y.

It would be desirable to have similar structures when $\mathcal{C}$ is not simplicial, at least at the level of homotopy categories. We will see in this chapter that for any model category $\mathcal{C}$, $\mathrm{Ho}(\mathcal{C})$ is a closed $\mathrm{Ho}(\mathrm{sSet})$–module. In other words, we can always achieve well-behaved homotopy mapping objects. This technique is known as *framings*. We give an outline of the constructions, results and properties using [Hov99, Chapter 5], [Hir03, Chapters 16 and 17] and [BR11]. We will furthermore see that for a *stable* model category $\mathcal{C}$, $\mathrm{Ho}(\mathcal{C})$ becomes a closed module over the stable homotopy category [Len12]. In particular, we obtain mapping spectra.

In [Dug01], Dugger creates a Quillen equivalence between any left proper and cellular (or combinatorial) model category and a simplicial model category. While these assumptions on the model category are often satisfied in practice, they are non-trivial nonetheless, whereas the method of framings we present in this chapter requires no such assumptions at all. Furthermore, the concept of framings classifies *all* Quillen adjunctions between simplicial sets and an arbitrary model category $\mathcal{C}$ up to homotopy, making simplicial sets "initial" among model categories.

The general idea of framings is the following. If $\mathcal{C}$ is not simplicial, then for $A, B \in \mathcal{C}$, the set $\mathcal{C}(A, B)$ is not necessarily a simplicial set in any meaningful way. However, we can "include" A and B into a bigger model category that is simplicial. We use the enrichment in this category to obtain a simplicial mapping object.

Recall the category Δ, whose objects are finite ordinals $[n] = \{0, 1, \cdots, n\}$, and whose morphisms are the order-preserving maps.

Definition 6.9.1 Let $\mathcal{C}$ be a category. The category of *cosimplicial objects* $\mathcal{C}^{\Delta}$ *in* $\mathcal{C}$ is defined as the category of functors

$$\Delta \longrightarrow \mathcal{C}.$$

Dually, the category of *simplicial objects* $\mathcal{C}^{\Delta^{op}}$ *in* $\mathcal{C}$ is defined as the category of functors

$$\Delta^{op} \longrightarrow \mathcal{C}.$$

Analogously to (co)simplicial sets, this means that a cosimplicial object $A^\bullet \in \mathcal{C}^{\Delta}$ (respectively, a simplicial object $B_\bullet \in \mathcal{C}^{\Delta^{op}}$) consists of an object A^n (resp. B_n) for each $n \in \mathbb{N}$ and structure maps (face and degeneracy maps) between those levels. Given $A^\bullet \in \mathcal{C}^{\Delta}$ and $B_\bullet \in \mathcal{C}^{\Delta^{op}}$, we write $A^\bullet([n]) = A^n$ and $B_\bullet([n]) = B_n$.

The category $\mathcal{C}^{\Delta}$ is very useful for describing adjunctions between simplicial sets sSet and $\mathcal{C}$, see [Hov99, Proposition 3.1.5].

Lemma 6.9.2 *Let $\mathcal{C}$ be a category containing all colimits. Then the category of cosimplicial objects $\mathcal{C}^{\Delta}$ is equivalent to the category of adjunctions*

$$F\colon \mathrm{sSet} \rightleftarrows \mathcal{C} : G.$$

Proof Given an adjunction

$$F\colon \mathrm{sSet} \rightleftarrows \mathcal{C} : G,$$

we obtain a cosimplicial object $X^\bullet$ in the following way. Let $\Delta[n] \in \mathrm{sSet}$ denote the standard n-simplex, that is, the functor given by

$$\Delta[n]\colon [k] \mapsto \Delta([k], [n]).$$

Then the composite

$$\Delta \longrightarrow \mathrm{sSet} \xrightarrow{F} \mathcal{C}$$

is the object of $X^\bullet$ of $\mathcal{C}^\Delta$, where the first functor is given by $[n] \mapsto \Delta[n]$, so $X^n = F(\Delta[n])$.

Conversely, given $X^\bullet \in \mathcal{C}^\Delta$, we would like to construct an adjunction

$$F_X\colon \mathrm{sSet} \rightleftarrows \mathcal{C} : G_X.$$

Let $K \in \mathrm{sSet}$. By ΔK, we denote the category of simplices in K: the objects are maps of simplicial sets

$$\Delta[n] \longrightarrow K,$$

and the morphisms are induced by the maps $[k] \longrightarrow [n]$ such that the resulting triangle commutes. The functor

$$\Delta K \longrightarrow \Delta, \quad (\Delta[n] \longrightarrow K) \mapsto [n]$$

induces a restriction functor $\mathcal{C}^\Delta \xrightarrow{\mathrm{res}} \mathcal{C}^{\Delta K}$, which we compose with the colimit over ΔK to obtain

$$F_X\colon \mathcal{C}^\Delta \xrightarrow{\mathrm{res}} \mathcal{C}^{\Delta K} \xrightarrow{\mathrm{colim}_{\Delta K}} \mathcal{C}, \quad \text{so } F_X(K) := (\operatorname*{colim}_{\Delta K} \circ \mathrm{res})(X^\bullet).$$

Because a morphism of simplicial sets $K \longrightarrow L$ induces

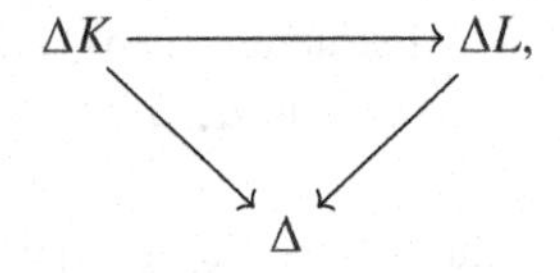

F_X is indeed functorial in K.

Before we examine the right adjoint of this functor F_X, we note that $\Delta(\Delta[n])$ has a cofinal object (namely, the identity of $[n]$), therefore the above construction gives

$$F_X(\Delta[n]) = X^n.$$

The right adjoint of F_X is given by

$$\mathcal{C}(X^\bullet, -)\colon \mathcal{C} \longrightarrow \mathrm{sSet}.$$

The nth level of the simplicial set $\mathcal{C}(X^\bullet, Y)$ is given by

$$\mathcal{C}(X^\bullet, Y)_n = \mathcal{C}(X^n, Y),$$

and the structure maps of $X^\bullet \in \mathcal{C}^\Delta$ induce the structure maps of $\mathcal{C}(X^\bullet, Y)_\bullet$ in sSet. The adjunction isomorphism is given by

$$\begin{aligned}\mathcal{C}(F_X(K), Y) &\cong \mathcal{C}(\operatorname*{colim}_{\Delta K} X^\bullet, Y)\\ &\cong \lim_{\Delta K} \mathcal{C}(X^\bullet, Y)\\ &\cong \lim_{\Delta K} \mathrm{sSet}(\Delta[-], \mathcal{C}(X^\bullet, Y))\\ &\cong \mathrm{sSet}(\operatorname*{colim}_{\Delta K} \Delta[-], \mathcal{C}(X^\bullet, Y))\\ &= \mathrm{sSet}(K, \mathcal{C}(X^\bullet, Y)).\end{aligned}$$

We now show that we have an equivalence of categories between adjunctions and $\mathcal{C}^\Delta$. Given a left adjoint $F\colon \mathrm{sSet} \longrightarrow \mathcal{C}$, the corresponding cosimplicial object is given by

$$X^n = F(\Delta[n]).$$

Constructing the functor $F_X\colon \mathrm{sSet} \longrightarrow \mathcal{C}$ out of $X^\bullet$ as before, we see that $F_X = F$, as both functors agree on $\Delta[n]$ and commute with colimits. Conversely, given $X^\bullet$, we construct F_X. This satisfies $F_X(\Delta[n]) = X^n$, therefore the cosimplicial object arising from F_X is again $X^\bullet$. This concludes our proof. □

We have seen in the proof of Lemma 6.9.2 that for $X^\bullet \in \mathcal{C}$, the thus constructed

$$F_X\colon \mathrm{sSet} \longrightarrow \mathcal{C}$$

has the following properties.

- $F_X(\Delta[n]) = X^n$.
- The right adjoint of F_X is $\mathcal{C}(X^\bullet, -)\colon \mathcal{C} \longrightarrow \mathrm{sSet}$.

The standard notation for $F_X(-)$ is $X^\bullet \otimes -$ as this functor will be the basis for defining our tensor (at the level of homotopy categories) of simplicial sets with $\mathcal{C}$.

Remark 6.9.3 An analogous proof shows that there is an equivalence between the category of simplicial objects $\mathcal{C}^{\Delta^{op}}$ and adjunctions

$$\mathrm{sSet}^{op} \underset{\longrightarrow}{\longleftarrow} \mathcal{C}.$$

Now that we have a convenient description of adjunctions between simplicial sets and a category $\mathcal{C}$, we can ask how to identify those adjunctions that are Quillen adjunctions in the case of $\mathcal{C}$ being a model category. In order to do so, we need to introduce a model structure in $\mathcal{C}^\Delta$.

Definition 6.9.4 A *Reedy category* is a small category I together with a function $d\colon \mathrm{Ob}(I) \longrightarrow \lambda$ for an ordinal λ, called a "degree function". Furthermore, there are two subcategories I_+ and I_- of I satisfying the following. The category I_+ consists of those morphisms of I raising degree, and the category I_- consists of those morphisms lowering degree. In addition, every $f \in I$ can be factored uniquely as $f = f_+ \circ f_-$, where f_+ raises degree and f_- lowers degree.

Reedy categories are useful for defining model structures on functor categories $\mathcal{C}^I$. In this chapter, we would like to apply this to $\mathcal{C}^\Delta$. The category Δ is a Reedy category, where Δ_+ are the injections and Δ_- are the surjections. Similarly, Δ^{op} is also a Reedy category, with I_+ and I_- swapped.

Definition 6.9.5 Let $\mathcal{C}$ be a category and I a Reedy category. Then for $i \in I$ we define a functor

$$L_i\colon \mathcal{C}^I \longrightarrow \mathcal{C}^{I_+} \longrightarrow \mathcal{C}^{(I_+)_i} \overset{\text{colim}}{\longrightarrow} \mathcal{C}.$$

Here, the category $(I_+)_i$ consists of those morphisms of I_+ with codomain i, and the first two functors are the restrictions induced by the inclusions

$$(I_+)_i \subset I_+ \subset I.$$

For $X \in \mathcal{C}^I$, the object L_iX is called the *latching object* of X at i.

Similarly, we have the following.

Definition 6.9.6 Let $\mathcal{C}$ be a category and I a Reedy category. Then for $i \in I$ we define a functor

$$M_i\colon \mathcal{C}^I \longrightarrow \mathcal{C}^{I_-} \longrightarrow \mathcal{C}^{(I_-)^i} \overset{\lim}{\longrightarrow} \mathcal{C}.$$

Here, the category $(I_-)^i$ consists of those morphisms of I_- with domain i, and the first two functors are the restrictions. For $X \in \mathcal{C}^I$, the object M_iX is called the *matching object* of X at i.

As all functors in the definition of the latching object factor over the evaluation functor at i, we have natural transformation $L_iX \longrightarrow X_i$ for every $i \in I$, where X_i denotes the value of X at $i \in I$. Analogously, the constant functor gives us a natural transformation $X_i \longrightarrow M_iX$ for each $i \in I$. For a morphism $f\colon X \longrightarrow Y$ in $\mathcal{C}^I$, we obtain pushout and pullback diagrams

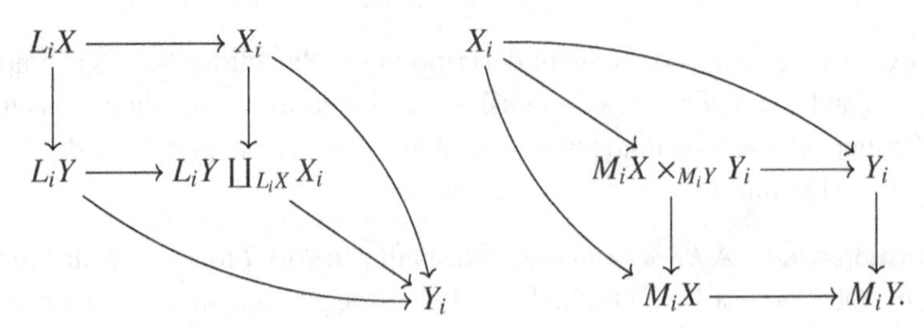

Now we have all the ingredients to describe a model structure on $\mathcal{C}^I$. The proof can be found, for example, in [Hov99, Theorem 5.2.5].

Theorem 6.9.7 *Let $\mathcal{C}$ be a model category and I a Reedy category. Then there is a model structure on $\mathcal{C}^I$ called the* Reedy model structure *with the following properties. A morphism $f: X \longrightarrow Y$ in $\mathcal{C}^I$ is a*

- *weak equivalence if $f_i: X_i \longrightarrow Y_i$ is a weak equivalence in $\mathcal{C}$ for all $i \in I$,*
- *cofibration if the map $L_iY \coprod_{L_iX} X_i \longrightarrow Y_i$ is a cofibration in $\mathcal{C}$ for all $i \in I$,*
- *fibration if the map $X_i \longrightarrow M_iX \times_{M_iY} Y_i$ is a fibration in $\mathcal{C}$ for all $i \in I$.*

Recall that we showed in Lemma 6.9.2 that for a category $\mathcal{C}$, the category of adjunctions

$$\mathrm{sSet} \rightleftarrows \mathcal{C}$$

is equivalent to the category $\mathcal{C}^\Delta$. If $\mathcal{C}$ is in addition a model category, we can now identify those adjunctions that are Quillen adjunctions.

Lemma 6.9.8 *Let $\mathcal{C}$ be a model category. An adjunction*

$$F: \mathrm{sSet} \rightleftarrows \mathcal{C} : G$$

is a Quillen adjunction if and only if the corresponding $X^\bullet \in \mathcal{C}^\Delta$ satisfies the following.

- *The map $L_nX^\bullet \longrightarrow X^n$ is a cofibration for all n.*
- *All structure maps in $X^\bullet$ are weak equivalences.*

In other words, $X^\bullet \in \mathcal{C}^\Delta$ is cofibrant as well as weakly constant.

Proof The category of simplicial sets sSet is cofibrantly generated (see Example A.6.8) with generating cofibrations

$$I = \{\partial\Delta[n] \longrightarrow \Delta[n] \mid n \in \mathbb{N}\}$$

and generating acyclic cofibrations

$$J = \{\Lambda^r[n] \longrightarrow \Delta[n] \mid 0 < n, 0 \leqslant r \leqslant n\}.$$

As F is a left adjoint, it commutes with colimits. Asking for F to be a left Quillen functor is equivalent to asking for F to send the elements of I to cofibrations in $\mathcal{C}$ and the elements of J to acyclic cofibrations in $\mathcal{C}$. The morphism

$$F(\partial\Delta[n]) \longrightarrow F(\Delta[n])$$

is identical to the map $L_nX^\bullet \longrightarrow X^n$ by [Hir03, Lemma 16.3.8]. By the characterisation of the Reedy model structure (see Theorem 6.9.7), asking for this map to be a cofibration for all n is equivalent to $X^\bullet$ being cofibrant in $\mathcal{C}^\Delta$.

If F is a left Quillen functor, then F sends the acyclic cofibrations

$$\Delta[0] \longrightarrow \Delta[n]$$

to acyclic cofibrations $X^0 \longrightarrow X^n$, which implies that all structure maps of $X^\bullet$ are weak equivalences.

Conversely, assume that all $X^0 \longrightarrow X^n$ are weak equivalences and that $X^\bullet$ is cofibrant. As stated at the beginning of the proof, the assumption that $X^\bullet$ is cofibrant means that F sends the generating cofibrations of sSet to cofibrations in $\mathcal{C}$ and therefore sends inclusions of simplicial sets to cofibrations. In particular, $X^0 \longrightarrow X^n$ is not just a weak equivalence, but also an acyclic cofibration, and

$$X^\bullet \otimes \Lambda^r[n] = F(\Lambda^r[n]) \longrightarrow F(\Delta[n]) = X^\bullet \otimes \Delta[n]$$

is a cofibration. We still need to show that it is also a weak equivalence. This is an inductive argument. By [Hir03, Lemma 16.4.10], there is a finite sequence of inclusions of simplicial sets

$$\Delta[0] = K_0 \longrightarrow K_1 \longrightarrow \cdots \longrightarrow K_m = \Lambda^r[n]$$

arising from pushouts

$$\begin{array}{ccc} \Lambda^{r_i}[n_i] & \longrightarrow & K_i \\ \downarrow & & \downarrow \\ \Delta[n_i] & \longrightarrow & K_{i+1}, \end{array}$$

for $i < r$ and $n_i < n$. We note that $\Lambda^0[1] \cong \Lambda^1[1] \cong \Delta[0]$, and that the functor $F = X^\bullet \otimes -$ commutes with pushouts. Thus, the assumption that $X^0 \longrightarrow X^1$ is a weak equivalence implies that $F(\Delta[0]) \longrightarrow F(\Lambda^i[2])$ is a weak equivalence, because this map arises as a finite sequence of pushouts of cofibrations. As the composite

$$F(\Delta[0]) \xrightarrow{\sim} F(\Lambda^i[2]) \longrightarrow F(\Delta[2])$$

is assumed to be a weak equivalence, the two-out-of-three axiom tells us that $F(\Lambda^i[2]) \longrightarrow F(\Delta[2])$ is one. Continuing this argument inductively completes our proof. □

We now have half of the structure we need.

Theorem 6.9.9 *Let $\mathcal{C}$ be a model category, and let $X \in \mathcal{C}$ a cofibrant object. Then there is a Quillen adjunction*

$$X \otimes -: \mathrm{sSet} \rightleftarrows \mathcal{C} : \mathrm{map}_l(X, -)$$

such that $X \otimes \Delta[0] \simeq X$.

Proof By Lemma 6.9.8, the statement is equivalent to creating a cosimplicial object in $\mathcal{C}$ which is weakly constant and whose zero degree object is weakly equivalent to X. Therefore, we can take a cofibrant replacement of the constant cosimplicial object on X. □

Again, we would like to stress that the notation $(X \otimes -, \mathrm{map}_l(X, -))$ is simply notation that is justified by some of the functors' properties. It is not implying that $\mathcal{C}$ is tensored or enriched over simplicial sets. We will discuss the difference later in this section and see the justification of the subscript in map_l.

Remark 6.9.10 The results outlined in this chapter so far also work in the pointed context. The left adjoint in the pointed analogue of Theorem 6.9.9 is usually denoted $X \wedge -$ and has the property that $X \wedge S^0 \cong X$. The pointed and unpointed functors are related via $X \wedge K_+ \cong X \otimes K$, where K is an unpointed simplicial set, and K_+ is K with a disjoint basepoint.

A very useful consequence of Theorem 6.9.9 is the following.

Corollary 6.9.11 *Let $\mathcal{C}$ be a model category, and let*

$$F, G\colon \mathrm{sSet} \longrightarrow \mathcal{C}$$

be left Quillen functors with $F(\Delta[0]) \cong G(\Delta[0])$. Then the left derived functors

$$\mathbb{L}F,\ \mathbb{L}G\colon \mathrm{Ho}(\mathrm{sSet}) \longrightarrow \mathrm{Ho}(\mathcal{C})$$

agree.

Proof Both F and G correspond to weakly constant cofibrant cosimplicial objects with the same object X in level zero. Any two such objects are weakly equivalent in $\mathcal{C}^{\Delta}$, as they are both weakly equivalent to the constant cosimplicial object on X. □

Example 6.9.12 Let $\mathcal{C}$ be a simplicial model category and $X \in \mathcal{C}$ be a cofibrant object. Then we can define the canonical cosimplicial object on X as

$$X^n := X \otimes \Delta[n],$$

where $\otimes$ is now the tensor that is part of the simplicial structure of $\mathcal{C}$. Thus, we see that the left Quillen functors

$$X \otimes -\colon \mathrm{sSet} \longrightarrow \mathcal{C}$$

given by either tensoring with X as part of the simplicial model category structure or by evaluating $X^\bullet$ induce the same left derived functors in the respective homotopy categories.

Remark 6.9.13 In [Hov99], the Quillen adjunction of Theorem 6.9.9 is assigned to an object in $\mathcal{C}$ *functorially*. This works because [Hov99] assumes that all model categories satisfy functorial factorisation, in particular, functorial cofibrant replacement. We do not assume this for this book. However, if $\mathcal{C}$ is cofibrantly generated, functorial cofibrant replacement holds by Corollary A.6.14. In this case, the resulting functor assigning a Quillen adjunction to an object in $\mathcal{C}$ is called a *framing functor*.

Note that sometimes the functor $X \mapsto X^\bullet$ is called a *cosimplicial frame*, and sometimes this term is used just for the object $X^\bullet$ itself.

We can apply this machinery to adjunctions

$$\mathrm{sSet}^{op} \rightleftarrows \mathcal{C}.$$

For a model category $\mathcal{C}$, one can identify which of those adjunctions are Quillen adjunctions using the model category $\mathcal{C}^{\Delta^{op}}$, rather than $\mathcal{C}^{\Delta}$ in Lemma 6.9.2. Furthermore, the analogue of Theorem 6.9.9 holds, namely, that for fibrant $X \in \mathcal{C}$ there is a Quillen adjunction

$$X^{(-)}\colon \mathrm{sSet}^{op} \rightleftarrows \mathcal{C} : \mathrm{map}_r(-, X)$$

with $X^{\Delta[0]} \cong X$. If we assume functorial factorisation for $\mathcal{C}$ (such as in the case of $\mathcal{C}$ being cofibrantly generated), these Quillen adjunctions can be assigned functorially, as before, which is called a *simplicial framing functor*.

Using both simplicial and cosimplicial framing functors together, we obtain bifunctors

$$\begin{array}{rcll} -\otimes- & : & \mathcal{C}\times \mathrm{sSet} \longrightarrow \mathcal{C}, & (A,K)\mapsto A^\bullet\otimes K \\ \mathrm{map}_l(-,-) & : & \mathcal{C}^{op}\times\mathcal{C} \longrightarrow \mathrm{sSet}, & (A,B)\mapsto \mathcal{C}(A^\bullet, B) \\ (-)^{(-)} & : & \mathrm{sSet}^{op}\times\mathcal{C} \longrightarrow \mathcal{C}, & (A,K)\mapsto A^K_\bullet \\ \mathrm{map}_r(-,-) & : & \mathcal{C}^{op}\times\mathcal{C} \longrightarrow \mathrm{sSet}, & (A,B)\mapsto \mathcal{C}(A, B_\bullet). \end{array}$$

The following is [Hov99, Theorem 5.4.9].

Theorem 6.9.14 *The bifunctors*

$$-\otimes-\colon \mathcal{C}\times\mathrm{sSet}\longrightarrow\mathcal{C} \ \ \textit{and}\ \ (-)^{(-)}\colon \mathrm{sSet}^{op}\times\mathcal{C}\longrightarrow\mathcal{C}$$

possess total left derived functors.

We mentioned before that in an ideal situation, the above bifunctors would satisfy the conditions of Definitions 6.1.25 and 6.1.27. Unfortunately, this fails in two significant aspects.

- The mapping spaces map_l and map_r do not generally agree – they only agree up to a zig-zag of weak equivalences [Hov99, Proposition 5.4.7].

- The functor $- \otimes -$ is not in general associative, that is, for an object $X \in \mathcal{C}$ and simplicial sets K and L, the objects $X \otimes (K \otimes L)$ and $(X \otimes K) \otimes L$ are not necessarily identical. In [Hov99, Theorem 5.5.3], Hovey provides an explicit associativity weak equivalence to relate them.

Thus, one can achieve the following.

Theorem 6.9.15 (Hovey) *Let $\mathcal{C}$ be a model category with functorial cofibrant replacement. Then* $\mathrm{Ho}(\mathcal{C})$ *is a closed* $\mathrm{Ho}(\mathrm{sSet})$ – *module category.*

We define

$$\mathrm{map}_{\mathcal{C}}(X, Y) = \mathbb{R}\,\mathrm{map}_l(X, Y) \cong \mathbb{R}\,\mathrm{map}_r(X, Y)$$

and call this the homotopy mapping space.

Note that this result also holds without the functorial factorisation discussed earlier. In the general case, there may not be a functorial way to assign to an $X \in \mathcal{C}$ a cofibrant, weakly constant $X^\bullet \in \mathcal{C}^\Delta$. However, any two such choices give the same result in the homotopy category, see Remark A.3.4.

Example 6.9.16 If $\mathcal{C}$ is a simplicial model category, then the simplicial structure makes $\mathrm{Ho}(\mathcal{C})$ into a closed $\mathrm{Ho}(\mathrm{sSet})$–module category by Theorem 6.1.29. We saw previously that we can also define a framing functor on $\mathcal{C}$ by sending a (cofibrant) $X \in \mathcal{C}$ to $X^\bullet$ with $X^n = X \otimes \Delta[n]$ in a way that is unique up to homotopy. Consequently, the $\mathrm{Ho}(\mathrm{sSet})$–module structure on $\mathrm{Ho}(\mathcal{C})$ from framings agrees with the $\mathrm{Ho}(\mathrm{sSet})$–module structure coming from the simplicial structure on $\mathcal{C}$.

In particular, the simplicial enrichment $\mathrm{hom}_{\mathcal{C}}(-,-)$ gives a suitable homotopy mapping space object, that is, if X is cofibrant and Y is fibrant,

$$\mathrm{map}_{\mathcal{C}}(X, Y) \simeq \mathrm{hom}_{\mathcal{C}}(X, Y).$$

In any case, our constructions give us the following.

- For an arbitrary model category $\mathcal{C}$, we have a very well-behaved notion of a homotopy mapping "space" functor.
- For any cofibrant object $X \in \mathcal{C}$, there is a left Quillen functor $F_X \colon \mathrm{sSet} \longrightarrow \mathcal{C}$ with $F_X(\Delta[0]) \cong X$. Thus, the model category of simplicial sets is initial among model categories in this sense.
- This left Quillen functor F_X is unique up to homotopy, meaning that any other left Quillen functor with the above properties has the same left derived functor.
- Therefore, a left Quillen functor $F \colon \mathrm{sSet} \longrightarrow \mathcal{C}$ is uniquely determined by $F(\Delta[0])$.

Lastly, we stress again that the results of this section all have pointed analogues. If $\mathcal{C}$ is a pointed model category, we replace sSet with sSet_* and see that a left Quillen functor $\mathrm{sSet}_* \longrightarrow \mathcal{C}$ is entirely determined by its value on the simplicial sphere S^0. This will be important in the next section.

We finish this subsection with three useful lemmas on homotopy mapping spaces and the tensor with simplicial sets in the case of a pointed model category. Recall that in the pointed case we write $\wedge$ instead of $\otimes$.

Lemma 6.9.17 *Let $\mathcal{C}$ be a pointed model category with X cofibrant and Y fibrant. Then in* $\mathrm{Ho}(\mathcal{C})$*, there are natural isomorphisms*

$$\Sigma X \cong S^1 \wedge X \qquad \Omega X \cong X^{S^1}.$$

Proof We may write S^1 as the pushout of the square

$$\begin{array}{ccc} \partial\Delta[1]_+ & \longrightarrow & \Delta[1]_+ \\ \downarrow & & \downarrow \\ * & \longrightarrow & S^1. \end{array}$$

Applying the left adjoint $X \wedge -$ gives a pushout square in $\mathcal{C}$. Since

$$X \wedge \partial\Delta[1]_+ \cong X \vee X$$

and the object $X \wedge \partial\Delta[1]_+$ is a cylinder object for X, the result follows.

The case of an Ω is dual. □

As a consequence, we obtain the following.

Lemma 6.9.18 *For X and Y in $\mathcal{C}$, there are natural isomorphisms in* $\mathrm{Ho}(\mathrm{sSet}_*)$

$$\mathrm{map}_{\mathcal{C}}(\Sigma X, Y) \cong \mathrm{map}_{\mathcal{C}}(X, \Omega Y) \cong \Omega\, \mathrm{map}_{\mathcal{C}}(X, Y).$$

Proof By the previous lemma, we can recognise Σ and Ω in terms of framings. The result follows from the associativity of the action of $\mathrm{Ho}(\mathrm{sSet}_*)$ on $\mathrm{Ho}(\mathcal{C})$. □

We can relate the homotopy groups of mapping spaces to maps in the homotopy category. In the following, we always take the basepoint of the homotopy groups to be the element arising from the zero map.

Lemma 6.9.19 *For X and Y in $\mathcal{C}$, there is a natural isomorphism*

$$\pi_n(\mathrm{map}_{\mathcal{C}}(X, Y)) \cong [\Sigma^n X, Y]^{\mathcal{C}}.$$

If $\mathcal{C}$ is stable then

$$\pi_n(\mathrm{map}_{\mathcal{C}}(X, \Sigma^m Y)) \cong [X, Y]^{\mathcal{C}}_{n-m}.$$

Proof By the previous lemmas, we have isomorphisms

$$\pi_n(\mathrm{map}_{\mathcal{C}}(X,Y)) = [S^n, \mathrm{map}_{\mathcal{C}}(X,Y)]^{\mathrm{sSet}_*} \cong [X \wedge S^n, Y]^{\mathcal{C}} \cong [\Sigma^n X, Y]^{\mathcal{C}}. \qquad \square$$

6.9.2 Stable Framings

As our focus generally lies on stable homotopy theory, we would like results analogous to the previous section for *stable* model categories $\mathcal{C}$. Specifically, we will show that for stable $\mathcal{C}$, the homotopy category $\mathrm{Ho}(\mathcal{C})$ is a closed module category over the stable homotopy category $\mathcal{SHC}$. Furthermore, we will study and classify all Quillen adjunctions

$$\mathcal{S} \rightleftarrows \mathcal{C},$$

which will lead to the notion that spectra (specifically, sequential spectra in simplicial sets, denoted $\mathcal{S}$) are initial among stable model categories. We will be closely following work of Lenhardt [Len12] in this chapter for our outline.

In this section, all categories $\mathcal{C}$ are pointed, and we work with sequential spectra in simplicial sets (see Subsection 5.5.1). Thus, a spectrum X consists of a sequence of pointed simplicial sets X_n, $n \in \mathbb{N}$, together with structure maps

$$\sigma_n^X : \Sigma X_n \longrightarrow X_{n+1},$$

where Σ is smashing with $S^1 = \Delta[1]/\partial\Delta[1]$, the simplicial circle. For every $n \in \mathbb{N}$, there is an adjunction

$$\mathrm{F}_n^{\mathbb{N}} : \mathrm{sSet}_* \rightleftarrows \mathcal{S} : \mathrm{Ev}_n^{\mathbb{N}},$$

where $\mathrm{Ev}_n^{\mathbb{N}}$ is the evaluation of a spectrum X in degree n. The left adjoint $\mathrm{F}_n^{\mathbb{N}}$ is the shifted suspension functor from Example 2.1.2.

Again, we start by describing all adjunctions between spectra and an arbitrary category. This can be related to (unstable) framings from the previous section by pre-composing an adjunction (L, R) between spectra and a category $\mathcal{C}$ with $(\mathrm{Ev}_n^{\mathbb{N}}, \mathrm{F}_n^{\mathbb{N}})$,

$$L_n := L \circ \mathrm{F}_n^{\mathbb{N}} : \mathrm{sSet}_* \rightleftarrows \mathcal{S} \rightleftarrows \mathcal{C} : \mathrm{Ev}_n^{\mathbb{N}} \circ R =: R_n$$

to obtain an adjunction between pointed simplicial sets and $\mathcal{C}$, which is precisely the subject of Lemma 6.9.2. Therefore, for an adjunction

$$F : \mathcal{S} \rightleftarrows \mathcal{C} : G,$$

we obtain a cosimplicial object X_n for every $n \in \mathbb{N}$. These objects $X_n \in \mathcal{C}^\Delta$ are linked as follows. By definition of the shifted suspension spectrum, there is a natural transformation $\mathrm{F}_n^{\mathbb{N}} \circ \Sigma \longrightarrow \mathrm{F}_{n-1}^{\mathbb{N}}$. This gives us natural transformations

$$\tau_n \colon L_n \circ \Sigma \longrightarrow L_{n-1}.$$

Translating this back into cosimplicial objects $\mathcal{C}^\Delta$ with Lemma 6.9.2 gives us a morphism $\Sigma X_n \longrightarrow X_{n-1}$ in $\mathcal{C}^\Delta$. More precisely, X_n is $(L \circ \mathrm{F}_n^{\mathbb{N}})(S^0)$, and so $\Sigma X_n \longrightarrow X_{n-1}$ is the canonical map

$$\Sigma L(\mathrm{F}_n^{\mathbb{N}}(S^0)) \longrightarrow L(\mathrm{F}_{n-1}^{\mathbb{N}}(S^0)).$$

Here, the suspension of a cosimplicial object is obtained by tensoring with the simplicial circle at every level, using pointed framings. Altogether, there is a name for this type of object arising from the above discussion.

Definition 6.9.20 Let $\mathcal{C}$ be a pointed category. A Σ-*cospectrum* is a sequence of objects $X_n \in \mathcal{C}^\Delta$, $n \in \mathbb{N}$, together with structure maps

$$\sigma \colon \Sigma X_n \longrightarrow X_{n-1}$$

for $n \geq 1$. A morphism of Σ-cospectra is a morphism in $\mathcal{C}^\Delta$ at each level n commuting with the structure maps. The resulting category of Σ-cospectra is denoted $C^\Delta(\Sigma)$.

Thus, we have already shown the following.

Lemma 6.9.21 *Let $\mathcal{C}$ be a pointed category with all colimits. Then the category of adjunctions*

$$\mathbb{S} \rightleftarrows \mathcal{C}$$

is equivalent to the category of Σ-cospectra $C^\Delta(\Sigma)$ in $\mathcal{C}^\Delta$. □

For a Σ-cospectrum X, we denote the corresponding adjunction by

$$X \wedge - \colon \mathbb{S} \rightleftarrows \mathcal{C} : \mathrm{Map}(X, -).$$

This is justified as $X \wedge \mathbb{S} = X$, and the right adjoint in level n is given by the simplicial set $\mathcal{C}(X_n, -)$. The adjoint structure maps of this mapping spectrum are induced by the cospectrum structure maps

$$\mathcal{C}(X_{n-1}, Y) \longrightarrow \mathcal{C}(\Sigma X_n, Y) \cong \Omega\mathcal{C}(X_n, Y).$$

Now if $\mathcal{C}$ is not just a category, but a model category, we can seek to identify those adjunctions that are Quillen adjunctions. For this, we need a model structure on Σ-cospectra. The following is [Len12, Theorem 4.6]. We use the Reedy model structure on $\mathcal{C}^{\Delta}$.

Theorem 6.9.22 *The category of* Σ*-cospectra* $C^{\Delta}(\Sigma)$ *admits a model structure called the* level model structure *with the following properties.*

- *A morphism* $f\colon X \longrightarrow Y$ *is a weak equivalence if* $f_n\colon X_n \longrightarrow Y_n$ *is a weak equivalence in* $\mathcal{C}^{\Delta}$ *for every* $n \in \mathbb{N}$.
- *A morphism* $f\colon X \longrightarrow Y$ *is a (trivial) cofibration if* $f_n\colon X_n \longrightarrow Y_n$ *is a (trivial) cofibration in* $\mathcal{C}^{\Delta}$ *for every* $n \in \mathbb{N}$.
- *A morphism* $f\colon X \longrightarrow Y$ *is a (trivial) fibration if* $f_0\colon X_0 \longrightarrow Y_0$ *is a (trivial) fibration in* $\mathcal{C}^{\Delta}$*, and*

$$X_n \longrightarrow Y_n \times_{\Omega Y_{n-1}} \Omega X_{n-1}$$

is a (trivial) fibration in $\mathcal{C}^{\Delta}$ *for all* $n \geqslant 1$.

Before we get to describe the Quillen adjunctions between sequential spectra and other stable model categories, we need the following.

Lemma 6.9.23 *Let* $\mathcal{C}$ *be a model category and* $G\colon \mathcal{C} \longrightarrow \mathcal{S}$ *be a functor. If* G *satisfies the following points*

1. *G preserves acyclic fibrations,*
2. *$G(Y)$ is an Ω-spectrum for all fibrant $Y \in \mathcal{C}$,*
3. *G sends fibrations between fibrant objects to level fibrations,*

then G *is a right Quillen functor.*

Proof We use Lemma A.4.3, which says that G is a right Quillen functor if and only if it preserves acyclic fibrations and fibrations between fibrant objects. Therefore, all that is left to do for us is to show that level fibrations between fibrant spectra are fibrations. Let $f\colon A \longrightarrow B$ be a level fibration between Ω-spectra. This means we have commutative squares (which are also homotopy pullback squares) for each $n \in \mathbb{N}$

$$\begin{array}{ccc} A_n & \twoheadrightarrow & B_n \\ \sim\downarrow & & \downarrow\sim \\ \Omega A_{n+1} & \twoheadrightarrow & \Omega B_{n+1}. \end{array}$$

By Proposition 2.3.10, this implies that f is a fibration in sequential spectra, which is what we wanted to prove. □

With this in mind, we are now in a position to describe Quillen adjunctions.

Lemma 6.9.24 *Let X be a Σ-cospectrum. Then the corresponding adjunction*

$$X \wedge -: \mathcal{S} \rightleftarrows \mathcal{C} : \mathrm{Map}(X, -)$$

is a Quillen adjunction if and only if all structure maps $\Sigma X_n \longrightarrow X_{n-1}$ of X are weak equivalences and all levels X_n are weakly constant and cofibrant in $\mathcal{C}^\Delta$.

Proof We prove that $\mathrm{Map}(X, -)$ satisfies the three points of Lemma 6.9.23:

1. $\mathrm{Map}(X, -)$ preserves acyclic fibrations,
2. $\mathrm{Map}(X, Y)$ is an Ω-spectrum for all fibrant $Y \in \mathcal{C}$,
3. $\mathrm{Map}(X, -)$ sends fibrations between fibrant objects to level fibrations.

We have that $\mathrm{Map}(X, Y)_n = \mathcal{C}(X_n, Y)$ with adjoint structure maps

$$\mathcal{C}(X_{n-1}, Y) \longrightarrow \mathcal{C}(\Sigma X_n, Y) \cong \Omega\mathcal{C}(X_n, Y).$$

Thus, the three points above are equivalent to the following:

1. $\mathcal{C}(X_n, -): \mathcal{C} \longrightarrow \mathrm{sSet}_*$ preserves acyclic fibrations for all n,
2. $\mathcal{C}(X_{n-1}, Y) \longrightarrow \Omega\mathcal{C}(X_n, Y)$ is a weak equivalence for all fibrant $Y \in \mathcal{C}$ and all n,
3. $\mathcal{C}(X_n, -)$ sends fibrations between fibrant objects to fibrations for all n.

Those three points are again equivalent to the following:

- $\mathcal{C}(X_n, -): \mathcal{C} \longrightarrow \mathrm{sSet}_*$ is a right Quillen functor for every n,
- $\Sigma X_n \longrightarrow X_{n-1}$ is a weak equivalence in $\mathcal{C}^\Delta$ for all n,

which is precisely our claim. □

Theorem 6.9.25 *Let $\mathcal{C}$ be a stable model category. Then for any fibrant and cofibrant $A \in \mathcal{C}$, there is a Quillen adjunction*

$$L: \mathcal{S} \rightleftarrows \mathcal{C} : R$$

with $L(\mathbb{S}) \simeq A$.

Proof We begin with a cosimplicial frame X_0 corresponding to

$$L \circ \mathrm{F}_0^{\mathbb{N}}: \mathrm{sSet}_* \longrightarrow \mathcal{C}.$$

We can take this to be a cofibrant replacement of a constant cosimplicial object on A. We furthermore take a fibrant replacement of this cosimplicial object so that we can use the following result of Schwede and Shipley

[SS02, Lemma 6.4]: If $Y \in \mathcal{C}^\Delta$ is fibrant and $\mathcal{C}$ is stable, then there is an $X \in \mathcal{C}^\Delta$ which is cofibrant and weakly constant such that

$$\Sigma X \longrightarrow Y$$

is a weak equivalence in $\mathcal{C}^\Delta$. Continuing this inductively gives us all the X_n to obtain a Σ-cospectrum that is a cosimplicial frame on every level and whose structure maps are weak equivalences. □

Theorem 6.9.25 gives us Quillen adjunctions $(X \wedge -, \mathrm{Map}(X, -))$ for every X. We would like to see what happens if we vary X.

Lemma 6.9.26 *Let $\mathcal{C}$ be a model category and $B \in \mathcal{S}$ be a cofibrant spectrum. Then the functor*

$$- \wedge B\colon \mathcal{C}^\Delta(\Sigma) \longrightarrow \mathcal{C}$$

is a left Quillen functor.

Proof We show that if $f\colon X \longrightarrow Y$ is a cofibration in Σ-cospectra and

$$g\colon A \longrightarrow B$$

is a cofibration in spectra, then the pushout product of f and g

$$X \wedge B \coprod_{X \wedge A} Y \wedge A \longrightarrow Y \wedge B$$

is a cofibration which is trivial if f is as well. It is enough to show the above for g being the generating cofibrations of sequential spectra in sSet_*

$$\mathrm{F}_m^{\mathbb{N}}\, \partial\Delta[n]_+ \longrightarrow \mathrm{F}_m^{\mathbb{N}}\, \Delta[n]_+,$$

see Subsection 5.5.1. Using that $(X \wedge -) \circ \mathrm{F}_m^{\mathbb{N}} = X_m$, the pushout product becomes

$$X_m^n \coprod_{X_m \wedge \partial\Delta[n]_+} (Y_m \wedge \partial\Delta[n]_+) \longrightarrow Y_m^n.$$

We have assumed that $f\colon X \longrightarrow Y$ was a (trivial) cofibration in Σ-cospectra. By Theorem 6.9.22, this means that $f_m\colon X_m \longrightarrow Y_m$ is a (trivial) cofibration in $\mathcal{C}^\Delta$, which by the Reedy model structure is precisely asking for the above pushout product map to be a (trivial) cofibration. Our claim now follows from taking $g\colon * \longrightarrow B$. □

Let X be an object of a stable model category $\mathcal{C}$. By ωX, we denote a Σ-cospectrum corresponding to a left Quillen functor $L\colon \mathcal{S} \longrightarrow \mathcal{C}$ sending the sphere spectrum to X. We call ωX a *stable framing* on X. Note that we say "a Σ-cospectrum" and "a Quillen functor" rather than "the Σ-cospectrum" and "the Quillen functor". This is because so far we have no reason to assume that either is unique in any sense. However, with a lot of attention to technical detail this issue can be resolved. We will not prove this theorem here, as it goes beyond the scope of this chapter.

Theorem 6.9.27 (Lenhardt) *Let $\mathcal{C}$ be a stable model category, $X \in \mathcal{C}$ cofibrant, and $Y \in \mathcal{C}$ fibrant and cofibrant. Let $\omega X \in \mathcal{C}^{\Delta}(\Sigma)$ be a stable framing on X, and let $\omega Y \in \mathcal{C}^{\Delta}(\Sigma)$ be a fibrant stable frame on Y.*

- *Any morphism $f\colon X \longrightarrow Y$ in $\mathcal{C}$ extends non-uniquely to a morphism*
$$F\colon \omega X \longrightarrow \omega Y$$
and therefore corresponds to a natural transformation
$$(X \wedge -) \longrightarrow (Y \wedge -)$$
such that the map
$$X \wedge \mathbb{S} \cong X \longrightarrow Y \wedge \mathbb{S} \cong Y$$
is again f.
- *If $f'\colon X \longrightarrow Y$ is homotopic to f as before, then the thus constructed*
$$F'\colon \omega X \longrightarrow \omega Y$$
is homotopic to F, and the induced derived natural transformations
$$(X \wedge^{\mathbb{L}} -) \longrightarrow (Y \wedge^{\mathbb{L}} -)$$
agree.
- *If $f\colon X \longrightarrow Y$ is a weak equivalence in $\mathcal{C}$, then the map*
$$F\colon \omega X \longrightarrow \omega Y$$
from the first point is a weak equivalence in $\mathcal{C}^{\Delta}(\Sigma)$.
- *Evaluation at level 0 and degree 0 induces an equivalence of categories*
$$\mathrm{Ho}(\mathcal{C}^{\Delta}(\Sigma)_{fr}) \longrightarrow \mathrm{Ho}(\mathcal{C}),$$
where $\mathrm{Ho}(\mathcal{C}^{\Delta}(\Sigma)_{fr})$ is the full subcategory of $\mathrm{Ho}(\mathcal{C}^{\Delta}(\Sigma))$ on those objects that are stable frames.

Because of the subtleties arising in the proof of this theorem, it would not be reasonable to aim for the four bifunctors between spectra and $\mathcal{C}$ we saw in the unstable case. However, the previous theorem tells us that whilst ω is not a functor from $\mathcal{C}$ to Σ-cospectra in $\mathcal{C}$, it at least gives us a functor
$$\omega\colon \mathrm{Ho}(\mathcal{C}) \longrightarrow \mathrm{Ho}(\mathcal{C}^{\Delta}(\Sigma)_{fr}).$$

The theorem also tells us that we have a bifunctor

$$\mathrm{Ho}(\mathcal{C}^{\Delta}(\Sigma)_{fr}) \times \mathcal{SHC} \longrightarrow \mathrm{Ho}(\mathcal{C}).$$

Combining this with the equivalence

$$\mathrm{Ho}(\mathcal{C}^{\Delta}(\Sigma)_{fr}) \longrightarrow \mathrm{Ho}(\mathcal{C})$$

in the last point of the theorem gives us a bifunctor

$$- \wedge^{\mathbb{L}} -\colon \mathrm{Ho}(\mathcal{C}) \times \mathcal{SHC} \longrightarrow \mathrm{Ho}(\mathcal{C}).$$

Theorem 6.9.28 (Lenhardt) *Let $\mathcal{C}$ be a stable model category. The bifunctor*

$$- \wedge^{\mathbb{L}} -\colon \mathrm{Ho}(\mathcal{C}) \times \mathcal{SHC} \longrightarrow \mathrm{Ho}(\mathcal{C})$$

makes $\mathrm{Ho}(\mathcal{C})$ *into a closed module over the stable homotopy category.*

In particular, for a stable model category, we have the nicely behaved homotopy mapping spectra which we will use in several places throughout this book.

Since symmetric spectra (in simplicial sets) is a symmetric monoidal model category, we can consider categories enriched (tensored and cotensored) over symmetric spectra.

Definition 6.9.29 Let $\mathcal{S}^{\Sigma}(\mathrm{sSet}_*)$ be the category of symmetric spectra in simplicial sets. A $\mathcal{S}^{\Sigma}(\mathrm{sSet}_*)$–model category is called a *spectral model category.*

Combining the previous two theorems gives the following result.

Corollary 6.9.30 *Let $\mathcal{C}$ be a spectral model category. Then the $\mathcal{SHC}$–module structure on* $\mathrm{Ho}(\mathcal{C})$ *from the derived spectral structure on $\mathcal{C}$ agrees with the $\mathcal{SHC}$–module structure from Theorem 6.9.28.* □

Additionally, stable frames induce another very useful piece of structure on a stable model category $\mathcal{C}$: using $- \wedge^{\mathbb{L}} -$, the stable homotopy groups of spheres $\pi_*(\mathbb{S}) = [\mathbb{S}, \mathbb{S}]_*$ act on morphisms in $\mathrm{Ho}(\mathcal{C})$ via

$$\pi_n(\mathbb{S}) \otimes [A, B]^{\mathcal{C}}_k \longrightarrow [A, B]^{\mathcal{C}}_{k+n}, \quad \alpha \otimes f = f \wedge^{\mathbb{L}} \alpha.$$

Now let $\mathcal{C}$ and $\mathcal{D}$ be stable model categories, and let $F\colon \mathcal{C} \longrightarrow \mathcal{D}$ be a left Quillen functor. By the uniqueness part of Theorem 6.9.27, we have a natural weak equivalence

$$F \circ (X \wedge -) \simeq F(X) \wedge -$$

for $X \in \mathcal{C}$. Applying this to the action of $\pi_*(\mathbb{S})$ yields that a left Quillen functor induces an exact functor

$$\mathbb{L}F\colon \mathrm{Ho}(\mathcal{C}) \longrightarrow \mathrm{Ho}(\mathcal{D}),$$

that is also "$\pi_*(\mathbb{S})$-linear" in the sense that we have a commutative diagram

$$\begin{array}{ccc} \pi_n(\mathbb{S}) \otimes [A,B]_k^{\mathcal{C}} & \longrightarrow & [A,B]_{n+k}^{\mathcal{C}} \\ \Big\downarrow {\scriptstyle \mathrm{Id}\otimes\mathbb{L}F} & & \Big\downarrow {\scriptstyle \mathbb{L}F} \\ \pi_n(\mathbb{S}) \otimes [\mathbb{L}F(A), \mathbb{L}F(B)]_k & \longrightarrow & [\mathbb{L}F(A), \mathbb{L}F(B)]_{n+k}^{\mathcal{C}}. \end{array}$$

This $\pi_*(\mathbb{S})$-linear structure can be useful in proving uniqueness results. For example, Schwede and Shipley apply it in [SS02, Theorem 3.2] to prove Margolis' unique conjecture.

Theorem 6.9.31 (Schwede–Shipley) *Let $\mathcal{T}$ be a symmetric monoidal triangulated category whose product $- \wedge -$ is exact in each variable. Furthermore, $\mathcal{T}$ should satisfy the following.*

- *$\mathcal{T}$ has infinite coproducts.*
- *The unit S of $- \wedge -$ is a compact generator of $\mathcal{T}$.*
- *There is an exact strong symmetric monoidal equivalence*

$$\Phi\colon \widehat{\mathcal{SW}}^f \longrightarrow \mathcal{T}^c,$$

where $\widehat{\mathcal{SW}}^f$ is the full subcategory of finite CW-complexes in $\widehat{\mathcal{SW}}$, and $\mathcal{T}^c$ is the full subcategory of compact objects in $\mathcal{T}$.

If there is a stable model category $\mathcal{C}$ and a $\pi_*(\mathbb{S})$-linear exact equivalence

$$\mathcal{T} \longrightarrow \mathrm{Ho}(\mathcal{C}),$$

then $\mathcal{C}$ is Quillen equivalent to $\mathcal{S}^{\mathbb{N}}$ and, in particular, $\mathcal{T} \cong \mathcal{SHC}$.

In other words, if there is a stable model category $\mathcal{C}$ whose homotopy category $\mathrm{Ho}(\mathcal{C})$ satisfies the bullet points of the theorem (which are the points from Margolis' original conjecture), then the $\pi_*(\mathbb{S})$-linear structure on $\mathrm{Ho}(\mathcal{C})$ is enough to prove that $\mathrm{Ho}(\mathcal{C})$ is in fact the stable homotopy category $\mathcal{SHC}$.

7

Left Bousfield Localisation

Bousfield localisation, or more specifically, *left* Bousfield localisation, is an established tool to formally add more weak equivalences to a model category. The most common setting is localisation of spaces or spectra with respect to a homology theory E_*: rather than the weak equivalences being π_*-isomorphisms, one constructs a model structure with the E_*-isomorphisms as the weak equivalences. As a consequence, the E_*-isomorphisms become strict isomorphisms in the corresponding homotopy category. Therefore, we can think of Bousfield localisation as a good formal framework for inverting maps in the homotopy category.

Typically, information is lost in this process, but some specific aspects may stand out clearer after localisation. We will see an example of this behaviour in the final section when we show that the p-local stable homotopy category has vast computational advances over working with the stable homotopy category itself. We will also see how Bousfield localisation can help us gain insight into the deeper structure of the stable homotopy category.

7.1 General Localisation Techniques

We start by giving some general definitions and examples before moving on to proving that, under some assumptions, we can form a "local" model structure from an existing one. When we work with stable model categories, we will see that the main existence result is a lot simpler to obtain than in the general case.

The aim of this localisation is to add weak equivalences to a given model structure. The first point to address is that adding a class of maps $\mathbb{W}$ to the weak equivalences of a model category $\mathcal{C}$ may make more maps than just $\mathbb{W}$ into weak equivalences.

Recall that $\mathrm{map}_{\mathcal{C}}(-,-)$ denotes the homotopy mapping space construction introduced in Section 6.9.

Definition 7.1.1 Let $\mathcal{C}$ be a model category and $\mathbb{W}$ be a class of maps.

- We say that an object $Z \in \mathcal{C}$ is $\mathbb{W}$-*local* if

$$\mathrm{map}_{\mathcal{C}}(f, Z)\colon\ \mathrm{map}_{\mathcal{C}}(B, Z) \longrightarrow \mathrm{map}_{\mathcal{C}}(A, Z)$$

  is a weak equivalence of simplicial sets for all $f\colon A \longrightarrow B$ in $\mathbb{W}$.
- A morphism $g\colon X \longrightarrow Y$ in $\mathcal{C}$ is a $\mathbb{W}$-*equivalence* if

$$\mathrm{map}_{\mathcal{C}}(g, Z)\colon\ \mathrm{map}_{\mathcal{C}}(Y, Z) \longrightarrow \mathrm{map}_{\mathcal{C}}(X, Z)$$

  is a weak equivalence of simplicial sets for all $\mathbb{W}$-local objects $Z \in \mathcal{C}$.
- An object $X \in \mathcal{C}$ is $\mathbb{W}$-*acyclic* if

$$\mathrm{map}_{\mathcal{C}}(X, Z) \simeq *$$

  for all $\mathbb{W}$-local objects Z.

We may think of this definition as giving a procedure to define the weak equivalences of $L_{\mathbb{W}}\mathcal{C}$ from $\mathbb{W}$. We start with a class of maps $\mathbb{W}$, from which we construct $\mathbb{W}$-local objects. We then use the $\mathbb{W}$-local objects to construct the $\mathbb{W}$-equivalences. In particular,

$$\mathbb{W} \subseteq \mathbb{W}\text{-equivalences}.$$

The process stabilises at this point, meaning that further steps will not change either the set of local objects or the set of equivalences. In other words,

$$(\mathbb{W}\text{-equivalences})\text{-equivalences} = \mathbb{W}\text{-equivalences}.$$

Some references including [Hir03] define a $\mathbb{W}$-local object to be a fibrant object of $\mathcal{C}$ satisfying the mapping space condition above. We find it more convenient to separate these two conditions.

We have the following lemma.

Lemma 7.1.2 *Every element of $\mathbb{W}$ and every weak equivalence in $\mathcal{C}$ is a $\mathbb{W}$-equivalence.*

If $g\colon Z \longrightarrow Z'$ is a weak equivalence in $\mathcal{C}$, then Z is $\mathbb{W}$-local if and only if Z' is $\mathbb{W}$-local.

A $\mathbb{W}$-equivalence $f\colon X \longrightarrow Y$ between $\mathbb{W}$-local objects is a weak equivalence of $\mathcal{C}$.

Proof The first two statements are a direct consequence of Definition 7.1.1. The third statement is a local version of the Whitehead Theorem and will be proved in a similar way to Proposition 1.2.3 and Lemma 5.3.13. Let

$$f\colon X \longrightarrow Y$$

be a $\mathbb{W}$-equivalence between $\mathbb{W}$-local objects. Then

$$f^*\colon \operatorname{map}_{\mathcal{C}}(Y, X) \longrightarrow \operatorname{map}_{\mathcal{C}}(X, X)$$

is a weak equivalence. In particular, there is a $g\colon Y \longrightarrow X$ with $g \circ f \cong \mathrm{Id}_X$ in $\mathrm{Ho}(\mathcal{C})$. We also have that

$$f^*\colon \operatorname{map}_{\mathcal{C}}(Y, Y) \longrightarrow \operatorname{map}_{\mathcal{C}}(X, Y)$$

is a weak equivalence which sends $f \circ g$ to $f \circ g \circ f$, which is isomorphic to f in $\mathrm{Ho}(\mathcal{C})$. This map also sends Id_Y to f, therefore $f \circ g \cong \mathrm{Id}_Y$ in $\mathrm{Ho}(\mathcal{C})$, meaning that f is a weak equivalence. □

The definitions also imply that only the equivalence class of $\mathbb{W}$ in the homotopy category of $\mathcal{C}$ matters. That is, if f and g are weakly equivalent maps (see Example A.7.2), then the maps $\operatorname{map}_{\mathcal{C}}(f, Z)$ and $\operatorname{map}_{\mathcal{C}}(g, Z)$ are weakly equivalent, so one is a weak equivalence if and only if the other is as well. Hence, one can replace $\mathbb{W}$ by a class of weakly equivalent maps. In particular, one may always assume that $\mathbb{W}$ consists of cofibrations between cofibrant objects.

We may now define left Bousfield localisations. The idea is to have a new model structure that focuses on the $\mathbb{W}$-local objects and $\mathbb{W}$-equivalences and has a suitable universal property, see Remark 7.1.7. Since the definition specifies a category, a class of cofibrations and a class of weak equivalences, we see that left Bousfield localisations are unique. Note that we do *not* claim that any given left Bousfield localisation exists.

Definition 7.1.3 Let $\mathcal{C}$ be a model category and $\mathbb{W}$ be a class of maps. The *left Bousfield localisation* of $\mathcal{C}$ is a model structure $L_{\mathbb{W}}\mathcal{C}$ on the same category as $\mathcal{C}$ with the weak equivalences given by the class of $\mathbb{W}$-equivalences and with the same cofibrations as $\mathcal{C}$.

The fibrations of $L_{\mathbb{W}}\mathcal{C}$ are usually quite complicated and often have no explicit characterisation outside of the right lifting property. However, in the case where $\mathcal{C}$ is left proper, we have a clear description of the fibrant objects of $L_{\mathbb{W}}\mathcal{C}$: they are the $\mathbb{W}$-local objects that are fibrant in $\mathcal{C}$ by [Hir03, Proposition 3.4.1].

This leads to the following definition.

Definition 7.1.4 Let $\mathcal{C}$ be a model category and $\mathbb{W}$ a class of maps. We say that a morphism $f: X \longrightarrow Y$ in $\mathcal{C}$ is a $\mathbb{W}$-*localisation* if

- the map f is a $\mathbb{W}$-equivalence,
- the object Y is $\mathbb{W}$-local.

In the above situation, Y satisfies the following universal property in $\mathrm{Ho}(\mathcal{C})$. If Z is another $\mathbb{W}$-local object and $g: X \longrightarrow Z$ is a map, Definition 7.1.1 implies that g uniquely factors over f in $\mathrm{Ho}(\mathcal{C})$, because

$$\mathrm{map}_{\mathcal{C}}(f, Z)\colon \mathrm{map}_{\mathcal{C}}(Y, Z) \longrightarrow \mathrm{map}_{\mathcal{C}}(X, Z)$$

is an isomorphism in $\mathrm{Ho}(\mathrm{sSet}_*)$. Therefore, it is common at this stage already to speak of Y as "the" $\mathbb{W}$-localisation of X and write $Y = L_{\mathbb{W}}X$, although we would like to specify $L_{\mathbb{W}}X$ further later in this chapter.

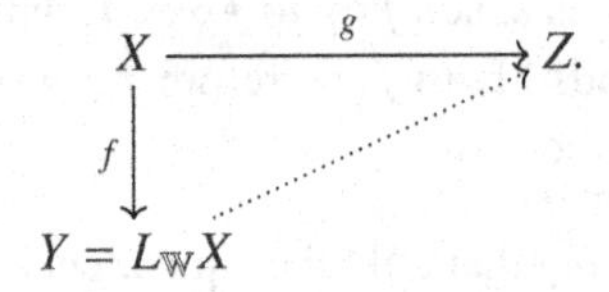

Example 7.1.5 The following example is known as *finite localisation* and was discussed by Miller in [Mil92]. This localisation can be described explicitly. Let $\mathcal{C}$ be a stable model category, and let $\mathbb{W}$ be a set of maps f in $\mathcal{C}$, where each f has a compact homotopy cofibre (see Definition 5.6.2). Let $\mathcal{A}$ denote the set of (cofibrant replacements of) these cofibres. Note that $\mathbb{W}$ is assumed to be a set rather than a class, but $\mathbb{W}$ does not have to be finite. Furthermore, assume that $\mathbb{W}$ and therefore $\mathcal{A}$ is closed under suspension and desuspension. The resulting localisation is often denoted by $L_{\mathbb{W}}X = L^f_{\mathcal{A}}X$ and called *finite* $\mathcal{A}$-*localisation* rather than $\mathbb{W}$-localisation. The choice of letter $\mathcal{A}$ refers to the fact that its elements are $\mathbb{W}$-acyclic.

Let X be a fibrant object of $\mathcal{C}$. We start with $X_0 := X$ and set

$$W_0 := \coprod_{A \in \mathcal{A}} \coprod_{f \in [A,X]^{\mathcal{C}}} A,$$

the coproduct over all the maps $A \longrightarrow X$ in $\mathrm{Ho}(\mathcal{C})$ and $A \in \mathcal{A}$. Choosing representatives for the maps in this coproduct, we obtain a map

$$W_0 \longrightarrow X_0$$

by sending the summand A corresponding to a representative $g: A \longrightarrow X$ to X via g. We now set X_1 to be the (fibrant replacement of the) cofibre of this map. Continuing, we set

$$W_1 := \coprod_{A \in \mathcal{A}} \coprod_{f \in [A,X_1]^{\mathcal{C}}} A,$$

which gives us a map

$$W_1 \longrightarrow X_1,$$

where we call the cofibre X_2 and so on. We thus obtain a cofibre sequence

$$W_i \longrightarrow X_i \xrightarrow{j_i} X_{i+1} \text{ for all } i \geqslant 0.$$

The maps j_i induce a map

$$\amalg X_i \xrightarrow{j} \amalg X_i.$$

We define X_∞ to be the homotopy cofibre of the map

$$\mathrm{Id} - j\colon \amalg X_i \longrightarrow \amalg X_i,$$

where Id denotes the identity of the coproduct. Our claim is that the canonical map

$$X \longrightarrow X_\infty$$

is a $\mathbb{W}$-localisation. For this, we have to show that

- the object X_∞ is $\mathbb{W}$-local,
- the map $X \longrightarrow X_\infty$ is a $\mathbb{W}$-equivalence.

For the first point, we check that $[A, X_\infty] = 0$ for all $A \in \mathcal{A}$. This is sufficient by Lemma 7.2.7 and by the long exact sequence of the cofibre sequence defining A, see Theorem 3.6.4.

Since X_∞ is defined as the cofibre of a map, there is a long exact sequence

$$\cdots \longrightarrow [A, \amalg X_i]_k \xrightarrow{(\mathrm{Id}-j)_*} [A, \amalg X_i]_k \longrightarrow [A, X_\infty]_k \longrightarrow \cdots.$$

Using compactness of A and injectivity of the maps $\mathrm{Id} - \oplus_i (j_i)_*$, the long exact sequence splits into short exact sequences

$$0 \longrightarrow \oplus_i [A, X_i]_k \xrightarrow{\mathrm{Id}-\oplus_i (j_i)_*} \oplus_i [A, X_i]_k \longrightarrow [A, X_\infty]_k \longrightarrow 0$$

for each $k \in \mathbb{Z}$, just as with Lemma 5.6.15. Hence,

$$[A, X_\infty]_*^{\mathcal{C}} \cong \mathrm{colim}_n [A, X_n]_*^{\mathcal{C}}.$$

Therefore, any map $f\colon A \longrightarrow X_\infty$ must factor over one of the X_n, that is,

$$f\colon A \xrightarrow{f_n} X_n \xrightarrow{i_n} X_\infty,$$

where i_n is the canonical map. By construction,

$$(j_n)_*\colon [A, X_n]^{\mathcal{C}} \longrightarrow [A, X_{n+1}]^{\mathcal{C}}$$

sends f_n to 0, so f must be trivial in the sequential colimit.

For the second point, we show that there is an exact triangle

$$\widetilde{W}_\infty \longrightarrow X \longrightarrow X_\infty \longrightarrow \Sigma\widetilde{W}_\infty$$

with $\widetilde{W}_\infty$ a $\mathbb{W}$-acyclic object. The long exact sequence of a cofibre will then imply that

$$[X_\infty, Z]_*^{\mathcal{C}} \longrightarrow [X, Z]_*^{\mathcal{C}}$$

is an isomorphism for all $\mathbb{W}$-local Z.

Since we assumed $\mathbb{W}$ to be closed under suspensions and desuspensions, we can use morphism sets in $\mathrm{Ho}(\mathcal{C})$ rather than homotopy mapping spaces to detect local objects, see Lemma 7.2.7. Hence, the above isomorphism will imply that $X \longrightarrow X_\infty$ is a $\mathbb{W}$-equivalence.

We have that

$$[A, Z]_*^{\mathcal{C}} = 0 \text{ for all } A \in \mathcal{A},$$

as all the A are acyclic and Z is $\mathbb{W}$-local. Therefore, we also have

$$[W_i, Z]_*^{\mathcal{C}} = 0 \text{ for all } i \geqslant 0,$$

as the W_i have been defined to be coproducts of the A.

Define $\widetilde{W}_i$ to be the fibre of $X = X_0 \longrightarrow X_i$. The octahedral axiom shows that $\widetilde{W}_2$ fits into an exact triangle with (suspensions of) W_1 and W_2. The Five Lemma gives that $\widetilde{W}_2$ is $\mathbb{W}$-acyclic (see Corollary 7.2.8). Continuing inductively, each $\widetilde{W}_i$ is $\mathbb{W}$-acyclic. The maps $j_i \colon X_i \longrightarrow X_{i+1}$ induce maps

$$h_i \colon \widetilde{W}_i \longrightarrow \widetilde{W}_{i+1}.$$

We can repeat our construction of X_∞ with the $\widetilde{W}_i$ and h_i to obtain $\widetilde{W}_\infty$. This gives the left-most column in the diagram below. The middle column below is another instance of this construction using the identity map of X. In this case, the cofibre is just X itself. The 3×3 Lemma, Lemma 4.1.13, implies that we have a commutative diagram with all rows and columns being part of exact triangles

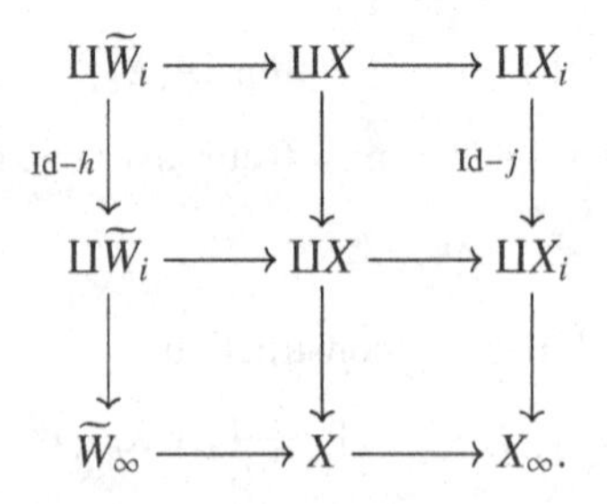

The object $\widetilde{W}_\infty$ is $\mathbb{W}$-acyclic, as it is part of an exact triangle with two other $\mathbb{W}$-acyclic objects. Moreover, $\widetilde{W}_\infty$ is constructed from the objects of $\mathcal{A}$ using (de)suspensions, coproducts and exact triangles.

The most commonly known existence result is the following [Hir03, Theorem 4.1.1]. Most notably, it requires $\mathbb{W}$ to be a *set* rather than a class. Recall that a cellular model category is a cofibrantly generated model category satisfying some additional conditions related to smallness, see [Hir03, Chapter 12].

Theorem 7.1.6 *Let $\mathcal{C}$ be a left proper, cellular model category, and let $\mathbb{W}$ be a set of morphisms in $\mathcal{C}$. Then there is a left proper, cellular model structure $L_{\mathbb{W}}\mathcal{C}$ on the category $\mathcal{C}$ such that*

- *a map $f\colon X \longrightarrow Y$ in $\mathcal{C}$ is a weak equivalence in $L_{\mathbb{W}}\mathcal{C}$ if and only if it is a $\mathbb{W}$-equivalence,*
- *a map $f\colon X \longrightarrow Y$ in $\mathcal{C}$ is a cofibration in $L_{\mathbb{W}}\mathcal{C}$ if and only if it is a cofibration in $\mathcal{C}$,*
- *the fibrant objects in $L_{\mathbb{W}}\mathcal{C}$ are precisely those objects in $\mathcal{C}$ that are fibrant in $\mathcal{C}$ and $\mathbb{W}$-local.*

In particular, the last point implies that fibrant replacement in the model structure $L_{\mathbb{W}}\mathcal{C}$ provides a $\mathbb{W}$-localisation. Another consequence is that the identity functor provides a Quillen adjunction

$$\mathrm{Id}\colon \mathcal{C} \rightleftarrows L_{\mathbb{W}}\mathcal{C} : \mathrm{Id},$$

which in turn induces an adjunction

$$\mathrm{Ho}(\mathcal{C}) \rightleftarrows \mathrm{Ho}(L_{\mathbb{W}}\mathcal{C}).$$

Hence, for X and Y in $\mathcal{C}$, there is a natural isomorphism

$$[X, L_{\mathbb{W}}Y]^{\mathcal{C}} \cong [X, Y]^{L_{\mathbb{W}}\mathcal{C}}.$$

The homotopy category of a model category $\mathcal{M}$ is equivalent to the full subcategory of cofibrant-fibrant objects in $\mathcal{M}$ and maps up to homotopy. Therefore, the above implies that the homotopy category of $L_{\mathbb{W}}\mathcal{C}$ is equivalent to the full subcategory of $\mathrm{Ho}(\mathcal{C})$ on the cofibrant-fibrant and $\mathbb{W}$-local objects in $\mathcal{C}$.

Remark 7.1.7 Our construction gives a *localisation functor* of categories in the following sense. Let

$$L\colon \mathrm{Ho}(\mathcal{C}) \longrightarrow \mathrm{Ho}(L_{\mathbb{W}}\mathcal{C})$$

be the left derived functor of the Quillen functor

$$\mathrm{Id}\colon \mathcal{C} \longrightarrow L_{\mathbb{W}}\mathcal{C}.$$

Then L takes the images of $\mathbb{W}$ in the homotopy category to isomorphisms in $\mathrm{Ho}(L_{\mathbb{W}}\mathcal{C})$ and furthermore satisfies the following universal property. Let $\mathcal{D}$ be another model category and

$$F\colon \mathcal{C} \longrightarrow \mathcal{D}$$

be a left Quillen functor such that the left derived functor

$$\mathbb{L}F\colon \mathrm{Ho}(\mathcal{C}) \longrightarrow \mathrm{Ho}(\mathcal{D})$$

takes the images of elements of $\mathbb{W}$ in $\mathrm{Ho}(\mathcal{C})$ to isomorphisms in $\mathrm{Ho}(\mathcal{D})$. Then F passes to a left Quillen functor

$$F'\colon L_{\mathbb{W}}\mathcal{C} \longrightarrow \mathcal{D}$$

(which on underlying categories is the same functor as F) making the diagram of Quillen functors below commute.

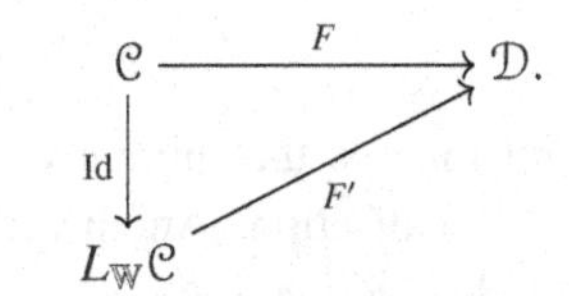

Further properties of $L_{\mathbb{W}}\mathcal{C}$ are given in [Hir03, Section 3.3], and further properties of $\mathrm{Ho}(L_{\mathbb{W}}\mathcal{C})$ in the stable setting are discussed in [HPS97]. As it is not our aim to provide a comprehensive reference for everything related to left Bousfield localisation, we will only touch on those things when we have a concrete need for them.

Remark 7.1.8 Note that the class of E_*-isomorphisms (or "E-equivalences") of spectra for a homology theory E_* does not form a set. However, it is possible (and highly non-trivial) to find a set J_E such that E_*-isomorphisms are precisely the J_E-equivalences, see Section 7.3. Therefore, we can apply Theorem 7.1.6 (or Theorem 7.2.17) to the set J_E to obtain a model structure on spectra where the weak equivalences are the E_*-isomorphisms.

In many cases, the main challenge to localising a model category lies in finding a set of maps, rather than a class of maps, which generates the desired equivalences that one would like to formally invert in the homotopy category.

Example 7.1.9 The stable model structures on $\mathcal{S}^{\mathbb{N}}$, $\mathcal{S}^{\Sigma}$ and $\mathcal{S}^{\mathcal{O}}$ are left Bousfield localisations of the levelwise model structures at the class of stable equivalences.

If we wanted to use Theorem 7.1.6 to construct the stable model structure, we would need to give a *set* of stable equivalences $\mathbb{W}$ such that the $\mathbb{W}$-equivalences are the class of all stable equivalences, similar to Remark 7.1.8. In this case, we can give an explicit set at which to localise. Choosing sequential spectra for definiteness, we claim that the stable model is the left Bousfield localisation of the levelwise model structure at the set of maps

$$\mathbb{W} = \{\lambda_n \colon \mathrm{F}^{\mathbb{N}}_{n+1} S^1 \longrightarrow \mathrm{F}^{\mathbb{N}}_n S^0 \mid n \in \mathbb{N}\}.$$

To see this, we note that we can use the enrichment over pointed topological spaces to give homotopy mapping objects, see Example 6.9.16. Since these shifted suspension spectra are cofibrant in the levelwise model structure and every object is fibrant, we see that X is $\mathbb{W}$-local if and only if the map

$$X_n \cong \mathcal{S}^{\mathbb{N}}(\mathrm{F}^{\mathbb{N}}_n S^0, X) \longrightarrow \mathcal{S}^{\mathbb{N}}(\mathrm{F}^{\mathbb{N}}_{n+1} S^1, X) \cong \Omega X_{n+1}$$

is a weak equivalence of spaces for all $n \in \mathbb{N}$. This means that the $\mathbb{W}$-local spectra are precisely the Ω-spectra, the fibrant objects of the stable model structure. The claim then follows.

7.2 Localisation of Stable Model Categories

It is not surprising that the proof of the existence result, Theorem 7.1.6 is very technical. However, a considerable amount of this complexity is removed if both the underlying model category is stable and the desired localisation is also "stable" in a certain sense. In this section, we examine how this stability can make localisations better behaved. The main result of this section is Theorem 7.2.17, a simpler existence result for left Bousfield localisations, which is a development of [BR13, Remark 4.12].

By Lemma 6.9.18, the homotopy mapping complexes interact with suspension and loop functors in the expected way. In particular,

$$\mathrm{map}_{\mathcal{C}}(\Sigma X, Y) \cong \mathrm{map}_{\mathcal{C}}(X, \Omega Y) \cong \Omega\, \mathrm{map}_{\mathcal{C}}(X, Y).$$

From this, we obtain the following.

Lemma 7.2.1 *Let $\mathcal{C}$ be a model category and $\mathbb{W}$ a class of maps in $\mathcal{C}$. Then the class of $\mathbb{W}$-local objects is closed under Ω, and the class of $\mathbb{W}$-equivalences is closed under Σ.*

Proof We have the following natural isomorphisms in $\mathrm{Ho}(\mathrm{sSet}_*)$

$$\Omega\, \mathrm{map}_{\mathcal{C}}(X, Z) \cong \mathrm{map}_{\mathcal{C}}(X, \Omega Z) \cong \mathrm{map}_{\mathcal{C}}(\Sigma X, Z).$$

Thus, if

$$\mathrm{map}_{\mathcal{C}}(f, Z)\colon\ \mathrm{map}_{\mathcal{C}}(B, Z) \longrightarrow \mathrm{map}_{\mathcal{C}}(A, Z)$$

is a weak equivalence for all $f\colon A \longrightarrow B$ in $\mathbb{W}$, then so is

$$\mathrm{map}_{\mathcal{C}}(f, \Omega Z)\colon\ \mathrm{map}_{\mathcal{C}}(B, \Omega Z) \longrightarrow \mathrm{map}_{\mathcal{C}}(A, \Omega Z).$$

Similarly, if

$$\mathrm{map}_{\mathcal{C}}(g, Z)\colon\ \mathrm{map}_{\mathcal{C}}(Y, Z) \longrightarrow \mathrm{map}_{\mathcal{C}}(X, Z)$$

is a weak equivalence of simplicial sets, then so is

$$\mathrm{map}_{\mathcal{C}}(\Sigma g, Z)\colon\ \mathrm{map}_{\mathcal{C}}(\Sigma Y, Z) \longrightarrow \mathrm{map}_{\mathcal{C}}(\Sigma X, Z). \qquad \square$$

If the class of $\mathbb{W}$-equivalences is not just closed under Σ, but also under Ω, we call a localisation *stable*.

Definition 7.2.2 Let $\mathcal{C}$ be a stable model category and $\mathbb{W}$ a class of maps in $\mathcal{C}$. We say that $\mathbb{W}$ is *stable* if one of the following two equivalent conditions hold.

- The class of $\mathbb{W}$-local objects is closed under Σ.
- The class of $\mathbb{W}$-equivalences is closed under Ω.

Stability of $\mathcal{C}$ together with the (Σ, Ω)–adjunction gives weak equivalences

$$\mathrm{map}_{\mathcal{C}}(\Omega X, Y) \simeq \mathrm{map}_{\mathcal{C}}(\Omega X, \Omega\Sigma Y) \simeq \mathrm{map}_{\mathcal{C}}(\Sigma\Omega X, \Sigma Y) \simeq \mathrm{map}_{\mathcal{C}}(X, \Sigma Y).$$

Hence, when $\mathbb{W}$ itself is closed under Ω (up to weak equivalence), an object $Y \in \mathcal{C}$ is $\mathbb{W}$-local if and only if ΣY is, in which case $\mathbb{W}$ is stable.

Lemma 7.2.3 *Let $\mathcal{C}$ be a stable model category, and let $\mathbb{W}$ be a class of maps such that the model category $L_{\mathbb{W}}\mathcal{C}$ exists. Then $L_{\mathbb{W}}\mathcal{C}$ is a stable model category if and only if $\mathbb{W}$ is a stable class of maps.*

Proof First, assume that $\mathbb{W}$ is stable, that is, the class of $\mathbb{W}$-local objects in $\mathrm{Ho}(L_{\mathbb{W}}\mathcal{C})$ is closed under suspension and desuspension. As mentioned earlier, the $\mathbb{W}$-local homotopy category $\mathrm{Ho}(L_{\mathbb{W}}\mathcal{C})$ is equivalent to the full subcategory of $\mathrm{Ho}(\mathcal{C})$ on its $\mathbb{W}$-local objects. Therefore, the adjunction

$$\Sigma\colon \mathrm{Ho}(\mathcal{C}) \rightleftarrows \mathrm{Ho}(\mathcal{C}) : \Omega$$

restricts to this full subcategory. The functor Ω on $\mathrm{Ho}(L_{\mathbb{W}}\mathcal{C})$ is fully faithful, as it is the restriction of such a functor. It is essentially surjective by stability of $\mathcal{C}$, hence, it is again an equivalence.

Now assume that $L_{\mathbb{W}}\mathcal{C}$ is stable and that $X \in \mathcal{C}$ is $\mathbb{W}$-local. Since Ω is an equivalence on $\mathrm{Ho}(L_{\mathbb{W}}\mathcal{C})$, there is a $\mathbb{W}$-local Y with $\Omega Y \cong X$ in $\mathrm{Ho}(L_{\mathbb{W}}\mathcal{C})$.

Since X and ΩY are $\mathbb{W}$-local, we see that $\Omega Y \cong X$ in $\mathrm{Ho}(\mathcal{C})$. By stability of $\mathcal{C}$, there are isomorphisms in $\mathrm{Ho}(\mathcal{C})$

$$Y \cong \Sigma\Omega Y \cong \Sigma X.$$

As Y is $\mathbb{W}$-local, so is ΣX. □

Example 7.2.4 The prime example of a stable localisation is localisation of spectra with respect to a homology theory. Let $\mathcal{S}$ denote a suitable category of spectra and E a (cofibrant) spectrum. Recall that an *E-equivalence* is defined to be an E_*-isomorphism, that is, a map of spectra that induces an isomorphism in the homology theory E_*. We want to localise $\mathcal{S}$ at the class

$$\mathbb{W} = E\text{-equivalences} = E_*\text{-isomorphisms}.$$

The model category $L_{\mathbb{W}}\mathcal{S}$ is more commonly called $L_E\mathcal{S}$. This choice of $\mathbb{W}$ is stable because for a morphism of spectra $f\colon X \longrightarrow Y$,

$$E_*(\Sigma f)\colon E_*(\Sigma X) \longrightarrow E_*(\Sigma Y)$$

is an isomorphism if and only if $E_*(f)$ is one. We will discuss further properties of this localisation in Sections 7.3 and 7.4.

In this case, we have that the class of $\mathbb{W}$-equivalences is exactly the class $\mathbb{W}$. That is, no further equivalences are introduced when we add the class $\mathbb{W}$ to the weak equivalences of $\mathcal{S}$. This follows as the class $\mathbb{W}$ is proven to be the J_E-equivalences for some set J_E, and the process of taking equivalences stabilises.

Example 7.2.5 The finite localisation discussed in Example 7.1.5 is stable by definition.

Example 7.2.6 Let $\mathcal{S}$ denote a suitable category of spectra, and let

$$f_k\colon \Sigma^\infty S^{k+1} \longrightarrow \Sigma^\infty D^{k+2}$$

denote the canonical inclusion of the $(k+1)$-sphere into the $(k+2)$–disc. By definition, a spectrum Z is f_k-local if and only if

$$\mathrm{map}_{\mathcal{S}}(f_k, Z)\colon\ \mathrm{map}_{\mathcal{S}}(\Sigma^\infty D^{k+2}, Z) \longrightarrow \mathrm{map}_{\mathcal{S}}(\Sigma^\infty S^{k+1}, Z)$$

is a weak equivalence of simplicial sets. As $\Sigma^\infty D^{k+2}$ is contractible, Z is f_k-local if and only

$$\pi_m(Z) = 0 \quad \text{for} \quad m \geqslant k+1.$$

Thus, for a spectrum X, its localisation

$$X \longrightarrow L_{f_k} X$$

is the kth Postnikov section of X. For f_k-local Z, we see that $\pi_m(\Sigma Z)$ is only guaranteed to be trivial for $m \geqslant k+2$. Therefore, the class of local objects is not closed under suspension, and thus $\mathbb{W} = \{f_k\}$ is not stable.

The definition of $\mathbb{W}$-local objects and $\mathbb{W}$-equivalences, Definition 7.1.1, is written for general model categories and uses the homotopy mapping space construction, $\mathrm{map}_{\mathcal{C}}(-,-)$. In the case of a stable model category and a stable localisation, an equivalent definition can be given in terms of $[-,-]_*^{\mathcal{C}}$. This formulation can often be more convenient.

Note that the statements of the next result still hold after replacing all graded morphism groups $[-,-]_*^{\mathcal{C}}$ with ungraded morphisms $[-,-]^{\mathcal{C}}$ because of the stability of $\mathbb{W}$.

Lemma 7.2.7 *Let $\mathcal{C}$ be a stable model category and $\mathbb{W}$ a stable class of morphisms in $\mathcal{C}$.*

- *An object $Z \in \mathcal{C}$ is $\mathbb{W}$-local if and only if*
$$[f,Z]_*^{\mathcal{C}} : [B,Z]_*^{\mathcal{C}} \longrightarrow [A,Z]_*^{\mathcal{C}}$$
is an isomorphism for all $f\colon A \longrightarrow B$ in $\mathbb{W}$.
- *A morphism $g\colon X \longrightarrow Y$ in $\mathcal{C}$ is a $\mathbb{W}$-equivalence if and only if*
$$[g,Z]_*^{\mathcal{C}} : [Y,Z]_*^{\mathcal{C}} \longrightarrow [X,Z]_*^{\mathcal{C}}$$
is an isomorphism for all $\mathbb{W}$-local objects $Z \in \mathcal{C}$.
- *An object $X \in \mathcal{C}$ is $\mathbb{W}$-acyclic if and only if*
$$[X,Z]_*^{\mathcal{C}} \simeq *$$
for all $\mathbb{W}$-local objects Z.

Proof We only show the first part as the others are very similar. By Lemma 6.9.19, for any $n,\ m \in \mathbb{N}$ we have
$$\pi_n(\mathrm{map}_{\mathcal{C}}(X,\Sigma^m Z)) \cong [X,Z]_{n-m}^{\mathcal{C}}$$
using the zero map as the basepoint. By the stability of $\mathcal{C}$ and $\mathbb{W}$, we have the "only if" direction.

For the converse, let $f\colon A \longrightarrow B$ in $\mathbb{W}$. Then Cf, the cofibre of f, is $\mathbb{W}$-acyclic. By Lemma 6.9.19,
$$0 = [Cf,Z]_n \cong \pi_n(\mathrm{map}_{\mathcal{C}}(Cf,Z))$$
for all $n \in \mathbb{N}$. As $\mathrm{map}_{\mathcal{C}}(Cf,Z)$ is weakly equivalent to the homotopy fibre of $\mathrm{map}_{\mathcal{C}}(f,Z)$, we see that
$$\pi_n(\mathrm{map}_{\mathcal{C}}(f,Z))\colon \pi_n(\mathrm{map}_{\mathcal{C}}(B,Z)) \longrightarrow \pi_n(\mathrm{map}_{\mathcal{C}}(A,Z))$$

is an isomorphism for all $n > 0$, using the zero map as the basepoint for the homotopy groups. Hence,

$$\mathrm{map}_{\mathcal{C}}(B, \Omega Z) \simeq \Omega\, \mathrm{map}_{\mathcal{C}}(B, Z) \longrightarrow \Omega\, \mathrm{map}_{\mathcal{C}}(A, Z) \simeq \mathrm{map}_{\mathcal{C}}(A, \Omega Z)$$

is a weak equivalence of pointed simplicial sets, and so ΩZ is $\mathbb{W}$-local. As $\mathbb{W}$ is stable, the local objects are closed under suspension. Since $\mathcal{C}$ is stable, we have the weak equivalence

$$\Sigma \Omega Z \xrightarrow{\sim} Z$$

which shows that Z is $\mathbb{W}$-local. □

Corollary 7.2.8 *Let $\mathcal{C}$ be a stable model category and $\mathbb{W}$ a stable class of morphisms. Furthermore, let*

$$A \longrightarrow B \longrightarrow C \longrightarrow \Sigma A$$

be an exact triangle in $\mathrm{Ho}(\mathcal{C})$. *If two out of the objects A, B and C are $\mathbb{W}$-local, then so is the third. If two out of the objects A, B and C are $\mathbb{W}$-acyclic, then so is the third.*

A map $X \longrightarrow Y$ is a $\mathbb{W}$-equivalence if and only if the homotopy cofibre (or homotopy fibre) is $\mathbb{W}$-acyclic. □

Corollary 7.2.9 *Let $\mathcal{C}$ be a stable model category, and let $\mathbb{W}$ be a stable class of morphisms such that $L_{\mathbb{W}}\mathcal{C}$ exists. For a diagram of $\mathbb{W}$-local objects*

$$X_1 \xleftarrow{f_1} X_2 \xleftarrow{f_2} X_3 \xleftarrow{f_3} \cdots$$

the homotopy limit $X = \mathrm{holim}_n X_n$ is $\mathbb{W}$-local.

Proof This follows from Corollary A.7.23 and Lemma 7.2.7. □

Stable localisations also interact well with pullbacks. Recall that the fibre of a map $f \colon X \longrightarrow Y$ in a model category $\mathcal{C}$ is the pullback of

$$X \xrightarrow{f} Y \longleftarrow *.$$

The homotopy fibre of f is the pullback of

$$Y' \xrightarrow{f} Y \longleftarrow *,$$

where Y' is given by factoring f into an acyclic cofibration and a fibration

$$X \overset{\sim}{\rightarrowtail} Y' \twoheadrightarrow Y.$$

If $\mathcal{C}$ is right proper and f is a fibration, then the fibre is weakly equivalent to the homotopy fibre Ff by Lemma A.7.20.

Lemma 7.2.10 *Let $\mathcal{C}$ be a stable model category, $\mathbb{W}$ be a stable set of maps and A, B, C and P be objects in $\mathcal{C}$. Furthermore, let*

$$\begin{array}{ccc} P & \xrightarrow{f} & A \\ {\scriptstyle i}\downarrow & & \downarrow{\scriptstyle j} \\ B & \xrightarrow[g]{} & C \end{array}$$

be a commutative square.

If the square is a homotopy pullback square and A, B and C are $\mathbb{W}$-local, then P is also $\mathbb{W}$-local.

If $\mathcal{C}$ is right proper, the square is a pullback square, g is a fibration and j is a $\mathbb{W}$-equivalence, then i is also a $\mathbb{W}$-equivalence.

Proof A homotopy pullback square gives rise to a long exact sequence

$$\cdots \longrightarrow [X, P]_n \longrightarrow [X, A]_n \oplus [X, B]_n \longrightarrow [X, C]_n \longrightarrow [X, P]_{n-1} \longrightarrow \cdots$$

for $X \in \mathcal{C}$ by Lemma A.7.21. If A, B and C are $\mathbb{W}$-local and X is $\mathbb{W}$-acyclic, then

$$[X, A]_n = [X, B]_n = [X, C]_n = 0$$

and thus also $[X, P]_n = 0$ for all n. This means that P is also $\mathbb{W}$-local as claimed.

For the second statement, as we have a pullback square, the fibres F_f and F_g of f and g are isomorphic. Moreover, the fibres are weakly equivalent to the homotopy fibres Ff and Fg.

This gives a map of homotopy fibre sequences in $\mathcal{C}$

$$\begin{array}{ccccc} Ff & \longrightarrow & P & \xrightarrow{f} & A \\ \downarrow{\scriptstyle \simeq} & & \downarrow{\scriptstyle i} & & \downarrow{\scriptstyle j} \\ Fg & \longrightarrow & B & \xrightarrow[g]{} & C. \end{array}$$

Applying $[-, Z]_*^{\mathcal{C}}$ for a $\mathbb{W}$-local object Z gives a map of long exact sequences. The Five Lemma shows that i is also a $\mathbb{W}$-equivalence. □

We now turn to our existence theorem for left Bousfield localisations in the stable context. We begin by constructing the maps which will be our generating acyclic cofibrations of the localised model structure.

Let $\mathcal{C}$ be a cofibrantly generated stable model category and $\mathbb{W}$ be a set of cofibrations between cofibrant objects. Let $f \colon X \longrightarrow Y$ be in $\mathbb{W}$ and

$i_n\colon \partial\Delta[n]_+ \to \Delta[n]_+$ be the standard inclusion of simplicial sets. The construction of framings from Section 6.9 gives us a bifunctor

$$- \wedge - \colon \mathcal{C} \times \mathrm{sSet}_* \longrightarrow \mathcal{C}.$$

In particular, f induces a natural transformation

$$f \wedge - \colon (X \wedge -) \longrightarrow (Y \wedge -)$$

of functors from pointed simplicial sets to $\mathcal{C}$ that is a cofibration in $\mathcal{C}^\Delta$. Then $f \,\square\, i_n$ is the indicated map out of the pushout

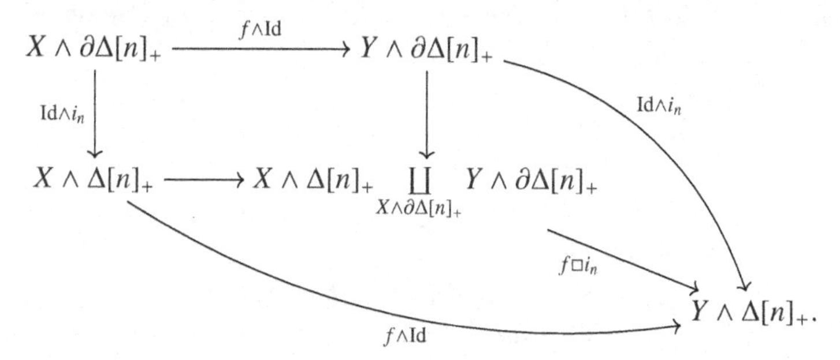

Recall from Lemma 6.9.8 that the map from the nth latching object of a cosimplicial frame on X to the nth level of that cosimplicial frame is exactly

$$X \wedge \partial\Delta[n]_+ \longrightarrow X \wedge \Delta[n]_+,$$

see also [Hir03, Lemma 16.3.8]. It follows that the map $f \,\square\, i_n$ is a cofibration between cofibrant objects by the definition of the model structure on cosimplicial frames, Theorem 6.9.7.

Definition 7.2.11 Let $\mathbb{W}$ be a set of cofibrations between cofibrant objects in a cofibrantly generated model category $\mathcal{C}$. We define a *set of horns on* $\mathbb{W}$ to be the set

$$\Lambda\mathbb{W} = \{f \,\square\, i_n \mid f \in \mathbb{W},\ n \in \mathbb{N}\},$$

where $i_n\colon \partial\Delta[n]_+ \to \Delta[n]_+$ is the standard inclusion.

The set of horns on $\mathbb{W}$ is strongly related to localisations.

Lemma 7.2.12 *Let $\mathbb{W}$ be a set of maps in a model category $\mathcal{C}$. Then the set of horns satisfies the following properties.*

- *Every element $f \,\square\, i_n$ of $\Lambda\mathbb{W}$ is a $\mathbb{W}$-equivalence.*
- *A fibrant object $Z \in \mathcal{C}$ is $\mathbb{W}$-local if and only if $Z \to *$ has the right lifting property with respect to the maps in $\Lambda\mathbb{W}$.*
- *If $\mathcal{C}$ is left proper and J is a set of acyclic cofibrations of $\mathcal{C}$, then every element of* $(J \cup \Lambda\mathbb{W})$-cell *is a cofibration and a $\mathbb{W}$-equivalence.*

As the proof of these statements is very set-theoretic, we refer to [Hir03, Propositions 3.2.10 and 3.2.11] and [Hir03, Propositions 4.2.3 and 4.2.4]. Note that the first two points do not require the model category to be left proper or cellular. The last point is essentially the statement that $\text{map}_r(-, Z)$ sends colimits to limits and $\mathbb{W}$-equivalences to weak equivalences of simplicial sets when Z is $\mathbb{W}$-local.

We give some technical lemmas needed for our existence theorem. These lemmas require that $\mathcal{C}$ is stable.

Lemma 7.2.13 *Let $\mathcal{C}$ be a stable model category and $\mathbb{W}$ a stable set of maps in $\mathcal{C}$. Assume we have a commutative triangle*

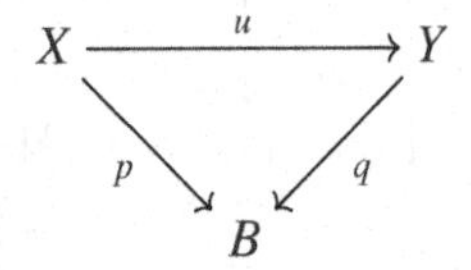

such that the homotopy fibres of p and q are $\mathbb{W}$-local. Then u is a $\mathbb{W}$-equivalence if and only if it is a weak equivalence in $\mathcal{C}$.

Proof We only need to prove one direction of the statement as the other follows straight from the definition. The commutative triangle gives a map of exact triangles in $\text{Ho}(\mathcal{C})$

$$\begin{array}{ccccccc}
\Omega B & \longrightarrow & Fp & \longrightarrow & X & \xrightarrow{p} & B \\
\| & & \downarrow{\scriptstyle v} & & \downarrow{\scriptstyle u} & & \| \\
\Omega B & \longrightarrow & Fq & \longrightarrow & Y & \xrightarrow{q} & B.
\end{array}$$

Applying the functor $[-, Z]_*$ for Z, a fibrant object, we obtain a diagram of long exact sequences

$$\begin{array}{ccccccccccc}
\cdots & \longrightarrow & [Fp, Z]_{n+1} & \longrightarrow & [B, Z]_n & \longrightarrow & [X, Z]_n & \longrightarrow & [Fp, Z]_n & \longrightarrow & \cdots \\
& & \downarrow{\scriptstyle v^*} & & \downarrow{\scriptstyle =} & & \downarrow{\scriptstyle u^*} & & \downarrow{\scriptstyle v^*} & & \\
\cdots & \longrightarrow & [Fq, Z]_{n+1} & \longrightarrow & [B, Z]_n & \longrightarrow & [Y, Z]_n & \longrightarrow & [Fq, Z]_n & \longrightarrow & \cdots .
\end{array}$$

Assume that u is a $\mathbb{W}$-equivalence. Then u^* is an isomorphism for each $\mathbb{W}$-local Z, and so is v^* by the Five Lemma. Lemma 7.2.7 then implies that v is a $\mathbb{W}$-equivalence. Since $\mathbb{W}$-equivalences between $\mathbb{W}$-local objects are weak equivalences by Lemma 7.1.2, it follows that v is a weak equivalence. Thus v^* is an isomorphism for any fibrant Z, and by the Five Lemma, u^* is an isomorphism for any fibrant Z. Therefore, u is a weak equivalence. □

The following definition is necessary for the section's main result.

Definition 7.2.14 We say that a model category $\mathcal{C}$ is *acceptable* if there are sets of generating cofibrations I and generating acyclic cofibrations J such that

1. the codomains of I are transfinitely small with respect to I,
2. the domains of J are cofibrant.

Examples 7.2.15 The levelwise and stable model structures on the various kinds of spectra we have studied are acceptable.

Other examples include chain complexes with the projective model structure, (pointed) topological spaces and (pointed) simplicial sets.

Note that being acceptable is a considerably weaker condition than being cellular. We are particularly interested in the following property: every cofibrant object in an acceptable model category is transfinitely small with respect to the class of cofibrations (see Lemma A.6.15).

We can now give our key lemma.

Lemma 7.2.16 *Let $\mathcal{C}$ be a proper, acceptable stable model category, and let $\mathbb{W}$ be a stable set of cofibrations between cofibrant objects in $\mathcal{C}$.*

If $f\colon X \longrightarrow Y$ is a fibration in $\mathcal{C}$, then the following statements are equivalent.

1. *The map f has the right lifting property with respect to $\Lambda\mathbb{W}$.*
2. *The fibre F of f is $\mathbb{W}$-local.*
3. *The homotopy fibre Ff of f is $\mathbb{W}$-local.*

Proof We show each condition implies the next.

$(1 \Rightarrow 2)$: Assume f has the right lifting property with respect to $\Lambda\mathbb{W}$. By the properties of pullbacks, the terminal map $F \longrightarrow *$ also has the right lifting property with respect to $\Lambda\mathbb{W}$. Hence, F is $\mathbb{W}$-local by Lemma 7.2.12.

$(2 \Rightarrow 3)$: Now assume that F, the fibre of f, is $\mathbb{W}$-local. Since f is a fibration and $\mathcal{C}$ is right proper, the strict fibre F is weakly equivalent to the homotopy fibre Ff. Hence, Ff is also $\mathbb{W}$-local.

$(3 \Rightarrow 1)$: Now assume that Ff, the homotopy fibre of f, is $\mathbb{W}$-local. As the domains of $J \cup \Lambda\mathbb{W}$ are cofibrant, they are transfinitely small with respect to the class of all cofibrations (see Lemma A.6.15), and hence transfinitely small with respect to $J \cup \Lambda\mathbb{W}$. By the small object argument (Proposition A.6.13), we may factor f as an element of $(J \cup \Lambda\mathbb{W})$-cell $j\colon X \longrightarrow B$ followed by a fibration $q\colon B \longrightarrow Y$ with the right lifting property with respect to $\Lambda\mathbb{W}$.

By the earlier parts of the proof, the homotopy fibre of q is also $\mathbb{W}$-local. By Lemma 7.2.12, j is a cofibration and $\mathbb{W}$-equivalence. Applying

Lemma 7.2.13, we see that j is a weak equivalence. Then we have a lifting square as on the left (with lift g), and hence a retract diagram as on the right.

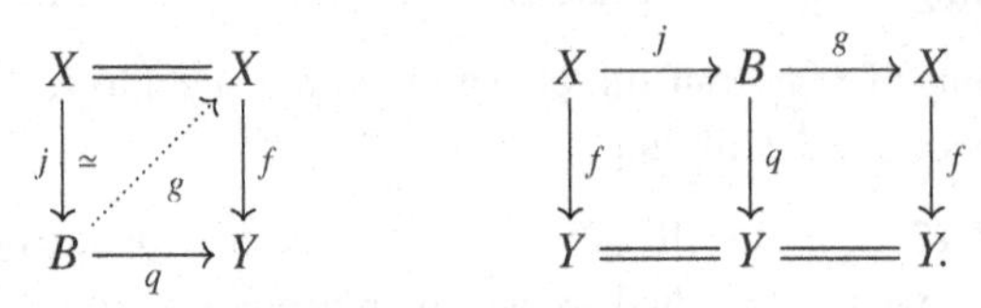

Since f is a retract of q, it also has the right lifting property with respect to $\Lambda\mathbb{W}$. □

We may now give an existence theorem for the $\mathbb{W}$-local model structure in the stable case. This theorem has two advantages over the unstable case. Firstly, it does not require the model category to be cellular and secondly, it provides explicit generating sets.

Theorem 7.2.17 *Let $\mathcal{C}$ be a proper, acceptable stable model category, and let $\mathbb{W}$ be a stable set of maps in $\mathcal{C}$.*

Then the left Bousfield localisation $L_{\mathbb{W}}\mathcal{C}$ of $\mathcal{C}$ at $\mathbb{W}$ from Theorem 7.1.6 exists and is a proper, acceptable stable model structure. The generating cofibrations are given by the generating cofibrations I of $\mathcal{C}$. The generating acyclic cofibrations are given by the set $J \cup \Lambda\mathbb{W}$, where J denotes the generating acyclic cofibrations of $\mathcal{C}$.

Proof We start by replacing $\mathbb{W}$ with a weakly equivalent set of cofibrations between cofibrant objects.

We follow the Recognition Theorem, Theorem A.6.9. The $\mathbb{W}$-equivalences satisfy the two-out-of-three property.

As for the smallness conditions, we have not changed the generating cofibrations I. Because every element of $J \cup \Lambda\mathbb{W}$ is a cofibration, every map in $(J \cup \Lambda\mathbb{W})$-cell is a cofibration. The domains of $J \cup \Lambda\mathbb{W}$ are cofibrant, hence, they are transfinitely small with respect to the class of cofibrations by Lemma A.6.15, hence, they are transfinitely small with respect to $(J \cup \Lambda\mathbb{W})$.

Now that the smallness properties have been dealt with, let us move on to the rest of the Recognition Theorem. Every element of $(J \cup \Lambda\mathbb{W})$-cell is a cofibration and a $\mathbb{W}$-equivalence by Lemma 7.2.12.

Let a morphism f in $\mathcal{C}$ have the right lifting property with respect to I. Then we know that f is a weak equivalence in $\mathcal{C}$ (hence, a $\mathbb{W}$-equivalence) and a fibration. Since the maps $J \cup \Lambda\mathbb{W}$ are cofibrations, f has the right lifting property with respect to $J \cup \Lambda\mathbb{W}$. Thus, we have shown that

$$I\text{-inj} \subseteq (J \cup \Lambda\mathbb{W})\text{-inj} \cap (\mathbb{W}\text{-equivalences}).$$

Let f have the right lifting property with respect to $J \cup \Lambda\mathbb{W}$. Then we know that f is a fibration in $\mathcal{C}$. In addition, by Lemma 7.2.16, the homotopy fibre Ff of f is $\mathbb{W}$-local. If we assume that f is also a $\mathbb{W}$-equivalence, then Ff is a $\mathbb{W}$-local object that is $\mathbb{W}$-equivalent to $*$ by Corollary 7.2.8. Hence, Ff is weakly equivalent to $*$ in $\mathcal{C}$. As $\mathcal{C}$ is stable, f is a weak equivalence in $\mathcal{C}$ and a fibration. Thus, f has the right lifting property with respect to I. Therefore, we have shown

$$(J \cup \Lambda\mathbb{W})\text{-inj} \cap (\mathbb{W}\text{-equivalences}) \subseteq I\text{-inj}.$$

We have proved that the model structure exists and is cofibrantly generated. It is an acceptable model category because we have not changed the set of generating cofibrations, and the domains of $J \cup \Lambda\mathbb{W}$ are cofibrant. Lemmas 7.2.3 and 7.2.10 imply that the model structure is stable and right proper. Left properness of $L_{\mathbb{W}}\mathcal{C}$ is a dual proof to Lemma 7.2.10. □

Remark 7.2.18 We have required $\mathcal{C}$ to be acceptable in order to give checkable conditions that $J \cup \Lambda\mathbb{W}$ will satisfy the conditions of the small object argument. The minimum requirement is that the domains of $J \cup \Lambda\mathbb{W}$ are transfinitely small with respect to $J \cup \Lambda\mathbb{W}$, see [BR13, Remark 4.12].

Since the standard examples of stable model categories (that is, the various model categories of spectra, and chain complexes with the projective model structure) are acceptable model categories, we find the requirements quite acceptable.

However, if we remove stability from the assumptions above, the result will fail, as with [BR13, Example 4.13].

Combining Corollary 7.2.8, Lemma 7.2.16 and Theorem 7.2.17 gives the following result.

Corollary 7.2.19 *Let $\mathcal{C}$ be a proper, acceptable stable model category, and let $\mathbb{W}$ be a stable set of maps in $\mathcal{C}$. Let $f\colon X \to Y$ be a fibration between $\mathbb{W}$-local objects. Then f has the right lifting property with respect to all cofibrations that are $\mathbb{W}$-equivalences.* □

Remark 7.2.20 Let $\mathbb{W}$ be a stable set of maps in a proper, acceptable stable model category $\mathcal{C}$. Then $L_{\mathbb{W}}\mathcal{C}$ exists and is cofibrantly generated with weak equivalences given by the $\mathbb{W}$-equivalences. It can be hard to give a good characterisation of the $\mathbb{W}$-equivalences beyond the definition and description in terms of maps in the homotopy category.

However, up to weak equivalence in $\mathcal{C}$, every $\mathbb{W}$-equivalence is a transfinite composition of pushouts of maps in $J \cup \Lambda\mathbb{W}$, where J is a set of generating acyclic cofibrations of $\mathcal{C}$. This is due to the fact that we may factor a

$\mathbb{W}$-equivalence f as $i \in (J \cup \Lambda\mathbb{W})$-cell followed by a fibration p. Since p is also a $\mathbb{W}$-equivalence, it is an acyclic fibration in $L_{\mathbb{W}}\mathcal{C}$. Consequently, p is an acyclic fibration of $\mathcal{C}$.

7.3 Localisation of Spectra with Respect to Homology Theories

Recall from Proposition 5.1.7 that $E \in \mathcal{SHC}$ defines a homology theory on $\mathcal{SHC}$ by

$$E_n(X) = \pi_n(E \wedge X).$$

A map $f\colon X \longrightarrow Y$ in $\mathcal{SHC}$ is called an E_*-isomorphism, or *E-equivalence*, if $E_*(f)$ is an isomorphism of graded abelian groups. In this section, we concentrate on localising the stable homotopy category at the E_*-isomorphisms for fixed $E \in \mathcal{SHC}$.

We work in the model category of orthogonal spectra for convenience. Let E be a cofibrant orthogonal spectrum. Then for any cofibrant spectrum X,

$$E_n(X) = \pi_n(X \wedge E) = \operatorname{colim}_a E_{n+a}(X_a),$$

see Corollary 6.4.12. Since E is cofibrant, Theorem 6.4.3 implies that E_* preserves all weak equivalences. By Example 7.2.4, we know that the E-equivalences are stable, hence, the resulting localised model structure on orthogonal spectra will also be a stable model category by Lemma 7.2.3.

Definition 7.3.1 We say that an orthogonal spectrum Z is *E-local* if for any E-equivalence $f\colon A \longrightarrow B$, the induced map

$$f^*\colon [B,Z] \longrightarrow [A,Z]$$

is an isomorphism. We say that an orthogonal spectrum A is *E-acyclic* if A is E-equivalent to $*$.

Because E-localisation is stable, we see that an orthogonal spectrum A is E-acyclic if and only if $[A,Z] = 0$ for every E-local Z.

Let $\mathcal{S}$ be any of our models for the stable homotopy category, that is, sequential spectra, symmetric spectra, or orthogonal spectra.

Definition 7.3.2 We define $L_E\mathcal{S}$ to be the localisation of $\mathcal{S}$ at the class of E-equivalences. We call this the *E-localisation* of spectra.

The existence of this model category does not follow from Theorem 7.2.17 as the E-equivalences are a proper class, not a set. Instead, we must prove that there is a set of maps $\mathbb{W}$ such that the class of $\mathbb{W}$-equivalences equals the E-equivalences. This is essentially a set-theoretic argument, and we follow the approach of [EKMM97, Chapter VIII], which is based upon [Bou75, Section 11]. It is also known as the Bousfield–Smith cardinality argument.

Theorem 7.3.3 *The left Bousfield localisation of orthogonal spectra at the class of E-equivalences exists.*

Proof Recall that the stable model structure on orthogonal spectra is cofibrantly generated by the following sets from Definition 5.2.15

$$\begin{aligned} I^{\mathcal{O}}_{\text{stable}} &= I^{\mathcal{O}}_{\text{level}} = \{\mathrm{F}^{\mathcal{O}}_d(S^{a-1}_+ \longrightarrow D^a_+) \mid a, d \in \mathbb{N})\} \\ J^{\mathcal{O}}_{\text{stable}} &= \{\mathrm{F}^{\mathcal{O}}_d(D^a_+ \longrightarrow (D^a \times [0,1])_+) \mid a, d \in \mathbb{N}\} \\ & \quad \cup \{k_n \square (S^{a-1}_+ \to D^a_+) \mid a, n \in \mathbb{N}\}. \end{aligned}$$

Let us call a map which is a cofibration and an E_*-isomorphism an *E-acyclic cofibration*, and let us call a map which has the right lifting property with respect to the E-acyclic cofibrations an *E–fibration.*

We will construct a set of generating E-acyclic cofibrations $J^{\mathcal{O}}_E$. This requires the transfinite small object argument, see Remark A.6.4 and Proposition A.6.13.

Let κ be a fixed infinite cardinal that is at least the cardinality of $E_*(\mathbb{S})$. Let $* \longrightarrow X$ be a map in orthogonal spectra which is in $I^{\mathcal{O}}_{\text{stable}}$-cell (a transfinite composition of pushouts of maps in $I^{\mathcal{O}}_{\text{stable}}$). We call X a *cell complex*. In this transfinite setting, we do not need to use coproducts due to [Hov99, Lemma 2.1.13].

Given a cell complex X, we regard each pushout over a map in $I^{\mathcal{O}}_{\text{stable}}$ as one *cell* and let $\#X$ be the cardinality of the set of cells of X. We define $J^{\mathcal{O}}_E$ to be the following set.

$$J^{\mathcal{O}}_E = \{f\colon X \longrightarrow Y \text{ in } I^{\mathcal{O}}_{\text{stable}}\text{-cell} \mid X \text{ is a cell complex}, \\ \#Y \leqslant \kappa \text{ and } f \text{ is an } E\text{-equivalence}\}$$

Note that each map in $J^{\mathcal{O}}_{\text{stable}}$ is in $J^{\mathcal{O}}_E$. We claim that the domains of $J^{\mathcal{O}}_E$ are κ-small with respect to $J^{\mathcal{O}}_E$. Let λ be a κ-filtered ordinal and

$$Y_0 \longrightarrow Y = \operatorname*{colim}_{\beta<\lambda} Y_\beta$$

a transfinite composition of pushouts of elements of $J^{\mathcal{O}}_E$, where the composition is indexed by λ. Consider a map $A \longrightarrow Y$ such that A is a cell complex with $\#A \leqslant \kappa$.

Since spheres are compact topological spaces, a map from $\mathrm{F}^{\mathcal{O}}_k S^n$ into Y lands in some finite subcomplex. That is, each cell of A must be mapped into some finite subcomplex of Y. Since there are at most κ-many cells, the image of

X must be mapped to some subcomplex of Y of size at most κ. This is precisely the required smallness condition.

We use the Recognition Theorem A.6.9 to show that $I^{\mathcal{O}}_{\text{stable}}$ and $J^{\mathcal{O}}_{E}$ are generating sets for a model structure on orthogonal spectra where the weak equivalences are the E_*-isomorphisms. The two-out-of-three property holds, and we have shown the smallness conditions. It remains to be shown that the lifting properties hold.

Since the maps in $J^{\mathcal{O}}_{E}$ are q-cofibrations, every map in $J^{\mathcal{O}}_{E}$-cell is a q-cofibration. The functor $E \wedge -$ commutes with colimits and sends E_*-acyclic cofibrations to acyclic cofibrations. Hence, Lemma 5.2.11 implies that E_*-acyclic cofibrations are preserved by pushouts and transfinite compositions. It follows that every map in $J^{\mathcal{O}}_{E}$-cell is an E_*-acyclic cofibration.

This shows that

$$J^{\mathcal{O}}_{E}\text{-cell} \subseteq \mathcal{W} \cap I^{\mathcal{O}}_{\text{stable}}\text{-cof}.$$

The maps in $J^{\mathcal{O}}_{E}$ are cofibrations, hence, every acyclic fibration has the right lifting property with respect to $J^{\mathcal{O}}_{E}$. Moreover, an acyclic fibration is a π_*-isomorphism and therefore an E_*-isomorphism. In other words,

$$I^{\mathcal{O}}_{\text{level}}\text{-inj} \subseteq \mathcal{W} \cap J^{\mathcal{O}}_{E}\text{-inj}.$$

It remains to be shown that any E-acyclic q-cofibration is in $J^{\mathcal{O}}_{E}$-cof. That is, any E-acyclic q-cofibration has the left lifting property with respect to $J^{\mathcal{O}}_{E}$-inj.

We show this in two steps. First, we show that every map in $J^{\mathcal{O}}_{E}$-inj has the right lifting property with respect to every inclusion of cell subcomplexes that is an E_*-isomorphism. Secondly, we show that every map in $J^{\mathcal{O}}_{E}$-inj has the right lifting property with respect to every E-acyclic q-cofibration.

We put the first step into Lemma 7.3.5. The second step we give below. Take a q-cofibration

$$i \colon X \longrightarrow Y$$

that is an E_*-isomorphism. Then we construct a cofibrant replacement

$$X' \to X$$

using the transfinite small object argument on $I^{\mathcal{O}}_{\text{stable}}$. By construction, X' is a cell complex. We then use the transfinite small object argument again to factor $X' \longrightarrow Y$ into a q-cofibration

$$i' \colon X' \longrightarrow Y'$$

followed by an acyclic fibration $Y' \longrightarrow Y$. The map i' is an E_*-acyclic inclusion of a cell subcomplex and hence has the left lifting property with respect to any map in $J_E^{\mathcal{O}}$-inj by the first step. We then use a technical result on left proper model categories [Hir03, Proposition 13.2.1]: if in a left proper model category a map has the right lifting property with respect to a cofibrant approximation of a map, then it also has the right lifting property with respect to the original map. This implies that i itself has the left lifting property with respect to any map in $J_E^{\mathcal{O}}$-inj and thus

$$I_{\text{stable}}^{\mathcal{O}}\text{-cof} \subseteq \mathcal{W} \cap J_E^{\mathcal{O}}\text{-cof}$$

which is the final condition required for the Recognition Theorem. □

Corollary 7.3.4 *In the E-localisation of orthogonal spectra, the class of E_*-isomorphisms is the class of $J_E^{\mathcal{O}}$-equivalences.* □

Since $J_E^{\mathcal{O}}$ consists of inclusions of cell complexes, the generating acyclic cofibrations of the E-local model structure have cofibrant domains and codomains.

Lemma 7.3.5 *Let $i\colon A \longrightarrow B$ be an E-acyclic inclusion of a cell subcomplex of orthogonal spectra. Then i has the left lifting property with respect to any map in $J_E^{\mathcal{O}}$-inj.*

Proof Let $f\colon X \longrightarrow Y$ be a map with the right lifting property with respect to $J_E^{\mathcal{O}}$ and consider a square

$$\begin{array}{ccc} A & \longrightarrow & X \\ {\scriptstyle i}\downarrow & & \downarrow{\scriptstyle f} \\ B & \longrightarrow & Y. \end{array}$$

Let T be the set of all pairs (B_t, k_t) with the following properties.

- B_t is a subcomplex of B containing A.
- $i_t\colon A \longrightarrow B_t$ is an E_*-isomorphism.
- The map $k_t\colon B_t \longrightarrow X$ provides a lift in the square

$$\begin{array}{ccccc} A & & \longrightarrow & & X \\ {\scriptstyle i_t}\downarrow & & {\scriptstyle k_t} \nearrow & & \downarrow{\scriptstyle f} \\ B_t & \longrightarrow & B & \longrightarrow & Y. \end{array}$$

We define a pre-order on T by $(B_s, k_s) < (B_t, k_t)$ if $B_s \subseteq B_t$ and the restriction of k_t to B_s is k_s. If a set $T' \subseteq T$ is totally ordered, define

$$B_u = \operatorname*{colim}_{t \in T'} B_t \qquad k_u = \operatorname*{colim}_{t \in T'} k_t.$$

It follows that (B_u, k_u) is in T and is an upper bound for T'.

This means that we can apply Zorn's Lemma, showing that there is a maximal element (B_m, k_m). We show by contradiction that $B_m = B$, hence, k_m is a lift in the original square, which would prove our claim. Assume B_m is not B, then choose a cell c of B not in B_m. There is a finite cell complex K_0 of B containing this cell.

Recall that κ is a fixed infinite cardinal that is at least the cardinality of $E_*(\mathbb{S})$. For each $x \in E_*(K_0)$ that is not in the image of $E_*(K_0 \cap B_m)$ (that is, an element of $E_*(K_0/(K_0 \cap B_m))$), we can attach finitely many cells of B to "kill" x, that is, make x trivial in the E–homology of the resulting pushout. Repeating for all such x gives a cell complex K_1, where we have added at most κ-many cells to K_0, as $E_*(K_0/(K_0 \cap B_m))$ has cardinality at most κ. For successor ordinals, we repeat the above process: for a limit ordinal λ we define

$$K_\lambda = \operatorname*{colim}_{\beta<\lambda} K_\beta.$$

Taking the colimit

$$K = \operatorname*{colim}_{\beta<\kappa} K_\beta$$

gives a subcomplex K of B such that

$$K \cap B_m \longrightarrow K$$

is E_*-acyclic and $\#K \leqslant \kappa$. The pushout of this map along $K \cap B_m \longrightarrow B_m$ gives an E_*-acyclic inclusion of subcomplexes $B_m \longrightarrow K \cup B_m$. This map has the left lifting property with respect to $J_E^{\mathcal{O}}$-inj (with a lift k), as it is the pushout of such a map. This means that we have

$$(K \cup B_m, k) > (B_m, k_m),$$

so B_m is not maximal. □

We can now prove directly what the fibrant objects in E-local spectra are.

Corollary 7.3.6 *The fibrant objects of $L_E\mathcal{S}^{\mathcal{O}}$ are the E-local Ω-spectra.*

Proof Let Z be fibrant in $L_E\mathcal{S}^{\mathcal{O}}$. Since

$$\mathrm{Id}\colon \mathcal{S}^{\mathcal{O}} \rightleftarrows L_E\mathcal{S}^{\mathcal{O}} : \mathrm{Id}$$

is a Quillen adjunction, Z is also fibrant in $\mathcal{S}^{\mathcal{O}}$. In particular, Z is an Ω-spectrum. It remains to be shown that Z is E-local.

As the model categories have the same cofibrations, the Quillen adjunction gives us an isomorphism

$$[A, Z]_* \cong [A, Z]_*^E$$

for any orthogonal spectrum A, where the right-hand side denotes the E-local stable homotopy category. Now if $f\colon A \longrightarrow B$ is an E-equivalence, it induces an isomorphism

$$[B, Z]_*^E \longrightarrow [A, Z]_*^E$$

and because of the previous isomorphism,

$$[B, Z]_* \xrightarrow{\cong} [A, Z]_*.$$

Thus, we have shown that Z is an E-local Ω-spectrum.

Now assume that X is an E-local Ω-spectrum. Taking a fibrant replacement in the E-local model structure gives an E-acyclic q-cofibration

$$q\colon X \longrightarrow L_E X$$

between E-local objects. By Lemma 7.1.2, q is an acyclic q-cofibration. Since q has the left lifting property with respect to $X \longrightarrow *$, it follows that the terminal map $X \longrightarrow *$ is a retract of $L_E X \longrightarrow *$. Consequently, X is a retract of an E–fibrant object and therefore itself E–fibrant. □

We now consider sequential spectra and symmetric spectra and show that they have E-local model structures, which are all Quillen equivalent to the E-local model structure on orthogonal spectra.

The following lemma is [Hir03, Theorem 3.3.20].

Lemma 7.3.7 *Let $F\colon \mathcal{C} \rightleftarrows \mathcal{D} :G$ be a Quillen equivalence of model categories. Let S be a set of maps in $\mathcal{C}$, and let $S' = \mathbb{L}F(S)$. Then, assuming both model structures exist, the adjunction*

$$F\colon L_S\mathcal{C} \rightleftarrows L_{S'}\mathcal{D} :G$$

is a Quillen equivalence.

Starting with a set of maps T in $\mathcal{D}$, we can obtain the analogous result, using the fact that for a Quillen equivalence $\mathbb{L}F(\mathbb{R}G(T)) \simeq T$, and that localisations only depend on the homotopy type of the maps.

Corollary 7.3.8 *Let $F\colon \mathcal{C} \rightleftarrows \mathcal{D} :G$ be a Quillen equivalence of model categories. Let T be a set of maps in $\mathcal{D}$, and let $T' = \mathbb{R}G(T)$. Then, assuming both model structures exist, the adjunction*

$$F\colon L_{T'}\mathcal{C} \rightleftarrows L_T\mathcal{D} :G$$

is a Quillen equivalence. □

Theorem 7.3.9 *There are E-local model structures on sequential spectra, symmetric spectra and orthogonal spectra with Quillen equivalences*

$$L_E \mathcal{S}^{\mathbb{N}} \underset{\mathbb{U}_{\mathbb{N}}^{\Sigma}}{\overset{\mathbb{P}_{\mathbb{N}}^{\Sigma}}{\rightleftarrows}} L_E \mathcal{S}^{\Sigma} \underset{\mathbb{U}_{\Sigma}^{\mathcal{O}}}{\overset{\mathbb{P}_{\Sigma}^{\mathcal{O}}}{\rightleftarrows}} L_E \mathcal{S}^{\mathcal{O}}.$$

Proof Starting with the set $J_E^{\mathcal{O}}$ in orthogonal spectra, we can use Theorem 7.2.17 with the sets $\mathbb{U}_{\Sigma}^{\mathcal{O}} J_E^{\mathcal{O}}$ for $\mathcal{S}^{\Sigma}$ and $\mathbb{U}_{\mathbb{N}}^{\mathcal{O}} J_E^{\mathcal{O}}$ for $\mathcal{S}^{\mathbb{N}}$ to obtain E-local model structures $L_E \mathcal{S}^{\Sigma}$ and $L_E \mathcal{S}^{\mathbb{N}}$ on symmetric spectra and sequential spectra. The result then follows from Corollary 7.3.8. □

Definition 7.3.10 The *E-local stable homotopy category* $\mathrm{Ho}(L_E \mathcal{S})$ is the homotopy category of the E-localisation of a category of spectra from Theorem 7.3.9.

The monoidal structure of the stable homotopy category passes to that of the E-local stable homotopy category.

Theorem 7.3.11 *Let E be a cofibrant orthogonal (respectively symmetric) spectrum. Then the E-local model structure on $\mathcal{S}^{\mathcal{O}}$ (respectively $\mathcal{S}^{\Sigma}$) is a symmetric monoidal model structure, and hence the E-local stable homotopy category is symmetric monoidal. The Quillen equivalence*

$$L_E \mathcal{S}^{\Sigma} \underset{\mathbb{U}_{\Sigma}^{\mathcal{O}}}{\overset{\mathbb{P}_{\Sigma}^{\mathcal{O}}}{\rightleftarrows}} L_E \mathcal{S}^{\mathcal{O}}$$

is a strong symmetric monoidal Quillen equivalence.

Proof Since the stable model structure on $\mathcal{S}^{\mathcal{O}}$ is monoidal and the cofibrations are unchanged, we only need to check that for a cofibration and E-equivalence $f\colon X \longrightarrow Y$ and a generating cofibration $g\colon A \longrightarrow B$ of orthogonal spectra, the pushout product map j

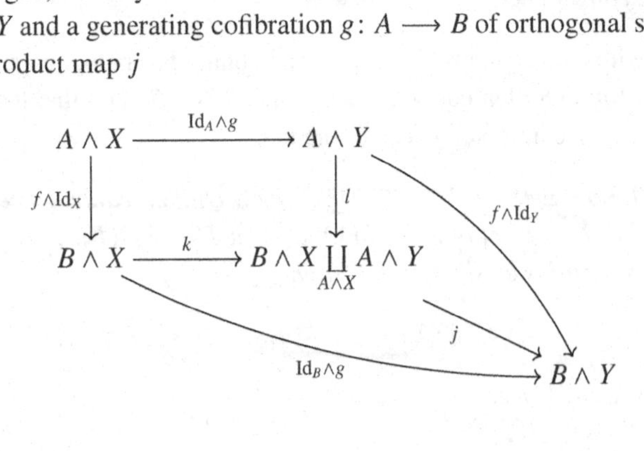

is an E-equivalence. Since A and B are cofibrant, the maps $\mathrm{Id}_A \wedge g$ and $\mathrm{Id}_B \wedge g$ are E-equivalences and cofibrations. Since pushouts preserve acyclic cofibrations in any model structure, the map k is an E-equivalence. Thus, j is an E-equivalence.

The unit condition also holds as before as cofibrant replacement in the E-local model structure is given by cofibrant replacement in the stable model structure.

The adjunction is a Quillen equivalence by Theorem 7.3.9. It is a strong symmetric monoidal adjunction, and cofibrant replacement has not changed, so it is a strong symmetric monoidal Quillen equivalence. □

We finish this section with a small result relating module structures to local spectra.

Proposition 7.3.12 *If E is a cofibrant ring spectrum, then every E–module is E-local.*

Proof Let E be a cofibrant orthogonal ring spectrum with unit map η, let X be a module over E in orthogonal spectra with E-action map ν, and let A be an orthogonal spectrum. Given any map $f\colon A \longrightarrow X$, we have a commutative square

$$\begin{array}{ccc} A & \xrightarrow{f} & X \\ {\scriptstyle \eta}\downarrow & & \uparrow{\scriptstyle \nu} \\ A \wedge E & \xrightarrow[f\wedge \mathrm{Id}]{} & X \wedge E. \end{array}$$

It follows that the composite

$$[A, X] \xrightarrow{-\wedge^{\mathbb{L}} E} [A \wedge^{\mathbb{L}} E, X \wedge^{\mathbb{L}} E] \xrightarrow{\eta^*} [A, X \wedge^{\mathbb{L}} E] \xrightarrow{\nu_*} [A, X]$$

is the identity. When A is E-acyclic, the second term is zero, so $[A, X] = 0$. Consequently, X is E-local. □

7.4 Examples of Left Localisation

In this section, we are going to look at some common examples of localisation of spectra. As all the calculations presented take place in the stable homotopy category, we do not have to worry about what kind of spectra we are using, so throughout the section, we let $\mathcal{S}$ denote a suitable category of spectra. Furthermore, the smash product $\wedge$ will denote the smash product in $\mathcal{SHC}$.

Many of the examples we will encounter satisfy the following property, which equips them with a wealth of technical advantages.

Definition 7.4.1 Let $\mathcal{S}$ be a category of spectra and $\mathbb{W}$ a stable set such that $L_{\mathbb{W}}\mathcal{S}$ exists. We say that the localisation at $\mathbb{W}$ is *smashing* if

$$X \longrightarrow X \wedge L_{\mathbb{W}}\mathbb{S}$$

is a $\mathbb{W}$-localisation, that is, the map is a $\mathbb{W}$-equivalence and $X \wedge L_{\mathbb{W}}\mathbb{S}$ is $\mathbb{W}$-local.

We could make this definition with any monoidal stable model category $\mathcal{C}$ such that the stable localisation $L_{\mathbb{W}}\mathcal{C}$ exists and replace the sphere in the definition with the monoidal unit. However, we chose to only discuss spectra in this section.

Example 7.4.2 We will encounter the following examples and counterexamples of localisation of spectra.

- Miller's finite localisation (Example 7.1.5) is smashing.
- Localisation of spectra at a set of primes is smashing. In particular, rationalisation is smashing, see Corollary 7.4.11.
- Completion at a prime p is *not* smashing, see Corollary 7.4.14.
- Localisation with respect to topological K-theory is smashing, see Theorem 7.4.37.
- Localisation at the Johnson–Wilson theories $E(n)$ is smashing, see Theorem 7.4.46.
- Localisation at the Morava K-theories $K(n)$ is *not* smashing, see Proposition 7.4.50.

In general, left localisation does not commute with sequential homotopy colimits. For example, the homotopy colimit of p-complete spectra is not necessarily p-complete itself. However, in the smashing case, these two operations do commute.

Proposition 7.4.3 *Let $\mathcal{S}$ be a category of spectra and $\mathbb{W}$ a stable set such that $L_{\mathbb{W}}\mathcal{S}$ exists. Then the localisation at $\mathbb{W}$ is smashing if and only if the (transfinite) sequential homotopy colimit of $\mathbb{W}$-local spectra is again $\mathbb{W}$-local.*

Proof We formulate the proof for orthogonal spectra. By the discussion above Lemma 6.4.13, we have

$$\text{hocolim}_n(X_n \wedge A) \cong (\text{hocolim}_n X_n) \wedge A$$

for a sequential diagram of spectra $f_n\colon X_n \longrightarrow X_{n+1}$, $n \in \mathbb{N}$. This extends to the transfinite case giving one direction of the result.

For the converse, assume that $\mathbb{W}$-localisation commutes with transfinite sequential homotopy colimits. Since we have a stable localisation,

$$\Sigma^m L_{\mathbb{W}} X \cong L_{\mathbb{W}} \Sigma^m X$$

for any $m \in \mathbb{Z}$ and any $X \in \mathcal{SHC}$. By transfinite induction and our assumption that $L_{\mathbb{W}}$ commutes with transfinite sequential (homotopy) colimits, it follows that $L_{\mathbb{W}}$ commutes with arbitrary (small) coproducts of cofibrant spectra.

Now consider the class of those $X \in \mathcal{SHC}$ such that $X \longrightarrow X \wedge L_{\mathbb{W}}\mathbb{S}$ is a $\mathbb{W}$-localisation. This class contains the sphere spectrum, is closed under suspensions, triangles and coproducts, and hence is the whole of the stable homotopy category by Corollary 5.6.8. □

As a consequence of the proof of Proposition 7.4.3, we have the following.

Corollary 7.4.4 *If localisation at $\mathbb{W}$ is smashing, then $L_{\mathbb{W}}$ commutes with coproducts and preserves compact objects.* □

Corollary 7.4.5 *Let $L_{\mathbb{W}}$ be a smashing localisation. Then the $\mathbb{W}$-local sphere $L_{\mathbb{W}}\mathbb{S}$ is a compact generator for* $\mathrm{Ho}(L_{\mathbb{W}}\mathcal{S})$.

Proof We have the following sequence of equivalences.

$$\begin{aligned}
& X \longrightarrow Y \text{ is a weak equivalence in } L_{\mathbb{W}}\mathcal{S}. \\
\iff & L_{\mathbb{W}}X \longrightarrow L_{\mathbb{W}}Y \text{ is a weak equivalence in } L_{\mathbb{W}}\mathcal{S}. \\
\iff & L_{\mathbb{W}}X \longrightarrow L_{\mathbb{W}}Y \text{ is a weak equivalence in } \mathcal{S} \text{ by Lemma 7.1.2}. \\
\iff & [\mathbb{S}, L_{\mathbb{W}}X]_* \longrightarrow [\mathbb{S}, L_{\mathbb{W}}Y]_* \text{ is an isomorphism}. \\
\iff & [L_{\mathbb{W}}\mathbb{S}, L_{\mathbb{W}}X]_* \longrightarrow [L_{\mathbb{W}}\mathbb{S}, L_{\mathbb{W}}Y]_* \text{ is an isomorphism}. \\
\iff & [L_{\mathbb{W}}\mathbb{S}, X]_*^{L_{\mathbb{W}}\mathcal{S}} \longrightarrow [L_{\mathbb{W}}\mathbb{S}, Y]_*^{L_{\mathbb{W}}\mathcal{S}} \text{ is an isomorphism}.
\end{aligned}$$

□

Remark 7.4.6 There are plenty more good properties of smashing localisations of spectra to study and exploit. We refer the interested reader to [EKMM97, Chapter VIII]. For example, Wolbert's Theorem says that if a localisation $L_{\mathbb{W}}$ is smashing, then $L_{\mathbb{W}}\mathcal{S}$ and the model category of modules over the $\mathbb{W}$-local sphere spectrum $L_{\mathbb{W}}\mathbb{S}$ are Quillen equivalent.

7.4.1 p-Localisation and p-Completion

Localisation and completion at a prime p or at a set of primes is often one of the first (left) Bousfield localisations one encounters. We will give some explicit constructions of p-localisation and p-completion as in [Bou79]. Then we will conclude the section by collecting some useful related results.

The spectra we will use for localisation and completion at a prime p are more general versions of the mod-n Moore spectra.

Example 7.4.7 For G an abelian group, the *Moore spectrum* $M(G)$ of G is constructed as follows. Assume that G is obtained via a short exact sequence

$$0 \longrightarrow F_2 \xrightarrow{\rho} F_1 \longrightarrow G \longrightarrow 0,$$

where F_1 is a free group on a generating set I_1 and F_2 is a free group on a generating set I_2. We think of I_1 as the generators of G and of $\rho(I_2)$ as the relations. We then construct a cofibre sequence

$$\bigvee_{I_2} \mathbb{S} \xrightarrow{r} \bigvee_{I_1} \mathbb{S} \longrightarrow M(G),$$

where r is a map satisfying $\pi_0(r) = \rho$. Such a map exists as $[\mathbb{S}, \mathbb{S}] = \mathbb{Z}$, and $[\mathbb{S}, -]$ commutes with coproducts by Corollary 5.6.8.

Applying homology to the above cofibre sequence gives rise to a long exact homology sequence which implies that $M(G)$ has the following homology groups

$$H_i(M(G)) = \begin{cases} G, & i = 0 \\ 0, & i \neq 0. \end{cases}$$

One can check that different choices of resolution give weakly equivalent spectra.

We would like to study the $M(G)$-localisation of a spectrum X. We start with the following computational tool.

Lemma 7.4.8 *Given a spectrum X and abelian group G, there is a homological universal coefficient short exact sequence*

$$0 \longrightarrow G \otimes \pi_*(X) \longrightarrow \pi_*(M(G) \wedge X) \longrightarrow \mathrm{Tor}^{\mathbb{Z}}(G, \pi_{*-1}(X)) \longrightarrow 0,$$

and a cohomological universal coefficient short exact sequence

$$0 \longrightarrow \mathrm{Ext}_{\mathbb{Z}}(G, \pi_{*+1}(X)) \longrightarrow [M(G), X]_* \longrightarrow \mathrm{Hom}_{\mathbb{Z}}(G, \pi_*(X)) \longrightarrow 0.$$

These sequences are natural in X.

Proof Following Example 7.4.7, we take a free resolution of G

$$0 \longrightarrow F_2 \xrightarrow{\rho} F_1 \longrightarrow G \longrightarrow 0,$$

where F_1 is a free group on a generating set I_1 and F_2 is a free group on a generating set I_2. We then construct a cofibre sequence

$$\bigvee_{I_2} \mathbb{S} \xrightarrow{r} \bigvee_{I_1} \mathbb{S} \longrightarrow M(G).$$

Applying the functor $[\mathbb{S}, - \wedge^{\mathbb{L}} X]_*$ to this cofibre sequence gives a long exact sequence

$$\cdots \longrightarrow \bigoplus_{I_2} \pi_n(X) \xrightarrow{r} \bigoplus_{I_1} \pi_n(X) \xrightarrow{s} \pi_n(M(G) \wedge^{\mathbb{L}} X) \xrightarrow{t} \bigoplus_{I_2} \pi_{n-1}(X) \longrightarrow \cdots.$$

We split this long exact sequence into the desired short exact sequence by taking the cokernel of s and the kernel of t.

A map of spectra $f\colon X \longrightarrow Y$ induces a natural transformation

$$f_*\colon [\mathbb{S}, - \wedge X]_* \longrightarrow [\mathbb{S}, - \wedge Y]_*,$$

so naturality of the short exact sequence follows.

The cohomological case is dual. □

Let us now turn to localisation with respect to Moore spectra.

Definition 7.4.9 Let P be a set of primes, and let

$$\mathbb{Z}_{(P)} = \{\frac{a}{b} \mid p \nmid b \text{ for all } p \in P\} \subseteq \mathbb{Q}.$$

Then the localisation $L_{M(\mathbb{Z}_{(P)})}X$ is called the *P-localisation* of X and is denoted $X_{(P)}$. The *P-local stable homotopy category* $\mathrm{Ho}(L_{M(\mathbb{Z}_{(P)})}\mathcal{S})$ is denoted $\mathrm{Ho}(\mathcal{S}_{(P)}) = \mathcal{SHC}_{(P)}$.

Quite often we only care about localisation at one prime at a time and write $X_{(p)}$ for $X_{(\{p\})}$. Note that for the case $P = \emptyset$, we obtain $\mathbb{Z}_{(P)} = \mathbb{Q}$, and thus we call this localisation *rationalisation*. In this case, we write $X_{\mathbb{Q}}$ for $X_{(\emptyset)}$.

The p-local stable homotopy category still has much the same "complexity" as the entire stable homotopy category itself. However, working in a p-local context often makes things easier in practice, such as when dealing with stable homotopy groups. Furthermore, there are powerful results about the structure of $\mathcal{SHC}_{(p)}$, which we will talk about in Subsection 7.4.3.

Proposition 7.4.10 *Let X be a spectrum and P a set of primes. Then the* P-localisation *of X is given by*

$$X_{(P)} = L_{M(\mathbb{Z}_{(P)})}X = X \wedge M(\mathbb{Z}_{(P)}).$$

Furthermore, $\pi_*(X_{(P)}) = \pi_*(X) \otimes \mathbb{Z}_{(P)}$.

Proof Since $\mathbb{Z}_{(P)}$ is torsion-free, the homological universal coefficient short exact sequence of Lemma 7.4.8 gives an isomorphism

$$\mathbb{Z}_{(P)} \otimes \pi_*(X) \xrightarrow{\cong} \pi_*(X \wedge M(\mathbb{Z}_{(P)})).$$

It follows that a $M(\mathbb{Z}_{(P)})$-equivalence is precisely a map of spectra inducing an isomorphism on P-local homotopy groups and that a spectrum A is $M(\mathbb{Z}_{(P)})$-acyclic if and only if its P-local homotopy groups are trivial.

We also see that $p \cdot \mathrm{Id}\colon X \longrightarrow X$ is a $M(\mathbb{Z}_{(P)})$-equivalence for a spectrum X and a prime p not in P. If X is a $M(\mathbb{Z}_{(P)})$-local spectrum, then $p \cdot \mathrm{Id}$ is a π_*-isomorphism. Hence, the homotopy groups of $M(\mathbb{Z}_{(P)})$-local spectra are uniquely p-divisible for each prime $p \notin P$.

We may assume that in the resolution of $\mathbb{Z}_{(P)}$ as a $\mathbb{Z}$–module, there is a generator corresponding to the unit 1. This gives a preferred map $\eta\colon \mathbb{S} \longrightarrow M(\mathbb{Z}_{(P)})$, and we claim that

$$\eta \wedge \mathrm{Id}\colon X \longrightarrow X \wedge M(\mathbb{Z}_{(P)})$$

is P-localisation. The map

$$(\eta \wedge \mathrm{Id})_*\colon \pi_*(X) \longrightarrow \mathbb{Z}_{(P)} \otimes \pi_*(X)$$

is induced by the unit map of the ring $\mathbb{Z}_{(P)}$, and hence, $\eta \wedge \mathrm{Id}$ is a $M(\mathbb{Z}_{(P)})$-equivalence.

It remains to be proven that $X \wedge M(\mathbb{Z}_{(P)})$ is $M(\mathbb{Z}_{(P)})$-local. Let $X \longrightarrow X_{(P)}$ be the fibrant replacement of X in the $M(\mathbb{Z}_{(P)})$-local model structure on spectra. Consider the commutative square below

$$\begin{array}{ccc} X & \xrightarrow{\simeq \mathbb{Z}_{(P)}} & X \wedge M(\mathbb{Z}_{(P)}) \\ {\scriptstyle \simeq \mathbb{Z}_{(P)}}\downarrow & & \downarrow{\scriptstyle \simeq} \\ X_{(P)} & \xrightarrow{\simeq \mathbb{Z}_{(P)}} & X_{(P)} \wedge M(\mathbb{Z}_{(P)}), \end{array}$$

where the arrows labelled $\simeq \mathbb{Z}_{(P)}$ are $\mathbb{Z}_{(P)} \otimes \pi_*$-isomorphisms. As the homotopy groups of $X_{(P)}$ and $X_{(P)} \wedge M(\mathbb{Z}_{(P)})$ are uniquely p-divisible, the bottom horizontal map is a π_*-isomorphism. Hence, $X \wedge M(\mathbb{Z}_{(P)})$ is $M(\mathbb{Z}_{(P)})$-local and $\pi_*(X_{(P)}) \cong \pi_*(X) \otimes \mathbb{Z}_{(P)}$. □

Corollary 7.4.11 *For a set of primes P, P-localisation is smashing, and the P-local sphere is $\mathbb{S}_{(P)} = M(\mathbb{Z}_{(P)})$.* □

Now we turn to the case $G = \mathbb{Z}/p$ for a prime p.

Definition 7.4.12 Let X be a spectrum and let p be a prime. Then the *p-completion* of X is given by $L_{M(\mathbb{Z}/p)}X$ and is denoted by $X_p^\wedge$. The *p-complete stable homotopy category* $\mathrm{Ho}(L_{M(\mathbb{Z}/p)}\mathbb{S})$ is denoted by $\mathrm{Ho}(\mathbb{S}_p^\wedge) = \mathcal{SHC}_p^\wedge$.

Proposition 7.4.13 *Let X be a spectrum and p be a prime. Then the p-completion of X is given by*

$$X_p^\wedge = L_{M(\mathbb{Z}/p)}X = \mathbb{R}\,\mathrm{Hom}(\Sigma^{-1}M(\mathbb{Z}/p^\infty), X),$$

where $\mathbb{R}\,\mathrm{Hom}$ denotes the function object in $\mathcal{SHC}$, and

$$\mathbb{Z}/p^\infty = \mathbb{Z}[\tfrac{1}{p}]/\mathbb{Z} = \mathrm{colim}(\mathbb{Z}/p \longrightarrow \mathbb{Z}/p^2 \longrightarrow \cdots).$$

Proof We start with the cohomological universal coefficient short exact sequence from Lemma 7.4.8

$$0 \longrightarrow \mathrm{Ext}_{\mathbb{Z}}(G, \pi_{*+1}(Y)) \longrightarrow [M(G), Y]_* \longrightarrow \mathrm{Hom}_{\mathbb{Z}}(G, \pi_*(Y)) \longrightarrow 0.$$

Let us put in $G = \mathbb{Z}/p$. If Y has uniquely p-divisible homotopy groups (i.e. for every $a \in \pi_*(Y)$ there is a $b \in \pi_*(Y)$ with $pb = a$), then both the Ext-term and the Hom-term are trivial, and therefore Y is $M(\mathbb{Z}/p)$-acyclic. We apply this to the following case. Let X be a spectrum. The exact triangle in $\mathcal{SHC}$

$$\Sigma^{-1}M(\mathbb{Z}/p^\infty) \longrightarrow \mathbb{S} \longrightarrow M(\mathbb{Z}[\tfrac{1}{p}]) \longrightarrow M(\mathbb{Z}/p^\infty)$$

induces an exact triangle of homotopy mapping spectra

$$\begin{aligned}\mathbb{R}\,\mathrm{Hom}(M(\mathbb{Z}[\tfrac{1}{p}]), X) &\longrightarrow \mathbb{R}\,\mathrm{Hom}(\mathbb{S}, X) \cong X \longrightarrow \\ &\longrightarrow \mathbb{R}\,\mathrm{Hom}(\Sigma^{-1}M(\mathbb{Z}/p^\infty), X) \longrightarrow \Sigma\mathbb{R}\,\mathrm{Hom}(M(\mathbb{Z}[\tfrac{1}{p}]), X).\end{aligned}$$

Let A be an $M(\mathbb{Z}/p)$-acyclic spectrum. Using exact triangles and induction,

$$A \wedge M(\mathbb{Z}/p) \simeq *$$

implies that $A \wedge M(\mathbb{Z}/p^n) \simeq *$ for all n and thus,

$$A \wedge M(\mathbb{Z}/p^\infty) \simeq *,$$

see the proof of Theorem 7.4.39 for the full argument. The spectrum

$$\mathbb{R}\,\mathrm{Hom}(\Sigma^{-1}M(\mathbb{Z}/p^\infty), X)$$

is $M(\mathbb{Z}/p)$-local because for A as above,

$$[A, \mathbb{R}\,\mathrm{Hom}(\Sigma^{-1}M(\mathbb{Z}/p^\infty), X)] = [A \wedge \Sigma^{-1}M(\mathbb{Z}/p^\infty), X] = [*, X] = 0.$$

The homotopy groups of $\mathbb{R}\operatorname{Hom}(M(\mathbb{Z}[\frac{1}{p}]), X)$ are uniquely p-divisible, hence, this mapping spectrum is $M(\mathbb{Z}/p)$ acyclic as explained above. Therefore, the map from the previous exact triangle

$$X \longrightarrow \mathbb{R}\operatorname{Hom}(\Sigma^{-1}M(\mathbb{Z}/p^\infty), X)$$

is a $M(\mathbb{Z}/p)$-equivalence and thus a $M(\mathbb{Z}/p)$-localisation as claimed. □

Corollary 7.4.14 *As a functor, p-completion is not smashing.*

Proof Proposition 7.4.13 says that

$$\mathbb{S}_p^\wedge = \mathbb{R}\operatorname{Hom}(\Sigma^{-1}M(\mathbb{Z}/p^\infty), \mathbb{S}).$$

Thus, for p-localisation to be smashing, we would have to have

$$\mathbb{R}\operatorname{Hom}(\Sigma^{-1}M(\mathbb{Z}/p^\infty), X) \cong X \wedge \mathbb{R}\operatorname{Hom}(\Sigma^{-1}M(\mathbb{Z}/p^\infty), \mathbb{S})$$

for any spectrum X. The above is equivalent to the spectrum $M(\mathbb{Z}/p^\infty)$ being strongly dualisable by Lemma 6.5.3. However, a spectrum is only strongly dualisable if and only if it is compact by Theorem 6.5.5. Since compact spectra have finitely generated homotopy groups in each degree by Corollary 5.6.14 and $\mathbb{Z}/p^\infty$ is not finitely generated over $\mathbb{Z}$, we conclude that $M(\mathbb{Z}/p^\infty)$ is not compact. □

Proposition 7.4.15 *For a spectrum X and a prime p, the p-completion of X is given by*

$$X_p^\wedge = \operatorname{holim}(M(\mathbb{Z}/p) \wedge X \longleftarrow M(\mathbb{Z}/p^2) \wedge X \longleftarrow M(\mathbb{Z}/p^3) \wedge X \longleftarrow \cdots),$$

where the maps are induced by the projections $\mathbb{Z}/p^n \longrightarrow \mathbb{Z}/p^{n-1}$.

Proof The Moore spectrum of $\mathbb{Z}/p^\infty$ is given by

$$M(\mathbb{Z}/p^\infty) = \operatorname{hocolim}(M(\mathbb{Z}/p) \longrightarrow M(\mathbb{Z}/p^2) \longrightarrow M(\mathbb{Z}/p^3) \longrightarrow \cdots),$$

where the maps are induced by multiplication by p on $\mathbb{Z}/p^{n-1} \longrightarrow \mathbb{Z}/p^n$.

Since the dual of $M(\mathbb{Z}/p^n)$ is $\Sigma^{-1}M(\mathbb{Z}/p^n)$, and these spectra are strongly dualisable,

$$M(\mathbb{Z}/p^n) \wedge X \cong \mathbb{R}\operatorname{Hom}(\mathbb{S}, M(\mathbb{Z}/p^n) \wedge X) \cong \mathbb{R}\operatorname{Hom}(\Sigma^{-1}M(\mathbb{Z}/p^n), X).$$

Consequently,

$$\begin{aligned} \operatorname{holim}_n(M(\mathbb{Z}/p^n) \wedge X) &\cong \operatorname{holim}_n \mathbb{R}\operatorname{Hom}(\Sigma^{-1}M(\mathbb{Z}/p^n), X) \\ &\cong \mathbb{R}\operatorname{Hom}(\operatorname{hocolim}_n \Sigma^{-1}M(\mathbb{Z}/p^n), X) \\ &\cong \mathbb{R}\operatorname{Hom}(\Sigma^{-1}M(\mathbb{Z}/p^\infty), X). \end{aligned}$$

□

We will need the following statement in Theorem 7.4.39.

Lemma 7.4.16 *Let X be a q-local spectrum for a prime q. Then, for any prime p different from q, the p-completion $X_p^\wedge$ is trivial.*

Proof By Proposition 7.4.15, we have that

$$X_p^\wedge = \operatorname{holim}(M(\mathbb{Z}/p) \wedge X \longleftarrow M(\mathbb{Z}/p^2) \wedge X \longleftarrow M(\mathbb{Z}/p^3) \wedge X \longleftarrow \cdots).$$

We will show that all terms in the homotopy limit are trivial. The universal coefficient short exact sequence tells us that

$$0 \longrightarrow \mathbb{Z}/p^n \otimes \pi_*(X) \longrightarrow \pi_*(M(\mathbb{Z}/p^n) \wedge X) \longrightarrow \operatorname{Tor}^{\mathbb{Z}}(\mathbb{Z}/p^n, \pi_{*-1}(X)) \longrightarrow 0$$

is exact. As X is q-local, all homotopy groups of X are of the form $G \otimes \mathbb{Z}_{(q)}$, see Proposition 7.4.10. This means that the first and third term of the short exact sequence are zero. Thus, $M(\mathbb{Z}/p^n) \wedge X \simeq *$ for all n. Therefore, by the properties of homotopy limits listed after Definition A.7.16, $X_p^\wedge \simeq *$. □

The homotopy groups of a p-completion have a nice algebraic characterisation as described below. However, it is not quite as straightforward as in the p-local case. We now have to make the extra assumption that the homotopy groups are finitely generated. By Corollary 5.6.14, this assumption holds for compact spectra.

Proposition 7.4.17 *Let X be a spectrum and p a prime. If $\pi_n(X)$ is finitely generated for all $n \in \mathbb{Z}$, then the homotopy groups of the p-completion of X are*

$$\pi_*(X_p^\wedge) = \pi_*(X) \otimes \mathbb{Z}_p^\wedge,$$

where $\mathbb{Z}_p^\wedge$ denotes the p-adic integers.

Proof By Proposition 7.4.13, the p-completion of X is

$$X_p^\wedge = \mathbb{R}\operatorname{Hom}(\Sigma^{-1} M(\mathbb{Z}/p^\infty), X).$$

The homotopy groups of $\mathbb{R}\operatorname{Hom}(\Sigma^{-1} M(\mathbb{Z}/p^\infty), X)$ are

$$[\mathbb{S}, \mathbb{R}\operatorname{Hom}(\Sigma^{-1} M(\mathbb{Z}/p^\infty), X)] \cong [\Sigma^{-1} M(\mathbb{Z}/p^\infty), X].$$

Thus, the short exact sequence

$$0 \longrightarrow \operatorname{Ext}_{\mathbb{Z}}(G, \pi_{*+1}(Y)) \longrightarrow [M(G), Y]_* \longrightarrow \operatorname{Hom}_{\mathbb{Z}}(G, \pi_*(Y)) \longrightarrow 0$$

gives us the following short exact sequence for $G = \mathbb{Z}/p^\infty$.

$$0 \longrightarrow \operatorname{Ext}_{\mathbb{Z}}(\mathbb{Z}/p^\infty, \pi_*(X)) \longrightarrow \pi_*(X_p^\wedge) \longrightarrow \operatorname{Hom}_{\mathbb{Z}}(\mathbb{Z}/p^\infty, \pi_{*-1}(X)) \longrightarrow 0$$

As the homotopy groups of X itself are finitely generated in each degree, the Hom-term of this sequence is trivial: a finitely generated abelian group is a finite coproduct of copies of cyclic groups (either infinite, of order p^k, or of order q^l for $q \neq p$), and in all those three cases $\mathrm{Hom}_{\mathbb{Z}}(\mathbb{Z}/p^\infty, -)$ is indeed zero. Thus,

$$\mathrm{Ext}_{\mathbb{Z}}(\mathbb{Z}/p^\infty, \pi_*(X)) \cong \pi_*(X_p^\wedge).$$

Let us calculate this Ext group. We have that $\mathbb{Z}/p^\infty = \mathrm{colim}_n\, \mathbb{Z}/p^n$, where the maps in the colimit are multiplication by p. Therefore, we get the short exact sequence

$$0 \longrightarrow \lim\nolimits_n^1 \mathrm{Hom}_{\mathbb{Z}}(\mathbb{Z}/p^n, \pi_*(X)) \longrightarrow \mathrm{Ext}_{\mathbb{Z}}(\mathbb{Z}/p^\infty, \pi_*(X)) \longrightarrow \lim\nolimits_n \mathrm{Ext}_{\mathbb{Z}}(\mathbb{Z}/p^n, \pi_*(X)) \longrightarrow 0.$$

The system

$$\cdots \longrightarrow \mathrm{Hom}_{\mathbb{Z}}(\mathbb{Z}/p^n, \pi_*(X)) \longrightarrow \mathrm{Hom}_{\mathbb{Z}}(\mathbb{Z}/p^{n-1}, \pi_*(X)) \longrightarrow \cdots$$

satisfies the Mittag–Leffler condition [Wei94, Proposition 3.5.7], which we can verify directly by either assuming $\pi_*(X) = \mathbb{Z}$, $\pi_*(X) = \mathbb{Z}/p^k$ or $\pi_*(X) = \mathbb{Z}/q^l$ for $q \neq p$. This implies that the $\lim^1$-term is zero, and consequently

$$\begin{aligned}\mathrm{Ext}_{\mathbb{Z}}(\mathbb{Z}/p^\infty, \pi_*(X)) &\cong \lim\nolimits_n \mathrm{Ext}_{\mathbb{Z}}(\mathbb{Z}/p^n, \pi_*(X)) \\ &= \lim\nolimits_n \pi_*(X)/(p^n) = \pi_*(X) \otimes \mathbb{Z}_p^\wedge.\end{aligned}$$ □

Sometimes it is also useful to have the following algebraic characterisation.

Proposition 7.4.18 *A spectrum X is p-complete if and only if $\pi_*(X)$ is* Ext-*p-complete, that is,*

$$\mathrm{Ext}_{\mathbb{Z}}(\mathbb{Z}[\tfrac{1}{p}], \pi_*(X)) = 0 = \mathrm{Hom}_{\mathbb{Z}}(\mathbb{Z}[\tfrac{1}{p}], \pi_*(X)).$$

Proof We use the universal coefficient short exact sequence again for $G = \mathbb{Z}[\frac{1}{p}]$,

$$0 \longrightarrow \mathrm{Ext}_{\mathbb{Z}}(\mathbb{Z}[\tfrac{1}{p}], \pi_{*+1}(X)) \longrightarrow [M(\mathbb{Z}[\tfrac{1}{p}]), X]_* \longrightarrow \mathrm{Hom}_{\mathbb{Z}}(\mathbb{Z}[\tfrac{1}{p}], \pi_*(X)) \longrightarrow 0.$$

If X is p-complete, then $X = \mathbb{R}\mathrm{Hom}(\Sigma^{-1}M(\mathbb{Z}/p^\infty, X)$ by Proposition 7.4.13. Thus,

$$\begin{aligned}[M(\mathbb{Z}[\tfrac{1}{p}]), X] &= [M(\mathbb{Z}[\tfrac{1}{p}]), \mathbb{R}\mathrm{Hom}(\Sigma^{-1}M(\mathbb{Z}/p^\infty, X)] \\ &\cong [\Sigma^{-1}M(\mathbb{Z}[\tfrac{1}{p}]) \wedge M(\mathbb{Z}/p^\infty), X].\end{aligned}$$

The spectrum $\Sigma^{-1}M(\mathbb{Z}[\frac{1}{p}]) \wedge M(\mathbb{Z}/p^\infty)$ is trivial by the universal coefficient short exact sequence, and therefore, $\pi_*(X)$ is Ext-p-complete.

Conversely, if the groups $\pi_*(X)$ are Ext-p-complete, the short exact sequence forces $[M(\mathbb{Z}[\frac{1}{p}]), X] = 0$ and consequently,

$$X \longrightarrow \mathbb{R}\operatorname{Hom}(\Sigma^{-1}M(\mathbb{Z}/p^\infty), X) = X_p^\wedge$$

is weak equivalence as in the proof of Proposition 7.4.13. □

We can also generalise completion in the following way. Let P be a set of primes again, and let $\mathbb{Z}/P := \bigoplus_{p \in P} \mathbb{Z}/p$. Then we can set

$$X_P^\wedge := L_{M(\mathbb{Z}/P)}X = \prod_{p \in P} X_p^\wedge.$$

Our results about localisation and completion now allow us to characterise $L_{M(G)}$ for all abelian G. Recall that an abelian group G is *uniquely p-divisible* if for all $g \in G$, there exists exactly one $h \in G$ with $ph = g$.

Definition 7.4.19 Let G_1 and G_2 be two abelian groups. Then we say that G_1 and G_2 are of the same *acyclicity type* if

- G_1 is torsion if and only if G_2 is torsion,
- for every prime p, G_1 is uniquely p-divisible if and only if G_2 is as well.

Every abelian group G is of the same acyclicity type as $\mathbb{Z}_{(P)}$ or $\mathbb{Z}/P$ for exactly one set of primes P. First, note that the groups $\mathbb{Z}/p$ and $\mathbb{Z}_{(p)}$ are uniquely q-divisible for all primes $q \neq p$. Therefore, if G is torsion and

$$P = \{\text{all primes } p \text{ for which } G \text{ is not uniquely } p\text{-divisible}\},$$

the group G has the same acyclicity type as $\mathbb{Z}/P$. Likewise, if G is not torsion, then it has the same acyclicity type as $\mathbb{Z}_{(P)}$. The term "acyclicity type" refers to the following.

Theorem 7.4.20 (Bousfield) *Let G_1 and G_2 be two abelian groups. Then G_1 and G_2 are of the same acyclicity type if and only if $M(G_1)$ and $M(G_2)$ give rise to the same Bousfield localisation, that is, a spectrum X is $M(G_1)$-local if and only if it is $M(G_2)$-local.*

Proof The full proof uses lengthy arguments involving the arithmetic of abelian groups, so we only provide an outline of [Bou74]. Let E be a connective spectrum. Then the class of groups

$$\mathcal{M}_E = \{\bigoplus_n \pi_n(X) \mid X \text{ is an } E\text{-acyclic spectrum}\}$$

satisfies the following.

- $0 \in \mathcal{M}_E$.
- $\mathcal{M}_E$ is closed under direct sums.
- Let $G_1 \to G_2 \to G_3 \to G_4 \to G_5$ be an exact sequence of abelian groups such that G_1, G_2, G_4 and G_5 are in $\mathcal{M}_E$, then so is G_3.

Then it is possible to prove that a class of groups $\mathcal{M}$ satisfying the points above can only be one of the two following cases.

1. $\mathcal{M}$ consists of all p-divisible groups, where P is a fixed set of primes and $p \in P$.
2. $\mathcal{M}$ consists of all P-torsion groups for a set of primes P. (A group G is P-torsion if for every $x \in G$ there is an n with $nx = 0$, where n is a product of elements of P.)

This means that all groups in $\mathcal{M}$ have the same acyclicity type. In the first case, $\mathcal{M}$ contains $\mathbb{Z}_{(Q)}$ for Q all primes not in P, in the second case, M contains $\mathbb{Z}/Q$. In particular, this holds for $\mathcal{M}_{M(G)}$ for any group G. □

Putting these pieces together, we see that P-localisation and P-completion describe all instances of localisation at $M(G)$.

Corollary 7.4.21 *Let X be a spectrum and G an abelian group. Then either*

$$L_{M(G)}X = L_{\mathbb{Z}_{(P)}}X = X_{(P)} \quad or \quad L_{M(G)}X = L_{\mathbb{Z}/P}X = X^{\wedge}_{P}$$

for a set of primes P. □

There is the following nice result for connective spectra.

Proposition 7.4.22 *Let X be a connective spectrum. Then X is $H\mathbb{Z}$-local.*

Proof The Postnikov tower of a connected spectrum X describes how a spectrum can be successively constructed from fibre sequences

$$K_i \longrightarrow X_i \longrightarrow X_{i-1}, \quad i \geqslant 1,$$

where the fibres K_i are Eilenberg–Mac Lane spectra. Any Eilenberg–Mac Lane spectrum HG is $H\mathbb{Z}$-local as HG can be constructed as a module over the ring spectrum $H\mathbb{Z}$, see Examples 6.6.4 and Proposition 7.3.12. Furthermore, Corollary 7.2.8 states that if two out of the three spectra in an exact triangle are $H\mathbb{Z}$-local, then so is the third.

In the Postnikov tower, X_0 is an Eilenberg–Mac Lane spectrum itself, as is K_1. Hence, X_1 is $H\mathbb{Z}$-local. Continuing inductively, we see that all the Postnikov sections X_i are $H\mathbb{Z}$-local. We then have $X = \operatorname{holim}_n X_n$, so by Corollary 7.2.9, X itself is $H\mathbb{Z}$-local. The result for connective spectra follows as a connective spectrum is a shift of a connected spectrum. □

By Lemma 7.1.2, we have the following.

Corollary 7.4.23 *A map between connective spectra that induces an isomorphism on integer homology groups (i.e. $H\mathbb{Z}$–homology) is a stable equivalence.* □

We are going to conclude this section by collecting a few more related results. As most of the proofs involve lengthy arithmetic arguments differentiating between various types of abelian groups, we are only going to include the statements of the results and refer to literature for proofs.

In [Bou79], Bousfield shows that for any spectra E and X, there are homotopy pullback squares

$$\begin{array}{ccc} L_E X & \longrightarrow & \prod_{p \text{ prime}} L_{E\wedge M\mathbb{Z}/p} X \\ \downarrow & & \downarrow \\ L_{\mathbb{Q}} L_E X & \longrightarrow & L_{\mathbb{Q}}\left(\prod_{p \text{ prime}} L_{E\wedge M\mathbb{Z}/p} X \right) \end{array}$$

as well as

$$\begin{array}{ccc} L_E X & \longrightarrow & L_{E\wedge \mathbb{Z}_{(P)}} X \\ \downarrow & & \downarrow \\ L_{E\wedge \mathbb{Z}_{(R)}} X & \longrightarrow & L_{\mathbb{Q}} L_E X, \end{array}$$

where P is a set of primes and R is the set of all primes not in P. These are often referred to as *Bousfield arithmetic squares*. These two squares can be used to deduce the following.

Proposition 7.4.24 (Bousfield) *Let E and X be spectra, and let G be an abelian group. If in addition G is either a torsion group or $E \wedge H\mathbb{Q} \not\simeq *$, then*

$$L_{E\wedge M(G)} X \cong L_{M(G)}(L_E X).$$

The arithmetic squares together with the results on acyclicity types can be combined to discuss all localisations with respect to connective spectra.

Theorem 7.4.25 (Bousfield) *Let E be a connective spectrum, and let $\pi_*(E) = \oplus_n \pi_n(E)$ be of either acyclicity type $G = \mathbb{Z}_{(P)}$ or $G = \mathbb{Z}/P$ for a set of primes P. Then for another connective spectrum X, the E-localisation of X is $L_E X = L_{M(G)} X$, that is, either*

$$L_E X = X_{(P)} \text{ or } L_E X = X^{\wedge}_{P}.$$

The following useful result arises from the proof of the previous theorem, see [Bou79].

Proposition 7.4.26 (Bousfield) *For any connective spectrum X and abelian group G, $L_{HG}X = L_{M(G)}X$.*

So, in particular, we have some understanding of localisation at Eilenberg–Mac Lane spectra. We can also consider localisation *of* Eilenberg–Mac Lane spectra at any other spectrum. While this is not immediately related to P-localisation and P-completion, it is nevertheless an interesting result. Details and applications can be found in [Gut10].

Theorem 7.4.27 (Gutiérrez) *Let E be any spectrum, and let G be an abelian group. Then $L_E HG$ is again related to a generalised Eilenberg–Mac Lane spectrum, that is, a product of wedges of Eilenberg–Mac Lane spectra. Specifically, let P be the set of primes p such that $E_*(H\mathbb{Z}/p) \neq 0$ and G is not uniquely p-divisible. Define*

$$A_p = \mathrm{Ext}_{\mathbb{Z}}(\mathbb{Z}/p^\infty, G) \text{ and } B_p = \mathrm{Hom}_{\mathbb{Z}}(\mathbb{Z}/p^\infty, G).$$

If $E_*(H\mathbb{Q}) = 0$, then

$$L_E HG \cong \prod_{p\in P} HA_p \vee \Sigma HB_p,$$

otherwise, there is a cofibre sequence of spectra

$$L_E HG \longrightarrow H(\mathbb{Q} \otimes G) \wedge \prod_{p\in P} HA_p \vee \Sigma HB_p \longrightarrow H\mathbb{Q} \wedge \prod_{p\in P} HA_p \vee \Sigma HB_p.$$

7.4.2 Localisation with Respect to K-Theory

In this section, we are going to describe Bousfield localisation with respect to topological K-theory. We could simply start with a definition of K-theory using nothing but spectra, but we feel that it would not be fitting to omit its geometric interpretation. For more details about K-theory, [Ati89] and [Hus94] are excellent sources, but we also think that [Gra75, Chapter 29] provides a very good introduction.

Background on K-Theory

The K-theory of a space is defined in terms of the vector bundles over that space. Let X be a compact Hausdorff topological space, and let k be either the field $\mathbb{R}$ or $\mathbb{C}$. By $\mathrm{Vect}_n(X)$, we denote the set of equivalence classes of n-dimensional k-vector bundles over X. Here, two vector bundles

$$\xi_1 \colon E_1 \longrightarrow X \text{ and } \xi_2 \colon E_2 \longrightarrow X$$

are equivalent if there is a map $f\colon E_1 \longrightarrow E_2$ inducing an isomorphism on each fibre and satisfying $\xi_2 \circ f = \xi_1$.

Take two vector bundles

$$\xi_1 \colon E_1 \longrightarrow X \text{ and } \xi_2 \colon E_2 \longrightarrow X.$$

The *Whitney sum* $\xi_1 \oplus \xi_2$ is given by the pullback of the diagonal of the product of ξ_1 and ξ_2, that is, it is the left vertical arrow in the pullback square

$$\begin{array}{ccc} \Delta^*(\xi_1 \times \xi_2) & \longrightarrow & E_1 \times E_2 \\ {\scriptstyle \pi}\downarrow & & \downarrow{\scriptstyle \xi_1\times\xi_2} \\ X & \xrightarrow{\Delta} & X \times X, \end{array}$$

where

$$\Delta^*(\xi_1 \times \xi_2) = \{(e_1, e_2, x) \in E_1 \times E_2 \times X \mid \xi_1(e_1) = \xi_2(e_2) = x\},$$

and

$$\pi \colon \Delta^*(\xi_1 \times \xi_2) \longrightarrow X$$

is the projection. This construction is well-defined on equivalence classes and therefore provides

$$\mathrm{Vect}(X) = \bigoplus_n \mathrm{Vect}_n(X)$$

with the structure of a semigroup. We now define $K_k(X)$ to be the *Grothendieck group* of the semigroup $(\mathrm{Vect}(X), \oplus)$. It has the universal property that if G is a group and $f\colon \mathrm{Vect}(X) \longrightarrow G$ a semigroup homomorphism, then f factors uniquely over $K_k(X)$,

$$\begin{array}{ccc} \mathrm{Vect}(X) & \xrightarrow{f} & G. \\ \downarrow & \nearrow & \\ K_k(X) & & \end{array}$$

More concretely, $K_k(X)$ is the quotient of the free group $F(\mathrm{Vect}(X))$ on $\mathrm{Vect}(X)$ by the subgroup generated by all $[\xi \oplus \eta] - [\xi] - [\eta]$. In other words, the Grothendieck construction formally adds inverses to a group in a "minimal" fashion.

If X is a pointed space, the information of the fibre over the basepoint is redundant, so one discards it by taking *reduced K-theory*

$$\tilde{K}_k(X) = \mathrm{coker}(K_k(*) \longrightarrow K_k(X)).$$

Now we can define the K-theory group of a pointed space X by

$$K^0(X) = \tilde{K}_{\mathbb{C}}(X) \text{ and } KO^0(X) = \tilde{K}_{\mathbb{R}}(X)$$

and for $n \in \mathbb{N}$,

$$K^{-n}(X) = \tilde{K}_{\mathbb{C}}(\Sigma^n X) \text{ and } KO^{-n}(X) = \tilde{K}_{\mathbb{R}}(\Sigma^n X).$$

We will soon see that for $n \in \mathbb{N}$, there are natural isomorphisms

$$K^{-n}(X) \cong K^{-n-2}(X) \text{ and } KO^{-n}(X) \cong KO^{-n-8}(X).$$

We use this periodicity to define K-theory for positive indices,

$$K^n(X) := K^{-n}(X) \text{ and } KO^n(X) := KO^{n-8k}(X),\ 8k \geqslant n.$$

In the first case, one speaks of *complex topological K-theory*, in the second case of *real topological K-theory*. For reasons of convenience, we will not decorate those K-theory groups with a tilde. In literature, K is sometimes denoted by KU. The groups $K^*(X)$ and $KO^*(X)$ form reduced cohomology theories, so we would like to look at the representing spectra.

It can be shown that for paracompact spaces, every n-dimensional vector bundle is a pullback of the canonical bundle ξ over $BU(n)$ in the complex case, respectively, $BO(n)$ in the real case. We see that every map $f: X \longrightarrow BU(n)$ gives us a vector bundle over X with total space

$$f^*(\xi) = \{(x, e) \in X \times EU \mid f(x) = \xi(e)\}.$$

A map homotopic to f gives an equivalent vector bundle. Conversely, constructing a bundle map from a vector bundle over X to the canonical bundle over $BU(n)$ is point-set topology. Thus, one obtains

$$\mathrm{Vect}_n(X) = \begin{cases} [X, BU(n)], & k = \mathbb{C} \\ [X, BO(n)], & k = \mathbb{R} \end{cases}$$

and after taking the colimit over n,

$$\mathrm{Vect}(X) = \begin{cases} [X, BU], & k = \mathbb{C} \\ [X, BO], & k = \mathbb{R}. \end{cases}$$

Therefore, a spectrum K representing K^* would have to satisfy $K_n = \Omega^n BU$, but how would we get its structure maps? The answer comes from the famous *Bott periodicity theorem.*

Theorem 7.4.28 (Bott) *There are homotopy equivalences $\Omega^2 BU \simeq BU \times \mathbb{Z}$ and $\Omega^8 BO \simeq BO \times \mathbb{Z}$.*

Thus, we have spectra K and KO via

$$K_n = \begin{cases} BU \times \mathbb{Z} & n \text{ even} \\ U & n \text{ odd} \end{cases}$$

and

$$KO_n = \Omega^i(BO \times \mathbb{Z}), \quad n = 8s + i.$$

So now for a topological space X, we have that

$$K^n(X) = [X, K_n] \text{ and } KO^n(X) = [X, KO_n].$$

As a consequence,

$$K^n(S^0) = \begin{cases} \mathbb{Z} & n \text{ even} \\ 0 & n \text{ odd} \end{cases} \qquad KO^n(S^0) = \begin{cases} \mathbb{Z} & n = 8s, n = 8s + 4 \\ \mathbb{Z}/2 & n = 8s + 1, n = 8s + 2 \\ 0 & \text{otherwise.} \end{cases}$$

The tensor product on vector bundles gives rise to an associative multiplication on Vect(X), which corresponds to the multiplicative structure on BU and BO. These multiplicative structures allow one to construct multiplications (cup products) on K^* and KO^*. Schwede [Sch07b, Example I.2.10] and Joachim [Joa01] give constructions of K and KU as commutative ring spectra which induce these multiplications.

From now on, when discussing K-theory, we will be talking only about K-theory of *spectra* instead of *spaces*, using the representing K-theory spectra to do so.

Because we would like to consider Bousfield localisation with respect to K-theory, the following lemma tells us that we do not have to worry about working over $\mathbb{R}$ or over $\mathbb{C}$.

Lemma 7.4.29 *Let X be a spectrum. Then $K^*(X) = 0$ if and only if $KO^*(X) = 0$. Equivalently, a spectrum X is K-local if and only if it is KO-local.*

Proof There is an exact triangle

$$\Sigma KO \xrightarrow{\mathrm{Id} \wedge \eta} KO \longrightarrow K \longrightarrow \Sigma^2 KO,$$

where $\eta \colon \mathbb{S}^1 \longrightarrow \mathbb{S}$ is the Hopf map, see [Ada74, p.206]. Consequently,

$$\Sigma KO \wedge X \xrightarrow{\mathrm{Id} \wedge \eta \wedge \mathrm{Id}} KO \wedge X \longrightarrow K \wedge X \longrightarrow \Sigma^2 KO \wedge X$$

is exact. Therefore, if $KO \wedge X \simeq *$ for a spectrum X, we must have $K \wedge X \simeq *$ too. Conversely, if $K \wedge X \simeq *$, the map

$$\Sigma KO \wedge X \xrightarrow{\mathrm{Id} \wedge \eta \wedge \mathrm{Id}} KO \wedge X$$

is an isomorphism in $\mathcal{SHC}$. However, it is also nilpotent as $\eta^4 = 0$, so it can only be an isomorphism if $KO \wedge X \simeq *$ as well. □

In [Ada66b, IV, Corollary 8], Adams showed that one can split up K-theory into smaller parts.

Theorem 7.4.30 (Adams) *Let X be a space, and let $K^*_{(p)}$ denote complex topological K-theory at the prime p, that is, $K_{(p)} = K \wedge \mathbb{S}_{(p)}$. Then there are idempotent maps*

$$E_\alpha \colon K^*_{(p)}(X) \longrightarrow K^*_{(p)}(X), \quad \alpha \in \mathbb{Z}/(p-1)$$

such that there are natural isomorphisms

$$K^*_{(p)}(X) = \bigoplus_\alpha E_\alpha(K^*_{(p)}(X)).$$

Furthermore, the functors $E_\alpha \circ K^*_{(p)}$ are cohomology theories and therefore are representable.

It was shown that the summands also satisfy the following. Let G denote the spectrum representing $E_1 \circ K^*_{(p)}$. Then

$$K^*_{(p)}(X) = \bigoplus_{0 \leqslant i \leqslant p-1} \Sigma^{2i} G^*(X),$$

that is, K-theory splits into shifted copies of G. This G is known as the *Adams summand*, and its homotopy groups are

$$\pi_*(G) = \mathbb{Z}[v_1, v_1^{-1}], \quad |v_1| = 2p-2.$$

The Adams summand also appears in literature denoted by L. A more contemporary notation for the Adams summand is $E(1)$ because it is the special case of some $E(n)$ for $n = 1$, as we will see in Subsection 7.4.3. Note that the prime p is absent from the notation, and for $p = 2$, $K_{(p)} = E(1)$. As a consequence of Adams' result we obtain the following.

Corollary 7.4.31 *A spectrum X is $K_{(p)}$-local if and only if it is $E(1)$-local.* □

K-Localisation

We can now study K-localisation of spectra. We will actually study localisation with respect to p-local K-theory, $K_{(p)} = K \wedge \mathbb{S}_{(p)}$. Throughout this section, p denotes a prime and $M = M(\mathbb{Z}/p)$ the mod-p Moore spectrum. Note that M is p-local.

In [Ada66a], Adams constructed $K_{(p)}$-equivalences

$$v_1^4 \colon \Sigma^8 M \longrightarrow M \ (p = 2) \ \text{ and } \ v_1 \colon \Sigma^{2p-2} M \longrightarrow M \ (p \text{ odd}).$$

In [Bou79], these maps are denoted by A_p. The notation v_1^4, respectively, v_1 is a more contemporary notation and refers to the element $v_1 \in E(1)_*$. The map in each case is called a v_1-*self-map*. In the case $p = 2$, v_1^4 is not, however, the fourth power of another map v_1. The notation merely alludes to the fact that it has degree 8. For convenience, in this section, we will write

$$v_1 : \Sigma^d M \longrightarrow M$$

to denote both instances.

The mod-p Moore spectrum plays a big role in understanding and constructing K-localisations. First, we define

$$M^\infty := \mathrm{hocolim}(M \xrightarrow{v_1} \Sigma^{-d} M \xrightarrow{v_1} \Sigma^{-2d} M \xrightarrow{v_1} \cdots).$$

Lemma 7.4.32 *The map $M \longrightarrow M^\infty$ is a $K_{(p)}$-equivalence.*

Proof We have that $K_{(p)*}(M^\infty) = \mathrm{colim}(K_{(p)*}(M) \xrightarrow{\cong} K_{(p)*}(M) \xrightarrow{\cong} \cdots)$, see Lemma 6.4.13, therefore, $K_{(p)*}(M) \longrightarrow K_{(p)*}(M^\infty)$ is an isomorphism. □

The homotopy groups of this spectrum have been calculated by Mahowald [Mah70] and Miller [Mil81].

Theorem 7.4.33 (Mahowald, Miller) *For $p = 2$, the order of the group $\pi_i(M^\infty)$ is*

$$\begin{array}{ll} 4, & i = 0 \; mod \; 8, \\ 8, & i = 1 \; mod \; 8, \\ 8, & i = 2 \; mod \; 8, \\ 4, & i = 3 \; mod \; 8, \\ 2, & i = 4 \; mod \; 8, \\ 1, & i = 5 \; mod \; 8, \\ 1, & i = 6 \; mod \; 8, \\ 2, & i = 7 \; mod \; 8. \end{array}$$

For odd primes, one has

$$\pi_i(M^\infty) = \mathbb{Z}/p \text{ for } i = 0, -1 \; mod \; (2p-2)$$

and $\pi_i(M^\infty) = 0$ otherwise.

To study K-localisation, we need to know the stable homotopy groups of the $K_{(p)}$-local sphere, which were computed in, for example, [Ada66a] and [Rav78]. The actual calculations and reasonings leading up to them are too extensive to include in this book.

Theorem 7.4.34 (Adams, Ravenel) *The stable homotopy groups $\pi_i(L_{K_{(2)}}\mathbb{S})$ of the $K_{(2)}$-local sphere are*

$$\begin{array}{ll} \mathbb{Z}_{(2)} \oplus \mathbb{Z}/2, & i = 0, \\ \mathbb{Z}/2^\infty, & i = -2, \\ \mathbb{Z}/2, & i = 0 \ mod\ 8, i \neq 0 \\ (\mathbb{Z}/2)^2, & i = 1 \ mod\ 8, \\ \mathbb{Z}/2, & i = 2 \ mod\ 8, \\ \mathbb{Z}/8, & i = 3 \ mod\ 8, \\ 0, & i = 4, 5, 6 \ mod\ 8, \\ \mathbb{Z}/2^\nu, & i = 7 \ mod\ 8, i + 1 = 2^{\nu-1}q, q \ odd. \end{array}$$

For odd primes, $\pi_i(L_{K_{(p)}}\mathbb{S})$ is given by

$$\begin{array}{ll} \mathbb{Z}_{(p)}, & i = 0, \\ \mathbb{Z}/p^\infty, & i = -2, \\ \mathbb{Z}/p^\nu, & i = -1 \ mod\ (2p-2),\ i \neq -1,\ i + 1 = (2p-2)qp^{\nu-1},\ p \nmid q, \\ 0, & else. \end{array}$$

Furthermore, the map $\pi_*(\mathbb{S}) \longrightarrow \pi_*(L_{K_{(p)}}\mathbb{S})$ is a split epimorphism in degrees greater than 1 for $p = 2$ and in non-negative degrees for p odd.

Since localisation at $K_{(p)}$ is the right derived functor of the identity functor from $L_{K_{(p)}}\mathcal{S}$ to $\mathcal{S}$, $L_{K_{(p)}}$ is an exact functor on $\mathcal{SHC}$ by Theorem 4.5.2. Hence, it preserves the cofibre sequence defining M

$$L_{K_{(p)}}\mathbb{S} \xrightarrow{p} L_{K_{(p)}}\mathbb{S} \longrightarrow L_{K_{(p)}}M,$$

and so we have an isomorphism in $\mathcal{SHC}$

$$L_{K_{(p)}}M \cong M \wedge L_{K_{(p)}}\mathbb{S}.$$

This cofibre sequence and our knowledge of $\pi_*(L_{K_{(p)}}\mathbb{S})$ allow us to calculate the order of the homotopy groups of $M \wedge L_{K_{(p)}}\mathbb{S}$. We see that the homotopy groups of M^∞ and $M \wedge L_{K_{(p)}}\mathbb{S}$ have the same order, a key step in the proof of the next proposition.

Let us consider the map

$$M \longrightarrow M \wedge L_{K_{(p)}}\mathbb{S}.$$

As $M \wedge L_{K_{(p)}}\mathbb{S}$ is $K_{(p)}$-local and $M \longrightarrow M^\infty$ is a $K_{(p)}$-equivalence by Lemma 7.4.32, the map $M \longrightarrow M \wedge L_{K_{(p)}}\mathbb{S}$ factors over M^∞ as follows.

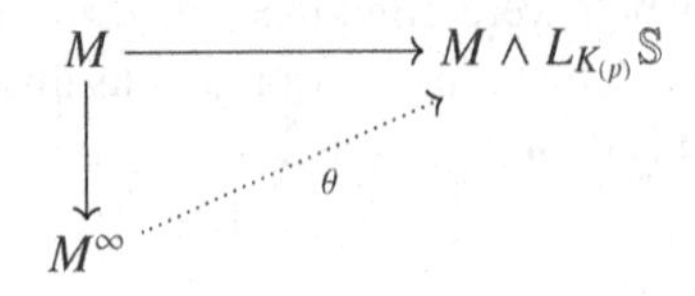

Lemma 7.4.35 *The map $\theta\colon M^\infty \longrightarrow M \wedge L_{K_{(p)}}\mathbb{S}$ is an isomorphism in $\mathcal{SHC}$.*

Proof We would like to show that

$$\pi_*(\theta)\colon \pi_*(M^\infty) \longrightarrow \pi_*(M \wedge L_{K_{(p)}}\mathbb{S})$$

is an isomorphism. We already remarked that the groups on either side have the same finite order in each degree, therefore, it is enough to prove that $\pi_*(\theta)$ is surjective. Theorem 7.4.34 says that

$$\pi_*(\mathbb{S}) \longrightarrow \pi_*(L_{K_{(p)}}\mathbb{S})$$

is a split epimorphism in degrees greater than 1. We have a commutative diagram of long exact sequences,

$$\begin{array}{ccccccccc}
\cdots \longrightarrow & \pi_i(\mathbb{S}) & \xrightarrow{p} & \pi_i(\mathbb{S}) & \longrightarrow & \pi_i(M) & \longrightarrow & \pi_{i-1}(\mathbb{S}) & \longrightarrow \cdots \\
& \downarrow & & \downarrow & & \downarrow & & \downarrow & \\
\cdots \longrightarrow & \pi_i(L_{K_{(p)}}\mathbb{S}) & \xrightarrow{p} & \pi_i(L_{K_{(p)}}\mathbb{S}) & \longrightarrow & \pi_i(M \wedge L_{K_{(p)}}\mathbb{S}) & \longrightarrow & \pi_{i-1}(L_{K_{(p)}}\mathbb{S}) & \longrightarrow \cdots .
\end{array}$$

The Five Lemma tells us that for $i > 2$, the map $\pi_i(M) \longrightarrow \pi_i(M \wedge L_{K_{(p)}}\mathbb{S})$ is also an epimorphism. We furthermore have a commutative diagram

$$\begin{array}{ccccccc}
M & \xrightarrow{v_1} & \Sigma^{-d}M & \xrightarrow{v_1} & \Sigma^{-2d}M & \xrightarrow{v_1} & \cdots \\
\downarrow & & \downarrow & & \downarrow & & \\
M \wedge L_{K_{(p)}}\mathbb{S} & \underset{\cong}{\xrightarrow{v_1}} & \Sigma^{-d}M \wedge L_{K_{(p)}}\mathbb{S} & \underset{\cong}{\xrightarrow{v_1}} & \Sigma^{-2d}M \wedge L_{K_{(p)}}\mathbb{S} & \longrightarrow & \cdots .
\end{array}$$

This means that we have a morphism between the homotopy colimits of the top and the bottom row. The homotopy colimit of the top row is again M^∞, and the homotopy colimit of the bottom row is $M \wedge L_{K_{(p)}}\mathbb{S}$, as all horizontal maps in the bottom row are isomorphisms in $\mathcal{SHC}$. The previous diagram induces

$$\begin{array}{ccccccc}
\pi_i(M) & \xrightarrow{v_1} & \pi_{i+d}(M) & \xrightarrow{v_1} & \pi_{i+2d}(M) & \xrightarrow{v_1} & \cdots \\
\downarrow & & \downarrow & & \downarrow & & \\
\pi_i(M \wedge L_{K_{(p)}}\mathbb{S}) & \underset{\cong}{\xrightarrow{v_1}} & \pi_{i+d}(M \wedge L_{K_{(p)}}\mathbb{S}) & \underset{\cong}{\xrightarrow{v_1}} & \pi_{i+2d}(M \wedge L_{K_{(p)}}\mathbb{S}) & \longrightarrow & \cdots .
\end{array}$$

As $i + kd$ will eventually be greater than 2 for any i, the vertical maps in this diagram will eventually become epimorphisms. Therefore, the map

$$\pi_i(M^\infty) \longrightarrow \pi_i(M \wedge L_{K_{(p)}}\mathbb{S})$$

is also an epimorphism, which is exactly what we wanted to prove. □

Together with Lemma 7.4.32, we obtain the following.

Corollary 7.4.36 *The map $M \longrightarrow M^\infty$ is a $K_{(p)}$-localisation.*

Proof We saw in Lemma 7.4.35 that $M^\infty \cong M \wedge L_{K_{(p)}}\mathbb{S}$. This means, in particular, that M^∞ is $K_{(p)}$-local. We already saw in Lemma 7.4.32 that

$$M \longrightarrow M^\infty$$

is also a $K_{(p)}$-equivalence. These two things together prove our claim. □

In fact, smashing with the $K_{(p)}$-local sphere provides a $K_{(p)}$-localisation of any spectrum. However, the proof of this requires extensive knowledge of the $K_{(p)}$-local sphere again. Therefore, we will just state the result.

Theorem 7.4.37 (Ravenel) *For any spectrum X, the map*

$$X \longrightarrow X \wedge L_{K_{(p)}}\mathbb{S}$$

is a $K_{(p)}$-equivalence, that is, $K_{(p)}$-localisation is smashing.

Corollary 7.4.38 *Let X be a rational spectrum, that is, $X = X_{\mathbb{Q}}$. Then X is also $K_{(p)}$-local.*

Proof The only torsion–free homotopy group of the $K_{(p)}$-local sphere is in degree 0, where it contains a copy of $\mathbb{Z}_{(p)}$, see Theorem 7.4.34. Hence,

$$\mathbb{S}_{\mathbb{Q}} = (L_{K_{(p)}}\mathbb{S})_{\mathbb{Q}},$$

and therefore,

$$L_{K_{(p)}}(X) = L_{K_{(p)}}(X_{\mathbb{Q}}) = X \wedge \mathbb{S}_{\mathbb{Q}} \wedge L_{K_{(p)}}\mathbb{S} = X \wedge \mathbb{S}_{\mathbb{Q}} = X,$$

which proves our claim. □

Finally, we have collected the necessary information to take on the main result of this section, namely, that $K_{(p)}$-localisation is detected by the v_1-self-map $v_1 \colon \Sigma^d M \longrightarrow M$ of the mod-p Moore spectrum.

Theorem 7.4.39 *Let X be a p-local spectrum for p a prime, and let M denote the mod-p Moore spectrum. The following are equivalent.*

1. *The spectrum X is $K_{(p)}$-local.*
2. *The map $v_1^* \colon [M, X]_* \longrightarrow [M, X]_{*+d}$ is an isomorphism.*
3. *The map $(v_1)_* \colon \pi_*(M \wedge X) \longrightarrow \pi_{*+d}(M \wedge X)$ is an isomorphism.*

Proof The second and third points are equivalent as M is strongly dualisable by Example 6.5.2 with $DM = \Sigma^{-1}M$.

If X is $K_{(p)}$-local, then

$$f^* \colon [B, X] \longrightarrow [A, X]$$

is an isomorphism for all $K_{(p)}$-equivalences $f\colon A \longrightarrow B$. In particular, it is an isomorphism for the $K_{(p)}$-equivalence

$$v_1 : \Sigma^d M \longrightarrow M.$$

This proves that the first point implies the second point.

Lastly, assume that $(v_1)_* : \pi_*(M \wedge X) \longrightarrow \pi_{*+d}(M \wedge X)$ is an isomorphism. Let X be a p-local spectrum. We first show that $M \wedge X$ is $K_{(p)}$-local. Consider the sequence

$$M \wedge X \xrightarrow{v_1 \wedge \mathrm{Id}} \Sigma^{-d} M \wedge X \xrightarrow{v_1 \wedge \mathrm{Id}} \Sigma^{-2d} M \wedge id \xrightarrow{v_1 \wedge \mathrm{Id}} \cdots .$$

Its homotopy colimit is

$$\mathrm{hocolim}(M \xrightarrow{v_1} \Sigma^{-d} M \xrightarrow{v_1} \cdots) \wedge X = M^{\infty} \wedge X,$$

see Lemma 6.4.13. However, as all the maps in the sequence are isomorphisms in $\mathcal{SHC}$ by assumption, the homotopy colimit is also isomorphic to its first term, namely, $M \wedge X$. Altogether, we have

$$M \wedge X = M^{\infty} \wedge X,$$

which, by Lemma 7.4.35 is also equal to

$$M \wedge L_{K_{(p)}} \mathbb{S} \wedge X = L_{K_{(p)}}(M \wedge X).$$

Thus, $M \wedge X$ is $K_{(p)}$-local.

We use induction to verify that $M(\mathbb{Z}/p^n) \wedge X$ is also $K_{(p)}$-local for all n. It is part of an exact triangle in $\mathcal{SHC}$,

$$M(\mathbb{Z}/p^{n-1}) \wedge X \longrightarrow M(\mathbb{Z}/p^n) \wedge X \longrightarrow M(\mathbb{Z}/p) \wedge X \longrightarrow \Sigma M(\mathbb{Z}/p^{n-1}) \wedge X.$$

To see this, we can use the octahedral axiom applied to the following diagram.

$$\begin{array}{ccccccc}
X & \xrightarrow{p^{n-1}} & X & \longrightarrow & M(\mathbb{Z}/p^{n-1}) \wedge X & \longrightarrow & \Sigma X \\
\| & & \downarrow{\scriptstyle p} & & \downarrow & & \| \\
X & \xrightarrow{p^{n}} & X & \longrightarrow & M(\mathbb{Z}/p^{n}) \wedge X & \longrightarrow & \Sigma X \\
 & & \downarrow & & \downarrow & & \\
 & & M \wedge X & = & M \wedge X & & \\
 & & \downarrow & & \downarrow & & \\
 & & \Sigma X & \longrightarrow & \Sigma M(\mathbb{Z}/p^{n-1}) \wedge X. & &
\end{array}$$

We know that the first two rows and the first column are exact triangles. The octahedral axiom tells us that so is the second column. We also know that if two out of three spectra in an exact triangle are $K_{(p)}$-local, then so is the third, see Corollary 7.2.8. Consequently, $M(\mathbb{Z}/p^n) \wedge X$ is $K_{(p)}$-local for all n.

By Proposition 7.4.15, we have

$$X_p^\wedge = \mathrm{holim}(M(\mathbb{Z}/p) \wedge X \longleftarrow M(\mathbb{Z}/p^2) \wedge X \longleftarrow M(\mathbb{Z}/p^3) \wedge X \longleftarrow \cdots).$$

We now know that all terms in the homotopy limit are $K_{(p)}$-local. By Corollary 7.2.9, the homotopy limit of $K_{(p)}$-local spectra is also $K_{(p)}$-local, so $X_p^\wedge$ is $K_{(p)}$-local.

Recall that we have the Bousfield arithmetic square,

$$\begin{array}{ccc} X & \longrightarrow & \prod_p X_p^\wedge \\ \downarrow & & \downarrow \\ X_{\mathbb{Q}} & \longrightarrow & (\prod_p X_p^\wedge)_{\mathbb{Q}}. \end{array}$$

As in our case, the spectrum X is p-local, all its q-completions are trivial except for $q = p$ (see Lemma 7.4.16). Therefore, the top right corner is simply $X_p^\wedge$, which we have just shown is $K_{(p)}$-local. We also know from Corollary 7.4.38 that the rational terms in the square are also $K_{(p)}$-local. Lemma 7.2.10 tells us that this means that X itself is $K_{(p)}$-local as well, which is what we wanted to prove. □

In particular, we see that Bousfield localisation with respect to p-local complex K-theory is the same as localisation with respect to just

$$v_1 : \Sigma^d M \longrightarrow M$$

and its (de)suspensions. Therefore, $K_{(p)}$-localisation is a special case of Miller's finite localisation $L_{\mathcal{A}}^f \mathcal{S}$, which we encountered at the beginning of Section 7.1, where in this case, $\mathcal{A}$ consists of the cofibre of

$$v_1 : \Sigma^d M \longrightarrow M$$

and its (de)suspensions.

A lovely feature of the K-local stable homotopy category at the prime $p = 2$ is that it is rigid in the sense of Theorem 5.7.1, see [Roi07].

Theorem 7.4.40 (Roitzheim) *Let $\mathcal{C}$ be a stable model category, and let $L_{K_{(2)}}\mathcal{S}^{\mathbb{N}}$ denote the model category of sequential spectra with the $K_{(2)}$-local model structure. If there is an equivalence of triangulated categories*

$$\Psi\colon \mathrm{Ho}(L_{K_{(2)}}\mathcal{S}) \longrightarrow \mathrm{Ho}(\mathcal{C}),$$

then $L_{K_{(2)}}\mathcal{S}^{\mathbb{N}}$ and $\mathcal{C}$ are Quillen equivalent.

The first challenge presents itself immediately, as the stable framings from Theorem 6.9.25 give us a Quillen adjunction

$$X \wedge -: \mathcal{S}^{\mathbb{N}} \rightleftarrows \mathcal{C} : \mathbb{R}\operatorname{Hom}(X, -)$$

between spectra and $\mathcal{C}$, but not necessarily between $K_{(2)}$-local spectra and $\mathcal{C}$. Therefore, the following has to be verified first.

Lemma 7.4.41 *Let $G\colon \mathcal{C} \longrightarrow \mathcal{D}$ be a right Quillen functor and $L_{\mathbb{W}}\mathcal{D}$ a left Bousfield localisation of $\mathcal{D}$. Then $G\colon \mathcal{C} \longrightarrow L_{\mathbb{W}}\mathcal{D}$ is a right Quillen functor if and only if $G(X)$ is $\mathbb{W}$-local for all fibrant $X \in \mathcal{C}$.*

Proof The functor G sends acyclic fibrations to acyclic fibrations, as $\mathcal{D}$ and $L_{\mathbb{W}}\mathcal{D}$ have the same cofibrations. If $X \in \mathcal{C}$ is fibrant, then $G(X)$ is fibrant. Furthermore, $G(X)$ is $\mathbb{W}$-local by assumption and therefore fibrant in $L_{\mathbb{W}}\mathcal{D}$. We still have to show that G sends fibrations to fibrations in $L_{\mathbb{W}}\mathcal{D}$. By Lemma A.4.3 it is sufficient to show that G sends fibrations between fibrant objects to fibrations.

Let $g\colon X \longrightarrow Y$ be a fibration in $\mathcal{C}$ between fibrant objects. This implies that

$$G(g)\colon G(X) \longrightarrow G(Y)$$

is a fibration in $\mathcal{D}$ between $L_{\mathbb{W}}\mathcal{D}$–fibrant objects. We can factor $G(g)$ in $L_{\mathbb{W}}\mathcal{D}$ as an acyclic cofibration followed by a fibration

$$G(X) \overset{i}{\underset{\sim}{\rightarrowtail}} C \overset{p}{\twoheadrightarrow} G(Y).$$

As C and $G(X)$ are fibrant in $L_{\mathbb{W}}\mathcal{D}$ and i is an $\mathbb{W}$-equivalence, i is also a weak equivalence in $\mathcal{D}$. As it is also a cofibration, it is thus an acyclic cofibration in $\mathcal{D}$. This implies that there is a lift H in the following diagram

$$\begin{array}{ccc} G(X) & = & G(X) \\ {\scriptstyle i}\downarrow & \overset{H}{\nearrow} & \downarrow{\scriptstyle G(g)} \\ C & \underset{p}{\longrightarrow} & G(Y). \end{array}$$

This means that there is a retract in $L_{\mathbb{W}}\mathcal{D}$

$$\begin{array}{ccccc} G(X) & \overset{i}{\longrightarrow} & C & \overset{H}{\longrightarrow} & G(X) \\ {\scriptstyle G(g)}\downarrow & & {\scriptstyle p}\downarrow & & \downarrow{\scriptstyle G(g)} \\ G(Y) & = & G(Y) & = & G(Y). \end{array}$$

Therefore, $G(g)$ is a retract of a fibration in $L_{\mathbb{W}}\mathcal{D}$ and thus is a fibration itself.

□

In the case of $K_{(2)}$-localisation, Theorem 7.4.39 gives us a criterion to check whether a spectrum is local. In our case, we take X a fibrant–cofibrant replacement of $\Psi(L_{K_{(2)}}\mathbb{S})$ with

$$\Psi\colon \operatorname{Ho}(L_{K_{(2)}}\mathcal{S}) \longrightarrow \operatorname{Ho}(\mathcal{C})$$

as in Theorem 7.4.40. A very laborious calculation in [Roi07] shows that for any object Y in $\mathcal{C}$, the mapping spectrum $\mathbb{R}\operatorname{Hom}(X, Y)$ is always $K_{(2)}$-local. This is done by examining the behaviour of v_1^4 on its mod-2 homotopy groups. Therefore, one obtains a Quillen adjunction

$$X \wedge -\colon L_{K_{(2)}}\mathcal{S}^{\mathbb{N}} \rightleftarrows \mathcal{C} : \mathbb{R}\operatorname{Hom}(X, -),$$

which one can now prove to be an equivalence.

The $K_{(2)}$-local sphere is a compact generator of $\operatorname{Ho}(L_{K_{(2)}}\mathcal{S}^{\mathbb{N}})$ as localisation with respect to $K_{(2)}$ is smashing. As outlined in Section 5.7, the $K_{(2)}$-local Rigidity Theorem can therefore be reduced to showing the following.

Proposition 7.4.42 *Let $F\colon \operatorname{Ho}(L_{K_{(2)}}\mathcal{S}) \longrightarrow \operatorname{Ho}(L_{K_{(2)}}\mathcal{S})$ be an exact endofunctor of the $K_{(2)}$-local stable homotopy category. Then*

$$F\colon \pi_n(L_{K_{(2)}}\mathbb{S}) \longrightarrow \pi_n(L_{K_{(2)}}\mathbb{S})$$

is an isomorphism for all n.

Like the proof of Theorem 5.7.1, the above statement in turn needs to be reduced to checking where F sends just a manageable list of generators of $\pi_n(L_{K_{(2)}}\mathbb{S})$. However, this reduction argument does not involve the Adams spectral sequence as in the non-local proof. In the $K_{(2)}$-local case, one manually verifies that F is an isomorphism in degrees -1 through to 9, using again that $2 \cdot \operatorname{Id}_M \neq 0$ and Toda bracket relations. The claim for the other degrees follows from using the 8-periodicity of the homotopy groups of $L_{K_{(2)}}M$ and homological algebra. Finally, one arrives at Theorem 7.4.40, namely, that all $K_{(2)}$-local higher homotopy information is encoded in the triangulated structure of the $K_{(2)}$-local stable homotopy category.

There is no such result at odd primes. In [Fra96], Franke constructs a stable model category whose homotopy category is triangulated equivalent to $\operatorname{Ho}(L_{K_{(p)}}\mathbb{S})$ for $p \geqslant 5$, but this model category is *not* Quillen equivalent to $K_{(p)}$-local spectra. (For example, one can show that the respective homotopy mapping spaces disagree.) It is baffling that one should obtain such different results at $p = 2$ and odd primes. This is partly due to the fact that $\pi_*(L_{K_{(2)}}\mathbb{S})$ is much more densely equipped with algebraic relations than its odd primary counterpart. The phenomenon still holds many open questions for future study.

7.4.3 A Brief Introduction to Chromatic Homotopy Theory

So far, we have only sporadically touched on the practical uses of left Bousfield localisation. In this section, we are going to give an overview of how localisation with respect to certain homology theories gives great insight into the overall structure of the stable homotopy category. This section's intention is to provide a first point of introduction to this type of result to a novice, proofs and details are left to the references, which we would encourage the reader to follow up.

For the rest of the section, we fix a prime p and work in the p-local stable homotopy category $\mathcal{SHC}_{(p)}$. Let us introduce our main tools. Ravenel [Rav92a] is a good introduction to these, but also see [Rav84].

Definition 7.4.43 For $n \geqslant 1$, the nth *Johnson–Wilson spectrum* $E(n)$ gives rise to a multiplicative homology theory $E(n)_*$ with

$$E(n)_* = \mathbb{Z}_{(p)}[v_1, v_2, \ldots, v_n, v_n^{-1}], \quad |v_i| = 2p^i - 2.$$

Definition 7.4.44 For $n \geqslant 1$, the nth *Morava K-theory spectrum* $K(n)$ gives rise to a multiplicative homology theory $K(n)_*$ with

$$K(n)_* = \mathbb{Z}/p\,[v_n, v_n^{-1}], \quad |v_n| = 2p^n - 2.$$

By convention, one sets $K(0) = E(0) = H\mathbb{Q}$ and $v_0 = p$. Also, note that the prime p is absent from notation – there is an $E(n)$ and a $K(n)$ for every p.

It is somewhat inappropriate to call the above "definitions", as it is by no means assumed that these are the only spectra with these properties. We will give a summary of the construction of the spectra $E(n)$ and $K(n)$.

Start with the complex cobordism spectrum MU, see [Ada74]. Its coefficients are

$$MU_* = \mathbb{Z}[x_1, x_2, \ldots], \quad |x_i| = 2i.$$

The p-localisation of MU splits into $p-1$ copies of a "smaller" spectrum, the *Brown–Peterson spectrum*,

$$MU_{(p)} = \bigvee_{i=0}^{p-2} \Sigma^{2i} BP,$$

where BP is a p-local spectrum with

$$BP_* = \mathbb{Z}_{(p)}[v_1, v_2, \cdots], \quad |v_i| = 2p^i - 2.$$

We can see from the notation that the rings $E(n)_*$ and $K(n)_*$ are modules over the ring BP_*. The spectra $E(n)$ and $K(n)$ are constructed as module spectra over BP to realise exactly this.

The Morava K-theories $K(n)$ are constructed from BP by first "killing" the unwanted elements $v_i, i \neq n$, of the homotopy groups by taking homotopy cofibres to obtain a spectrum $k(n)$ with $k(n)_* = \mathbb{Z}/p[v_n]$. Then one inverts v_n by taking a "telescope" to obtain

$$K(n) = \operatorname{colim}(k(n) \xrightarrow{v_n} \Sigma^{-2p^n-2}k(n) \xrightarrow{v_n} \Sigma^{-2(2p^n-2)}k(n) \xrightarrow{v_n} \cdots),$$

see, for example, [Rav86, Chapter 4.2]. Similarly, one obtains a spectrum $BP\langle n\rangle$ with

$$BP\langle n\rangle_* = \mathbb{Z}_{(p)}[v_1, \ldots, v_n],$$

where one formally inverts v_n to arrive at

$$E(n)_* := \operatorname{colim}(BP\langle n\rangle \xrightarrow{v_n} \Sigma^{-2p^n-2}BP\langle n\rangle \xrightarrow{v_n} \Sigma^{-2(2p^n-2)}BP\langle n\rangle \xrightarrow{v_n} \cdots).$$

Despite the similarity of their construction, the spectra $E(n)$ and $K(n)$ exhibit very different behaviour in places. For a spectrum X, we obtain a BP_*–module via

$$E(n)_*(X) := BP_*(X) \otimes_{BP_*} E(n)_*.$$

However, it is not at all clear for which BP_*–modules M the functor

$$M_*(X) := BP_*(X) \otimes_{BP_*} M$$

actually defines a homology theory. The *Landweber exact functor theorem* gives conditions on M making M_* a homology theory. The coefficient ring $E(n)_*$ satisfies Landweber exactness, see, for example, [Rav86]. This gives an alternative way of constructing $E(n)$ as the representing spectrum of that cohomology theory.

Contrastingly, $K(n)_*$ is not Landweber exact. Nonetheless, $K(n)_*$ has several useful algebraic properties. For example, it is a graded field in the sense that every module over it is free. Furthermore, it comes with a *Künneth isomorphism* for any spectra X and Y,

$$K_*(X) \otimes_{K(n)_*} K_*(Y) \cong K_*(X \wedge Y).$$

For $n = 1$, $E(1)$ and $K(1)$ are closely related to K-theory (see Subsection 7.4.2). Comparing their coefficient rings, we see that

$$E(1)_* = \mathbb{Z}_{(p)}[v_1, v_1^{-1}] \text{ and } K(1)_* = \mathbb{Z}/p\,[v_1, v_1^{-1}],$$

so in fact $K(1) = E(1) \wedge M(\mathbb{Z}/p)$. The spectrum $E(1)$ is the Adams summand of p-local K-theory, see Theorem 7.4.30, and therefore, $L_{E(1)} = L_{K_{(p)}}$. Thus, by Proposition 7.4.24, for a spectrum X we have

$$L_{K(1)}X = L_{E(1)\wedge M(\mathbb{Z}/p)}X = (L_{E(1)}(X))^\wedge_p = (L_{K_{(p)}}(X))^\wedge_p,$$

that is, localisation with respect to $E(1)$ is K-localisation, and localisation with respect to $K(1)$ is K-localisation followed by p-completion.

Definition 7.4.45 Localisation with respect to $E(n)$ is denoted by $L_n := L_{E(n)}$.

In particular, $L_1 = L_{K_{(p)}}$. Localisation with respect to $E(n)$ enjoys the following property, see [Rav92a, Theorem 7.5.6].

Theorem 7.4.46 (Ravenel) *Localisation with respect to $E(n)$ is smashing.*

We saw that as a consequence of Theorem 7.4.39, L_1 is a finite localisation as introduced at the beginning of Section 7.1, that is, it can be constructed as the localisation at a set of maps with finite cofibre. However, it has been unclear for decades if the same can be said about L_n. The conjecture that L_n is a finite localisation is known as the *Telescope Conjecture*. It was once thought to be disproven by [Rav92b], but the proof has since been found to be insufficient and the conjecture therefore remains open.

The following result relates the $E(n)$ and $K(n)$ in a very strong way, see [Rav84, Theorem 2.1.(d)].

Theorem 7.4.47 *A spectrum is $E(n)$-local if and only if it is local with respect to $(K(0) \vee K(1) \vee \cdots \vee K(n))$.*

Corollary 7.4.48 *If a spectrum is $E(n)$-local, it is also $E(n+1)$-local. Consequently, there is a natural transformation $L_{n+1} \longrightarrow L_n$.*

If a spectrum is $K(n)$-local, it is also $E(n)$-local. Consequently, there is a natural transformation $L_n \longrightarrow L_{K(n)}$.

Theorem 7.4.47 tells us that the "difference" between $E(n)$-localisation and $E(n-1)$-localisation is governed by $K(n)$. This has been made precise by the *chromatic square*, see Dwyer [Dwy04, Section 3.9]: for every p-local spectrum X, there is a homotopy pullback square

$$\begin{array}{ccc} L_n X & \longrightarrow & L_{K(n)} X \\ \downarrow & & \downarrow \\ L_{n-1} X & \longrightarrow & L_{n-1} L_{K(n)} X. \end{array}$$

The left vertical arrow and the top arrow are given by the natural transformations from Corollary 7.4.48. The right vertical arrow is localisation with respect to $E(n-1)$, and the bottom arrow is L_{n-1} applied to the localisation $X \longrightarrow L_{K(n)} X$.

The following theorem says that for growing n, $E(n)$-localisation provides a better and better "approximation" of the p-local stable homotopy category itself. It also justifies the term "chromatic homotopy theory", as it hints at the idea of decomposing $\mathcal{SHC}_{(p)}$ into "chromatic layers" using the L_n – we will see more of this later on.

Theorem 7.4.49 (Ravenel) *Let X be the p-localisation of a finite CW-spectrum. Then*

$$X \simeq \text{holim}(L_0X \longleftarrow L_1X \longleftarrow L_2X \longleftarrow \cdots).$$

This is known as the *Chromatic Convergence Theorem*. Note that the finiteness assumption is actually necessary. For example, Eilenberg–Mac Lane spectra are not finite. By Theorem 7.4.27, the $E(n)$-localisation of an Eilenberg–Mac Lane spectrum is its rationalisation, that is, for all n,

$$L_nHG \cong HG_{\mathbb{Q}}.$$

Therefore, the homotopy limit of the chromatic tower would also be $HG_{\mathbb{Q}}$, but not every Eilenberg–Mac Lane spectrum is rational.

Returning to the Morava K-theories, we see that localisation with respect to $K(n)$ behaves very differently to the case of $E(n)$-localisation.

Proposition 7.4.50 *Localising with respect to $K(n)$ is not smashing.*

Proof To show this, we need a spectrum E satisfying the following.

- E is $K(n)$-local.
- $\pi_*(E \wedge H\mathbb{Q}) \neq 0$, that is, the rationalisation of E is non-trivial.

Then, if $K(n)$-localisation was smashing, we would have

$$E = L_{K(n)}E = E \wedge L_{K(n)}\mathbb{S}$$

and also $L_{K(n)}\mathbb{S} \wedge H\mathbb{Q} = L_{K(n)}H\mathbb{Q}$. The latter is trivial, because the $K(n)$-localisation of an Eilenberg–Mac Lane spectrum is trivial by Theorem 7.4.27. Combining these properties, we obtain

$$0 \neq \pi_*(E \wedge H\mathbb{Q}) = \pi_*(E \wedge L_{K(n)}\mathbb{S} \wedge H\mathbb{Q}) = \pi_*(E \wedge *) = 0,$$

which is a contradiction. For the above, we have to know that such a spectrum E exists, which is highly non-trivial. For example, the nth Morava E-theory (also known as Lubin–Tate theory) satisfies the conditions, see, for example, Rezk [Rez98]. □

Despite this deficit, $K(n)$-localisation is still a powerful tool. For example, it helps us answer the following questions.

- Does a spectrum X carry a non-trivial nilpotent self-map, that is, a map that eventually becomes trivial when iterated?
- Does a spectrum X carry a non-trivial self-map that never becomes trivial when iterated?

Definition 7.4.51 Let $n \in \mathbb{N}$, and let X be a spectrum such that $K(m)_*(X) = 0$ for $m < n$. Then a map

$$\alpha\colon \Sigma^d X \longrightarrow X$$

is called a v_n*–self map* if

- $K(n)_*(\alpha)$ is multiplication by v_n^k for some k,
- $K(m)_*(\alpha) = 0$ for $m > n$.

It is worth remarking (and non-trivial) that $K(n)_*(X) = 0$ for some n implies that $K(n-1)_*(X) = 0$ [Rav84, Theorem 2.11]. Because of the first point in the definition, a v_n-self-map in particular never becomes trivial when iterated by composition.

Note that we already encountered a v_1-self-map on the mod-p Moore spectrum at the beginning of Subsection 7.4.2. It is an intriguing question whether such maps exist on a given spectrum. This has been answered by Ravenel [Rav92a, Theorem 1.5.4 and Chapter 6], see also Hopkins and Smith [HS98, Theorem 9].

Theorem 7.4.52 (Periodicity Theorem) *Let X be the p-localisation of a finite CW-spectrum. If n is the largest integer such that $K(m)_*(X) = 0$ for $m < n$, then X has a v_n-self-map.*

Let us now look at the other extreme.

Definition 7.4.53 A map $f\colon X \longrightarrow Y$ of spectra is *smash nilpotent* if there is an n such that $f^{\wedge n}\colon X^{\wedge n} \longrightarrow Y^{\wedge n}$ is nullhomotopic.

A self-map $f\colon \Sigma^d X \longrightarrow X$ of a spectrum X is *nilpotent* if

$$f^n = f \circ \cdots \circ f\colon \Sigma^{dn} X \longrightarrow X$$

is nullhomotopic for some n.

It can be shown that the Morava K-theories answer the question as to when a map is nilpotent in those ways, see [HS98, Theorem 3].

Theorem 7.4.54 (Nilpotence Theorem) *Let X be the p-localisation of a finite CW-spectrum, and let Y be a p-local spectrum. Then a map $f\colon X \longrightarrow Y$ is smash nilpotent if and only if*

$$K(n)_*(f) = 0 \ \textit{for all} \ 0 \leqslant n \leqslant \infty$$

(where $K(\infty) = H\mathbb{Z}/p$). A map $f\colon \Sigma^d X \longrightarrow X$ is nilpotent if and only if

$$K(n)_*(f) = 0 \ \textit{for all} \ 0 \leqslant n < \infty.$$

In fact, a spectrum E detects smash nilpotence as in the above theorem if and only if $K(n)_*(E) \neq 0$ for $0 \leqslant n \leqslant \infty$.

Let us now look at how the $K(n)$ are vital for understanding the overall structure of $\mathcal{SHC}_{(p)}$.

Recall the following definition. Let $\mathcal{T}$ be a triangulated category, and let $\mathcal{F}$ be a full triangulated subcategory of $\mathcal{T}$. Then $\mathcal{F}$ is *thick* if $\mathcal{F}$ is closed under retracts, that is, if

$$A \xrightarrow{i} X \xrightarrow{p} A$$

such that $p \circ i = \mathrm{Id}$ and $X \in \mathcal{F}$, then A is also in $\mathcal{F}$.

Note that if a coproduct of objects of $\mathcal{T}$ is contained in $\mathcal{F}$, then all its summands are in $\mathcal{F}$ too.

Definition 7.4.55 Let $\mathcal{T}$ be a tensor-triangulated category (see Remark 6.1.15), and let $\mathcal{F}$ be a thick subcategory of $\mathcal{T}$. Then $\mathcal{F}$ is an *ideal* in $\mathcal{T}$ if for $X \in \mathcal{T}$ and $Y \in \mathcal{F}$, $X \wedge Y$ is in $\mathcal{F}$.

Recall that a finite p-local spectrum is a spectrum which is isomorphic to the p-localisation of a finite CW-spectrum.

We let $\mathcal{SHC}^{\mathrm{fin}}_{(p)}$ denote the full triangulated subcategory of $\mathcal{SHC}_{(p)}$ of finite p-local spectra.

We can make a non-trivial example of a thick subcategory that is also an ideal in $\mathcal{SHC}^{\mathrm{fin}}_{(p)}$ from these spectra.

$$\mathcal{F}_n = \{X \text{ a finite } p\text{-local spectrum} \mid K(n-1)_*(X) = 0\}.$$

The $\mathcal{F}_n$ form a descending chain of ideals in $\mathcal{SHC}^{\mathrm{fin}}_{(p)}$,

$$\mathcal{F}_1 \subset \mathcal{F}_2 \subset \cdots \subset \mathcal{F}_{n-1} \subset \mathcal{F}_n \subset \cdots .$$

In fact, the $\mathcal{F}_n$ as defined above give us all thick subcategories of $\mathcal{SHC}^{\mathrm{fin}}_{(p)}$, see [Rav92a, Theorem 3.4.3], [HS98, Theorem 7] and also [HPS97, Section 5.2]. This means that the $K(n)$-acyclic spectra form the "atomic" part of $\mathcal{SHC}^{\mathrm{fin}}_{(p)}$.

Theorem 7.4.56 (Thick Subcategory Theorem) *Let $\mathcal{F}$ be a non-trivial thick subcategory of the homotopy category of finite p-local spectra. Then $\mathcal{F} = \mathcal{F}_n$ for some $n \geqslant 1$.*

The Chromatic Convergence Theorem 7.4.49, the chromatic square and the Thick Subcategory Theorem allow us to think of the stable homotopy category as follows. For each prime p, the p-local stable homotopy category can be thought of as a building with infinitely many floors. The nth floor is described by $L_{K(n)}$, and the first n floors together are described by $L_{E(n)}$.

Visualising Ho($\mathcal{S}$) *in relation to* Ho($L_n\mathcal{S}$):

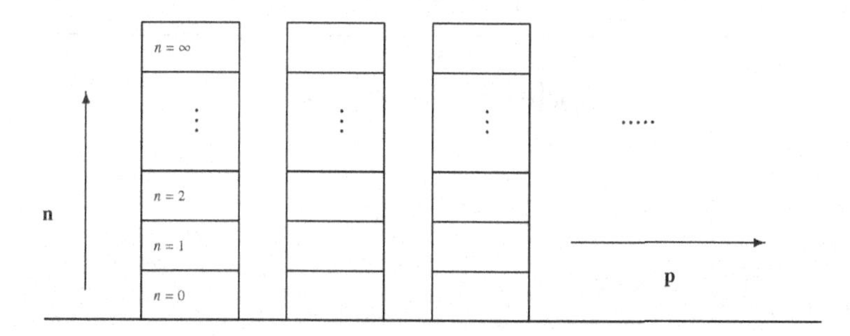

The "ground floor" is rationalisation, and the first floor is governed by K-theory. The second floor is related to *elliptic cohomology theories*, which we will not discuss in this book – we recommend Hopkins and Mahowald [HM14] for the interested reader. This second chromatic layer is already very difficult to describe explicitly or to use for calculations, and beyond this even less is known. The recently emerging *transchromatic homotopy theory* is aiming to shed more light on how the different chromatic layers are related.

Because we have the notion of an ideal in a tensor-triangulated category, we can also say what a prime ideal should be.

Definition 7.4.57 An ideal $\mathcal{F}$ is *prime* if $X \wedge Y \in \mathcal{F}$ implies that either $X \in \mathcal{F}$ or $Y \in \mathcal{F}$.

The notion of ideals and prime ideals in a tensor-triangulated category can be combined with techniques from commutative algebra in order to study the structure of other interesting homotopy categories as well. The set of thick prime ideals in a tensor-triangulated category is known as the *Balmer spectrum*

$$\mathrm{Spc}(\mathcal{T}) = \{\mathcal{P} \mid \mathcal{P} \subsetneq \mathcal{T} \text{ is a thick prime ideal in } \mathcal{T}\},$$

see Balmer and Sanders [BS17]. It can be given a Zariski-style topology by asking for the sets

$$\mathrm{supp}(X) = \{\mathcal{P} \in \mathrm{Spc}(\mathcal{T}) \mid X \notin \mathcal{P}\}, \quad X \in \mathcal{T},$$

to be closed. For example, the Balmer spectrum of the homotopy category of G-spectra for finite groups G is a particularly interesting object of study.

Appendix

Model Categories

In this Appendix, we give a short account of model categories, summarising the definitions and results that we use in this book. With some exceptions, we will not include proofs, but instead provide references where those can be found if desired.

A.1 Basic Definitions

A model category is a category with a notion of homotopy between morphisms. The ideas of a homotopy between continuous functions of topological spaces and a chain homotopy between chain maps of chain complexes are well established. We will see that they are not just "morally" similar, but that they are special cases arising from the same formal definition, see [DS95].

Recall that a morphism $f\colon X \longrightarrow Y$ is a *retract* of $g\colon U \longrightarrow V$, if there is a commutative diagram

$$\begin{array}{ccccc} X & \xrightarrow{i} & U & \xrightarrow{r} & X \\ {\scriptstyle f}\downarrow & & {\scriptstyle g}\downarrow & & \downarrow{\scriptstyle f} \\ Y & \xrightarrow[i']{} & V & \xrightarrow[r']{} & Y, \end{array}$$

such that $r \circ i$ and $r' \circ i'$ are the respective identities.

Definition A.1.1 A *model category* is a category $\mathcal{C}$ equipped with three distinguished classes of morphisms, which are closed under composition and all contain the identities:

- weak equivalences, denoted $\xrightarrow{\sim}$
- cofibrations, denoted $\rightarrowtail$
- fibrations, denoted $\twoheadrightarrow$.

This information satisfies the following axioms.

(MC1) The category $\mathcal{C}$ has all finite limits and colimits.

(MC2) **(2-out-of-3)** Let f and g be composable morphisms in $\mathcal{C}$. If two out of f, g and $g \circ f$ are weak equivalences, then so is the third.

(MC3) **(Retracts)** Weak equivalences, cofibrations and fibrations are closed under retracts.

(MC4) **(Lifts)** Assume there is a commutative square in $\mathcal{C}$

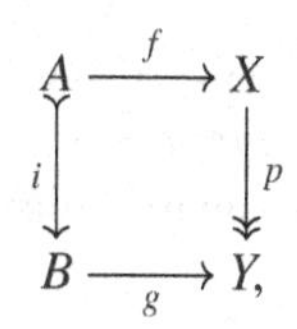

where i is a cofibration and p is a fibration. If in addition, either i or p is a weak equivalence, then there is a lift $h\colon B \longrightarrow X$ in the square, that is, $h \circ i = f$ and $p \circ h = g$ as below.

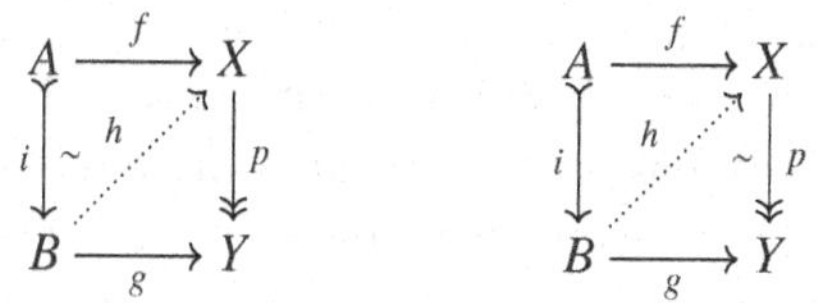

(MC5) **(Factorisation)** Every morphism $f\colon X \longrightarrow Y$ can be factored as a cofibration that is a weak equivalence followed by a fibration and as a cofibration followed by a fibration that is a weak equivalence, that is,

$$f\colon X \overset{\sim}{\rightarrowtail} U \twoheadrightarrow Y, \qquad f\colon X \rightarrowtail V \overset{\sim}{\twoheadrightarrow} Y.$$

Some sources ask for the factorisations in (MC5) to be functorial. We do not make this assumption in general, but it does hold for many of the model categories we encounter in practice, see Corollary A.6.14.

The information of weak equivalences, cofibrations and fibrations together with axioms (MC2) – (MC5) is often called a *model structure* on the category $\mathcal{C}$. We will see that there can be more than one different model structure on the same underlying category.

First, some terminology for convenience's sake.

Definition A.1.2 A morphism that is both a cofibration and a weak equivalence is called an *acyclic cofibration* or *trivial cofibration* and is denoted by $\overset{\sim}{\rightarrowtail}$. A morphism that is both a fibration and a weak equivalence is called an *acyclic fibration* or *trivial fibration* and is denoted by $\overset{\sim}{\twoheadrightarrow}$.

Definition A.1.3 Let $\mathcal{C}$ be a category. We say that a morphism $i\colon A \longrightarrow B$ has the *left lifting property* (LLP) with respect to another morphism $p\colon X \longrightarrow Y$ if for any commutative diagram

$$\begin{array}{ccc} A & \xrightarrow{f} & X \\ {\scriptstyle i}\downarrow & \sim & \downarrow{\scriptstyle p} \\ B & \xrightarrow[g]{} & Y \end{array}$$

there is a morphism $h\colon B \longrightarrow X$, such that $h \circ i = f$ and $p \circ h = g$. Similarly, $p\colon X \longrightarrow Y$ has the *right lifting property* (RLP) with respect to $i\colon A \longrightarrow B$ if such a lift h exists for every commutative diagram of the above form.

Let us list some consequences of the definition of a model category. Proofs can be found in [DS95], [Hov99] and [Qui67].

- Axiom (MC1) implies that a model category $\mathcal{C}$ has finite products and coproducts as well as an initial object $\emptyset$ and a terminal object $*$.
- Axiom (MC4) says that cofibrations have the left lifting property with respect to acyclic fibrations, acyclic cofibrations have the left lifting property with respect to fibrations, acyclic fibrations have the right lifting property with respect to cofibrations and fibrations have the right lifting property with respect to acyclic cofibrations. However, one can show that these conditions are also necessary, that is, a morphism is a cofibration *if and only if* it has the right lifting property with respect to all acyclic fibrations and so on.
- The above remark tells us that the full axioms of a model category contain more information than strictly necessary, as the fibrations could be defined as those maps having the right lifting property with respect to all acyclic cofibrations, and the cofibrations could be defined as precisely those maps having the left lifting property with respect to all acyclic fibrations. In other words, the weak equivalences and the cofibrations determine the fibrations, and the weak equivalences and the fibrations determine the cofibrations.
- Cofibrations and acyclic cofibrations are closed under pushouts.
- Fibrations and acyclic fibrations are closed under pullbacks.

Remark A.1.4 We usually assume a stronger condition than axiom (MC1), namely, that our model categories have all small limits and colimits. Recall that a small (co)limit is a (co)limit over a diagram whose object class is a set.

There is a wealth of naturally arising model structures. We will list some of the most common ones below. In each case, it is a non-trivial effort to prove that the model category axioms are satisfied. We leave this to the references.

Example A.1.5 There are many meaningful examples of model structures on the category of topological spaces Top. A commonly used one (which we are also using in this book) is the *Serre model structure*. In this model structure, a continuous map $f: X \longrightarrow Y$ is

- a weak equivalence if f is a weak homotopy equivalence,
- a cofibration if f is a retract of a map $X \longrightarrow Y'$, where Y' is obtained from X by attaching cells (i.e. (X, Y') is a relative CW-complex),
- a fibration if f is a Serre fibration, that is, f has the right lifting property with respect to the inclusion

$$A \times \{0\} \longrightarrow A \times [0, 1]$$

for any CW-complexes A.

The *Hurewicz model structure* on Top is given by the following. A continuous map $f: X \longrightarrow Y$ is

- a weak equivalence if f is a homotopy equivalence,
- a cofibration if f has the left lifting property with respect to all fibrations that are also homotopy equivalences,
- a fibration if f has the right lifting property with respect to the inclusion

$$A \times \{0\} \longrightarrow A \times [0, 1]$$

for any topological space A. This is called a *Hurewicz fibration*.

In this book, the cofibrations in the Serre model structure are also called *q-cofibrations*, and the cofibrations in the Hurewicz model structure are called *h-cofibrations*.

Example A.1.6 Let Δ denote the category of finite ordered sets

$$[n] = \{0, 1, \cdots, n\}, \quad n \in \mathbb{N},$$

with the morphisms being the order-preserving maps, and let Set denote the category of sets. The category of *simplicial sets* sSet is the category of functors $\Delta^{op} \longrightarrow \text{Set}$. There is an adjunction

$$|-|: \text{sSet} \rightleftarrows \text{Top} : \text{sing},$$

where $|-|$ denotes the geometric realisation functor, and where for a space $A \in \text{Top}$, the set $\text{sing}(A)_n$ is the set of continuous functions from the standard n-simplex $\Delta_n \in \mathbb{R}^n$ into A. A model structure on sSet is given by the following data. A map $f: X \longrightarrow Y$ of simplicial sets is

- a weak equivalence if $|f|\colon |X| \longrightarrow |Y|$ is a weak homotopy equivalence in Top,
- a cofibration if for every $n \in \mathbb{N}$, the map $f([n])\colon X([n]) \longrightarrow Y([n])$ is a monomorphism,
- a fibration if f has the right lifting property with respect to all acyclic cofibrations.

The fibrations in this model structure are known as *Kan fibrations*.

Example A.1.7 Important algebraic examples of model categories come from chain complexes. Let us denote the category of non-negatively graded chain complexes of R–modules by $\mathrm{Ch}(R)_+$. We say that a chain map $f\colon C_* \longrightarrow D_*$ is

- a weak equivalence if $H_*(f)\colon H_*(X) \longrightarrow H_*(Y)$ is an isomorphism of graded abelian groups, that is, f is a homology isomorphism,
- a cofibration if for each degree n, the map $f_n\colon C_n \longrightarrow D_n$ is injective with a projective cokernel,
- a fibration if for each $n \geqslant 1$, the map $f_n\colon C_n \longrightarrow D_n$ is surjective.

This is known as the *projective model structure* on $\mathrm{Ch}(R)_+$.

We can define another model structure on $\mathrm{Ch}(R)_-$, the non-positively graded chain complexes of R–modules by asking for a chain map $f\colon C_* \longrightarrow D_*$ to be

- a weak equivalence if f is a homology isomorphism,
- a cofibration if for each degree $n \leqslant -1$, the map $f_n\colon C_n \longrightarrow D_n$ is injective,
- a fibration if for each n, the map $f_n\colon C_n \longrightarrow D_n$ is surjective with injective kernel.

This is known as the *injective model structure* on $\mathrm{Ch}(R)_-$. There are also projective and injective model structures on unbounded chain complexes, although they are not quite as straightforward to describe. We will do so in Section A.6.

Definition A.1.8 An object X in a model category $\mathcal{C}$ is *cofibrant* if the morphism $\emptyset \longrightarrow X$ out of the initial object is a cofibration. An object X in a model category is called *fibrant* if the unique morphism $X \longrightarrow *$ to the terminal object is a fibration.

Examples A.1.9 We see that in the Serre model structure on topological spaces Top defined earlier, a retract of a CW-complex is cofibrant. Meanwhile, every simplicial set is cofibrant.

In the projective model structure on $\mathrm{Ch}(R)_+$, a chain complex is cofibrant if and only if it is degreewise projective, and all chain complexes are fibrant. In the injective model structure on $\mathrm{Ch}(R)_-$, every chain complex is cofibrant, and a chain complex is fibrant if and only if it is degreewise injective.

For many properties and proofs, it is necessary to assume that an object is fibrant or cofibrant. However, this is not always a major restriction as any object can be replaced by a fibrant and a cofibrant one.

Definition A.1.10 Let $\mathcal{C}$ be a model category and X an object in $\mathcal{C}$. We say that an object Y is a *cofibrant replacement* for X if Y is cofibrant and there is a weak equivalence $Y \longrightarrow X$. An object Z is a *fibrant replacement* for X if Z is fibrant and there is a weak equivalence $X \longrightarrow Z$.

Fibrant and cofibrant replacements exist for every $X \in \mathcal{C}$ by applying (MC5) to the morphisms $\emptyset \longrightarrow X$ and $X \longrightarrow *$, respectively. They are unique up to weak equivalence, therefore we can write X^{cof} and X^{fib} for a cofibrant, respectively, fibrant replacement of X even if we have not assumed functorial factorisation in our model category as in practice, proofs will not depend on this choice.

Examples A.1.11 In Top with the Serre model structure, cofibrant replacement is given by CW-approximation.

Let A be an R–module, and let $A[0] \in \mathrm{Ch}(R)_+$ be the chain complex consisting of A in degree 0 and zero elsewhere. In the projective model structure on $\mathrm{Ch}(R)_+$, a cofibrant replacement of $A[0]$ is a projective resolution of A. If we consider $A[0]$ as an object of $\mathrm{Ch}(R)_-$ with the injective model structure, a fibrant replacement is an injective resolution.

A.2 Homotopies

In a model category, we can recreate the geometric notion of homotopy without having a "unit interval" in our category.

Definition A.2.1 Let $\mathcal{C}$ be a model category and $X \in \mathcal{C}$. A *cylinder object* for X is an object $\mathrm{Cyl}(X)$ together with a diagram

$$X \coprod X \xrightarrow{(i_0, i_1)} \mathrm{Cyl}(X) \xrightarrow[\sim]{r} X,$$

which factors the fold map $(\mathrm{Id}, \mathrm{Id})\colon X \coprod X \longrightarrow X$.

We have to make some important remarks concerning this definition. Firstly, the cylinder object is not just the object in the middle of this factorisation, but the entire diagram, although we are also going to abuse this wording. By the axiom (MC5), every object has (at least) one cylinder object, and it can be chosen so that (i_0, i_1) is a cofibration, i_0 and i_1 are weak equivalences and r an acyclic fibration. This is known in the literature as a *very good cylinder object*. Furthermore, if X is cofibrant, then i_0 and i_1 are acyclic cofibrations [DS95, Lemma 4.1]. For this book, we will assume all our cylinder objects to be very good. For X a topological space, an actual cylinder $X\times[0,1]$ provides a cylinder object in topological spaces. For a chain complex A, a cylinder object is given by

$$\mathrm{Cyl}(A_*)_n = A_n \oplus A_{n-1} \oplus A_n, \quad \partial(a,b,c) = (d(a)+b, -d(b), d(c)-b)$$

together with the evident inclusions and projection map.

Given a morphism $f\colon X \longrightarrow Y$, we have the following commutative diagram.

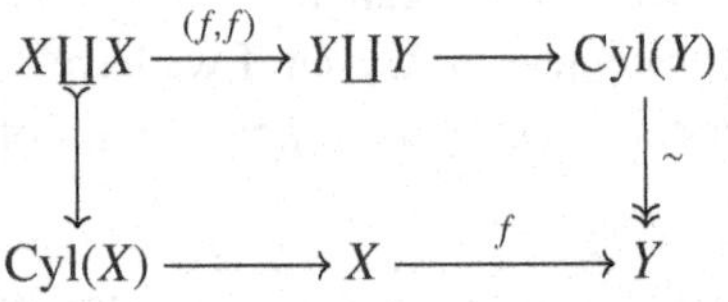

$$\begin{array}{ccccc} X\coprod X & \xrightarrow{(f,f)} & Y\coprod Y & \longrightarrow & \mathrm{Cyl}(Y) \\ \downarrow & & & & \downarrow \sim \\ \mathrm{Cyl}(X) & \longrightarrow & X & \xrightarrow{f} & Y \end{array}$$

Axiom (MC4) says that there is a lift $\mathrm{Cyl}(X) \longrightarrow \mathrm{Cyl}(Y)$, giving us a commutative diagram

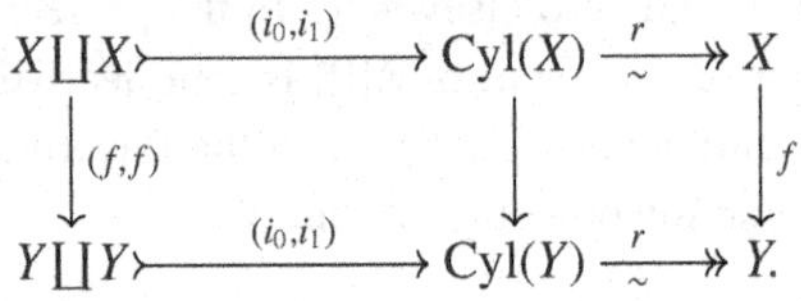

$$\begin{array}{ccccc} X\coprod X & \xrightarrow{(i_0,i_1)} & \mathrm{Cyl}(X) & \xrightarrow[\sim]{r} & X \\ \downarrow (f,f) & & \downarrow & & \downarrow f \\ Y\coprod Y & \xrightarrow{(i_0,i_1)} & \mathrm{Cyl}(Y) & \xrightarrow[\sim]{r} & Y. \end{array}$$

In particular, we see that two cylinder objects of the same object are weakly equivalent.

Definition A.2.2 Let $f, g\colon X \longrightarrow Y$ be two morphisms in a model category $\mathcal{C}$. A *left homotopy* between f and g is a morphism

$$H\colon \mathrm{Cyl}(X) \longrightarrow Y,$$

such that $H \circ i_0 = f$ and $H \circ i_1 = g$ for some cylinder object of X. We say that f and g are *left homotopic*.

This recovers the notion of homotopy between two continuous maps of CW-complexes in topological spaces, and one can check that this also recovers the definition of chain homotopy in chain complexes.

Dually, we can make the following definitions, where analogous remarks apply.

Definition A.2.3 Let $\mathcal{C}$ be a model category, and let $Y \in \mathcal{C}$. A *path object* for Y is an object $PY \in \mathcal{C}$ together with a diagram

$$Y \xrightarrow[\sim]{s} PY \xrightarrow{(e_0,e_1)} Y \times Y,$$

which factors the diagonal map $\Delta\colon Y \longrightarrow Y \times Y$.

Again, when we write "path object" in this book, we will actually mean a *very good path object* where s is an acyclic cofibration, e_0 and e_1 are weak equivalences and (e_0, e_1) is a fibration. If Y is fibrant, then e_0 and e_1 can be chosen to be acyclic fibrations [DS95, Lemma 4.14].

Definition A.2.4 Let $f, g\colon X \longrightarrow Y$ be two morphisms in a model category $\mathcal{C}$. A *right homotopy* between f and g is a morphism

$$H\colon X \longrightarrow PY,$$

such that $e_0 \circ H = f$ and $e_1 \circ H = g$ for some path object of Y. We say that f and g are *right homotopic*.

Definition A.2.5 Two morphisms f and g in a model category $\mathcal{C}$ are *homotopic* if they are both left and right homotopic. We denote this by $f \simeq g$.

We will now list some properties of left and right homotopies. The proofs of these can be found in [DS95, Section 4], except for (4.) which we prove below.

1. If X is cofibrant, then "being left homotopic" is an equivalence relation on $\mathcal{C}(X, Y)$.
2. If Y is fibrant, then "being right homotopic" is an equivalence relation on $\mathcal{C}(X, Y)$.
3. If X is cofibrant and Y is fibrant, then "being homotopic" is an equivalence relation on $\mathcal{C}(X, Y)$. An equivalence class is called a *homotopy class* of a morphism f, denoted $[f]$.
4. If X is cofibrant and Y is fibrant, then f and g are left homotopic if and only if they are right homotopic.
5. If X, Y and Z are both fibrant and cofibrant, then composition induces a well-defined map on homotopy classes

$$(\mathcal{C}(X, Y)/\sim) \times (\mathcal{C}(Y, Z)/\sim) \longrightarrow \mathcal{C}(X, Z)/\sim, \quad ([f], [g]) \mapsto [g \circ f].$$

6. If X and Y are both fibrant and cofibrant, then a map $f\colon X \longrightarrow Y$ is a weak equivalence if and only if it has a homotopy inverse.

Lemma A.2.6 *Let $f, g\colon X \longrightarrow Y$ be two morphisms in a model category $\mathcal{C}$. Furthermore, assume that X is cofibrant and Y fibrant. Then f and g are left homotopic if and only if they are also right homotopic.*

Proof We will only show one direction, as the other is very similar. Let $H\colon Y \longrightarrow PA$ be a right homotopy between $f = e_0 \circ H$ and $g = e_1 \circ H$. Consider the cylinder object

$$Y \coprod Y \xrightarrow{(i_0,i_1)} \mathrm{Cyl}(Y) \xrightarrow{r} Y$$

and the path object

$$A \xrightarrow{s} PA \xrightarrow{(e_0,e_1)} A \times A.$$

Because of our assumptions on X and Y, we can pick (i_0, i_1) to be a cofibration and e_0 to be an acyclic fibration, so there is a lift Φ in the diagram

$$\begin{array}{ccccc} Y\coprod Y & & \xrightarrow{(s\circ e_0 \circ H, H)} & & PA \\ \downarrow{\scriptstyle (i_0,i_1)} & & \nearrow{\scriptstyle \Phi} & & \downarrow{\scriptstyle e_0} \\ \mathrm{Cyl}(Y) & \xrightarrow{r} & Y & \xrightarrow{e_0\circ H} & A. \end{array}$$

The map $\bar{H} := e_1 \circ \Phi\colon \mathrm{Cyl}(Y) \longrightarrow A$ is now a left homotopy between f and g because

$$e_1 \circ \Phi \circ i_0 = e_1 \circ s \circ e_0 \circ H = e_0 \circ H$$

(because $e_1 \circ s = \mathrm{Id}$), and

$$e_1 \circ \Phi \circ i_1 = e_1 \circ H. \qquad \square$$

Definition A.2.7 The map Φ in the above proof is called a *correspondence* between the right homotopy H and the left homotopy $\bar{H}$.

As with topological spaces, we may concatenate (glue together) compatible homotopies.

Let $\alpha, \beta\colon \mathrm{Cyl}(A) \longrightarrow X$ be two left homotopies with $\alpha \circ i_1 = \beta \circ i_0$. This means that we can form the following pushout.

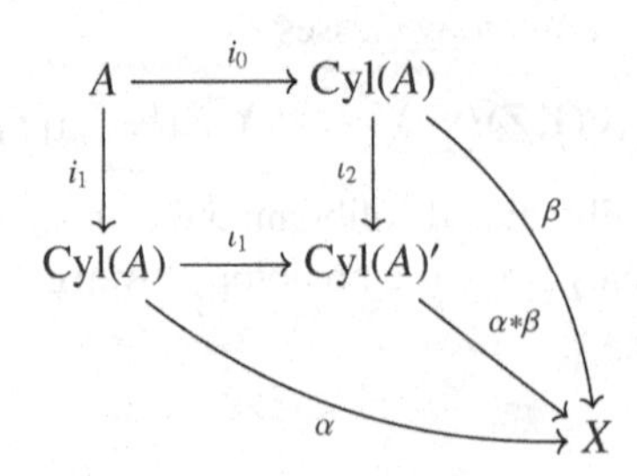

Definition A.2.8 The *concatenation* of the homotopies α and β is given by $\alpha * \beta$ in the above diagram.

Note that we read the notation $\alpha*\beta$ from left to right. The object $\mathrm{Cyl}(A)'$ is a cylinder object for A with $i_0' \colon A \longrightarrow \mathrm{Cyl}(A)'$ given by $\iota_1 \circ i_0$, and $i_1' \colon A \longrightarrow \mathrm{Cyl}(A)'$ given by $\iota_2 \circ i_1$.

If A is cofibrant, then i_0 and i_1 are acyclic cofibrations, thus ι_1 and ι_2 are, and therefore

$$(i_0', i_1') \colon A \coprod A \longrightarrow \mathrm{Cyl}(A)'$$

is an acyclic cofibration. The structure map $\mathrm{Cyl}(A)' \longrightarrow A$ is given by the pushout of the structure map $r \colon \mathrm{Cyl}(A) \longrightarrow A$ with itself. While the pushout of two fibrations is not necessarily a fibration, we can just factor $r * r$ into an acyclic cofibration with a fibration, and this factorisation gives us a very good cylinder object. However, we will simply assume without loss of generality that $\mathrm{Cyl}(A)'$ is itself very good.

Thus, we have

$$(\alpha * \beta) \circ i_0' = (\alpha * \beta) \circ \iota_1 \circ i_0 = \alpha \circ i_0$$

and

$$(\alpha * \beta) \circ i_1' = (\alpha * \beta) \circ \iota_2 \circ i_1 = \beta \circ i_1.$$

For a map $f \colon X \longrightarrow Y$ we get

$$f \circ (\alpha * \beta) = (f \circ \alpha) * (f \circ \beta).$$

We can of course perform a dual construction for right homotopies using pullbacks of path objects. These constructions satisfy some useful properties. We omit the details of the proofs, which can be found in [Qui67, Chapter 1.1].

1. Two maps $\alpha, \beta \colon \mathrm{Cyl}(A) \longrightarrow PB$ can be considered to be left homotopies as well as right homotopies. The concatenation $\alpha * \beta$ as left homotopies is not the same morphism as $\alpha * \beta$ concatenated as right homotopies, but the results are homotopic if A and B are both fibrant and cofibrant.
2. Let Φ^α be a correspondence between $\alpha \colon \mathrm{Cyl}(A) \longrightarrow B$ and some right homotopy in the sense of Definition A.2.7 and Φ^β a correspondence between $\beta \colon \mathrm{Cyl}(A) \longrightarrow B$ and some other right homotopy. Then the correspondence of their concatenation is homotopic to the concatenation of correspondences, that is,

$$\Phi^\alpha * \Phi^\beta \simeq \Phi^{\alpha*\beta}.$$

3. For a morphism $f: A \longrightarrow B$, let

$$\mathrm{const}_f := r \circ \mathrm{Cyl}(f) \colon \mathrm{Cyl}(A) \longrightarrow B$$

denote the constant homotopy. Then, for a left homotopy $H\colon \mathrm{Cyl}(A) \longrightarrow B$, we have

$$H * \mathrm{const}_{H \circ i_1} \simeq H \simeq \mathrm{const}_{H \circ i_0} * H.$$

An analogous statement holds for right homotopies.

4. Given $H\colon \mathrm{Cyl}(A) \longrightarrow B$, we can define the reverse $\bar{H}$ of H as follows. Consider the cylinder object

$$A \coprod A \xrightarrow{\tau} A \coprod A \xrightarrow{(i_0, i_1)} \mathrm{Cyl}(A) \xrightarrow{r} A,$$

where τ is the switch morphism. This cylinder now satisfies

$$i_0^{new} = i_1^{old} \circ \tau \text{ and } i_1^{new} = i_0^{old} \circ \tau.$$

Now $\bar{H}$ is the same morphism as H, but considered on this new cylinder object. Therefore, if H is a homotopy between $f: A \longrightarrow B$ and $g: A \longrightarrow B$, then $\bar{H}$ is a homotopy between

$$\bar{H} \circ i_0^{new} = H \circ i_1^{old} = g \text{ and } \bar{H} \circ i_1^{new} = H \circ i_0^{old} = f.$$

The respective concatenations between those two homotopies satisfy, for fibrant and cofibrant A and B,

$$H * \bar{H} \simeq \mathrm{const}_f \text{ and } \bar{H} * H \simeq \mathrm{const}_g.$$

An analogous statement holds for right homotopies.

5. As taking pushouts (resp. pullbacks) is associative, the concatenation of left (resp. right) homotopies is associative up to homotopy.
6. The previous three points can be summarised by saying that for cofibrant and fibrant A and B, the set of morphisms from A to B are the objects of a groupoid, with morphisms given by homotopies of maps (up to homotopy).

A.3 The Homotopy Category

We are now going to make the following important definition. Recall that for an object X in a model category, X^{cof} denotes the cofibrant replacement and X^{fib} the fibrant replacement.

Definition A.3.1 Let $\mathcal{C}$ be a model category. The *homotopy category* of $\mathcal{C}$, denoted $\mathrm{Ho}(\mathcal{C})$, is defined as follows:

- the objects of $\mathrm{Ho}(\mathcal{C})$ are the objects of $\mathcal{C}$,
- the morphisms of $\mathrm{Ho}(\mathcal{C})$ are given by

$$\mathrm{Ho}(\mathcal{C})(X, Y) = \mathcal{C}((X^{cof})^{fib}, (Y^{cof})^{fib})/\sim,$$

that is, by the homotopy classes between the respective fibrant-cofibrant replacements.

If there is no ambiguity over which model category we are working in, we will denote $\mathrm{Ho}(\mathcal{C})(X, Y)$ by $[X, Y]$.

Example A.3.2 An example of morphisms in the homotopy category comes from chain complexes $\mathrm{Ch}(R)_+$. Let $A[m]$ and $B[n]$ be those chain complexes which consist of the R–modules A concentrated in degree m and B in degree n, respectively, and which are zero elsewhere. Then, by applying the previous definitions to this specific example, one can show that

$$\mathrm{Ho}(\mathrm{Ch}(R)_+)(A[m], B[n]) = \mathrm{Ext}_R^{n-m}(A, B).$$

This is [DS95, Proposition 7.3], and it beautifully highlights the interaction between homotopy theory and algebra.

Further technical properties about homotopy categories can be found in [DS95, Section 5], as well as [Hov99, Chapter 1.2], including the following.

Proposition A.3.3 *A morphism* $[f] \in \mathrm{Ho}(\mathcal{C})$ *is an isomorphism if and only if* f *is a weak equivalence in* $\mathcal{C}$.

Therefore, $\mathrm{Ho}(\mathcal{C})$ is equivalent to the category-theoretic localisation $\mathcal{C}[\mathcal{W}^{-1}]$ of $\mathcal{C}$ at the class of weak equivalences $\mathcal{W}$, which is the free category on the arrows in $\mathcal{C}$ together with added reversed arrows of the weak equivalences. It is not initially clear why such a construction would result in a category with morphism *sets* between objects rather than classes of morphisms, but our construction using homotopies guarantees that this is in fact true in the model category case.

Before we move on to functors, we have to make the following important remark.

Remark A.3.4 The lifting axiom (MC3) for model categories does not make any claims about the uniqueness of such a lift. However, two lifts in a square, such as

$$\begin{array}{ccc} A & \xrightarrow{f} & X \\ {\scriptstyle i}\big\downarrow{\scriptstyle \sim} & \nearrow & \big\downarrow{\scriptstyle p} \\ B & \xrightarrow{g} & Y \end{array}$$

are identical in the homotopy category: assume that there are two lifts h_1 and h_2 in the square above. Then

$$[h_1] \circ [i] = [h_2] \circ [i] = [f].$$

But as i is a weak equivalence, $[i]$ is an isomorphism in $\mathrm{Ho}(\mathcal{C})$, thus $[h_1] = [h_2]$. This means, in particular, that in $\mathrm{Ho}(\mathcal{C})$, the map

$$\mathrm{Cyl}(X) \longrightarrow \mathrm{Cyl}(Y)$$

from earlier in Section A.2 is unique. Similarly, choices of left (or right) homotopies between morphisms and correspondences between left and right homotopies from Definition A.2.7 are unique in the homotopy category.

A.4 Quillen Functors

Now that we have discussed some properties of model categories, it makes sense to consider functors that respect this structure. We give a very simplified outline of the basics—more details and proofs of the statements can be found in, for example, [Hov99, Chapter 1.3].

Definition A.4.1 Let $\mathcal{C}$ and $\mathcal{D}$ be model categories. A functor

$$F\colon \mathcal{C} \longrightarrow \mathcal{D}$$

is a *left Quillen functor* if F preserves cofibrations and acyclic cofibrations. A functor

$$G\colon \mathcal{D} \longrightarrow \mathcal{C}$$

is a *right Quillen functor* if it preserves fibrations and acyclic fibrations. An adjunction of functors

$$F\colon \mathcal{C} \rightleftarrows \mathcal{D} : G$$

is a *Quillen adjunction* if F is a left Quillen functor and G is a right Quillen functor.

Example A.4.2 The adjunction consisting of geometric realisation and the singular complex functor

$$|-|\colon \mathrm{sSet} \rightleftarrows \mathrm{Top} : \mathrm{sing}$$

from Example A.1.6 is a Quillen adjunction.

Lemma A.4.3 *Let $\mathcal{C}$ and $\mathcal{D}$ be model categories and*

$$F\colon \mathcal{C} \rightleftarrows \mathcal{D} : G$$

an adjunction of functors. Then the following are equivalent.

- *F preserves cofibrations and G preserves fibrations.*
- *F is a left Quillen functor.*
- *G is a right Quillen functor.*
- *(F, G) is a Quillen adjunction.*
- *F preserves acyclic cofibrations and F preserves cofibrations between cofibrant objects.*
- *G preserves acyclic fibrations and G preserves fibrations between fibrant objects.*

For the proof that the last two points are equivalent to the others, see [Hir03, Proposition 8.5.4].

The following result has been found to be quite useful. Note that it does not require the functor F be a left or right Quillen functor.

Lemma A.4.4 (Ken Brown's Lemma) *Let $\mathcal{C}$ and $\mathcal{D}$ be model categories, and let $F\colon \mathcal{C} \longrightarrow \mathcal{D}$ be a functor.*

- *If F takes acyclic cofibrations between cofibrant objects to weak equivalences, then F takes all weak equivalences between cofibrant objects to weak equivalences.*
- *If F takes acyclic fibrations between fibrant objects to weak equivalences, then F takes all weak equivalences between fibrant objects to weak equivalences.*

The primary reason for Quillen functors to be the "correct" notion of structure-preserving functors between model categories is that they induce functors on the respective homotopy categories.

Definition A.4.5 Let $F\colon \mathcal{C} \longrightarrow \mathcal{D}$ be a left Quillen functor between model categories $\mathcal{C}$ and $\mathcal{D}$. Then the *left derived functor*

$$\mathbb{L}F\colon \operatorname{Ho}(\mathcal{C}) \longrightarrow \operatorname{Ho}(\mathcal{D})$$

is given by $\mathbb{L}F(X) := F(X^{cof})$. Dually, if $G\colon \mathcal{D} \longrightarrow \mathcal{C}$ is a right Quillen functor, then its *right derived functor*

$$\mathbb{R}G\colon \operatorname{Ho}(\mathcal{D}) \longrightarrow \operatorname{Ho}(\mathcal{C})$$

is given by $\mathbb{R}G(Y) := G(Y^{fib})$.

As mentioned, this is an extremely simplified version of the real picture. Non-trivial work goes into showing that the above actually describes well-defined functors with the desired universal properties. We leave this to [Hov99, Chapter 1.3.2] and [DS95, Section 9]. They also show the following important result.

Theorem A.4.6 *Let $\mathcal{C}$ and $\mathcal{D}$ be model categories and*

$$F\colon \mathcal{C} \rightleftarrows \mathcal{D} : G$$

a Quillen adjunction. Then the left and right derived functors

$$\mathbb{L}F\colon \mathrm{Ho}(\mathcal{C}) \rightleftarrows \mathrm{Ho}(\mathcal{D}) : \mathbb{R}G$$

form an adjunction.

Definition A.4.7 A Quillen adjunction

$$F\colon \mathcal{C} \rightleftarrows \mathcal{D} : G$$

is called a *Quillen equivalence*, if

$$\mathbb{L}F\colon \mathrm{Ho}(\mathcal{C}) \rightleftarrows \mathrm{Ho}(\mathcal{D}) : \mathbb{R}G$$

is an adjoint equivalence of categories.

The following characterisation is frequently used, see [Hov99, Corollary 1.3.16].

Proposition A.4.8 *Let $\mathcal{C}$ and $\mathcal{D}$ be model categories, and let*

$$F\colon \mathcal{C} \rightleftarrows \mathcal{D} : G$$

be a Quillen adjunction. Then the following are equivalent.

- *(F, G) is a Quillen equivalence.*
- *For every fibrant $Y \in \mathcal{D}$, the composite map*

$$F(G(Y)^{cof}) \longrightarrow F(G(Y)) \longrightarrow Y$$

is a weak equivalence in $\mathcal{D}$, and whenever $f\colon A \longrightarrow B$ is a morphism in $\mathcal{C}$ between cofibrant objects, such that

$$F(f)\colon F(A) \longrightarrow F(B)$$

is a weak equivalence in $\mathcal{D}$, then f is a weak equivalence in $\mathcal{C}$.

- *For every cofibrant $X \in \mathcal{C}$, the composite map*

$$X \longrightarrow G(F(X)) \longrightarrow G(F(X)^{fib})$$

is a weak equivalence in $\mathcal{C}$, and whenever $g: A \longrightarrow B$ is a morphism in $\mathcal{D}$ between fibrant objects, such that

$$G(g): G(A) \longrightarrow G(B)$$

is a weak equivalence in $\mathcal{C}$, then g is a weak equivalence in $\mathcal{D}$.

Let us continue with some examples.

Example A.4.9 The adjunction

$$|-|: \mathrm{sSet} \rightleftarrows \mathrm{Top} : \mathrm{sing}$$

from Example A.1.6 is a Quillen equivalence. Therefore, we can consider simplicial sets to be a "model" for topological spaces in this context.

Example A.4.10 Example A.6.10 gives a projective model structure and an injective model structure on the category $\mathrm{Ch}(R)$ of unbounded chain complexes extending the bounded case. The identity functor $\mathrm{Id}: \mathrm{Ch}(R) \longrightarrow \mathrm{Ch}(R)$ from the projective model structure to the injective model structure is the left adjoint of a Quillen equivalence.

The examples show the importance of Quillen equivalences: they allow us to use different model categories to model the same homotopy category. This is particularly useful when those models have technical advantages over each other. As a simple example, every simplicial set is cofibrant, while not every topological space is cofibrant. As a more involved example, the model category of sequential spectra is not a monoidal model category, but Chapters 5 and 6 discuss Quillen equivalent model categories with far better monoidal properties.

Quillen equivalences are more structured than equivalences of homotopy categories. In general, one can use the slogan *"Quillen equivalent model categories have the same homotopy theory"*. We will see instances of this phenomenon all over this book.

A.5 Homotopy Cofibrations

A useful notion in a model category is that of h-cofibrations, which is a generalisation of h-cofibrations of topological spaces. We give a general definition and some properties. Further references can be found in Cole [Col06], May [May99a] and May and Ponto [MP12].

Definition A.5.1 Let $\mathcal{C}$ be a model category. A map $i\colon A \longrightarrow X$ is a *h-cofibration* if for any map $f\colon X \longrightarrow Y$ and any homotopy $H\colon \mathrm{Cyl}(A) \longrightarrow Y$, such that the following square commutes

$$\begin{array}{ccc} A & \xrightarrow{i_0} & \mathrm{Cyl}(A) \\ {\scriptstyle i}\downarrow & & \downarrow{\scriptstyle H} \\ X & \xrightarrow[f]{} & Y, \end{array}$$

there is a homotopy $G\colon \mathrm{Cyl}(X) \longrightarrow Y$ with a map $\mathrm{Cyl}(A) \longrightarrow \mathrm{Cyl}(X)$, such that the following diagram commutes.

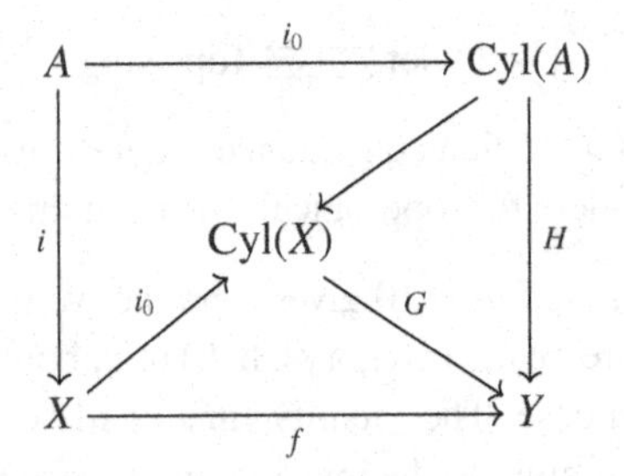

This condition on $i\colon A \longrightarrow X$ is also referred to as the homotopy extension property.

Example A.5.2 The h-cofibrations of pointed spaces are the Hurewicz cofibrations. The functor $A \wedge -$ preserves h-cofibrations and thus sends q-cofibrations to h-cofibrations for any pointed space A.

We may rewrite the definition of the homotopy extension property as a lifting diagram for well-behaved topological model categories (including sequential, symmetric and orthogonal spectra).

Lemma A.5.3 *Let $\mathcal{C}$ be a pointed topological model category, such that*

$$Y \otimes -\colon \mathrm{Top}_* \longrightarrow \mathcal{C} \quad \text{and} \quad Y^{(-)}\colon \mathrm{Top}_*^{op} \longrightarrow \mathcal{C}$$

send homotopy equivalences to weak equivalences for any $Y \in \mathcal{C}$.

Then a map $i\colon A \longrightarrow X$ in $\mathcal{C}$ is a h-cofibration if and only if every square of the following type admits a lift

$$\begin{array}{ccc} A & \xrightarrow{H} & Y^{I_+} \\ {\scriptstyle i}\downarrow & \nearrow & \downarrow{\scriptstyle p_0} \\ X & \xrightarrow[f]{} & Y. \end{array}$$

Proof The assumption implies that $(-) \otimes I_+$ and $(-)^{I_+}$ are functorial cylinder and path objects (though not necessarily very good ones). It follows that a homotopy $A \otimes I_+ \longrightarrow Y$ corresponds to a homotopy $A \longrightarrow Y^{I_+}$, without requiring cofibrancy and fibrancy assumptions. □

Using this description of h-cofibrations in terms of lifting properties, we have the following.

Corollary A.5.4 *Let $\mathcal{C}$ be a pointed topological model category satisfying the assumptions of Lemma A.5.3. The class of h-cofibrations in $\mathcal{C}$ is closed under pushouts, coproducts, retracts and (countable) sequential colimits.* □

There is a universal test case for a map to be a h-cofibration.

Lemma A.5.5 *Let $\mathcal{C}$ be a model category. A map $i\colon A \longrightarrow X$ is a h-cofibration if and only if it has the homotopy extension property with respect to the inclusion of X into the mapping cylinder of i*

$$X \longrightarrow \mathrm{Cyl}(A) \amalg_A X = Mi.$$

In this case, Mi is a retraction of $\mathrm{Cyl}(X)$.

The following result is well established. A proof can be found in [May99a, Section 8.4]. The proof given there can be extended to any topological model category. We may think of this as saying that when i is a h-cofibration, X/A has the "correct" homotopy type.

Lemma A.5.6 *Let $i\colon A \longrightarrow X$ be a h-cofibration of pointed spaces. Then the pushout X/A of i over a point is naturally homotopy equivalent to the homotopy cofibre Ci of i.*

A related statement for a Serre fibration $p\colon E \longrightarrow B$ of pointed spaces also holds: there is a weak homotopy equivalence between the pre-image of the basepoint and the homotopy fibre

$$p^{-1}(b_0) \simeq Fp.$$

We also need a technical result about h-cofibrations and compact spaces. For the proof, see [Hov99, Proposition 2.4.2] and [May99a, Problem 1, Page 48], which relates h-cofibrations to closed inclusions. Recall that a *closed inclusion* is an injective continuous map $f\colon A \longrightarrow B$ with closed image such that U is open in A if and only if $U = f^{-1}(V)$ for some open V in B.

Lemma A.5.7 *Compact spaces are small with respect to the class of h-cofibrations.*

A.6 Cofibrantly Generated Model Categories

In many situations, it is very helpful to know that the cofibrations and acyclic cofibrations in a model category are generated by specified *sets* in a certain sense rather than knowing them as classes of maps. This often allows one to prove technical results by checking statements on these generating sets. Details can be found in [Hov99, Section 2.1].

We begin with some notation.

Definition A.6.1 Let $\mathcal{C}$ be a category with all small colimits, and let I be a set of morphisms in $\mathcal{C}$. Then I-inj is the class of morphisms in $\mathcal{C}$ that have the right lifting property with respect to all elements of I. Furthermore, I-cof is the class of morphisms in $\mathcal{C}$ that have the left lifting property with respect to all morphisms in I-inj.

The class of morphisms I-cell consists of all sequential colimits of pushouts of elements of I. Hence, a map $f\colon A \longrightarrow X$ is in I-cell if there is a sequence

$$A = X_0 \xrightarrow{f_0} X_1 \xrightarrow{f_1} X_2 \xrightarrow{f_2} \cdots$$

such that the canonical map $A = X_0 \longrightarrow \operatorname{colim}_n X_n = X$ is the map f, and for each $n \in \mathbb{N}$, there is a pushout square

$$\begin{array}{ccc} \coprod_\alpha C_\alpha & \longrightarrow & X_n \\ {\scriptstyle i_\alpha}\downarrow & & \downarrow{\scriptstyle f_n} \\ \coprod_\alpha D_\alpha & \longrightarrow & X_{n+1} \end{array}$$

with $i_\alpha \in I$ and α in some small indexing set.

An I-cell *complex* X is an object of $\mathcal{C}$ whose map from the initial object is in I-cell.

Since I-cof is defined by a lifting property, it is closed under coproducts, pushouts and sequential colimits, hence, we have the following.

Lemma A.6.2 *Let $\mathcal{C}$ be a category with all small colimits, and let I be a set of morphisms in $\mathcal{C}$. Then,*

$$I\text{-cell} \subseteq I\text{-cof}. \qquad \square$$

We can now define a class of objects that interacts well with I-cell. This is a generalisation of the relationship between compact spaces and CW-complexes.

Definition A.6.3 Let $\mathcal{C}$ be a category with all small colimits, and let I be a class of morphisms in $\mathcal{C}$. We say that Z is *small with respect to* I if for any map $i: A \longrightarrow X = \operatorname{colim}_n X_n$ in I-cell, the map of sets

$$\operatorname{colim}_n \mathcal{C}(Z, X_n) \longrightarrow \mathcal{C}(Z, \operatorname{colim}_n X_n)$$

is an isomorphism.

Taking Z and $X = \operatorname{colim}_n X_n$ as in the definition, we see that a map from $Z \longrightarrow X$ factors as a map $Z \longrightarrow X_n$ followed by the map into the colimit $X_n \longrightarrow X$.

Remark A.6.4 We note that [Hov99, Chapter 2.1] calls such objects "ω–small", and [MMSS01] calls such objects "compact". The first reference considers smallness with respect to more general cardinals (see also [Hir03, Chapter 10]) and introduces the notion of transfinite composition (a generalisation of sequential colimits) to define I-cell for a set of maps I. We call this more general notion of smallness *transfinitely small.* We use transfinite composition in Sections 7.2 and 7.3, while for the earlier chapters, we only need sequential colimits.

In particular, our levelwise and stable model structures on spectra (and categories of rings, modules and commutative rings) only use sequential colimits in the proofs of the model structures.

Small objects in the sense of the previous definition are preserved by general colimits, see [Hir03, Propositon 10.4.8].

Lemma A.6.5 *Let $\mathcal{C}$ be a category with all small colimits, and let I be a class of morphisms in $\mathcal{C}$. If P is the pushout of a diagram of objects*

$$B \longleftarrow A \longrightarrow C$$

that are small with respect to I, then P is also small with respect to I.

We can now give the main definition of this section.

Definition A.6.6 A model category $\mathcal{C}$ is *cofibrantly generated* if there are sets I and J such that the following hold.

- The domains of I are small with respect to I.
- The domains of J are small with respect to J.
- The fibrations in $\mathcal{C}$ are precisely J-inj.
- The acyclic fibrations in $\mathcal{C}$ are precisely I-inj.

It can be shown that in a cofibrantly generated model category, the cofibrations are exactly I-cof and the acyclic cofibrations are exactly J-cof. Therefore, we often speak of I as the generating cofibrations of $\mathcal{C}$ and of J as the generating acyclic cofibrations of $\mathcal{C}$.

Example A.6.7 In the Serre model structure for topological spaces (Example A.1.5), the generating cofibrations are

$$I = \{S^{n-1} \longrightarrow D^n \mid n \in \mathbb{N}\},$$

where $S^{-1} = \emptyset$. The generating acyclic cofibrations are

$$J = \{D^n \longrightarrow D^n \times [0,1] \mid n \in \mathbb{N}\}.$$

The generating sets for the Serre model structure on pointed spaces Top_* are given by adding a disjoint basepoint to the maps in these sets.

Example A.6.8 Simplicial sets as in Example A.1.6 form a cofibrantly generated model category. Let us recall that the standard n-simplex is the functor

$$\Delta[n] \colon \Delta^{op} \longrightarrow \mathrm{Set},$$

sending $[k]$ to the set of order-preserving injections $\Delta([k],[n])$. The simplicial set $\partial\Delta[n]$ is the functor $\Delta^{op} \longrightarrow \mathrm{Set}$ sending $[k]$ to the set of order-preserving non-identity injections $[k] \longrightarrow [n]$.

For $0 \leqslant r \leqslant n$, the r-horn $\Lambda^r[n]$ of $\Delta[n]$ is the functor $\Delta^{op} \longrightarrow \mathrm{Set}$, which sends $[k]$ to the order-preserving injections $[k] \longrightarrow [n]$, but excluding both the identity $[n] \longrightarrow [n]$ and the map $d^r \colon [n-1] \longrightarrow [n]$, which avoids r. For the geometric realisations, we can picture $\partial\Delta[n]$ as the geometric n-simplex with its interior removed and the horn $\Lambda^r[n]$ as the geometric n-simplex without its interior and the $(n-1)$-face opposite the vertex r.

The generating cofibrations of sSet are the inclusions

$$I = \{\partial\Delta[n] \longrightarrow \Delta[n] \mid n \in \mathbb{N}\},$$

and the generating acyclic cofibrations are the inclusions

$$J = \{\Lambda^r[n] \longrightarrow \Delta[n] \mid n > 0,\ 0 \leqslant r \leqslant n\}.$$

Adding a disjoint basepoint gives the generating sets for a model structure on pointed simplicial sets.

If we are given sets of morphisms I and J in a category $\mathcal{C}$, we would like to know if I and J make $\mathcal{C}$ into a cofibrantly generated model category. The following Recognition Theorem [Hir03, Theorem 11.3.1], [Hov99, Theorem 2.1.19] gives a highly useful answer, which we use in several places in this book.

Theorem A.6.9 *Let $\mathcal{C}$ be a category with all small limits and colimits. Let $\mathcal{W}$ be a class of morphisms which is closed under composition and contains all the identity morphisms. Further, let I and J be sets of morphisms in $\mathcal{C}$. Assume that*

- *$\mathcal{W}$ satisfies the 2-out-of-3 property,*
- *the domains of I are small with respect to I,*
- *the domains of J are small with respect to J,*
- J-cell $\subseteq \mathcal{W} \cap I$-cof,
- I-inj $\subseteq \mathcal{W} \cap J$-inj,
- *either* $\mathcal{W} \cap I$-cof $\subseteq J$-cof *or* $\mathcal{W} \cap J$-inj $\subseteq I$-inj.

Then $\mathcal{C}$ can be given a cofibrantly generated model structure with $\mathcal{W}$ being the weak equivalences, I the set of generating cofibrations and J the set of generating acyclic cofibrations.

Example A.6.10 Let Ch(R) be the category of (unbounded) chain complexes over a ring R. By S^n, we denote the chain complex that is R in degree n and zero elsewhere. By D^n, we denote the chain complex which is R in degrees $n-1$ and n and zero elsewhere, with the identity differential between degrees n and $n-1$. Let I and J be the following sets of chain maps,

$$I = \{S^{n-1} \longrightarrow D^n \mid n \in \mathbb{Z}\} \text{ and } J = \{0 \longrightarrow D^n \mid n \in \mathbb{Z}\}.$$

Then $\mathcal{W} = (H_*\text{-isomorphisms})$, I and J satisfy the conditions of Theorem A.6.9 and therefore define a cofibrantly generated model structure on Ch(R), namely, the *projective model structure*. With some extra work (see [Hov99, Chapter 2.3]), one can show that this model structure satisfies the following.

- The weak equivalences are the H_*-isomorphisms.
- The fibrations are the surjections.
- A cofibration is an injection with a cofibrant cokernel.

Note that while a cofibrant chain complex is projective in each degree, this is not a sufficient condition.

We could have restricted ourselves to non-negatively graded chain complexes Ch$(R)_+$ and used

$$I_+ = \{0 \longrightarrow S^0\} \cup \{S^{n-1} \longrightarrow D^n \mid n \geqslant 1\} \text{ and } J_+ = \{0 \longrightarrow D^n \mid n \geqslant 1\}.$$

This would recover the first part of Example A.1.7, where the characterisation of cofibrations is more straightforward. The analogous injective model structures on Ch(R) and Ch$(R)_-$ are also cofibrantly generated, but their generating cofibrations and acyclic cofibrations are very difficult to make explicit, see [Hov99, Definition 2.3.14].

The next result is sometimes useful in verifying the conditions of the Recognition Theorem.

Lemma A.6.11 *Let $\mathcal{C}$ be a category with all small colimits and sets of maps I and J. Assume that I*-inj $\subset J$-inj. *Then J*-cell $\subset I$-cof.

Proof We know that J-cell $\subset J$-cof. The assumption I-inj $\subset J$-inj implies J-cof $\subset I$-cof. □

The following result is often known as the *lifting lemma*. It is a highly useful method of creating new model structures from an adjunction and a cofibrantly generated model structure. Lifting results with the adjunction in the other direction can be found in work of Hess et al. [HKRS17] and Garner et al. [GKR18].

Lemma A.6.12 *Let $\mathcal{C}$ be a cofibrantly generated model category and $\mathcal{D}$ a category with all small limits and colimits. Given an adjunction*

$$F\colon \mathcal{C} \rightleftarrows \mathcal{D} : G,$$

assume that

- *the domains of FI are small with respect to FI,*
- *the domains of FJ are small with respect to FJ,*
- *the right adjoint G sends maps in FJ*-cell *to weak equivalences in $\mathcal{C}$.*

Then there is a model structure on $\mathcal{D}$ with fibrations and weak equivalences defined via G, that is, a map f in $\mathcal{D}$ is a weak equivalence or a fibration if and only if $G(f)$ is so in $\mathcal{C}$. We call this model structure on $\mathcal{D}$ the lifted model structure.

The next proposition is a very powerful tool when it comes to constructing model structures and is often used in the context of cofibrantly generated model categories. It works just as well in the case of transfinitely small objects and transfinite composition.

Proposition A.6.13 (The Small Object Argument) *Let $\mathcal{C}$ be a category with all small colimits, and let I be a set of maps in $\mathcal{C}$. Assume that the domains of the elements of I are small relative to I*-cell. *Then there is a functorial factorisation*

$$X \xrightarrow{\;i\;} Z \xrightarrow{\;q\;} Y$$

of every map $f\colon X \longrightarrow Y$ in $\mathcal{C}$, where i is in I-cell *and q is in I*-inj.

As a consequence, we have the following.

Corollary A.6.14 *Let $\mathcal{C}$ be a cofibrantly generated model category. Then the factorisations of axiom (MC5) can be assigned functorially.*

The following is a consequence of cofibrant generation, see [Hir03, Section 11.2], but requires the notion of transfinitely small objects and transfinite composition for its cofibrant generation. We find it useful in Section 7.2.

Lemma A.6.15 *Let $\mathcal{C}$ be a cofibrantly generated model category with generating cofibrations I. If K is an object that is transfinitely small relative to I, then it is transfinitely small relative to all cofibrations. If the codomains of the elements of I are transfinitely small relative to I, then every cofibrant object is transfinitely small relative to all cofibrations.*

A.7 Homotopy Limits and Colimits

One general problem with limits and colimits is that they are not as homotopy invariant as one would ideally like. For example, the pushout of the diagram

$$* \longleftarrow S^n \longrightarrow *$$

in topological spaces is trivial. Meanwhile, the pushout of

$$D^{n+1} \longleftarrow S^n \longrightarrow D^{n+1}$$

is S^{n+1}, despite the two diagrams being termwise homotopy equivalent. Homotopy limits and colimits are one method of bringing some order to this problem. We will give an overview of the techniques which we hope to be sufficient for understanding the instances arising in this book. In some cases, we will provide references rather than explicit proofs.

Throughout this section let $\mathcal{C}$ be a model category and I a category that is also a poset. Furthermore, we require I to be *simple*, meaning that the function

$$d\colon I \longrightarrow \mathbb{N} \cup \infty$$

with

$$d(i) = \{\sup(n) \mid \text{there are non-identity maps } x_0 \to x_1 \to \cdots \to x_n = i\}$$

is required to only take finite values. The key examples we have in mind are the categories

$$I = (2 \longleftarrow 1 \longrightarrow 3)$$

and

$$I = \mathbb{N} = (0 \longrightarrow 1 \longrightarrow 2 \longrightarrow \cdots).$$

The following is [Hov99, Theorem 5.1.3].

Theorem A.7.1 *Let $\mathcal{C}$ be a model category and I a simple category. Then there is a model structure on $\mathcal{C}^I$ called the* projective model structure *such that a morphism $f: X \longrightarrow Y$ is a weak equivalence (respectively fibration) if and only if $f_i: X_i \longrightarrow Y_i$ is a weak equivalence (respectively fibration) for every $i \in I$.*

Example A.7.2 Let $\mathcal{C}$ be a model category and $I = (1 \longrightarrow 2)$. Then $\mathcal{C}^I$ is the category of morphisms in $\mathcal{C}$, often called the *arrow category*. A map $\alpha: f \longrightarrow g$ in $\mathcal{C}^I$ is a commutative square

$$\begin{array}{ccc} A & \xrightarrow{f} & B \\ {\scriptstyle \alpha_1}\downarrow & & \downarrow{\scriptstyle \alpha_2} \\ X & \xrightarrow[g]{} & Y. \end{array}$$

Theorem A.7.1 gives a model structure on this category and specifies that a map α between morphisms in $\mathcal{C}$ is a weak equivalence if α_1 and α_2 are weak equivalences in $\mathcal{C}$.

Definition A.7.3 The *homotopy colimit* of a diagram $X \in \mathcal{C}^I$ is given by the colimit of a cofibrant replacement X^{cof} of X in $\mathcal{C}^I$ with the projective model structure. For

$$I = (2 \longleftarrow 1 \longrightarrow 3),$$

the homotopy colimit is the *homotopy pushout*, for $I = \mathbb{N}$, we speak of the *sequential homotopy colimit*.

This construction enjoys the following properties.

- There is a natural comparison map $\operatorname{hocolim} X \longrightarrow \operatorname{colim} X$.
- If $F: X \longrightarrow Y \in \mathcal{C}^I$ is a vertexwise weak equivalence of diagrams, then F induces a weak equivalence $\operatorname{hocolim} X \longrightarrow \operatorname{hocolim} Y$.
- If our model category has functorial factorisation, then we have a functor $\operatorname{hocolim}: \mathcal{C}^I \longrightarrow \mathcal{C}$. Any other functor $\mathcal{C}^I \longrightarrow \mathcal{C}$ with the previous properties factors through hocolim.

The following result helps to make the construction of homotopy colimits explicit, see [Str11, Theorem 6.36]. As our indexing category I is a poset, we can define $I_{<i}$ to be the full subcategory of I on objects j with $j < i$.

Theorem A.7.4 *Let $\mathcal{C}$ be a model category and I be a simple category. Assume that $X \in \mathcal{C}^I$ is vertexwise cofibrant. If for all $i \in I$, the map*

$$\operatorname*{colim}_{j \in I_{<i}} X_j \longrightarrow X_i$$

is a cofibration in $\mathcal{C}$, then X is cofibrant in $\mathcal{C}^I$.

Example A.7.5 We will take a look at what the conditions of Theorem A.7.4 translate to in the example of

$$I = (2 \longleftarrow 1 \longrightarrow 3),$$

that is, a homotopy pushout. Let

$$X_2 \longleftarrow X_1 \longrightarrow X_3$$

be a diagram of cofibrant objects in $\mathcal{C}$. Then, for $i = 1$, Theorem A.7.4 asks for X_1 to be cofibrant, which we assumed anyway. For $i = 2$ and $i = 3$, we obtain that $X_1 \longrightarrow X_2$ and $X_1 \longrightarrow X_3$ are supposed to be cofibrations. Thus, we can construct the cofibrant replacement of a diagram by replacing all vertices with cofibrant objects and replace the "legs" of the diagram with cofibrations.

In the case of a left proper model category (see Definition 5.4.1), it is sufficient to replace just one of the legs in a pullback square with a cofibration.

Lemma A.7.6 *Let $\mathcal{C}$ be a left proper model category and $X \in \mathcal{C}^I$ for*

$$I = (2 \longleftarrow 1 \longrightarrow 3).$$

Then, if just one of the legs of the diagram X is a cofibration between cofibrant objects of $\mathcal{C}$, then its pushout is its homotopy pushout.

Proof Let

$$C \leftarrowtail A \xrightarrow{f} B$$

be a diagram in $\mathcal{C}$. We can write down another diagram that is vertexwise weakly equivalent to this diagram and where both morphisms are cofibrations by factoring f into a cofibration followed by an acyclic fibration.

$$\begin{array}{ccccc} C & \leftarrowtail & A & \rightarrowtail & Z \\ \| & & \| & & \downarrow{\sim} \\ C & \leftarrowtail & A & \xrightarrow{f} & B \end{array}$$

As both diagrams are vertexwise weakly equivalent, their homotopy pushouts agree. Furthermore, the homotopy pushout of the top row is just its pushout. Therefore, if we can show that the pushout of the bottom row is weakly equivalent to the pushout of the top row, we have shown that the pushout of the bottom row is also its homotopy pushout as claimed. Now let P be the pushout of the top row. We thus have a diagram

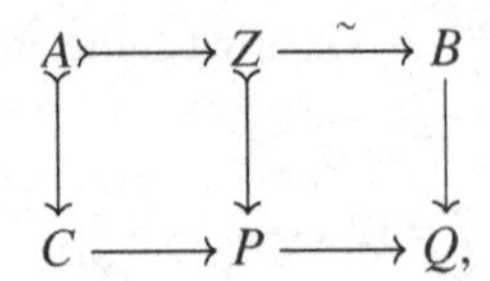

where both squares are pushout squares. Therefore, the outer square of the diagram is a pushout, that is, Q is the pushout of

$$C \leftarrowtail A \xrightarrow{f} B.$$

But as $\mathcal{C}$ is left proper, the map $P \longrightarrow Q$ is a weak equivalence (it is the pushout of a weak equivalence along a cofibration), which is what we wanted to show. □

The following is not a statement about homotopy colimits, but it uses the projective model structure of diagrams, which is why we found it fitting to include it in this place.

Lemma A.7.7 (The Patching Lemma) *Consider a diagram*

$$\begin{array}{ccccc} B & \longleftarrow & A & \longrightarrow & C \\ \downarrow & & \downarrow & & \downarrow \\ Y & \longleftarrow & X & \longrightarrow & Z, \end{array}$$

such that

$$C \longrightarrow Z \quad \text{and} \quad B \amalg_A X \longrightarrow Y$$

are cofibrations (respectively acyclic cofibrations). Then the map

$$B \amalg_A C \longrightarrow Y \amalg_X Z$$

is a cofibration (respectively acyclic cofibration).

Proof We may write the map in question as the composite

$$B \amalg_A C \longrightarrow B \amalg_A Z \cong (B \amalg_A X) \amalg_X Z \longrightarrow Y \amalg_X Z.$$

By the assumptions, these maps are colimits of cofibrations in the projective model structure for pushout diagrams of Theorem A.7.1. □

We give a sketch proof of the following.

Lemma A.7.8 *Let $\mathcal{C}$ be a stable model category, and let*

$$\begin{array}{ccc} A & \xrightarrow{i} & B \\ {\scriptstyle j}\downarrow & & \downarrow{\scriptstyle g} \\ C & \xrightarrow[f]{} & P \end{array}$$

be a homotopy pushout in $\mathcal{C}$. Then,

$$A \xrightarrow{i-j} B \amalg C \xrightarrow{(f,g)} P \longrightarrow \Sigma A$$

is an exact triangle in $\mathrm{Ho}(\mathcal{C})$.

Proof The homotopy pushout commutes with cofibres in the following sense. Consider a commutative diagram

$$\begin{array}{ccccc} X_2 & \longleftarrow & X_1 & \longrightarrow & X_3 \\ {\scriptstyle f_2}\downarrow & & {\scriptstyle f_1}\downarrow & & {\scriptstyle f_3}\downarrow \\ Y_2 & \longleftarrow & Y_1 & \longrightarrow & Y_3. \end{array}$$

Let P_X be the homotopy pushout of the top row and P_Y the homotopy pushout of the bottom row. Then the cofibre of the induced map $P_X \longrightarrow P_Y$ is the homotopy pushout of the cofibres

$$Cf_2 \longleftarrow Cf_1 \longrightarrow Cf_3.$$

This is [Str11, Corollary 7.28], which follows from various results about manipulating and "composing" squares, which we do not wish to discuss in detail. We will apply the above statement to the diagram

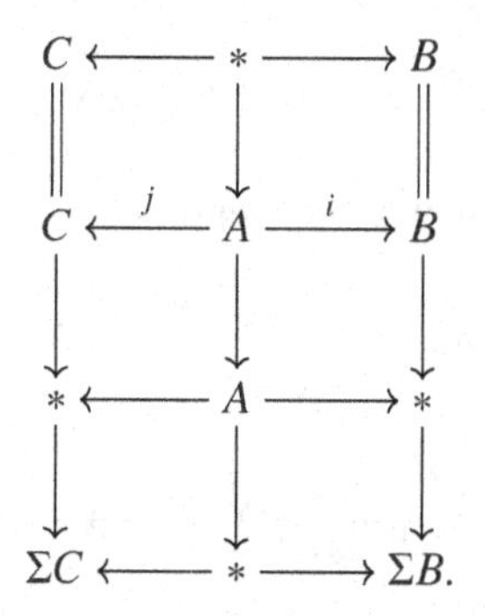

The three columns are exact triangles, therefore their homotopy pushouts form an exact triangle too. This is precisely

$$B \amalg C \xrightarrow{(f,g)} P \longrightarrow \Sigma A \xrightarrow{\Sigma(i-j)} \Sigma(B \amalg C). \qquad \square$$

Let us now take a closer look at sequential homotopy colimits.

Example A.7.9 Let

$$X = (X_0 \xrightarrow{f_0} X_1 \xrightarrow{f_1} X_2 \xrightarrow{f_2} \cdots)$$

be a diagram in $\mathcal{C}^{\mathbb{N}}$. Theorem A.7.4 says that if each f_i is a cofibration between cofibrant objects, then X is cofibrant. We can define a cofibrant replacement X' of X via the following inductive procedure. Assume X'_i has already been constructed for $i \leqslant n$ such that the squares

$$\begin{array}{ccc} X'_{i-1} & \overset{f'_{i-1}}{\rightarrowtail} & X'_i \\ {\scriptstyle p_{i-1}}\downarrow\!\!\!\downarrow{\scriptstyle \sim} & & {\scriptstyle p_i}\downarrow\!\!\!\downarrow{\scriptstyle \sim} \\ X_{i-1} & \underset{f_{i-1}}{\longrightarrow} & X_i \end{array}$$

commute, the X'_i are cofibrant, the vertical arrows are acyclic fibrations and the f'_i are cofibrations. Then, for the next step, we factor $f_n \circ p_n$ in the diagram below

$$\begin{array}{ccccc} X'_{n-1} & \overset{f'_{n-1}}{\rightarrowtail} & X'_n & & \\ {\scriptstyle p_{n-1}}\downarrow\!\!\!\downarrow{\scriptstyle \sim} & & {\scriptstyle p_n}\downarrow\!\!\!\downarrow{\scriptstyle \sim} & & \\ X_{n-1} & \underset{f_{n-1}}{\longrightarrow} & X_n & \underset{f_n}{\longrightarrow} & X_{n+1} \end{array}$$

as a cofibration followed by an acyclic fibration,

$$X'_n \overset{f'_n}{\rightarrowtail} X'_{n+1} \overset{p_{n+1}}{\underset{\sim}{\twoheadrightarrow}} X_{n+1},$$

which completes the next step of the diagram

$$\begin{array}{ccccc} X'_{n-1} & \overset{f'_{n-1}}{\rightarrowtail} & X'_n & \overset{f'_n}{\rightarrowtail} & X'_{n+1} \\ {\scriptstyle p_{n-1}}\downarrow\!\!\!\downarrow{\scriptstyle \sim} & & {\scriptstyle p_n}\downarrow\!\!\!\downarrow{\scriptstyle \sim} & & {\scriptstyle p_{n+1}}\downarrow\!\!\!\downarrow{\scriptstyle \sim} \\ X_{n-1} & \underset{f_{n-1}}{\longrightarrow} & X_n & \underset{f_n}{\longrightarrow} & X_{n+1} \end{array}$$

as desired.

The explicit construction of the sequential homotopy colimit gives the following result.

Corollary A.7.10 *In the category of pointed topological spaces with the Serre model structure, we have the following.*

- *If X is a sequential diagram of pointed spaces with non-degenerate base-points where all maps in X are h-cofibrations, then the homotopy colimit is homotopy equivalent to the colimit.*
- *The homotopy groups of a homotopy colimit of pointed spaces are the colimit of the homotopy groups of the spaces in the diagram.*
- *There is a weak equivalence*

$$\operatorname{hocolim}_n \Omega X_n \longrightarrow \Omega \operatorname{hocolim}_n X_n.$$

Proof The assumptions of the first point are precisely saying that the diagram is cofibrant in the Hurewicz model structure.

For the second point, let

$$X_0 \xrightarrow{f_0} X_1 \xrightarrow{f_1} X_2 \xrightarrow{f_2} \cdots$$

be a cofibrant diagram of pointed topological spaces. For $x_0 \in X_0$, let

$$x_n = (f_n \circ f_{n-1} \circ \cdots \circ f_0)(x_0) \in X_n.$$

Let $x_\infty \in \operatorname{colim}_n X_n$ be the image in the colimit. We must prove

$$\operatorname{colim}_n \pi_k(X_n, x_n) \cong \pi_k(\operatorname{colim}_n X_n, x_\infty).$$

We can assume that the X_n are CW-complexes and that the maps f_n are inclusions of CW-subcomplexes. Choose a map

$$\alpha\colon S^k \longrightarrow \bigcup_{n=0}^{\infty} X_n = \operatorname{colim}_n X_n \simeq \operatorname{hocolim}_n X_n.$$

As S^k is compact, its image under α lies in a finite subcomplex of the codomain and hence lies in some X_n.

The third point follows from the second. □

Example A.7.11 We consider how coequaliser diagrams interact with the conditions of Theorem A.7.4. A coequaliser has the shape

$$I = (1 \rightrightarrows 2).$$

Let

$$X = (X_1 \overset{f}{\underset{g}{\rightrightarrows}} X_2)$$

be a diagram in $\mathcal{C}$. Then Theorem A.7.4 asks for X_1 to be cofibrant and the map induced by f and g

$$X_1 \coprod X_1 \xrightarrow{(f,g)} X_2$$

to be a cofibration (which implies that f and g are cofibrations).

Since the coequaliser of f and g may be constructed as the pushout of the diagram

$$X_1 \xleftarrow{\text{fold}} X_1 \amalg X_1 \xrightarrow{(f,g)} X_2,$$

we may then construct the homotopy coequaliser using our construction of homotopy pushouts.

When $\mathcal{C}$ is left proper, we may construct the homotopy coequaliser by replacing the fold map by the map to a cylinder object $X_1 \longrightarrow \text{Cyl}(X_1)$ and taking the pushout.

Example A.7.12 The homotopy coequaliser gives an alternative construction of the sequential homotopy colimit. Let

$$X = (X_0 \xrightarrow{f_0} X_1 \xrightarrow{f_1} X_2 \xrightarrow{f_2} \cdots)$$

be a diagram in $\mathcal{C}^{\mathbb{N}}$. The colimit is exactly the coequaliser of the maps

$$\text{Id}, \coprod_{n \geqslant 0} f_n \colon \coprod_{n \geqslant 0} X_n \longrightarrow \coprod_{n \geqslant 0} X_n.$$

It follows that the sequential homotopy colimit is (weakly equivalent to) the homotopy coequaliser.

It may be helpful for the reader to sketch the homotopy pushout construction of a homotopy coequaliser and the homotopy colimits of the previous example in the case of topological spaces. Further details can be found in [MP12, Section 2.2].

Lemma A.7.13 *In a stable model category, the homotopy coequaliser of $f,\ g \colon X \longrightarrow Y$ is weakly equivalent to the cofibre of $g - f \colon X \longrightarrow X$.*

Proof By Lemma A.7.8, the homotopy pushout construction of the homotopy coequaliser $CE(f,g)$ gives an exact triangle

$$X \amalg X \xrightarrow{(-\text{fold}, f \amalg g)} X \amalg Y \longrightarrow CE(f,g) \longrightarrow \Sigma(X \amalg X).$$

Since the fold map is the sum of the two projections $X \amalg X \longrightarrow X$, we have a commutative diagram

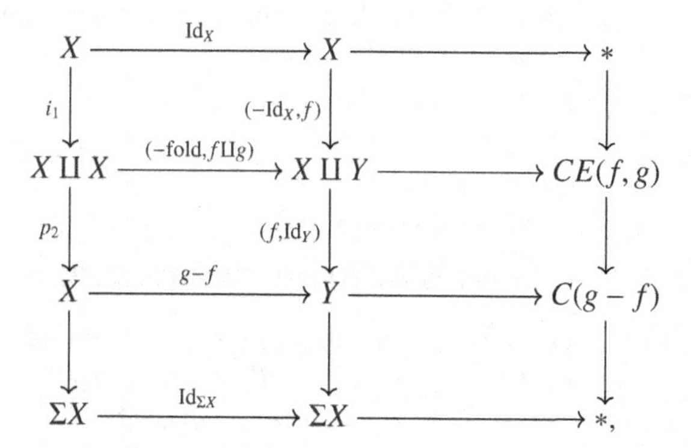

where $C(g - f)$ is the cofibre of $g - f$.

The middle column is an exact triangle, as it is isomorphic to the sum of two exact triangles

$$X \longrightarrow X \longrightarrow * \longrightarrow \Sigma X \text{ and } * \longrightarrow Y \longrightarrow Y \longrightarrow *,$$

as the diagram below demonstrates.

$$\begin{array}{ccccccc}
X & \xrightarrow{i_1} & X \amalg Y & \xrightarrow{p_2} & Y & \longrightarrow & \Sigma X \\
\downarrow{\scriptstyle \mathrm{Id}_X} & & \downarrow{\scriptstyle (-\mathrm{Id}_X, f) \oplus \mathrm{Id}_Y} & & \downarrow{\scriptstyle \mathrm{Id}_Y} & & \downarrow{\scriptstyle \mathrm{Id}_{\Sigma X}} \\
X & \xrightarrow[(-\mathrm{Id}_X, f)]{} & X \amalg Y & \xrightarrow[(f, \mathrm{Id}_Y)]{} & Y & \longrightarrow & \Sigma X
\end{array}$$

The result then follows from the 3×3 Lemma, Lemma 4.1.13. □

Corollary A.7.14 *Let $\mathcal{C}$ be a stable model category and consider a diagram*

$$X_0 \xrightarrow{f_0} X_1 \xrightarrow{f_1} X_2 \xrightarrow{f_2} \cdots .$$

Then, for any $Y \in \mathcal{C}$, there is a natural short exact sequence

$$* \longrightarrow \lim_n^1 [\Sigma X_n, Y] \longrightarrow [\mathrm{hocolim}_n X_n, Y] \longrightarrow \lim_n [X_n, Y] \longrightarrow *.$$

Proof Using the constructions of Example A.7.12 and Lemma A.7.13, we may write the homotopy colimit as the cofibre of

$$\mathrm{Id} - \coprod_{n \geqslant 0} f_n : \coprod_{n \geqslant 0} X_n \longrightarrow \coprod_{n \geqslant 0} X_n.$$

Applying the functor $[-, Y]$ to this exact triangle and splitting the resulting long exact sequences gives the result. See [HPS97, Proposition 2.2.4 (d)] or [MP12, Section 2.2] for details on the $\lim^1$ term. □

Unsurprisingly, the set-up for homotopy limits is dual to that of the previous section. For the rest of this section, I will denote an *opposite-simple* poset, that is, its opposite category I^{op} is a simple category. The main examples that we have in mind are

$$I = (2 \longrightarrow 1 \longleftarrow 3)$$

and

$$\mathbb{N}^{op} = (0 \longleftarrow 1 \longleftarrow 2 \longleftarrow \cdots).$$

The following is the second part of [Hov99, Theorem 5.1.3].

Theorem A.7.15 *Let $\mathcal{C}$ be a model category and I an opposite-simple category. Then there is a model structure on $\mathcal{C}^I$ called the* injective model structure *such that a morphism $f\colon X \longrightarrow Y$ is a weak equivalence (respectively cofibration) if and only if $f_i\colon X_i \longrightarrow Y_i$ is for every $i \in I$.*

Definition A.7.16 Let X be an object in $\mathcal{C}^I$, where $\mathcal{C}$ is a model category and I is opposite-simple. Then the *homotopy limit* $\operatorname{holim} X$ of X is the limit of a fibrant replacement of X in the injective model structure. In the case of

$$I = (2 \longrightarrow 1 \longleftarrow 3),$$

we talk about a homotopy pullback, and in the case of

$$\mathbb{N}^{op} = (0 \longleftarrow 1 \longleftarrow 2 \longleftarrow \cdots),$$

we speak of a sequential homotopy limit.

The homotopy limit satisfies the following properties.

- There is a natural comparison morphism $\lim X \longrightarrow \operatorname{holim} X$.
- If $F\colon X \longrightarrow Y \in \mathcal{C}^I$ is a vertexwise weak equivalence of diagrams, then F induces a weak equivalence $\operatorname{holim} X \longrightarrow \operatorname{holim} Y$.
- If our model category has functorial factorisation, then we have a functor $\operatorname{holim}\colon \mathcal{C}^I \longrightarrow \mathcal{C}$. Any other functor $\mathcal{C}^I \longrightarrow \mathcal{C}$ with the above properties factors over holim.

Again, we can characterise fibrant diagrams, see [Str11, Theorem 6.79]. For an opposite-simple category I, we define $I_{>i}$ as the full subcategory on all objects $j \in I$ with $j > i$.

Theorem A.7.17 *Let $\mathcal{C}$ be a model category and I an opposite-simple category. Assume that $X \in \mathcal{C}^I$ is vertexwise fibrant. If for all $i \in I$ the map*

$$X_i \longrightarrow \lim_{j \in I_{>i}} X_j$$

is a fibration in $\mathcal{C}$, then X is a fibrant diagram.

Example A.7.18 In a very similar way to homotopy pushouts, if in a diagram

$$X_2 \longrightarrow X_1 \longleftarrow X_3$$

all the X_i are fibrant and the arrows are fibrations, then the pullback of the diagram is a homotopy pullback.

Example A.7.19 The case of sequential homotopy limits is, unsurprisingly, dual to homotopy colimits, see Example A.7.9. A diagram

$$X = (X_0 \xleftarrow{g_0} X_1 \xleftarrow{g_1} X_2 \xleftarrow{g_2} \cdots)$$

is fibrant if all the g_i are fibrations between fibrant objects. In a general diagram, the g_i can inductively be replaced with fibrations between fibrant objects analogously to Example A.7.9.

The first lemma below says that if $\mathcal{C}$ is right proper, then for the homotopy pullback, it is sufficient to replace just one leg of the diagram with a fibration. As the following lemmas are dual to the corresponding statements about homotopy colimits, we will not include their proofs.

Lemma A.7.20 *Let $\mathcal{C}$ be a right proper model category and*

$$X_2 \longrightarrow X_1 \longleftarrow X_3$$

be a diagram in $\mathcal{C}$. If either $X_2 \longrightarrow X_1$ or $X_3 \longrightarrow X_1$ is a fibration between fibrant objects, then the pullback of the diagram is a homotopy pullback.

Lemma A.7.21 *Let $\mathcal{C}$ be a stable model category. Then a homotopy pullback*

$$\begin{array}{ccc} P & \xrightarrow{p} & B \\ {\scriptstyle q}\downarrow & & \downarrow{\scriptstyle s} \\ C & \xrightarrow[r]{} & A \end{array}$$

gives rise to an exact triangle in $\mathrm{Ho}(\mathcal{C})$

$$P \xrightarrow{(p,q)} B \times C \xrightarrow{r-s} A \longrightarrow \Sigma P.$$

Remark A.7.22 The two types of exact triangles arising from Lemmas A.7.8 and A.7.21 are isomorphic as finite coproducts and finite products agree in the homotopy category of a stable model category, see Lemma 4.4.4.

This leads to the slogan "homotopy pushouts and homotopy pullbacks agree in a stable model category". However, one must take care in going from a triangle of the form

$$P \xrightarrow{(p,q)} B \times C \xrightarrow{r-s} A \longrightarrow \Sigma P$$

in the homotopy category to a (strictly commuting) homotopy pullback square in the model category.

The duals of Example A.7.12 and Lemma A.7.13 also hold, giving the dual to Corollary A.7.14.

Corollary A.7.23 *Let $\mathcal{C}$ be a stable model category and consider a diagram*

$$X_0 \xleftarrow{f_0} X_1 \xleftarrow{f_1} X_2 \xleftarrow{f_2} \cdots .$$

Then for any $A \in \mathcal{C}$, there is a natural short exact sequence

$$* \longrightarrow \lim_n^1[\Sigma A, X_n] \longrightarrow [A, \operatorname{holim}_n X_n] \longrightarrow \lim_n[A, X_n] \longrightarrow *.$$

References

[Ada60] J. F. Adams. On the non-existence of elements of Hopf invariant one. *Ann. of Math. (2)*, 72:20–104, 1960.

[Ada66a] J. F. Adams. On the groups $J(X)$. IV. *Topology*, 5:21–71, 1966.

[Ada66b] J. F. Adams. *Stable homotopy theory*, volume 1961 of *Second revised edition. Lectures delivered at the University of California at Berkeley*. Springer-Verlag, Berlin and New York, 1966.

[Ada74] J. F. Adams. *Stable homotopy and generalised homology*. University of Chicago Press, Chicago, IL, 1974. Chicago Lectures in Mathematics.

[Ada78] J. F. Adams. *Infinite loop spaces*, volume 90 of *Annals of Mathematics Studies*. Princeton University Press, Princeton, NJ; University of Tokyo Press, Tokyo, 1978.

[Ati89] M. F. Atiyah. *K-theory*. Advanced Book Classics. Addison-Wesley Publishing Company, Advanced Book Program, Redwood City, CA, second edition, 1989. Notes by D. W. Anderson.

[BBD82] A. A. Beĭlinson, J. Bernstein, and P. Deligne. Faisceaux pervers. In *Analysis and topology on singular spaces, I (Luminy, 1981)*, volume 100 of *Astérisque*, pages 5–171. Soc. Math. France, Paris, 1982.

[Ben98] D. J. Benson. *Representations and cohomology. I*, volume 30 of *Cambridge Studies in Advanced Mathematics*. Cambridge University Press, Cambridge, second edition, 1998. Basic representation theory of finite groups and associative algebras.

[BF78] A. K. Bousfield and E. M. Friedlander. Homotopy theory of Γ-spaces, spectra, and bisimplicial sets. In *Geometric applications of homotopy theory (Proc. Conf., Evanston, Ill., 1977), II*, volume 658 of *Lecture Notes in Math.*, pages 80–130. Springer, Berlin, 1978.

[Bor94] F. Borceux. *Handbook of categorical algebra. 2*, volume 51 of *Encyclopedia of Mathematics and Its Applications*. Cambridge University Press, Cambridge, 1994. Categories and structures.

[Bou74] A. K. Bousfield. Types of acyclicity. *J. Pure Appl. Algebra*, 4:293–298, 1974.

[Bou75] A. K. Bousfield. The localization of spaces with respect to homology. *Topology*, 14:133–150, 1975.

[Bou79] A. K. Bousfield. The localization of spectra with respect to homology. *Topology*, 18(4):257–281, 1979.

[BR04] A. Baker and B. Richter, editors. *Structured ring spectra*, volume 315 of *London Mathematical Society Lecture Note Series*. Cambridge University Press, Cambridge, 2004.

[BR11] D. Barnes and C. Roitzheim. Local framings. *New York J. Math.*, 17:513–552, 2011.

[BR13] D. Barnes and C. Roitzheim. Stable left and right Bousfield localisations. *Glasg. Math. J.*, 56:13–42, 2014.

[Bre97] G. E. Bredon. *Topology and geometry*, volume 139 of *Graduate Texts in Mathematics*. Springer-Verlag, New York, 1997. Corrected third printing of the 1993 original.

[Bru09] R. R. Bruner. An Adams spectral sequence Primer. Available via the author's website, www.rrb.wayne.edu, 2009.

[BS17] P. Balmer and B. Sanders. The spectrum of the equivariant stable homotopy category of a finite group. *Invent. Math.*, 208(1):283–326, 2017.

[Coh68] J. M. Cohen. The decomposition of stable homotopy. *Ann. of Math. (2)*, 87:305–320, 1968.

[Col06] M. Cole. Many homotopy categories are homotopy categories. *Topology Appl.*, 153(7):1084–1099, 2006.

[CS98] J. D. Christensen and N. P. Strickland. Phantom maps and homology theories. *Topology*, 37(2):339–364, 1998.

[Day70] B. Day. On closed categories of functors. In *Reports of the midwest category seminar, IV*, Lecture Notes in Mathematics, Vol. 137, pages 1–38. Springer, Berlin, 1970.

[DP80] A. Dold and D. Puppe. Duality, trace, and transfer. In *Proceedings of the International Conference on Geometric Topology (Warsaw, 1978)*, pages 81–102. PWN, Warsaw, 1980.

[DS95] W. G. Dwyer and J. Spaliński. Homotopy theories and model categories. In *Handbook of algebraic topology*, pages 73–126. North-Holland, Amsterdam, 1995.

[Dug01] D. Dugger. Replacing model categories with simplicial ones. *Trans. Amer. Math. Soc.*, 353(12):5003–5027, 2001.

[Dwy04] W. G. Dwyer. Localizations. In *Axiomatic, enriched and motivic homotopy theory*, volume 131 of *NATO Sci. Ser. II Math. Phys. Chem.*, pages 3–28. Kluwer Acad. Publ., Dordrecht, 2004.

[EKMM97] A. D. Elmendorf, I. Kriz, M. A. Mandell, and J. P. May. *Rings, modules, and algebras in stable homotopy theory*, volume 47 of *Mathematical Surveys and Monographs*. American Mathematical Society, Providence, RI, 1997. With an appendix by M. Cole.

[Fra96] J. Franke. Uniqueness theorems for certain triangulated categories possessing an Adams spectral sequence. www.math.uiuc.edu/K-theory/0139/, 1996.

[GKR18] R. Garner, M. Kędziorek, and E. Riehl. Lifting accessible model structures. arXiv: 1802.09889, 2018.

[Gra75] B. Gray. *Homotopy theory*. Academic Press [Harcourt Brace Jovanovich, Publishers], New York and London, 1975. An introduction to algebraic topology, Pure and Applied Mathematics, Vol. 64.

[Gut10] J. J. Gutiérrez. Homological localizations of Eilenberg–Mac Lane spectra. *Forum Math.*, 22(2):349–356, 2010.

[Har09] J. E. Harper. Homotopy theory of modules over operads in symmetric spectra. *Algebr. Geom. Topol.*, 9(3):1637–1680, 2009.

[Har15] J. E. Harper. Corrigendum to "Homotopy theory of modules over operads in symmetric spectra" [MR2539191]. *Algebr. Geom. Topol.*, 15(2):1229–1237, 2015.

[Hat02] A. Hatcher. *Algebraic topology*. Cambridge University Press, Cambridge, 2002.

[Hir03] P. S. Hirschhorn. *Model categories and their localizations*, volume 99 of *Mathematical Surveys and Monographs*. American Mathematical Society, Providence, RI, 2003.

[HKRS17] K. Hess, M. Kędziorek, E. Riehl, and B. Shipley. A necessary and sufficient condition for induced model structures. *J. Topol.*, 10(2):324–369, 2017.

[HM14] M. J. Hopkins and M. Mahowald. From elliptic curves to homotopy theory. In *Topological modular forms*, volume 201 of *Math. Surveys Monogr.*, pages 261–285. American Mathematical Society, Providence, RI, 2014.

[Hov99] M. Hovey. *Model categories*, volume 63 of *Mathematical Surveys and Monographs*. American Mathematical Society, Providence, RI, 1999.

[Hov01a] M. Hovey. Model category structures on chain complexes of sheaves. *Trans. Amer. Math. Soc.*, 353(6):2441–2457 (electronic), 2001.

[Hov01b] M. Hovey. Spectra and symmetric spectra in general model categories. *J. Pure Appl. Algebra*, 165(1):63–127, 2001.

[HPS97] M. Hovey, J. H. Palmieri, and N. P. Strickland. Axiomatic stable homotopy theory. *Mem. Amer. Math. Soc.*, 128(610):x+114, 1997.

[HS98] M. J. Hopkins and J. H. Smith. Nilpotence and stable homotopy theory. II. *Ann. Math. (2)*, 148(1):1–49, 1998.

[HS99] M. Hovey and N. P. Strickland. Morava K-theories and localisation. *Mem. Amer. Math. Soc.*, 139(666):viii+100, 1999.

[HSS00] M. Hovey, B. Shipley, and J. Smith. Symmetric spectra. *J. Amer. Math. Soc.*, 13(1):149–208, 2000.

[Hus94] D. Husemoller. *Fibre bundles*, volume 20 of *Graduate Texts in Mathematics*. Springer-Verlag, New York, third edition, 1994.

[IX15] D. C. Isaksen and Z. Xu. Motivic stable homotopy and the stable 51 and 52 stems. *Topology Appl.*, 190:31–34, 2015.

[Jam99] I. M. James, editor. *History of topology*. North-Holland, Amsterdam, 1999.

[Joa01] M. Joachim. A symmetric ring spectrum representing KO-theory. *Topology*, 40(2):299–308, 2001.

[Kel05] G. M. Kelly. Basic concepts of enriched category theory. *Repr. Theory Appl. Categ.*, (10):vi+137 pp. (electronic), 2005. Reprint of the 1982 original [Cambridge University Press, Cambridge; MR0651714].

[Kra07] H. Krause. Derived categories, resolutions, and Brown representability. In *Interactions between homotopy theory and algebra*, volume 436 of *Contemporary Mathematics*, pages 101–139. American Mathematical Society, Providence, RI, 2007.

[Kro07] T. Kro. Model structure on operads in orthogonal spectra. *Homol. Homotopy Appl.*, 9(2):397–412, 2007.

[Len12] F. Lenhardt. Stable frames in model categories. *J. Pure Appl. Algebra*, 216(5):1080–1091, 2012.

[Lew91] L. G. Lewis, Jr. Is there a convenient category of spectra? *J. Pure Appl. Algebra*, 73(3):233–246, 1991.

[Lim59] E. L. Lima. The Spanier–Whitehead duality in new homotopy categories. *Summa Brasil. Math.*, 4:91–148 (1959), 1959.

[LMSM86] L. G. Lewis, Jr., J. P. May, M. Steinberger, and J. E. McClure. *Equivariant stable homotopy theory*, volume 1213 of *Lecture Notes in Mathematics*. Springer-Verlag, Berlin, 1986. With contributions by J. E. McClure.

[Mac71] S. Mac Lane. *Categories for the working mathematician*. Springer-Verlag, New York, 1971. Graduate Texts in Mathematics, Vol. 5.

[Mah70] M. Mahowald. The order of the image of the J-homomorphisms. *Bull. Amer. Math. Soc.*, 76:1310–1313, 1970.

[Mar83] H. R. Margolis. *Spectra and the Steenrod algebra*, volume 29 of *North-Holland Mathematical Library*. North-Holland Publishing Co., Amsterdam, 1983. Modules over the Steenrod algebra and the stable homotopy category.

[May70] J. P. May. A general algebraic approach to Steenrod operations. In *The Steenrod algebra and its applications (Proceedings of a conference to celebrate N. E. Steenrod's sixtieth birthday, Battelle Memorial Institute, Columbus, OH, 1970)*, Lecture Notes in Mathematics, Vol. 168, pages 153–231. Springer, Berlin, 1970.

[May77] J. P. May. *E_∞ ring spaces and E_∞ ring spectra*. Lecture Notes in Mathematics, Vol. 577. Springer-Verlag, Berlin and New York, 1977. With contributions by Frank Quinn, Nigel Ray, and Jørgen Tornehave.

[May96] J. P. May. *Equivariant homotopy and cohomology theory*, volume 91 of *CBMS Regional Conference Series in Mathematics*. Published for the Conference Board of the Mathematical Sciences, Washington, DC, 1996. With contributions by M. Cole, G. Comezaña, S. Costenoble, A. D. Elmendorf, J. P. C. Greenlees, L. G. Lewis, Jr., R. J. Piacenza, G. Triantafillou, and S. Waner.

[May99a] J. P. May. *A concise course in algebraic topology*. Chicago Lectures in Mathematics. University of Chicago Press, Chicago, IL, 1999.

[May99b] J. P. May. Stable algebraic topology, 1945–1966. In *History of topology*, pages 665–723. North-Holland, Amsterdam, 1999.

[May01] J. P. May. The additivity of traces in triangulated categories. *Adv. Math.*, 163(1):34–73, 2001.

[May09a] J. P. May. What are E_∞ ring spaces good for? In *New topological contexts for Galois theory and algebraic geometry (BIRS 2008)*, volume 16 of *Geometry & Topology Monographs*, pages 331–365. Geometry & Topology Publications, Coventry, 2009.

[May09b] J. P. May. What precisely are E_∞ ring spaces and E_∞ ring spectra? In *New topological contexts for Galois theory and algebraic geometry (BIRS 2008)*, volume 16 of *Geometry & Topology Monographs*, pages 215–282. Geometry & Topology Publications, Coventry, 2009.

[McC01] J. McCleary. *A user's guide to spectral sequences*, volume 58 of *Cambridge Studies in Advanced Mathematics*. Cambridge University Press, Cambridge, second edition, 2001.

[Mil58] J. W. Milnor. The Steenrod algebra and its dual. *Ann. Math. (2)*, 67:150–171, 1958.

[Mil81] H. R. Miller. On relations between Adams spectral sequences, with an application to the stable homotopy of a Moore space. *J. Pure Appl. Algebra*, 20(3):287–312, 1981.

[Mil92] H. R. Miller. Finite localizations. *Bol. Soc. Mat. Mexicana (2)*, 37(1-2):383–389, 1992. Papers in honor of José Adem (Spanish).

[MM65] J. W. Milnor and J. C. Moore. On the structure of Hopf algebras. *Ann. Math. (2)*, 81:211–264, 1965.

[MM02] M. A. Mandell and J. P. May. Equivariant orthogonal spectra and S-modules. *Mem. Amer. Math. Soc.*, 159(755):x+108, 2002.

[MMSS01] M. A. Mandell, J. P. May, S. Schwede, and B. Shipley. Model categories of diagram spectra. *Proc. London Math. Soc. (3)*, 82(2):441–512, 2001.

[Moo58] J. C. Moore. Semi-simplicial complexes and Postnikov systems. In *Symposium internacional de topología algebraica International symposium on algebraic topology*, pages 232–247. Universidad Nacional Autónoma de México and UNESCO, Mexico City, 1958.

[MP12] J. P. May and K. Ponto. *More concise algebraic topology*. Chicago Lectures in Mathematics. University of Chicago Press, Chicago, IL, 2012. Localization, completion, and model categories.

[MSS07] F. Muro, S. Schwede, and N. Strickland. Triangulated categories without models. *Invent. Math.*, 170(2):231–241, 2007.

[MV99] F. Morel and V. Voevodsky. $\mathbf{A}^1$-homotopy theory of schemes. *Inst. Hautes Études Sci. Publ. Math.*, 90:45–143 (2001), 1999.

[Qui67] D. G. Quillen. *Homotopical algebra*. Lecture Notes in Mathematics, No. 43. Springer-Verlag, Berlin, 1967.

[Rav78] D. C. Ravenel. A novice's guide to the Adams-Novikov spectral sequence. In *Geometric applications of homotopy theory (Proc. Conf., Evanston, Ill., 1977), II*, volume 658 of *Lecture Notes in Math.*, pages 404–475. Springer, Berlin, 1978.

[Rav84] D. C. Ravenel. Localization with respect to certain periodic homology theories. *Amer. J. Math.*, 106(2):351–414, 1984.

[Rav86] D. C. Ravenel. *Complex cobordism and stable homotopy groups of spheres*, volume 121 of *Pure and Applied Mathematics*. Academic Press, Inc., Orlando, FL, 1986.

[Rav92a] D. C. Ravenel. *Nilpotence and periodicity in stable homotopy theory*, volume 128 of Annals of Mathematics Studies. Princeton University Press, Princeton, NJ, 1992. Appendix C by Jeff Smith.

[Rav92b] D. C. Ravenel. Progress report on the Telescope Conjecture. In *Adams memorial symposium on algebraic topology, 2 (Manchester, 1990)*, volume 176 of *London Mathematical Society Lecture Note Series*, pages 1–21. Cambridge University Press, Cambridge, 1992.

[Rez98] C. Rezk. Notes on the Hopkins–Miller theorem. In *Homotopy theory via algebraic geometry and group representations (Evanston, IL, 1997)*, volume 220 of *Contemporary Mathematics*, pages 313–366. American Mathematical Society, Providence, RI, 1998.

[Rob87] A. Robinson. The extraordinary derived category. *Math. Z.*, 196(2):231–238, 1987.

[Rog12] J. Rognes. The Adams spectral sequence. Available via the author's website, https://folk.uio.no/rognes/home.html, 2012.

[Roi07] C. Roitzheim. Rigidity and exotic models for the K-local stable homotopy category. *Geom. Topol.*, 11:1855–1886, 2007.

[Rot88] J. J. Rotman. *An introduction to algebraic topology*, volume 119 of Graduate Texts in Mathematics. Springer-Verlag, New York, 1988.

[RS17] B. Richter and B. Shipley. An algebraic model for commutative $H\mathbb{Z}$-algebras. *Algebr. Geom. Topol.*, 17(4):2013–2038, 2017.

[Sch97] S. Schwede. Spectra in model categories and applications to the algebraic cotangent complex. *J. Pure Appl. Algebra*, 120(1):77–104, 1997.

[Sch99] S. Schwede. Stable homotopical algebra and Γ-spaces. *Math. Proc. Cambridge Philos. Soc.*, 126(2):329–356, 1999.

[Sch01a] S. Schwede. S-modules and symmetric spectra. *Math. Ann.*, 319(3):517–532, 2001.

[Sch01b] S. Schwede. The stable homotopy category has a unique model at the prime 2. *Adv. Math.*, 164(1):24–40, 2001.

[Sch07a] S. Schwede. The stable homotopy category is rigid. *Ann. Math. (2)*, 166(3):837–863, 2007.

[Sch07b] S. Schwede. An untitled book project about symmetric spectra. www.math.uni-bonn.de/people/schwede/SymSpec.pdf, 2007.

[Sch08] S. Schwede. On the homotopy groups of symmetric spectra. *Geom. Topol.*, 12(3):1313–1344, 2008.

[Sch18] S. Schwede. *Global homotopy theory*, volume 34. Cambridge University Press, Cambridge, 2018.

[Seg74] G. Segal. Categories and cohomology theories. *Topology*, 13:293–312, 1974.

[Ser53] J.-P. Serre. Groupes d'homotopie et classes de groupes abéliens. *Ann. Math. (2)*, 58:258–294, 1953.

[Shi01] B. Shipley. Monoidal uniqueness of stable homotopy theory. *Adv. Math.*, 160(2):217–240, 2001.

[Shi04] B. Shipley. A convenient model category for commutative ring spectra. In *Homotopy theory: Relations with algebraic geometry, group cohomology, and algebraic K-theory*, volume 346 of *Contemporary Mathematics*, pages 473–483. American Mathematical Society, Providence, RI, 2004.

[Shi07] B. Shipley. $H\mathbb{Z}$-algebra spectra are differential graded algebras. *Amer. J. Math.*, 129(2):351–379, 2007.

[Spa81] E. H. Spanier. *Algebraic topology*. Springer-Verlag, New York and Berlin, 1981. Corrected reprint.

[SS00] S. Schwede and B. Shipley. Algebras and modules in monoidal model categories. *Proc. Lond. Math. Soc. (3)*, 80(2):491–511, 2000.

[SS02] S. Schwede and B. Shipley. A uniqueness theorem for stable homotopy theory. *Math. Z.*, 239(4):803–828, 2002.

[SS03a] S. Schwede and B. Shipley. Equivalences of monoidal model categories. *Algebr. Geom. Topol.*, 3:287–334, 2003.

[SS03b] S. Schwede and B. Shipley. Stable model categories are categories of modules. *Topology*, 42(1):103–153, 2003.

[Ste62] N. E. Steenrod. *Cohomology operations*. Lectures by N. E. Steenrod written and revised by D. B. A. Epstein. Annals of Mathematics Studies, No. 50. Princeton University Press, Princeton, NJ, 1962.

[Ste67] N. E. Steenrod. A convenient category of topological spaces. *Michigan Math. J.*, 14:133–152, 1967.

[Str11] J. Strom. *Modern classical homotopy theory*, volume 127 of Graduate Studies in Mathematics. American Mathematical Society, Providence, RI, 2011.

[SW55] E. H. Spanier and J. H. C. Whitehead. Duality in homotopy theory. *Mathematika*, 2:56–80, 1955.

[Tod62] H. Toda. *Composition methods in homotopy groups of spheres*. Annals of Mathematics Studies, No. 49. Princeton University Press, Princeton, NJ, 1962.

[Vog70] R. Vogt. *Boardman's stable homotopy category*. Lecture Notes Series, No. 21. Matematisk Institut, Aarhus Universitet, Aarhus, 1970.

[Wei94] C. A. Weibel. *An introduction to homological algebra*, volume 38 of Cambridge Studies in Advanced Mathematics. Cambridge University Press, Cambridge, 1994.

[Whi62] G. W. Whitehead. Generalized homology theories. *Trans. Amer. Math. Soc.*, 102:227–283, 1962.

Index

3×3 Lemma, **137**, 245, 322, 407

abelian category, 94, 101, 128, 134
absolute stable model structure, **218**
acceptable model category, **333**, 333–335
acyclic cofibration, **377**
acyclic fibration, **377**
acyclic object, **318**, 320, 322, 323, 328–330, 349, 350, 374
acyclicity type, **353**, 353–355
Adams filtration, **90**, 91, 232, 233
Adams spectral sequence, 38, 71, **86**, 87, 88, 91, 92, 222, 368
Adams summand, **360**, 370
Adams tower, **88**, 88–91
Adams–Novikov spectral sequence, 91
additive category, **102**, 104, 129, 169, 249
additive functor, **102**, 104, 233
Adem relations, **73**, 77, 78, 81, 83, 92
adjunction of two variables, **250**
admissible monomial, **77**, 77–80, 84–86
antipode, **81**, 82, 84, 86
arrow category, **400**
augmentation, **81**

Balmer spectrum, **375**
bialgebra, 82
biproduct, **102**
Bockstein homomorphism, 73, 75
Bott periodicity, 25, 49, **358**
boundary map, **114**, 114–116, 124, 125, 127, 138–145, 147, 149, 150, 244
Bousfield arithmetic square, **355**, 366
Bousfield localisation, *see* localisation
Bousfield–Smith cardinality argument, 337
BP, 90, **369**, 370
Brown Representability, **22**, 71, 172, 283
Brown–Peterson spectrum, *see BP*

$\mathcal{C}$–model category, **250**, 251, 315
$\mathcal{C}$–module, **249**, 250, 251, 298, 307, 309, 315
candidate triangle, **161**, 162
  contractible, **162**
  exact, **162**
canonical action, 174, 187
Cartan formula, **73**, 74, 77, 79, 83
cell complex, 270, 337, 338, **394**
cellular, 220, 298, 323, 332
chain complex, 93, 97, 101, 152, 211, 236, 241, 248, 292, 297, 298, 333, 335, 376, 380–382, 391, 397
  bounded, 380, 381, 387, 397
change of rings, **288**, 289, 290
Chromatic Convergence Theorem, **372**, 374
chromatic homotopy theory, 371
chromatic square, **371**, 374
closed $\mathcal{C}$–module, *see* $\mathcal{C}$–module
closed inclusion, 8, **393**
coend, **257**
cofibrant implies flat, **273**, 288
cofibrant object, **380**
cofibrant replacement, **381**
cofibrantly generated, 40, 64, 183, 189–191, 203, 213, 215, 242, 287, 293, 295, 303, 306, 335, **395**, 396–399
cofibre, **105**
cofibre sequence, **113**, 149
cohomological functor, **134**, 135
cohomology operation, **72**
cohomology theory
  reduced, 19, **20**, 20–28, 30, 32, 33, 35, 36, 71, 72, 170, 171, 358

combinatorial model category, 298
commutative product, **235**
commutative ring spectrum, *see* ring spectrum, commutative
compact, **222**, 223, 225, 227–230, 280, 282, 283, 320, 321
compactly generated, **8**, 236
complex cobordism, *see* MU
concatenation of homotopies, 103, 111, 112, **385**, 386
connected spectrum, **69**, 354
connective spectrum, **69**, 70, 217–219, 354–356
convolution product, **258**, 259, 260
correspondence, 100, 103, 120, 121, 123, 124, **384**, 385, 388
cosimplicial frame, **306**
cosimplicial object, **299**, 301, 305, 310, 312
cotensor, 58, 177, 189, **249**, 250, 286, 298
CW-spectrum, 36, **46**, 66, 159, 228–230, 372–374
cylinder object, 94, **381**

decomposable, **80**, 81
deformation retract, 14, 19, 62, 180, 193
degree of a map, **101**
diagram spectrum, 217, 218
direct sum, **102**, 129
dual Steenrod algebra, **84**, 86, 88

E-acyclic, **336**, 339
E-acyclic cofibration, **337**, 338, 341
E-equivalence, **324**, 327, 336, 337, 339
E-fibration, **337**
E-local, **336**, 339–343
E-local stable homotopy category, **342**
E-localisation, **336**, 337, 339, 342, 355
$E(n)$, 90, 344, 360, **369**, 369–372
E_*-isomorphism, *see* E-equivalence
Eilenberg–Mac Lane object, **88**
Eilenberg–Mac Lane space, 25, 293
Eilenberg–Mac Lane spectrum, **33**, 71, 88, 188, 284, 285, 297, 298, 354, 356, 372
Eilenberg–Steenrod axioms, 19
elliptic cohomology, 375
end, **257**
endomorphism ring spectrum, **284**, 297
enrichment, 57–59, 102, 217, 218, **249**, 250, 253–259, 261, 263–267, 269, 286, 287, 297, 298, 305, 307, 315, 325
equivariant map, **174**
evaluation functor, **67**, **176**, 177, 186, **189**, 209, 268, 309
exact couple, 89
exact functor, **152**, 281, 362
exact triangle, **129**, 230, 281, 406
exotic object, 162, **164**, 165

$\mathscr{F}$-space, **218**, 219
fibrant object, **380**
fibrant replacement, **381**
fibre, **105**
fibre sequence, **113**, 149
finite localisation, **320**, 327, 344, 366, 371
Five Lemma, 55, 56, 134, 135, 143, 150, 223, 322, 330, 332, 363
forgetful functor, 176, 177, 187, 192, 207, 216, 219, 269, 275, 286, 291, 292, 294, 295
framing, 231, 297, 298, **306**, 306–310, 312, 331
free algebra on a spectrum, **291**
Freudenthal Suspension Theorem, 7, 12, **13**, 14, 15, 18
functorial factorisation, 68, 306, **377**, 381, 398–400, 408

G-space, **174**, 221
G-spectrum, 221
generator, **222**, 223, 226, 227, 297
  compact, 222–225, 227, 232, 233, 297, 316, 345, 368
  homotopy generator, **222**
geometric realisation, 188, 213, 214, 248, 296, 379, 388, 396
graded maps, **101**
Grothendieck group, 357

h-cofibration, 51, 54, 55, 61, 174, 177–179, 181, 183, 184, 200–202, 270, 273, 274, 379, 391, **392**, 393
homological degree, **16**, 16–18
homology isomorphism, **380**
homology theory, 317, 324
  reduced, **19**, 20, 27, 172, 173, 278, 279
homotopy cartesian square, **60**
homotopy category, **386**
homotopy coequaliser, **406**
homotopy cofibre, 11, **51**, 51–54, 150, 178, 179, 200, 205, 271, 274, 329, 393
homotopy cofibre sequence, **51**, 52
homotopy colimit, 15, 16, 35, 68, 69, 180, 196, 197, 202, 230, 270, 344, 363, 365, **400**
  sequential, 16, 29, 68, 230, 279, 344, 345, **400**, 404–406

homotopy excision, 10, 11, **392**
homotopy fibre, 10, **51**, 53, 54, 65, 150, 329, 330, 332, 333, 335, 393
homotopy fibre sequence, **51**, 330
homotopy group of a spectrum, 37, **47**, 48, 49, 51, 52, 54, 65, 68–70, 87, 179, 195, 197, 206, 273, 277, 278
homotopy limit, 351, 366, 372, **408**
  sequential, 329, **408**, 409
homotopy mapping space, **307**
homotopy pullback, 59–61, 64, 205, 214, 216, 330, 355, 371, **408**, 409
homotopy pushout, **400**, 401, 403, 404, 409
Hopf algebra, **81**, 82, 84, 86
Hopf element, 92, 158, 232, 233
Hopf map, 14, 92, 101, 158–160, **162**, 163, 232, 233, 359
hopfian, 162, **163**, 163–165
horn, 331, **396**
Hurewicz fibration, 51, **379**
Hurewicz map, 88, 189
Hurewicz model structure, 9, 55, 211, 241, **379**, 405

I-cell, **394**
I-cof, **394**
I-inj, **394**
ideal, **374**, 375
indecomposable, **80**, 91
indeterminacy, **157**, 158
infinite loop space, **69**, 70, 219
injective model structure
  on chain complexes, 98, 101, 211, 241, **380**, 381, 397
  on diagrams, **408**
internal function object, 8, 27, 172, 196, 237, 238, 251, 257, **266**, 267, 281, 288

Johnson-Wilson spectrum, *see* $E(n)$

k-connected, **10**, 10–14, 18
k-equivalence, **10**, 10–14
K-localisation, 344, **356**, 359–361, 363, 364, 366, 368, 371
K-theory, 25, 32, 49, 70, **356**, 358–360, 375
$K(n)$, 344, **369**, 369–374
Kan fibration, 214, 216, **380**
Kelly product topology, 9, 236
Ken Brown's Lemma, 185, 208, **389**
Künneth isomorphism, 74, 76, 79, 370

Landweber exactness, 370
latching object, **302**
left derived functor, **389**, 390
left homotopy, **382**
left Kan extension, **258**, 258
left lifting property, **378**
left localisation, *see* localisation
levelwise model structure, 37, 38, **40**, 42, 43, 46, 57–60, 62, 66, 67, 174, **185**, 189, **190**, 192–194, 196, 199, 201, 209, 213, 215, 220, 269, 270, 274, 324, 325
lifted model structure, *see* lifting lemma
lifting lemma, 183, 184, 287, 291, **398**
local object, **318**, 318–323, 325–333, 335, 341, 367
localisation, 57, 189, 192, 220, 317, **319**, 319–321, 323–325, 334, 337, 345, 356, 366, 367
  localisation functor, 323
  stable, **326**, 326–328, 334, 336, 344
localising subcategory, **226**
loop functor, 66, **95**, 98

mapping cone, **11**, 23, 147, 163, 180
mapping cylinder, **11**, 12, 16, 61, 177, 178, 180, 193, 200, 393
mapping space, 298
Margolis' Uniqueness Conjecture, **316**
matching object, **302**
model category, **376**
  pointed, **94**
  stable, **101**
module over a ring spectrum, 234, 283, **285**, 285–290, 295–297, 343, 345
  free, **285**, 286, 288
moment, **78**
monoid axiom, 273, **274**
monoidal product, **235**
monoidal unit, **235**
Moore spectrum, 49, 75, 159, 163, 233, 281, **346**, 350, 360, 361, 363, 364, 368, 373
Morava K-theory, *see* $K(n)$
MU, 90, 369

$\mathbb{N}$, 6
nilpotent, 372, **373**, 374
non-degenerate basepoint, **9**, 12–15, 20, 405
v_1-self map, **361**, 364, 368
v_n-self map, **373**

octahedral axiom, **131**, 143, 322, 365
Ω-spectrum, **33**, 34, 36, 38, 57, 66, 68, 69, 184, 186, 189, 192–195, 197–200, 202, 203, 205, 208–210, 213, 216, 230, 271, 325, 340, 341

op-lax symmetric monoidal functor, **246**
opposite-simple category, **408**
orthogonal sequence, **255**
orthogonal space, **255**
orthogonal spectrum, 166, **174**, 174–179, 181–186, 195, 207, 209, 210, 230, 234, 252, 261–263, 266–269, 271, 273–280, 283, 286–293, 295, 336, 337, 339, 341, 342

p-complete stable homotopy category, **349**
p-completion, 87, 344, 346, **349**, 349–354, 356, 371
p-local stable homotopy category, **347**, 369
p-localisation, 344, 346, **347**, 348, 351, 354, 356
parallelisable, 92
Patching Lemma, 62, 145, **402**
path object, 94, **383**
phantom maps, 173, 283
π_*-isomorphism, 37, **47**, 47–50, 53–57, 61, 62, 64–68, 166, 170, 178–186, 192, 195, 197, 199, 200, 206, 208–210, 213, 214, 218, 219, 224, 229, 269–272, 277, 278, 288
positive stable model structure, 219, 234, 293, **294**, 295, 296
Postnikov section, 328, 354
Postnikov tower, 354
prime ideal, **375**
projective model structure
  on chain complexes, 98, 101, 211, 241, 248, 297, 298, 333, 335, **380**, 381, 397
  on diagrams, **400**, 402
prolongation functor, **176**, 177, 180, 183, 185, 207, 208, 219, 268, 275, 292, 295
proper, 210, **211**, 287, 293, 297, 298, 332, 333, 335
  left, **211**, 220, 319, 323, 335, 339, 341
  right, **211**, 212, 329, 330, 333, 335
Puppe sequence, 10, 23, 31, 51, 52, 89, 114, **125**, 127, 159, 173, 201
pushout product, 191, **236**, 237, 239, 250, 275, 313, 342
pushout product axiom, 59, **239**, 240, 242, 269, 270, 273–275, 288

q-cofibration, 9, **40**, 40–47, 58–62, 64–66, 180, 181, **183**, 183–185, **189**, 190, 191, 193, 195, 200, 202–204, 274, 338, 341, **379**
Quillen equivalence, **390**
Quillen functor, **388**

rationalisation, 344, **347**, 364, 372, 375
Recognition Theorem, 40, 190, 203, 334, 339, 396, **397**
Reedy category, **301**, 302, 303
Reedy model structure, **303**, 311, 313
representation sphere, 221
retract, 226, 374, **376**, 377
right derived functor, **389**, 390
right homotopy, **383**
right lifting property, **378**
rigid, **231**, 233, 366, 368
ring spectrum, 234, 283, **284**, 284–287, 289–292, 297, 343
  commutative, **284**, 285, 292–295

S-modules, 219
sequential homotopy limit, *see* homotopy limit, sequential
sequential space, **255**
sequential spectrum, 37, **38**, 38–40, 42, 46, 50, 51, 53, 57, 58, 60, 62, 64–69, 97, 101, 102, 152, 166, 167, 169, 170, 172, 174, 176, 178–180, 182, 184, 186, 187, 189, 192, 193, 207, 209, 210, 214, 216–218, 223, 224, 228, 230, 231, 250–252, 261, 263, 268, 276, 279, 280, 336, 341, 342, 366, 391
  in simplicial sets, **212**, 212–214, 216, 309, 311
Serre cofibration, *see* q-cofibration
Serre fibration, 40, 52, **379**, 393
Serre model structure, 9, 41, 43, 57, 67, 209, 211, 212, 241, 270, **379**, 380, 381, 396, 405
set of horns, **331**
shifted suspension, **39**, 48, 67, 69, **176**, 177, 186, **188**, 189, 209, 223, 263, 268, 270, 287, 309, 310, 325
Σ-cospectrum, **310**, 310–313
simple category, **399**, 400, 408
simplicial framing, **306**
simplicial model category, **251**, 296–299, 305, 307
simplicial object, **299**
simplicial set, 97, 186, 188, 211, 212, 236, 238, 241, 251, 298–301, 303–305, 307–310, 318, 331, 333, **379**, 380, 388, 391, 396
singular cohomology, 25, 72, 87
singular complex functor, 213, 214, 296, 388
singular homology, 20, 253
small (co)limit, **378**
small object argument, 183, 337, 338, **398**

small with respect to I, 40, 41, 190, 203, 287, 333–335, 337, 393, **395**
smash nilpotent, **373**
smash product, 34, 39, 262, 266
smash product of spectra, 37, 57, 71, 234, 251–253, **262**, 262–264, 267–270, 276, 277, 279, 280, 294, 295
smashing localisation, **344**, 345, 348, 350, 364, 368, 371, 372
Spanier–Whitehead category, **28**, 28–30, 32, 36, 168, 251, 280
Spanier–Whitehead dual, 28, 280, **281**, 283
spectral model category, **315**
spectrum, 24, **33**, 34
spectrum of finite type, **87**
sphere spectrum, **33**, 35, 39, 48, 75, 88, 163, **175**, 185, **187**, 224, 228–230, 232, 251–253, **259**, 260, 262, 278, 281, 284, 285, 291, 294, 313, 344, 345, 348, 361, 364, 368
split exact triangle, **136**, 137
stable cell, **46**, 159, 228, 229
stable equivalence, **193**, 210, 216
stable fibration, **64**, 183, 204
stable framing, 231, **313**, 314, 315, 367
stable homotopy category, 7, 26–28, 32, 33, 35–38, **66**, 67, 71, 72, 128, 155, 159, 166–169, 172, 174, 185, 186, 192, 207, 209, 212, 217, 219, 224, 228, 231, 232, 234, 235, 251, 252, 276, 280, 298, 309, 315–317, 336, 342, 343, 347, 369, 374
stable homotopy class, **18**
stable homotopy groups, **15**, 18, 20, 25, 26, 29–31, 35, 37, 47, 53, 182, 347
  of spheres, 16–18, 38, 86, 158, 222, 232, 315
  of the K-local sphere, 362
stable localisation, *see* localisation, stable
stable model category, *see* model category, stable
stable model structure on spectra, 37, 38, 47, 57, 59, 61, **62**, 64–68, 166, 174, **178**, 180, 182–184, 186, 189, 192, 193, 195, 200, **203**, 207–211, 214–216, 218, 220, 224, 273, 274, 276, 294, 324, 325, 333, 337, 343
Steenrod algebra, 71, **72**, 72–74, 76, 77, 81–84, 86, 87, 91
Steenrod square, **73**, 74
strongly dualisable, **281**, 281–283, 350, 364
structure maps
  of a sequential spectrum, **38**
  of a sequential spectrum in simplicial sets, **212**
  of a spectrum, **22**, 33
  of a symmetric spectrum, **187**
  of an orthogonal spectrum, **174**
suspension functor, 66, **94**, 95, 98
suspension spectrum, **33**, 101, 167, 171, 173, 175, 188
symmetric monoidal category, **235**, 235–238, 243, 245, 246, 255–258, 262, 266, 267, 269, 276, 283, 316, 342
  closed, 196, 197, 234, 236, **237**, 237–240, 242, 249–251, 257, 267, 280
symmetric monoidal functor, 296, 316
  lax, **246**, 246–248
  op-lax, **246**
  strong, **246**, 248, 249, 251, 268, 276, 343
symmetric monoidal model category, 234, **240**, 240–243, 245, 247, 248, 250–252, 269, 270, 274, 276, 278, 315, 342
symmetric monoidal Quillen adjunction
  strong, 220, **247**, 248, 275, 292, 342
  weak, **247**, 248
symmetric monoidal Quillen equivalence, 277
  weak, 297
symmetric sequence, **255**
symmetric space, **255**
symmetric spectrum, **186**, 187, 189–193, 195, 196, 198–201, 203, 204, 207, 209, 210, 216, 230, 234, 252, 260–263, 266–269, 273–276, 280, 283, 286–293, 295, 297, 298, 336, 341, 342
  in simplicial sets, **215**, 216, 276, 296, 315
Telescope Conjecture, **371**
tensor, 57, 177, 178, 183, 189, 191, 200, **249**, 250, 286, 298, 301, 305, 308, 310
thick subcategory, **226**, 227–230, 283, 374, 375
Thick Subcategory Theorem, **374**
Toda bracket, 92, 128, 155, **156**, 157, 158, 232, 233, 368
topological model category, **251**, 392, 393
topological triangulated category, *see* triangulated category, topological
transfinitely small, 333–335, 337, 338, **395**, 398, 399

triangle, **129**
triangulated category, 30, 66, 128, **129**, 132–134, 136–138, 146, 152, 155, 156, 161, 162, 164, 171, 222, 223, 225–227, 234, 242, 243, 316, 374
tensor-triangulated category, **245**, 276, 374, 375
topological, **161**, 163, 165
triangulated subcategory, **132**
trivial cofibration, *see* acyclic cofibration
trivial fibration, *see* acyclic fibration
twist functor, **235**

$\mathbb{W}$-acyclic object, *see* acyclic object
$\mathbb{W}$-equivalence, **318**, 318–323, 325–336
$\mathbb{W}$-local object, *see* local object
$\mathbb{W}$-localisation, *see* localisation
$\mathscr{W}$-spectrum, **217**, 219
weak Hausdorff, **8**, 236
weak symmetric monoidal functor, *see* symmetric monoidal functor, lax
Whitney sum, **357**
Wolbert's Theorem, **345**

Yoneda Lemma, 103, 113, 136, 193, 223, 253, **258**, 264
Yoneda product, 87

Zariski topology, 375
zero map, **94**, 99, 103, 105, 106, 117
Zorn's Lemma, 340

Printed in the United States
by Baker & Taylor Publisher Services